Methods in Enzymology

Volume 77
DETOXICATION AND DRUG METABOLISM:
CONJUGATION AND RELATED SYSTEMS

METHODS IN ENZYMOLOGY

EDITORS-IN-CHIEF

Sidney P. Colowick   Nathan O. Kaplan

Methods in Enzymology

Volume 77

Detoxication and Drug Metabolism: Conjugation and Related Systems

EDITED BY

William B. Jakoby

NATIONAL INSTITUTE OF ARTHRITIS,
DIABETES, DIGESTIVE, AND KIDNEY DISEASES
NATIONAL INSTITUTES OF HEALTH
BETHESDA, MARYLAND

1981

ACADEMIC PRESS

A Subsidiary of Harcourt Brace Jovanovich, Publishers

New York   London   Toronto   Sydney   San Francisco

ACADEMIC PRESS, INC.
111 Fifth Avenue, New York, New York 10003

United Kingdom Edition published by
ACADEMIC PRESS, INC. (LONDON) LTD.
24/28 Oval Road, London NW1 7DX

Library of Congress Cataloging in Publication Data
Main entry under title:

Detoxication and drug metabolism.

   (Methods in enzymology; v. 77)
   Includes bibliographical references and index.
   1. Enzymes. 2. Drugs--Metabolism. 3. Drugs--
Metabolic detoxication.   I. Jakoby, William B., Date.
II. Series. [DNLM: 1. Drugs--Metabolism. 2. Metabolic
detoxication, Drug. 3. Enzymes. W1 ME9615K v. 77 /
QV 38 D482]
QP601.M49 vol. 77   574.1'925s   [615'.7]   81-14932
ISBN 0-12-181977-9   AACR2

PRINTED IN THE UNITED STATES OF AMERICA

81 82 83 84   9 8 7 6 5 4 3 2 1

Table of Contents

Section I. Animal Organ and Cell Preparations
A. General Methods

B. Organ Perfusion

C. Cells

Section II. Enzyme Preparations

Section III. Assay Systems

Section IV. Synthesis

Contributors to Volume 77

Article numbers are in parentheses following the names of contributors.
Affiliations listed are current.

THEODORUS P. M. AKERBOOM (48), *Institut für Physiologische Chemie I, Universität Düsseldorf, D-4000-Düsseldorf-1, Federal Republic of Germany*

IRWIN M. ARIAS (23), *Liver Research Center, Department of Medicine, Albert Einstein College of Medicine, Bronx, New York 10461*

ANNE-CHARLOTTE ARONSSON (39), *Department of Biochemistry, Arrhenius Laboratory, University of Stockholm, S-10691 Stockholm, Sweden*

KENT AXELSSON (36), *Department of Biochemistry, Arrhenius Laboratory, University of Stockholm, S-10691 Stockholm, Sweden*

JOHN R. BEND (8, 13), *Laboratory of Pharmacology, National Institute of Environmental Health Sciences, National Institutes of Health, Research Triangle Park, North Carolina 27709*

PHILIP BENTLEY (46), *Ciba-Geigy Limited, CH-4000 Basel, Switzerland*

JOHANNES BIRCHER (1), *Department of Clinical Pharmacology, University of Berne, CH-3010 Berne, Switzerland*

NORBERT BLANCKAERT (50), *Laboratory of Hepatology, Department of Medical Research, Catholic University of Louvain, B-3000 Louvain, Belgium*

RONALD T. BORCHARDT (34), *Department of Biochemistry, University of Kansas, Lawrence, Kansas 66044*

RICHARD V. BRANCHFLOWER (7), *Laboratory of Chemical Pharmacology, National Heart, Lung, and Blood Institute, National Institutes of Health, Bethesda, Maryland 20205*

BRIAN BURCHELL (20, 22), *Department of Biochemistry, Medical Sciences Institute, The University, Dundee DD1 4HN, Scotland*

MILTON T. BUSH (47), *Department of Pharmacology, Vanderbilt University School of Medicine, Nashville, Tennessee 37232*

LEE J. CHEN (26), *Departments of Internal Medicine and Biological Chemistry, University of California, Davis, California 95616*

J. ROY CHOWDHURY (23), *Department of Medicine, Albert Einstein College of Medicine, Bronx, New York 10461*

FRANS COMPERNOLLE (54), *Laboratory of Hepatology, Department of Medical Research, Catholic University of Louvain, B-3000 Louvain, Belgium*

THEODORA R. DEVEREUX (17), *Laboratory of Pharmacology, National Institute of Environmental Health Sciences, National Institutes of Health, Research Triangle Park, North Carolina 27709*

W. DIMPFEL (19), *Experimentell-Medizinische Forschung/Pharmakologie, E. Merck, D-6100 Darmstadt, Federal Republic of Germany*

MICHAEL W. DUFFEL (24), *Section on Enzymes and Cellular Biochemistry, National Institute of Arthritis, Diabetes, Digestive, and Kidney Diseases, National Institutes of Health, Bethesda, Maryland 20205*

ERICH E. DUMELIN (2), *Maple Leaf Monarch Company, Windsor, Ontario, Canada*

G. J. DUTTON (49), *Department of Biochemistry, Medical Sciences Institute,*

University of Dundee, Dundee DD1 4HN, Scotland

JAMES R. FOUTS (17), *Laboratory of Pharmacology, National Institute of Environmental Health Sciences, National Institutes of Health, Research Triangle Park, North Carolina 27709*

JEFFREY R. FRY (15), *Department of Physiology and Pharmacology, Medical School, Queen's Medical Centre, Nottingham NG7 2UH, United Kingdom*

PETER GOLDMAN (6), *Department of Pharmacology, Harvard Medical School, Boston, Massachusetts 02115*

OWEN W. GRIFFITH (9, 30), *Department of Biochemistry, Cornell University Medical College, New York, New York 10021*

THOMAS M. GUENTHNER (46), *Section on Biochemical Pharmacology, Institute of Pharmacology, University of Mainz, D-6500 Mainz, Federal Republic of Germany*

CLAES GUTHENBERG (28), *Department of Biochemistry, Arrhenius Laboratory, University of Stockholm, S-10691 Stockholm, Sweden*

WILLIAM H. HABIG (27, 51), *Bureau of Biologics, Food and Drug Administration, Bethesda, Maryland 20205*

KAREL P. M. HEIRWEGH (50), *Laboratory of Hepatology, Department of Medical Research, Catholic University of Louvain, B-3000 Louvain, Belgium*

EBERHARD HEYMANN (45, 52), *Biochemisches Institut, Universität Kiel, D-2300 Kiel, Federal Republic of Germany*

JERRY B. HOOK (12), *Center for Environmental Toxicology, Michigan State University, East Lansing, Michigan 48824*

WILLIAM B. JAKOBY (24, 25, 27, 32, 33, 51), *Section on Enzymes, National Institute of Arthritis, Diabetes, Digestive, and Kidney Diseases, National Institutes of Health, Bethesda, Maryland 20205*

REBECCA JARABAK (38), *Department of Biochemistry, University of Chicago, Chicago, Illinois 60637*

DEAN P. JONES (16), *Department of Biochemistry, Emory University, Atlanta, Georgia 30322*

KATJA KEULEMANS (11), *Department of Pharmacology, State University of Groningen, Groningen, The Netherlands*

PAUL G. KILLENBERG (41, 57), *Department of Medicine, Duke University, Durham, North Carolina 27710*

CHARLES M. KING (35), *Department of Chemical Carcinogenesis, Michigan Cancer Foundation, Detroit, Michigan 48201*

MARTTI KOIVUSALO (42, 43), *Department of Medical Chemistry, University of Helsinki, SF-00170 Helsinki 17, Finland*

KERSTIN LARSON (36, 55), *Department of Biochemistry, Arrhenius Laboratory, University of Stockholm, S-10691 Stockholm, Sweden*

BENGT LARSSON (10), *Department of Toxicology, University of Uppsala, S-751 23 Uppsala, Sweden*

J. E. A. LEAKEY (49), *Department of Biochemistry, Medical Sciences Institute, University of Dundee, Dundee DD1 4HN, Scotland*

ELLEN SUE LYON (25, 33), *Section on Enzymes, National Institute of Arthritis, Diabetes, Digestive, and Kidney Diseases, National Institutes of Health, Bethesda, Maryland 20205*

MARTHA A. MCLAFFERTY (6), *Department of Pharmacology, Harvard Medical School, Boston, Massachusetts 02115*

BENGT MANNERVIK (28, 36, 39, 55), *Department of Biochemistry, Arrhenius Laboratory, University of Stockholm, S-10691 Stockholm, Sweden*

CAROL J. MARCUS (25), *Laboratory of Experimental Pathology, National Institute of Arthritis, Diabetes, Digestive, and Kidney Diseases, National Institutes of Health, Bethesda, Maryland 20205*

EWA MARMSTÅL (39), *Department of Biochemistry, Arrhenius Laboratory, University of Stockholm, S-10691 Stockholm, Sweden*

HAZEL B. MATHEWS (5), *National Institute of Environmental Health Sciences, National Institutes of Health, Research Triangle Park, North Carolina 27709*

DIRK K. F. MEIJER (4, 11), *Department of Pharmacology and Pharmacotherapeutics, Faculty of Pharmacy, State University of Groningen, Groningen, The Netherlands*

ALTON MEISTER (30), *Department of Biochemistry, Cornell University Medical Center, New York, New York 10021*

ROLF MENTLEIN (45, 52), *Biochemisches Institut, Universität Kiel, D-2300 Kiel, Federal Republic of Germany*

GERARD J. MULDER (4, 11), *Department of Pharmacology, State University of Groningen, Groningen, The Netherlands*

JOHN F. NEWTON, JR. (12), *Department of Pharmacology and Toxicology, Center for Environmental Toxicology, Michigan State University, East Lansing, Michigan 48824*

FRANZ OESCH (46), *Section on Biochemical Pharmacology, Institute of Pharmacology, University of Mainz, D-6500 Mainz, Federal Republic of Germany*

KARI ORMSTAD (16), *Department of Forensic Medicine, Karolinska Institute, S-104 01 Stockholm 60, Sweden*

STEN ORRENIUS (16), *Department of Forensic Medicine, Karolinska Institute, S-104 01 Stockholm, Sweden*

LAWRENCE M. PINKUS (18), *Department of Medicine, State University of New York at Stony Brook, Stony Brook, New York, and Nassau County Medical Center, East Meadow, New York 11554*

JOHN L. PLUMMER (8), *Department of Anesthesia and Intensive Care, Flinders Medical Center, Bedford Park, SA 5042, Australia*

LANCE R. POHL (7), *Laboratory of Chemical Pharmacology, National Heart, Lung, and Blood Institute, National Institutes of Health, Bethesda, Maryland 20205*

M. R. POLLARD (49), *Department of Biochemistry, Medical Sciences Institute, University of Dundee, Dundee DD1 4HN, Scotland*

RUDOLF PREISIG (1), *Department of Clinical Pharmacology, University of Berne, CH-3010 Berne, Switzerland*

HELLA RIX (52), *Biochemisches Institut, Universität Kiel, D-2300 Kiel, Federal Republic of Germany*

EGBERT SCHOLTENS (4), *Department of Pharmacology, State University of Groningen, Groningen, The Netherlands*

RONALD D. SEKURA (24, 53), *Bureau of Biologics, Food and Drug Administration, Bethesda, Maryland 20205*

HIROTOSHI SHIMIZU (31), *Department of Biochemistry, Nippon Roche Research Center, 200 Kajiwara, Kamakura City, Kanagawa Pref. 247, Japan*

HELMUT SIES (3, 8, 48), *Institut für Physiologische Chemie I, Universität Düsseldorf, D-4000-Düsseldorf-1, Federal Republic of Germany*

PETER C. SIMONS (29), *Friedrich Miescher Institute, CH-4002, Basel, Switzerland*

BRIAN R. SMITH (8, 13), *School of Pharmacy, University of New Mexico, Albuquerque, New Mexico 87131*

ALBERT E. SPAETH (14), *Laboratory of Nutrition and Endocrinology, National Institute of Arthritis, Diabetes, Digestive, and Kidney Diseases, National Institutes of Health, Bethesda, Maryland 20205*

SURESH S. TATE (30), *Department of Biochemistry, Cornell University Medical College, New York, New York 10021*

MITSURU TATEISHI (31), *Department of Biochemistry, Nippon Roche Research Center, 200 Kajiwara, Kamakura City, Kanagawa Pref. 247, Japan*

THOMAS R. TEPHLY (21), *The Toxicology Center, Department of Pharmacology, University of Iowa, Iowa City, Iowa 52242*

GUDRUN TIBBELIN (39), *Department of Biochemistry, Arrhenius Laboratory, University of Stockholm, S-10691 Stockholm, Sweden*

ROBERT H. TUKEY (21), *Developmental Pharmacology Branch, National Institute of Child Health and Human Development, National Institutes of Health, Bethesda, Maryland 20205*

SVEN ULLBERG (10), *Department of Toxicology, University of Uppsala, S-751 23 Uppsala, Sweden*

LASSE UOTILA (42, 43, 56), *Department of Medical Chemistry, University of Helsinki, SF-00170 Helsinki 17, Finland*

DAVID L. VANDER JAGT (29), *Department of Biochemistry, School of Medicine, The University of New Mexico, Albuquerque, New Mexico 87131*

JUN-LAN WANG (25), *Department of Agricultural Chemistry, Oregon State University, Corvallis, Oregon 97330*

PHILIP WEATHERILL (20), *Tenovus Institute of Cancer Research, Welsh National School of Medicine, Heath Park, Cardiff, CF 4 4XN, Wales*

WENDELL W. WEBER (35), *Department of Pharmacology, University of Michigan, Ann Arbor, Michigan 48109*

LESLIE T. WEBSTER, JR. (40, 57), *Department of Pharmacology, School of Medicine, Case Western Reserve University, Cleveland, Ohio 44106*

RICHARD A. WEISIGER (32), *Department of Medicine, University of California, San Francisco, California 94143*

ALBRECHT WENDEL (2, 44), *Physiologisch-Chemisches Institut, Universität Tübingen, D-7400 Tübingen, Federal Republic of Germany*

JOHN WESTLEY (37), *Department of Biochemistry, University of Chicago, Chicago, Illinois 60637*

HERBERT G. WINDMUELLER (14), *Laboratory of Nutrition and Endocrinology, National Institute of Arthritis, Diabetes, Digestive, and Kidney Diseases, National Institutes of Health, Bethesda, Maryland 20205*

Preface

The area covered in this volume, the enzymes of detoxication and techniques of study related to them, has received little attention in this series. This was not a matter of neglect but largely reflected the unavailability, prior to the last decade, of highly purified enzyme preparations. The enzymes of detoxication catalyze a large variety of reactions in which xenobiotics, i.e., foreign compounds, are oxidized, reduced, hydrolyzed, and conjugated. Xenobiotics appear to be their natural substrates. With minor exceptions, the common feature of such enzymes may well be that each is characterized by a very broad specificity for lipophilic substrates.

The thirty or so catalytic proteins that fall into this category have been reviewed as a group (W. B. Jakoby, ed., "The Enzymatic Basis of Detoxication," Volumes I and II, Academic Press, 1980). Methods for most of the enzymes responsible for oxidation and reduction reactions have received attention in this series. The cytochrome P-450 system and other oxygenases have been the subject of special treatment in Volume LII (Biomembranes, Part C). The present volume concentrates on the reactions catalyzing conjugation and hydrolysis, steps in the process of preparing foreign compounds for elimination.

In addition to the enzymes, their substrates, and the assay systems for them, a number of other techniques have been included that may be of value to investigators in this field. Such procedures are mainly oriented toward higher levels of organization, including the use of the whole animals, perfusion of an organ, and the behavior of cells in culture. The latter two techniques have undergone major development during the last decade and form a vital link in the spectrum of methods for the study of detoxication. Such methods now employ the entire range from the intact animal to those "interesting artifacts," the homogeneous enzymes.

WILLIAM B. JAKOBY

METHODS IN ENZYMOLOGY

EDITED BY

Sidney P. Colowick and Nathan O. Kaplan

VANDERBILT UNIVERSITY
SCHOOL OF MEDICINE
NASHVILLE, TENNESSEE

DEPARTMENT OF CHEMISTRY
UNIVERSITY OF CALIFORNIA
AT SAN DIEGO
LA JOLLA, CALIFORNIA

METHODS IN ENZYMOLOGY

EDITORS-IN-CHIEF

Sidney P. Colowick   Nathan O. Kaplan

VOLUME VIII. Complex Carbohydrates
Edited by ELIZABETH F. NEUFELD AND VICTOR GINSBURG

VOLUME IX. Carbohydrate Metabolism
Edited by WILLIS A. WOOD

VOLUME X. Oxidation and Phosphorylation
Edited by RONALD W. ESTABROOK AND MAYNARD E. PULLMAN

VOLUME XI. Enzyme Structure
Edited by C. H. W. HIRS

VOLUME XII. Nucleic Acids (Parts A and B)
Edited by LAWRENCE GROSSMAN AND KIVIE MOLDAVE

VOLUME XIII. Citric Acid Cycle
Edited by J. M. LOWENSTEIN

VOLUME XIV. Lipids
Edited by J. M. LOWENSTEIN

VOLUME XV. Steroids and Terpenoids
Edited by RAYMOND B. CLAYTON

VOLUME XVI. Fast Reactions
Edited by KENNETH KUSTIN

VOLUME XVII. Metabolism of Amino Acids and Amines (Parts A and B)
Edited by HERBERT TABOR AND CELIA WHITE TABOR

VOLUME XVIII. Vitamins and Coenzymes (Parts A, B, and C)
Edited by DONALD B. MCCORMICK AND LEMUEL D. WRIGHT

VOLUME 61. Enzyme Structure (Part H)
Edited by C. H. W. HIRS AND SERGE N. TIMASHEFF

VOLUME 62. Vitamins and Coenzymes (Part D)
Edited by DONALD B. McCORMICK AND LEMUEL D. WRIGHT

VOLUME 63. Enzyme Kinetics and Mechanism (Part A: Initial Rate and Inhibitor Methods)
Edited by DANIEL L. PURICH

VOLUME 64. Enzyme Kinetics and Mechanism (Part B: Isotopic Probes and Complex Enzyme Systems)
Edited by DANIEL L. PURICH

VOLUME 65. Nucleic Acids (Part I)
Edited by LAWRENCE GROSSMAN AND KIVIE MOLDAVE

VOLUME 66. Vitamins and Coenzymes (Part E)
Edited by DONALD B. McCORMICK AND LEMUEL D. WRIGHT

VOLUME 67. Vitamins and Coenzymes (Part F)
Edited by DONALD B. McCORMICK AND LEMUEL D. WRIGHT

VOLUME 68. Recombinant DNA
Edited by RAY WU

VOLUME 69. Photosynthesis and Nitrogen Fixation (Part C)
Edited by ANTHONY SAN PIETRO

VOLUME 70. Immunochemical Techniques (Part A)
Edited by HELEN VAN VUNAKIS AND JOHN J. LANGONE

VOLUME 71. Lipids (Part C)
Edited by JOHN M. LOWENSTEIN

VOLUME 72. Lipids (Part D)
Edited by JOHN M. LOWENSTEIN

VOLUME 73. Immunochemical Techniques (Part B)
Edited by JOHN J. LANGONE AND HELEN VAN VUNAKIS

Methods in Enzymology

Volume 77
DETOXICATION AND DRUG METABOLISM:
CONJUGATION AND RELATED SYSTEMS

Section I

Animal Organ and Cell Preparations

A. General Methods
Articles 1 through 10

B. Organ Perfusion
Articles 11 through 14

C. Cells
Articles 15 through 19

[1] Exhalation of Isotopic CO₂

By JOHANNES BIRCHER and RUDOLF PREISIG

The collection of $^{14}CO_2$ derived from appropriately labeled test compounds has several methodological advantages. The procedure is noninvasive and, therefore, does not disturb the organism being tested. The technical realization is quite easy, requiring no sophisticated equipment other than a liquid scintillation counter. The method is extremely versatile because it can be applied to a wide range of test compounds, at pharmacological or tracer doses, in many species and in isolated organs or cells. However, the processes leading from the administration of the test compound to the formation of $^{14}CO_2$ are complex.[1] An optimal experimental design, therefore, requires an understanding of the mechanisms involved and of the limitations related to the many technical aspects that need to be considered.

Chemical and Biochemical Basis of $^{14}CO_2$ Breath Tests

The most frequently used $^{14}CO_2$ breath tests have been demethylation reactions (see table).[2-15] For this purpose, the test compounds must be labeled at a methyl group that is demethylated within the organisms, e.g., by the mixed function oxidases of the liver. In most instances, demethylation results in CO_2 only after several further oxidation steps with formaldehyde, formic acid, and carbonic acids as intermediates. Thus, the label

$$R-{}^{14}CH_3 \;\longrightarrow\; H_2{}^{14}C{=}O \;\longrightarrow\; H{}^{14}C{\overset{\displaystyle O}{\underset{\displaystyle OH}{}}} \;\longrightarrow\; H_2{}^{14}CO_3 \;\rightleftharpoons\; {}^{14}CO_2 \uparrow$$

| ^{14}C-Methylated Substrate | Formaldehyde | Formic Acid | Carbonic Acid | Carbon Dioxide |

[1] J. Bircher, *in* "Principles of Radiopharmacology" (L. G. Colombetti, ed.), Vol. 3, p. 179. CRC Press, Boca Raton, Florida, 1979.

[2] B. H. Lauterburg and J. Bircher, *J. Pharmacol. Exp. Ther.* **196**, 501 (1976).

[3] S. Hottinger, W. Röllinghoff, H. Wietholtz, and R. Preisig, *Biochem. Pharmacol.* (submitted for publication).

[4] S. Hottinger, W. Röllinghoff, and R. Preisig, *in* "The Liver: Quantitative Aspects of Structure and Function" (R. Preisig and J. Bircher, eds.), p. 155. Editio Cantor, Aulendorf, West Germany, 1979.

[5] G. W. Hepner and E. S. Vesell, *Clin. Pharmacol. Ther.* **20**, 654 (1976).

[6] J. Bircher, A. Küpfer, I. Gikalov, and R. Preisig, *Clin. Pharmacol. Ther.* **20**, 484 (1976).

METHODS IN ENZYMOLOGY, VOL. 77

SOME TEST COMPOUNDS AND SPECIES IN WHICH $^{14}CO_2$ BREATH TESTS HAVE BEEN CARRIED OUT

Test compound	Investigated reaction	Species	Experimental design[a]	References
[dimethylamine-^{14}C]Aminopyrine	Demethylation	Rat	OP, IO, HM	2,3
[dimethylamine-^{14}C]Aminopyrine	Demethylation	Chicken	OP, HM	Fig. 3; 4
[dimethylamine-^{14}C]Aminopyrine	Demethylation	Pig	SA, IO	Unpublished
[dimethylamine-^{14}C]Aminopyrine	Demethylation	Turkey	SA, IO	Unpublished
[dimethylamine-^{14}C]Aminopyrine	Demethylation	Horse	SA, IO	Unpublished
[dimethylamine-^{14}C]Aminopyrine	Demethylation	Man	SA, IO	Fig. 4; 5,6
[methoxy-^{14}C]Glycodiazine	Demethylation	Man	SA, IO	Fig. 4; 7
[methyl-^{14}C]Monomethylaminoantipyrine	Demethylation	Guinea pig	OP, IO	8
[dimethyl-^{14}C]Dimethylaminobenzene	Demethylation	Rat	SA, IO	9
[1-methyl-^{14}C]Diazepam	Demethylation	Man	SA, IO	10
[3-methyl-^{14}C]Mephenytoin	Demethylation	Dog	SA, IO	11
[7-methyl-^{14}C]Caffeine	Demethylation	Man	SA, IO	Fig. 4
L-[U-^{14}C]Ornithine	Decarboxylation	Rat	OP, HM	Fig. 3; 3
[6-^{14}C]Uric acid	Decarboxylation	Rat	OP, IO, HM	Fig. 3; 3,12
[acetyl-^{14}C]Phenacetine	Deacetylation	Rat	OP, IO	13
D-[U-^{14}C]Galactose	Galactokinase	Rat	OP, HM	Fig. 3
D-[U-^{14}C]Galactose	Galactokinase	Man	SA, IO	Fig. 4; 14
[26-^{14}C]Cholesterol	Side chain oxidation	Rat	OP, HM	Fig. 3; 15

[a] OP, Output measurement; SA, specific activity measurement; HM, hepatocyte monolayers in primary culture; IO, intact organism *in vivo*.

has to pass through the pools of several physiological intermediary metabolites resulting in changes of specific activity and in losses to other routes of metabolism. Under these circumstances the rate of $^{14}CO_2$ formation reflects the rate of metabolism of the administered test compound only if the modifications occurring at the level of the intermediary metabolites are predictable, i.e., remain the same for all the investigated experimental conditions. Furthermore, it is necessary that the initial demethylation reaction be the rate-limiting step and that all further oxidations are much more rapid.[16,17]

Fortunately, it is now well established that the preceding conditions are fulfilled to a large measure for many test compounds and experimental circumstances. For instance, the use of [*dimethylamine*-^{14}C]aminopyrine for assessing microsomal demethylation in the rat, in healthy human volunteers, and in patients with various liver diseases has been validated by independent measurements of the fate of the test compound in blood.[2,5,6] If, however, the procedure is to be applied to other experimental systems, the assumptions underlying the interpretation of the results have to be reevaluated.

Apart from demethylation reactions, the technique may be applied to study deacetylations,[13] decarboxylations,[12] and other rate limiting enzymes such as galactokinase.[14] In each instance, attention to the involved physiological processes and biochemical reactions is needed in order to realize an experimental design that assures that the rate of $^{14}CO_2$ exhalation really reflects the process to be studied. For instance, in the case of galactose in man, it was necessary to administer the sugar in sufficient quantities to saturate the galactokinase enzyme, thereby realizing first order galactose elimination for the first 40 to 60 min of the experiment.

[7] R. Platzer, R. L. Galeazzi, G. Karlaganis, and J. Bircher, *Eur. J. Clin. Pharmacol.* **14**, 293 (1978).

[8] I. Roots, S. Nigam, S. Gramatzki, G. Heinemeyer, and A. G. Hildebrandt, *Naunyn-Schmiedeberg's Arch. Pharmacol.* **313**, 175 (1980).

[9] G. W. Hepner and E. P. Peiken, *Gastroenterology* **76**, 267 (1979).

[10] G. W. Hepner, E. S. Vesell, A. Lipton, H. A. Harvey, G. R. Wilkinson, and S. Schenker, *J. Lab. Clin. Med.* **90**, 440 (1977).

[11] A. Küpfer and J. Bircher, *J. Pharmacol. Exp. Ther.* **209**, 190 (1979).

[12] B. Lauterburg, V. Sautter, R. Herz, J. P. Colombo, F. Roch-Ramel, and J. Bircher, *J. Lab. Clin. Med.* **90**, 92 (1977).

[13] P. V. Desmond, R. A. Branch, I. Calder, and S. Schenker, *Proc. Soc. Exp. Biol. Med.* **164**, 173 (1980).

[14] L. Grimm, J. Bircher, and R. Preisig, *Z. Gastroenterol.* **18**, 45 (1980).

[15] H. Wietholtz, S. Hottinger, and R. Preisig, *Proc. Int. Bile Acid Meet., 6th, 1981* (in press).

[16] H. Waydhas, K. Weigh, and H. Sies, *Eur. J. Biochem.* **89**, 143 (1978).

[17] C. Waydhas, H. Sies, and E. L. R. Stokstad, *FEBS Lett.* **103**, 366 (1979).

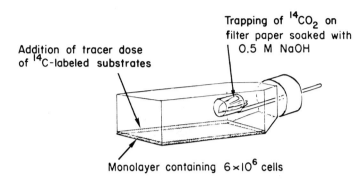

FIG. 1. Test system used for quantitative collection of CO_2 yield in tissue culture. Through the screw cap of a tissue culture flask, a plastic dipper is mounted to hold a Whatman GFA filter paper soaked with 200 μl 0.5 N NaOH. Following addition of tracer doses of specifically labeled substrates, $^{14}CO_2$ can be collected without disturbing the integrity of the hepatocyte monolayer.

This experimental design was reflected in the exhaled air by a linear rise in the specific activity of $^{14}CO_2$ for 40 to 60 min.[14]

The Techniques of $^{14}CO_2$ Measurements and Their Implications

If cell cultures, isolated organs, or small animals are investigated, it is ideal to sample all exhaled air continuously and to measure total $^{14}CO_2$ output per unit of time. $^{14}CO_2$ produced by hepatocyte monolayers can, for instance, be sampled simply by suspending an alkalinized filter paper within the cell culture flask (Fig. 1). At appropriate intervals the filter paper is transferred to a counting vial containing a scintillation fluid that will mix with aqueous $NaHCO_3$.[3] Simple experimental systems, e.g., a series of gas wash bottles, are also sufficient for small animals.[2] By choosing appropriate doses and sampling intervals, the time course of $^{14}CO_2$ production can be followed as accurately and as long as needed. Alternatively, similar measurements may be achieved within an ionization chamber.[8]

If $^{14}CO_2$ breath tests are applied to man and to larger animals, continuous output measurements are impractical. Exhaled air is, therefore, collected intermittently and, instead of output, the specific activity of $^{14}CO_2$ is assessed. Sample collection in man (Fig. 2) has now become so easy that patient acceptance is excellent, and technician time in the laboratory is minimal.[18] For large animals exhaled air is aspirated from the animal box (which must be closed for 5 min prior to sampling) or through a funnel, which is held in front of the nose.[11] The air is first humidified in a

[18] S. Scherrer, B. Haldimann, A. Küpfer, F. Reubi, and J. Bircher, *Clin. Sci.* **54**, 133 (1978).

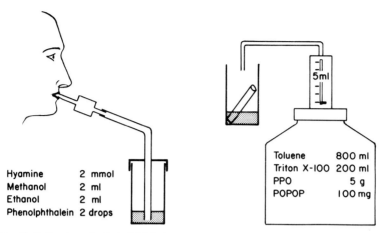

FIG. 2. Collection of exhaled $^{14}CO_2$ in man. The valve attached to the straw is essential, because it prevents aspiration of hyamine, which is caustic to the oral mucosa. The subject has to blow his exhaled air into the hyamine solution until the pink color of the added phenolphthalein has disappeared (approximately 1 to 2 min). At that moment, exactly 2 mmol of CO_2 have been collected. The straw is then cut, the scintillation fluid added and the sample counted. The Triton X-100 content of the scintillation cocktail has rendered precautions to avoid humidity in the sample unnecessary.[18]

gas wash bottle containing ethanol and then led through a vial containing 2 mmol of hyamine and phenolphthalein as in Fig. 2.

Figure 4 shows data on the specific activity of $^{14}CO_2$. When only the specific activity of exhaled $^{14}CO_2$ is measured, it is important to realize that the endogenously produced ^{12}CO is taken as the standard of refer-

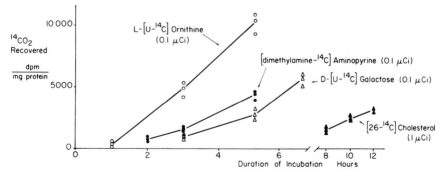

FIG. 3. $^{14}CO_2$ "exhalation" breath tests in tissue culture. Three adult rat hepatocyte monolayers, cultured for 2 days, were each incubated with one of the different labeled substrates for a period of time, and $^{14}CO_2$ was collected by use of NaOH-soaked filter paper suspended with a plastic dipper in the atmosphere of the culture flask.

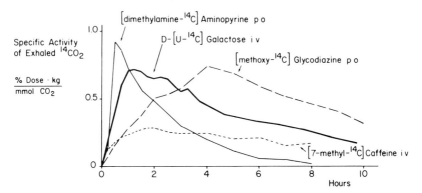

FIG. 4. Specific activity of exhaled $^{14}CO_2$ in different human subjects tested with four substances. The ordinate has been corrected for the dose and body weight of the subjects. It is obvious that the time course of $^{14}CO_2$ exhalation is different for each of the test compounds.

ence. In interpreting the data, consideration must be given to changes in $^{12}CO_2$ output as a result of many processes that are entirely unrelated to the specific metabolic reaction to be investigated. For example, endogenous CO_2 production is decreased by sleep, hypothermia, and hypothyroidism and increased by physical exertion, meals, fever, and hyperthyroidism. It is also smaller in adults and in larger species than in newborns and in smaller species. When designing experiments based on intermittent determination of the specific activity of exhaled $^{14}CO_2$, these variables need to be carefully controlled. In practice, patients should remain seated or be confined to bed; they should not be walking. After arrival at the laboratory, they should rest 30 min before starting the breath test in order to assure basal conditions.[19]

Pharmacokinetic Interpretation of Breath Test Results

When considering the results from $^{14}CO_2$ breath tests, it is appropriate to apply relatively simple pharmacokinetic principles only under optimal circumstances. It was possible, for instance, to model the pharmacokinetics of glycodiazine and to show that the terminal disappearance of the drug from plasma was identical to the terminal elimination of $^{14}CO_2$ from exhaled air.[7] In the case of aminopyrine, there are two labeled methyl groups with different rates of demethylation. If caffeine is used as substrate, the labeled methyl group may either be demethylated from caffeine itself or from one of its metabolites, which may be formed before demethylation of the label has occurred. Thus, loss of the labeled ^{14}C atom

[19] H. S. Winchell, H. Stahelin, and N. Kusubov, *J. Nucl. Med.* **12,** 711 (1970).

from the main metabolic pathway with reentry at a later time must always be considered. Such losses and reentry pathways may also occur with formaldehyde, formic acid, and carbonic acid. Because the label requires some time to travel through alternate pathways, the latter are likely to become more and more important with increasing time. From the foregoing it is evident that interpretation of the $^{14}CO_2$ breath test results are easier if only the early portion of the $^{14}CO_2$ exhalation time curves is considered.

$^{13}CO_2$ Collection

When ^{13}C-labeled test compounds are used, it obviously is necessary to assess $^{13}CO_2$ in exhaled air. The technique differs markedly from collection of $^{14}CO_2$.[20] Therefore, the experimental design requires changes.

The natural abundance of ^{13}C atoms is about 1%. Thus an instrumentation should be available that will detect small increases in the $^{13}CO_2/^{12}CO_2$ ratio of exhaled air. So far the method of choice has been single ion monitoring in a mass spectrometer.[20] Alternatively, infrared spectroscopic techniques may become available in the future[21] and would have the advantage of easier sample preparation.

Unfortunately, the natural abundance of ^{13}C atoms may not be stable. Carbohydrates from different sources can have different $^{13}C/^{12}C$ ratios. Exhaled $^{13}CO_2$ will, therefore, not only be influenced by a ^{13}C-labeled test compound, but also by food intake. In practice, the individuals under investigation should fast during the test.[22] This requirement, however, drastically limits the duration of a $^{13}CO_2$ exhalation breath test.

The lack of radiation involved in ^{13}C-labeled test compounds is a decided advantage, particularly for investigations in pediatric age groups. The limitations imposed by the $^{13}CO_2$ detection techniques, however, require that a much larger mass of the test compound be administered. Because larger amounts are used, the pharmacological consequences and potential adverse effects of the administered test compounds cannot be neglected.

Acknowledgment

This work was supported by the Swiss National Science Foundation.

[20] D. A. Schoeller, J. F. Schneider, N. W. Solomos, J. B. Watkins, and P. D. Klein, *J. Lab. Clin. Med.* **90**, 412 (1977).
[21] S. Hirano, T. Kanamatsu, Y. Takagi, and T. Abei, *Anal. Biochem.* **96**, 64 (1979).
[22] J. F. Schneider, D. A. Schoeller, B. Nemchansky, J. L. Boyer, and P. D. Klein, *Clin. Chim. Acta* **84**, 153 (1978).

[2] Hydrocarbon Exhalation

By ALBRECHT WENDEL and ERICH E. DUMELIN

Lipid peroxidation encounters increasing interest as a basic pathophysiological phenomenon responsible for cell damage at the molecular level.[1,2] Several *in vitro* methods are available for the estimation of lipid peroxidation: measurement of malondialdehyde formation by the thiobarbituric acid reaction; spectrophotometry of conjugated dienes at 232 nm; and determination of fluorescent products with different excitation/emission wavelength couples depending on the biological source. Research in food chemistry provided the basis of a new technique for measurement of lipid peroxidation in living animals by quantitative gas chromatographic determination of exhaled saturated short-chain hydrocarbons.[3] The overall reaction is the following:

$$R{=}CH{-}CH_2{-}CH{=}CH{-}(CH_2)_n{-}CH_3 \xrightarrow[\text{decomposition}]{\text{peroxidation}}$$
$$H_2(CH_2)_{n+1} + R{=}CH{-}CH{=}CH{-}CHO$$

R = carboxyl end of a polyunsaturated fatty acid
$n = 1$: ω 3-fatty acids,　　(linolenic acid family) $\rightarrow$ ethane
$n = 4$: ω 6-fatty acids,　　(linoleic acid family) $\rightarrow$ pentane

The mechanism is most likely to include abstraction of a hydrogen of the unsaturated carbon closest to the methyl end; isomerization to a diene radical; addition of oxygen to form a hydroperoxide; and β-scission to a hydroxyl and an alkoxy radical, the latter of which forms the corresponding alkane by hydrogen atom addition.[4,5]

Quantitative Analysis of Hydrocarbons from Breath

Method A[6,7]

RESPIRATORY CHAMBER

A mouse is placed within a cylindrical Plexiglas chamber (10 cm in diameter, 7.5 cm in height) with a screw top. A rubber ring between the

[1] G. L. Plaa and H. Witschi, *Annu. Rev. Pharmacol. Toxicol.* **16,** 125 (1976).
[2] M. Wolman, *Isr. J. Med. Sci.* **11,** 1 (1975).
[3] C. Riely, G. Cohen, and M. Lieberman, *Science* **183,** 208 (1974).
[4] C. D. Evans, G. R. List, A. Dolev, D. G. McConell, and R. L. Hoffmann, *Lipids* **2,** 432 (1967).
[5] D. H. Donovan and D. B. Menzel, *Experientia* **34,** 775 (1978).
[6] A. Wendel and S. Heidinger, *Res. Commun. Chem. Pathol. Pharmacol.* **28,** 473 (1980).
[7] A. Wendel and S. Feuerstein, *Biochem. Pharmacol.* **30,** 2513–2520 (1981).

lower and the upper part of the chamber seals the system. The rubber ring and the threads are greased with starch–glycerol (10 g in 100 ml, heated to 160° and cooled). Inside the Plexiglas cylinder, a stainless steel net, which is centrally held by two 1-cm Plexiglas rings attached to the bottom and to the top of the vessel, forms a second cylinder with a 7-cm diameter. The space between the Plexiglas and stainless steel net is filled with 120 g of a CO_2 adsorber and a desiccant (e.g., Sodasorb). The animal sits on a sieve disc 1 cm above the chamber bottom. The cover contains two stainless steel inlets with stopcocks. Once the animal is inside, the chamber can be flushed with synthetic air of the purest grade. This may be necessary in urban centers with polluted air. One inlet is connected to the oxygen supply: the simplest way is to attach a moderately inflated rubber balloon; the best way is to connect it to a spirometer filled with O_2. A calibration curve for every hydrocarbon gas is obtained by flushing the chamber with a premixed atmosphére of hydrocarbon gases in the range of 0.5–1.0 ppm (approximately 10–40 pmol hydrocarbon per ml), followed by sampling and analysis.

AIR SAMPLING

1. Direct Sampling from a Closed System. Single animals are allowed to breath in the chamber for several hours. At appropriate periods, the oxygen supply is closed, a gas-tight 20-ml glass syringe with a Teflon plunger (e.g., Hamilton Nr. 1020) is pushed back and forth several times for mixing, and approximately 10 ml of air are withdrawn. The outlet is closed and the oxygen supply is immediately reopened. The syringes must be free of ail lubricants! At this stage, samples may be stored for several hours before analyzing the contents. It should be noted that any lubricant with an oil base, especially in fittings or mechanical pumps, will absorb hydrocarbon gases and complicate analysis.

Gas Chromatographic Analysis. A 3-ml gas sampling valve attached to a gas chromatograph is flushed and filled by applying a 10-ml gas sample. A Carlo Erba Fractovap Model 2150 with flame ionization detection has been used. The system is equipped with a 5 m × 4 mm stainless steel column filled with Porasil C (80–100 mesh). Nitrogen of the highest grade of purity is used as carrier gas. Before entering the column, it passes a precolumn (2.5 × 25 cm) filled with $Na_2SO_4 \cdot 12H_2O$ in order to establish constant moisture; this is absolutely essential for good performance. Analyses are run isothermally at 70° with a nitrogen flow of 30 ml/min. The air flow rate through the flame ionization detector is 570 ml/min synthetic air and 23 ml/min hydrogen, both of the purest grade. Electrometer sensitivity is set to 0.1 pA/mV. Under these conditions, the elution pattern illustrated in Fig. 1 with a standard hydrocarbon mixture (A) and from an

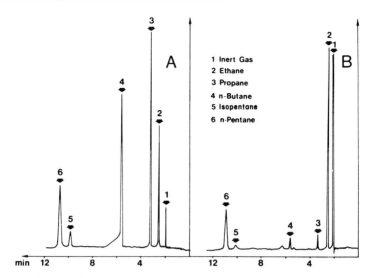

FIG. 1. High-sensitivity gas chromatogram of hydrocarbons on Porasil C. (A) Standard calibration mixture (amount/retention time): ethane 48.9 pmol/154 sec; propane 67.0 pmol/ 216 sec; n-butane 55.1 pmol/350 sec; isopentane 7.5 pmol/605 sec; n-pentane 40.8 pmol/660 sec. (B) Breath exhaled within 4 hr in a closed 500-ml respiration chamber by a male 36-g selenium-deficient mouse injected with 500 mg/kg paracetamol. Analyzed volume: 3 ml of gas.

experiment with a mouse (B) are obtained. After 50 consecutive analyses it is necessary to bake the column at 170° for at least 30 min.

2. Sampling for Concentration from an Open System. If very low amounts of hydrocarbons are to be measured, as is the case for humans, breath samples have to be collected over a period of time. Patients exhale breath into a plastic bag or into larger devices of known volume, e.g., spirometers. The breath is absorbed onto a short precolumn attached to a gas chromatograph. Actually, a 3-ml gas sampling valve may be filled with 1 g of Porapak Q (110–120 mesh) and the breath sucked through the column at –98° (methanol/liquid N₂) by a membrane pump. The loop is switched into the carrier gas stream and flash-heated to 100°. It is essential that elution be carried out in the direction opposite from that for absorption of the hydrocarbon gases. Further analysis is as described.

CALCULATION

Hydrocarbon exhalation is expressed in nmol gas per kg body weight of the animal per time. Several corrections are necessary according to the following expression:

$$\text{nmol gas/kg} = \left[a_i(V_\text{G} - V_\text{A}) + \sum_{i=0}^{n} a_i V_\text{S} \right] W_\text{A}^{-1}$$

where V_G, free gas volume of the respiration chamber (ml); V_A, volume of the animal determined postmortem (ml); W_A, weight of the animal (g); V_S, total sample volume withdrawn from the chamber (ml); i, number of samples withdrawn previously; and a_i, hydrocarbon concentration calculated from integrated calibration peaks (pmol/ml).

The rate of hydrocarbon exhalation of animals upon oxidative or xenobiotic-induced stress is usually not linear with time. After a lag phase of 30 min, a quasi-linear phase is observed that usually ends in respiratory depression before death. Under such conditions, the total amount expired over the entire time period must be given. At low xenobiotic doses or for the estimation of spontaneous exhalation, the value for nmol $\times$ kg body weight$^{-1} \times$ hr^{-1} is adequate.

SENSITIVITY

With this method, 1 pmol of hydrocarbon per sample can be accurately determined. For quantitation of smaller amounts, a larger sampling loop can be used, or samples may be concentrated before analysis (Method A,2).

Method B[8]

OPEN-SYSTEM-RESPIRATION CHAMBER

A rat is placed in a glass chamber with its head protruding into a stainless steel chamber through two collars: one is a rubber gasket and the other a smaller diameter rigid Teflon collar. A stream of hydrocarbon-free air flowing at 120–180 ml/min passes the rat's head, is dried as it passes through a tube of indicating Drierite, and is then split to obtain a flow rate of 60–90 ml/min.

SAMPLE COLLECTION

A 200–400 ml sample of the air-breath stream is absorbed on a stainless column (20 $\times$ 0.3 cm) partially filled with activated alumina (80–100 mesh) immersed into an ethanol–liquid nitrogen slush ($-130°$) and connected to a six-way gas sampling valve. The outlet is connected to a 1-liter vacuum flask by means of which quantitative sampling can be done. Immediately after removal of the trap from the cold slush, the sample is injected into the gas chromatograph *via* the gas sampling valve while the trap is heated by immersion into hot water.

ANALYSIS

A Varian gas chromatograph (model 1520 or 3700) is equipped with a column (150 to 300 $\times$ 0.3 cm) that contains activated alumina (80–100

[8] C. J. Dillard, E. E. Dumelin, and A. L. Tappel, *Lipids* 12, 109 (1977).

mesh). Analyses are performed with a nitrogen carrier flow rate of 25–40 ml/min, a hydrogen flow rate of 30 ml/min, and an air flow rate of 300 ml/min. A variety of temperature programs can be used. One program is 50° for 1 min, 20° rise/min for 7 min to 190°, hold for 3 min, followed by a 20° rise/min to 250°. The injector temperature is 165°, detector 265°, and the electrometer setting 2–16 pA/mV.

CALCULATIONS

The volume of air collected is corrected to standard temperature and pressure and the following calculation is applied:

pmol hydrocarbon gas/kg $\times$ min

$$= \frac{\text{incoming flow rate in ml/min} \times \text{pmol gas}}{\text{volume of air collected in ml} \times \text{body weight} \times 10^{-3}}$$

The value for pmol of gas is calculated by comparison with standards injected into the gas chromatograph.

Comments

TECHNICAL COMMENTS

Several alternative methods for collection and measurement of volatile hydrocarbons have been described. The variants include columns with Porapak,[3,9–11] silica gel,[12] and Carbosieve B.[13] With respect to sensitivity, low bleeding and high resolution Porasil columns (Method A) and activated alumina (Method B) appear to be the best choice.

The method of sample collection depends on the problem under investigation. It is convenient to work with a closed system, but the following reservations have to be made: (1) It is difficult to inject the animals during the experiment. (2) Rodents metabolize hydrocarbons. The following apparent half-lives have been found with mice: ethane, not significantly metabolized within 4 hr; ethylene, 7.6 min/kg; n-propane, 3.1 min/kg; n-butane, 2.5 min/kg; isopentane, 2.1 min/kg; n-pentane, 1.8 min/kg.

GENERAL COMMENTS

The pattern of hydrocarbons exhaled by mammals is subject to numerous influences (diet, age, sex, diurnal rhythm), only a few of which[13–15] have been investigated. In addition, the fate of hepatic lipid

[9] D. G. Hafemann and W. G. Hoekstra, *J. Nutr.* **107**, 656 (1977).
[10] A. Wendel, S. Feuerstein, and K. H. Konz, *Biochem. Pharmacol.* **28**, 2051 (1979).
[11] R. F. Burk and J. M. Lane, *Toxicol. Appl. Pharmacol.* **50**, 467 (1979).
[12] U. Köster, D. Albrecht, and H. Kappus, *Toxicol. Appl. Pharmacol.* **41**, 639 (1977).
[13] D. Lindstrom and M. W. Anders, *Biochem. Pharmacol.* **27**, 563 (1978).
[14] C. J. Dillard, R. E. Litov, and A. L. Tappel, *Lipids* **13**, 396 (1978).
[15] M. Sagai and T. Ichinose, *Life Sci.* **27**, 731 (1980).

hydroperoxide, its breakdown to a volatile hydrocarbon, and its transport to the lung alveoli have received only limited study. The isolated perfused rat liver shows an immediate and reversible release of hydrocarbons upon infusion of hydroperoxides or drug substrates.[16] In an *in vitro* model system, balance studies showed that, depending on the fatty acid composition, the catalyst, and the calculation basis used, up to 1.3 mol % of pentane, 4.3% ethane, and 10.6% ethylene were formed from 100% polyunsaturated fatty acid.[17] Figure 1B shows that a mouse also exhales low amounts of propane, *n*-butane, and isopentane. Not only are their origin and significance unknown, but it cannot be assumed that production of large amounts implies major biological significance.

Acknowledgment

This work was supported by the Deutsche Forschungsgemeinschaft Grant We 686/6.

[16] A. Müller, P. Graf, A. Wendel, and H. Sies, *FEBS Lett.* **126**, 241 (1981).
[17] E. E. Dumelin and A. L. Tappel, *Lipids* **12**, 894 (1977).

[3] Measurement of Hydrogen Peroxide Formation *in Situ*

By HELMUT SIES

The production of hydrogen peroxide in intact cells and tissues is now well established as a metabolic event occurring in a variety of subcellular localizations.[1] H$_2$O$_2$ is formed by reduction of oxygen either directly in a two-electron transfer reaction, often catalyzed by flavoproteins, or via an initial one-electron step to O$_2^-$ followed by dismutation to H$_2$O$_2$. Hydrogen peroxide fulfills a number of useful functions in metabolic detoxication as well as in some metabolic processes: in phagocytosis; ethanol and methanol oxidation; thyroid hormone production; during fertilization; and apparently even as second messenger in the insulin response in adipocytes. However, H$_2$O$_2$ is also known as a potentially dangerous oxidant produced, among other sources, during detoxication reactions catalyzed by monooxygenase systems of the endoplasmic reticulum. This calls for the existence of powerful defense systems in aerobic cells capable of controlling the steady state concentration at very low levels. Hydrogen peroxide concentrations are about 10^{-8} *M* in intact bacterial cells (*Micrococcus lysodeikticus*)[2] as well as in intact liver.[3,4]

[1] B. Chance, H. Sies, and A. Boveris, *Physiol. Rev.* **59**, 527 (1979).
[2] B. Chance, *Science* **116**, 202 (1952).
[3] H. Sies and B. Chance, *FEBS Lett.* **11**, 172 (1970).
[4] N. Oshino, B. Chance, H. Sies, and T. Bücher, *Arch. Biochem. Biophys.* **154**, 117 (1973).

The H_2O_2-metabolizing enzymes, hydroperoxidases,[5] fall into two main categories, catalases and peroxidases, with glutathione peroxidase as the most abundant and probably most important one in the latter group. It is noteworthy that, apart from the catalatic reaction, catalases have the capability of catalyzing peroxidatic reactions (the "coupled oxidations"[6]), thus being a member of both groups.

Quantitative detection of intracellular H_2O_2 production is based either (1) on measurement of the steady state level of the catalase–H_2O_2 intermediate (Compound I), or (2) on assaying flux through the peroxidatic reaction as catalyzed by a peroxidase or by catalase. Other indicator systems at present are of minor importance, as is measurement of H_2O_2 that might, under some conditions, escape from intact cells.

Much previous work was carried out with tissue homogenates[7] or with such subcellular fractions as mitochondria[8-10] or microsomes[11-13]; these will not to be discussed here in detail. Because the detection within intact cells and tissues is dependent upon subcellular localization of the indicator system, assignment of a given H_2O_2 production rate to a subcellular space must take into account both the distribution of the indicator system and the diffusion properties of H_2O_2.

Monitoring of the Steady State Level of Catalase–H_2O_2 (Compound I)

The direct optical measurement of Compound I has both quantitative and qualitative advantages for indicating H_2O_2 concentrations and production rates. Compound I is monitored specifically at the wavelength pair of 660–640 nm.[3,14] For quantitation, the method is based on the fact that the heme occupancy of catalase, i.e., the fraction of catalase heme present as Compound I, depends on the rate of H_2O_2 formation, $d[H_2O_2]/dt$, on the total catalase heme concentration, [Cat], and on the concentration of a hydrogen donor A for the peroxidatic reaction, [A]. The hydrogen donor concentration required for half-maximal steady state occupancy, $[A]_{1/2}$, is related in a simple way to the other parameters[14,15]:

$$d[H_2O_2]/dt = k \times [Cat] \times [A]_{1/2}$$

[5] H. Theorell, in "The Enzymes" (J. B. Sumner and K. Myrbäck, eds.), Vol. 2, p. 397. Academic Press, New York, 1951.
[6] D. Keilin and E. F. Hartree, Biochem. J. 39, 293 (1945).
[7] F. Portwich and H. Aebi, Helv. Physiol. Acta 18, 312 (1960).
[8] B. Chance and N. Oshino, Biochem. J. 122, 225 (1971).
[9] G. Loschen, L. Flohe, and B. Chance, FEBS Lett. 18, 261 (1971).
[10] A. Boveris, N. Oshino, and B. Chance, Biochem. J. 128, 617 (1972).
[11] J. R. Gillette, B. B. Brodie, and B. N. Ladu, J. Pharmacol. Exp. Ther. 119, 532 (1957).
[12] R. G. Thurman, H. G. Ley, and R. Scholz, Eur. J. Biochem. 25, 420 (1972).
[13] A. G. Hildebrandt and I. Roots, Arch. Biochem. Biophys. 171, 385 (1975).
[14] H. Sies, T. Bücher, N. Oshino, and B. Chance, Arch. Biochem. Biophys. 154, 106 (1973).
[15] B. Chance and N. Oshino, Biochem. J. 131, 564 (1973).

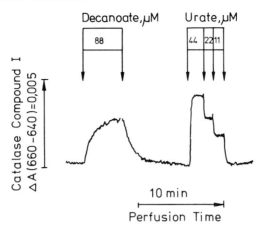

FIG. 1. Quantitation of H_2O_2 production in intact perfused rat liver during decanoate oxidation. Catalase heme occupancy is monitored continuously against time by organ spectrophotometry of Compound I. Calibration of the decanoate response is performed against the urate response. With the known stoichiometry of urate : H_2O_2 of 1 and measurement of the rate of urate removal in effluent perfusate samples, the decanoate response is quantitated to indicate an extra H_2O_2 production of 80 nmol/min per g wet weight. Trace was obtained with organ spectrophotometry attachment to Sigma ZWS11 dual wavelength photometer (Biochem. Co., Munich). For details, see Foerster et al.[18]

For the particular condition of A being methanol or ethanol and for rat liver, k has the value[15] of 32×10^3.

Experimentally, $[A]_{1/2}$ is determined by steady state titration of Compound I in the perfused organ,[4,14] in the exposed organ in situ,[16] or in isolated cells.[17] Methanol is the preferable hydrogen donor since ethanol may complicate matters because of side reactions. Catalase heme can be quantitated spectrophotometrically as the cyanide complex.[14,16] Data are usually expressed as H_2O_2 production in nanomoles per minute per gram wet weight or per 10^6 cells. Since those cellular compartments not equipped with catalase usually do not contribute to the signal, these data may represent an underestimate. Because catalase is located predominantly within the peroxisome in rat liver, this method indicates peroxisomal H_2O_2 production.

A simplified use of catalase Compound I spectrophotometry, independent of the determination of the value of the constant in the preceding equation and of the catalase concentration, is based on the comparison of the response of a test compound or test condition with that obtained by external addition of urate (Fig. 1).[18] In the presence of a low and constant

[16] N. Oshino, D. Jamieson, T. Sugano, and B. Chance, Biochem. J. 146, 67 (1975).

[17] D. P. Jones, H. Thor, B. Andersson, and S. Orrenius, J. Biol. Chem. 253, 6031 (1978).

[18] E. C. Foerster, T. Fährenkemper, U. Rabe, P. Graf, and H. Sies, Biochem. J., 196, 705 (1981).

concentration of methanol, the rise of the steady state level of Compound I upon addition of urate at appropriate concentrations is compared to that obtained by a test compound, e.g., a substrate for peroxisomal β-oxidation such as decanoate. The stoichiometric conversion of urate to allantoin and H_2O_2 by peroxisomal urate oxidase is detected as Compound I.[19] Thus only a simple measurement of urate oxidation in the extracellular fluid[20] is required for calibration.

Assay of Flux through Peroxidatic Reaction to Indicate H_2O_2 Production

The measurement of the rate of hydrogen donor oxidation is applicable to intact organs and cells. Methanol and formate have been employed as substrates for catalase, and glutathione as substrate for glutathione peroxidase.

Peroxidatic Reaction of Catalase Compound I: Methanol and Formate

Scheme 1 shows that the metabolism of C_1 substrates may occur via peroxidatic reaction but that this is not the only possible pathway. Therefore, conditions increasing the selectivity must be chosen. One crucial point resides in the competition between the catalatic and peroxidatic

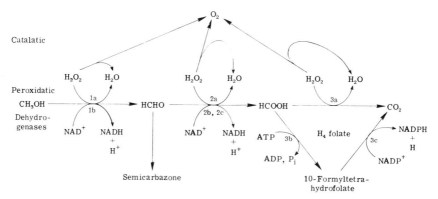

SCHEME 1. Simplified metabolic scheme of C_1 compounds in relation to catalase. ^{14}C-labeled methanol can be carried to formaldehyde or through to [14]CO_2 along the reactions from left to right. The reactions of catalase are catalatic (top) or peroxidatic (middle); the metabolic significance of 2a and 3a is still unclear. Competing reactions are (1b) alcohol dehydrogenase (methylpyrazole sensitive); (2b) formaldehyde dehydrogenase (GSH); (2c) aldehyde dehydrogenase; (3b) 10-formyltetrahydrofolate synthetase; (3c) 10-formyltetra-hydrofolate dehydrogenase (pathway 3b, 3c can be stimulated by L-methionine).

[19] N. Oshino, D. Jamieson, and B. Chance, *Biochem. J.* **146**, 53 (1975).
[20] P. Scheibe, E. Bernt, and H. U. Bergmeyer, "Methoden der enzymatischen Analyse" (H. U. Bergmeyer, ed.), 3rd ed., p. 1999. Verlag Chemie, Weinheim, 1974.

pathways of catalase, practically ruling out integer values of stoichiometric recovery in the peroxidatic reaction. The efficiency of the peroxidatic reaction, i.e., the ratio of the rate through the peroxidatic indicator reaction to the rate of H$_2$O$_2$ production, is increased by raising the hydrogen donor concentration.[21] It should be noted, however, that the efficiency also is inversely related to the rate of H$_2$O$_2$ production (i.e., the steady state turnover number of catalase), requiring redetermination of the efficiency if there are large fluctuations in H$_2$O$_2$ production. Competition with endogenous hydrogen donor further should be considered but usually presents little problem because of the high external hydrogen donor concentration employed.

Further side reactions regarding these C$_1$ substrates may require attention, depending on the nature of the particular biological sample studied. Reactivity of alcohol dehydrogenase toward methanol at the high concentrations used is comparable to the flux through the peroxidatic reaction in rat liver, making inhibition of the dehydrogenase desirable, e.g., by pyrazole derivatives (4-methylpyrazole, 50 μM). The formaldehyde generated may be directly trapped with semicarbazide and then quantitated either with isotopic methods[22] or by the Nash procedure[23] as recently applied also to intact hepatocytes[24] and liver homogenates.[24a] Alternatively, formaldehyde may be further oxidized to yield carbon dioxide as the final product. The first step, oxidation to formate, is predominantly carried out by GSH-dependent formaldehyde dehydrogenase[17] and is highly effective in tissues containing their normal level of glutathione.[17,25] The final step, oxidation of formate to carbon dioxide, is more prone to fluctuations because the 10-formyltetrahydrofolate pathway may be critically altered in its capacity, depending on the availability of an easily washed-out component, L-methionine.[25–27] To ensure high recovery of carbon from methanol in the form of (isotopically labeled) carbon dioxide as the final product, the addition of 0.2 mM methionine is useful.[25] However, if formate is added as external hydrogen donor instead of methanol, the addition of methionine will obviously decrease the indicator efficiency for the peroxidatic reaction of catalase.

[21] N. Oshino, R. Oshino, and B. Chance, *Biochem. J.* **131**, 555 (1973).

[22] E. Feytmans and F. Leighton, *Biochem. Pharmacol.* **22**, 349 (1972).

[23] T. Nash, *Biochem. J.* **55**, 416 (1953).

[24] G. P. Mannaerts, L. J. Debeer, J. Thomas, and P. J. De Schepper, *J. Biol. Chem.* **254**, 4585 (1979).

[24a] N. C. Inestrosa, M. Bronfman, and F. Leighton, *Biochem. J.* **182**, 779 (1979).

[25] C. Waydhas, K. Weigl, and H. Sies, *Eur. J. Biochem.* **89**, 143 (1978).

[26] H. A. Krebs, R. Hems, and B. Tyler, *Biochem. J.* **158**, 341 (1976).

[27] C. Waydhas, H. Sies, and E. L. R. Stokstad, *FEBS Lett.* **103**, 366 (1979).

Methanol[19,24,28,29] and formate[7,30–32] have been applied in studies of the peroxidatic reaction of catalase in cells, organs, and organisms. The studies with adipocytes[31,32] have suggested H_2O_2 as a second messenger in the insulin response.

Peroxidatic Reaction of Catalase Compound I: Aminotriazole

A special case of a peroxidatic reaction, the oxidation of 3-amino-1,2,4-triazole by Compound I, leads to selective alkylation of a histidine and, consequently, to inactivation of catalase.[33] Because the inactivation depends on H_2O_2 production, the measurement of the loss of catalase activity has been used to indicate H_2O_2 formation in intact cells, e.g., erythrocytes,[34,35] and in rat brain.[36]

Glutathione Peroxidase: GSSG Formation and GSSG Efflux

Although reactive not only with H_2O_2 but also with organic hydroperoxides, GSH peroxidase may serve as a tool to detect intracellular H_2O_2 formation. The formation of glutathione disulfide, GSSG, in the peroxidatic reaction can be followed either by GSSG assay in tissue or cell samples[37] or by measuring GSSG efflux from the cells. GSSG efflux is dependent on the rate of H_2O_2 infusion[38] and was found to be increased upon intracellular H_2O_2 formation, e.g., by benzylamine oxidation by monoamine oxidase.[39,40]

Acknowledgment

Supported by the Deutsche Forschungsgemeinschaft.

[28] T. R. Tephly, R. E. Parks, and G. J. Mannering, *J. Pharmacol. Exp. Ther.* **143**, 292 (1964).

[29] D. R. van Harken, T. R. Tephly, and G. J. Mannering, *J. Pharmacol. Exp. Ther.* **149**, 36 (1965).

[30] H. Aebi, F. Frei, R. Knab, and P. Siegenthaler, *Helv. Physiol. Pharmacol. Acta* **15**, 150 (1957).

[31] S. P. Mukherjee, R. H. Lane, and W. S. Lynn, *Biochem. Pharmacol.* **27**, 2589 (1978).

[32] J. M. May and C. deHaen, *J. Biol. Chem.* **254**, 2212 (1979).

[33] E. Margoliash, A. Novogrodsky, and A. Schejter, *Biochem. J.* **74**, 339 (1969).

[34] G. Cohen and P. Hochstein, *J. Pharmacol. Exp. Ther.* **147**, 139 (1965).

[35] R. E. Heikkila and G. Cohen, *Experientia* **28**, 1197 (1972).

[36] P. M. Sinet, R. E. Heikkila, and G. Cohen, *J. Neurochem.* **34**, 1421 (1980).

[37] T. P. M. Akerboom and H. Sies, this volume.

[38] H. Sies and K. H. Summer, *Eur. J. Biochem.* **57**, 503 (1975).

[39] N. Oshino and B. Chance, *Biochem. J.* **162**, 509 (1977).

[40] H. Sies, G. M. Bartoli, R. F. Burk, and C. Waydhas, *Eur. J. Biochem.* **89**, 113 (1978).

[4] Collection of Metabolites in Bile and Urine from the Rat

By GERARD J. MULDER, EGBERT SCHOLTENS, and DIRK K. F. MEIJER

Administration of Drugs

The conjugation pattern of a xenobiotic is dependent not only on its chemical structure, but also on the route of administration and the dose. For example, when a drug is given orally, either as a single dose or with food, it may be subject to first-passage conjugation during absorption in the gut mucosa and to first-passage metabolism in the liver. When a drug is given intravenously, however, it may have a first-passage effect in the lung.[1] Because the ratio between various competing enzyme activities for the same substrate varies greatly with the specific organ, the conjugation may be profoundly altered.

The dose is important because competing conjugation enzymes very likely will have different affinities for the same substrate, and hence the conjugation pattern of a drug is expected to be dose dependent.[2] Therefore, the same dose administered as a bolus injection need not have the same metabolic pattern as an infusion over a longer period. The same applies to the addition of a drug to food or drinking water for oral administration; it is particularly important with food, because its composition can influence the rate of drug entry into the systemic circulation. Because the oral route generally leads to lower plasma levels than does intravenous administration, and because the rate of drug entry may be different for various organs in the body, a varying pattern of conjugation may be found when oral and iv doses are compared. All of these factors should be assessed before considering the route of administration in any protocol involving experimental animals or human volunteers.

Bile Collection

Many details of operative procedures and equipment can be found in "Surgery of the Digestive System in the Rat," by R. Lambert.[3] It should be noted that the rat is a rather unique mammal in that it does not have a gall bladder.

[1] M. K. Cassidy and J. B. Houston, *J. Pharm. Pharmacol.* **32,** 57 (1980).
[2] J. M. van Rossum, C. A. M. van Ginneken, P. T. Henderson, H. C. J. Ketelaars, and T. B. Vree, *in* "Kinetics of Drug Action" (J. M. van Rossum, ed.), p. 125. Springer-Verlag, Berlin and New York, 1977.
[3] R. Lambert, "Surgery of the Digestive System in the Rat." Thomas, Springfield, Illinois, 1965.

METHODS IN ENZYMOLOGY, VOL. 77

Operative Techniques

ANESTHETIZED RATS

An intraperitoneal injection of pentobarbital (Nembutal, about 50 to 80 mg/kg) is normally used to induce anesthesia in rats. When required, additional anesthetic (15 to 20 mg/kg for example) may be administered later on for maintenance of anesthesia. If oxidative metabolism has been induced in the animals, as with prior treatment with phenobarbital, a greater does of pentobarbital will be required. When the rat is not stressed by the injection procedure, a deep anesthesia develops, usually within 4 to 8 min. Pentobarbital, especially at higher doses, may decrease blood pressure somewhat.

An alternative anesthetic is urethane (1 g/kg bodyweight; intraperitoneally). Although urethane may cause liver damage in chronic usage and is mildly carcinogenic, it is a convenient anesthetic for long-term anesthesia. It takes at least 10 to 15 min before sufficiently deep anesthesia is reached. Urethane should be injected in a rather large volume, e.g., 6 to 10 ml/kg bodyweight, because otherwise the injected solution is too hypertonic and may cause a shift of water from blood to the peritoneal cavity.

The anesthetized rat preparation may be used for as long as 8 hr if sufficient measures are taken to prevent dehydration; a slow infusion of Krebs–bicarbonate buffer containing albumin is effective. Blood pressure usually decreases after 2 to 3 hr and body temperature drops. The temperature of the animal should be maintained between 37.5 and 38.5° by means of a heat lamp or by an electrical heated pad[4] under the rat. Body temperature may be monitored rectally with a small thermometer or with a thermocouple. This is an important parameter: the body temperature of the rat drops after the operation, decreasing metabolism in general and diminishing bile flow.

The operation procedure begins with the catheterization of the external jugular vein for intravenous drug administration, either by infusion or by injection as a bolus. The compound may be dissolved in saline and may have albumin added to improve the dissolution of substances that are poorly water-soluble. Only small volumes, up to 3 ml per kg of body weight of unbuffered solutions of pH 10 or 3, may be injected. Next, the carotid artery is catheterized in order to obtain arterial blood samples. For both catheters, polyethylene tubing (medical quality) with inner diameter of 0.5 mm and outer diameter of 1.0 mm can be used. Blood samples are

---

[4] For example, type HN 8 from Koninklijke Fabriek Inventum N.V., Postbus 4, 3720 AA Bilthoven, The Netherlands.

usually collected in heparinized microtubes that can be centrifuged to furnish plasma.[5]

The arterial blood pressure, normally approximately 110 mm Hg, can be measured in either the carotid or femoral artery with a manometer.[6] It is advisable to check the effect of a test compound on blood pressure, because decreased blood pressure may induce lowered blood flow to the liver and other organs, thereby affecting metabolism. The rate and the extent of conjugation may be altered because of changes in liver blood flow when competing eliminating organs are operating.

Although many substances are devoid of pharmacological effects on respiration, a few have such activity and alter the condition of the preparation. Such effects may occur with very high doses of drugs that otherwise leave respiration unaffected as well. Therefore, it is advisable to routinely apply artificial respiration. This can be achieved simply with the use of a bifurcated tube (Y) secured in the trachea. The Y tube, in turn, is connected to a respiratory pump[7] that delivers approximately 2.1 ml of air at a frequency of 70 to 90 times per minute for a 250-g rat. A correct respiratory state can usually be maintained by checking blood pH at intervals after beginning artificial respiration. Although the breathing pattern of the anesthetized rat is controlled by a respiratory pump, the anesthetic will eventually be decreased and the rat will start to breath with its natural frequency, resulting in an irregular breathing pattern against the enforced artificial respiration. This is a sign that additional pentobarbital should be administered. Irregular breathing may also result from inefficient oxygenation of the rat, e.g., due to saliva and mucous secretion in the respiratory tract or obstruction of the Y tube of the respirator; the tube and part of the trachea should be cleaned out with a small plug of cotton wool. In special cases, it may be appropriate to administer atropine to decrease formation of saliva.

For cannulation of the bile duct, a 2-cm incision is made across the abdomen just below the liver. The overlying liver lobes and gut are gently moved out of the way, and the duodenum is gently pulled. The bile duct is noted as a thin vessel that lies between the hilus, where it leaves the liver, and the point at which it joins the duodenum. The upper part of the bile duct is free and should be used for catheterization at about 1 cm from the liver hilus. In that case the pancreatic ducts are left intact and mixing of bile with pancreatic secretions is avoided.

[5] Caraway Micro Blood Collecting tubes, Sherwood Medical Industries Inc., St. Louis, Missouri.

[6] For instance, the HSE Elektro-manometer, Hugo Sachs Elektronik, D-7801 March-Hugstetten, W. Germany.

[7] Respirator V5KG from Narco Biosystems Inc., Houston, Texas.

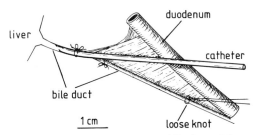

FIG. 1. Position of a cannula in the bile duct of the rat.

A loose surgical wire knot is placed around the bile duct, just above its juncture with the duodenum, and the wire is kept taught so as to straighten the bile duct (see Fig. 1). The bile duct should preferably not be dissected free from the peritoneal folds, in order to prevent damage to small blood vessels; a cannula can be applied without cleaning the bile duct. A small incision is made in the duct, approximately 10 mm below the hilus of the liver, and a cannula is inserted until it is approximately 2 to 3 mm from the hilus. It is fastened to the duct by a knot with surgical wire, and the original loose knot is released. Very little or no tissue injury need occur during the procedure. When, directly below the incision, the bile duct is ligated, the flow from the pancreatic ducts will still be led down the original bile duct into the gut lumen (Fig. 1). Eventually, the arteries of the kidney can be ligated to prevent urinary excretion of drug metabolites. The abdomen is closed with surgical clamps or suture.

Bile production of the rat is approximately 0.8 to 1.2 ml per hour in a 300-g rat for the first few hours. Because of a slow decrease in the bile salt pool size, bile flow decreases with time. Variations among strains will be found, and dietary or drug treatment may influence the amount of bile produced. The entire operation as just described (blood vessels, trachea, bile duct) takes only 10 min for a practiced operator.

FREELY MOVING RATS

In order to collect bile for a prolonged period, e.g., several days, a permanent biliary cannula is fixed to the skull. The bile is collected external to the animal and to the cage (Fig. 2) by way of a swivel joint (Fig. 3) that is connected to polyethylene tubing.[8,9] A much improved version of the swivel joint has been described recently.[10] Under anesthesia induced by diethyl ether, the bile duct of a rat is cannulated in a manner similar to that described in the preceding section. However, in this case, a midline

[8] J. H. Strubbe, *Physiol. Behav.* **12,** 317 (1974).

[9] R. J. Vonk, A. B. D. van Doorn, and J. H. Strubbe, *Clin. Sci.* **55,** 253 (1978).

[10] C. Darracq, P. Gonzalez, and C. Balabaud, *Physiol. Behav.* **25,** 327 (1980).

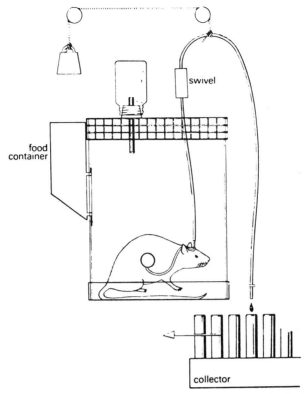

FIG. 2. Continuous collection of bile in the freely moving rat. (Reprinted with permission from Vonk *et al.* [9] and the Biochemical Society, London.)

incision is made to permit better positioning of the cannula. Flexible silicone tubing (0.5 mm i.d. and 1.0 mm o.d.) must be used for the biliary cannula instead of the rigid polyethylene tubing. Obviously, it is essential to leave the lower bile duct intact to ensure an unimpaired flow of pancreatic secretions. The silicone bile cannula is led under the skin to the head, and is attached to a metal cylinder (obtained from an injection needle) bent at a 90° angle. This cannula is fixed with screws and glue[11] to the skull of the rat. The other opening of the metal cylinder is connected by silicone or polyethylene tubing to a second metal cylinder fixed next to the first on the skull (Fig. 4). The second cylinder is connected to silicone tubing that leads bile back to the gut; it is secured in the gut wall of the duodenum so that the bile is delivered into the lumen of the duodenum just above the normal place of entry of the bile duct. This recycling of bile

[11] Ivoclar Resin Cement, from Ivoclar A.G., Schaan, Lichtenstein.

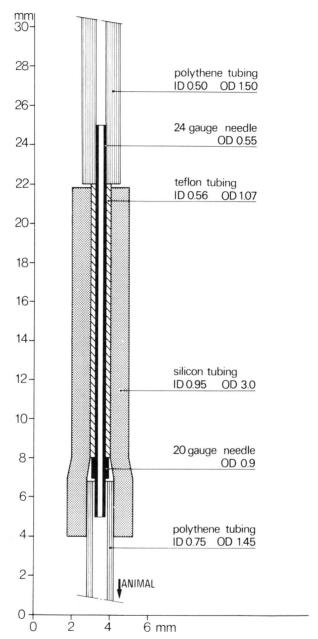

FIG. 3. Swivel joint required for the continuous collection of bile in the freely moving rat. This swivel joint is connected to the tubing outside the cage (see Fig. 2), and prevents torsion of the tubing consequent to movement of the rat. (Reprinted with permission from Strubbe,[8] and Pergamon Press Ltd.)

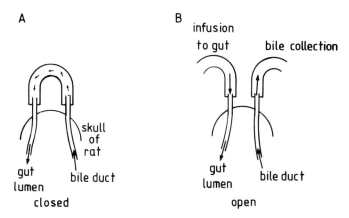

A

B

infusion

to gut   bile collection

skull
of
rat

gut
lumen   bile duct

closed

gut
lumen   bile duct

open

FIG. 4. Closed (A) and open (B) position of the bile duct–gut cannulae fixed to the skull of the freely moving rat.

ensures the conservation of bile and allows for both continuous and intermittent sampling. In the first case, a swivel joint is used and the cannula leading to the gut is closed; in the second, bile is sampled intermittently. The cannula leading to the gut can also be used to administer drugs directly into the lumen of the gut, thereby avoiding the complications of oral administration and the low stomach pH.

Because bile ceases to flow at a pressure exceeding 11 to 15 cm H_2O, care must be taken to ensure bile flow when the biliary cannula is extended outside the cage. To overcome this problem, polyethylene tubing (containing the swivel joint) is filled with water and is connected to the biliary cannula at the skull; the outlet, however, is kept a few centimeters below the level of the cage so as to create a siphoning effect, allowing for the flow of bile (Fig. 2). If required, bile from a second rat may be delivered through the skull cannula of the first rat to the gut, in order to replenish bile that is being collected from the recipient.

An alternative method is to implant a reservoir into the peritoneal cavity of the rat for the collection of bile.[3,12] A side arm of the collection vessel is introduced through the dorsal body wall; through the side arm, samples of bile can be withdrawn from the outside, and the rat can remain conscious and freely moving while bile is being collected. Replenishment of bile salts is not possible unless an intravenous infusion, or an infusion via the gut, is applied.

CONJUGATION *in Vivo* IN LIVER

In order to study the role of the liver in the conjugation of xenobiotics, two other operative procedures may be useful. A bypass may be used to

[12] P. Johnson and P. A. Rising, *Xenobiotica* **8**, 27 (1978).

temporarily exclude the liver from the circulation.[13,14] Obviously, the liver can be excluded from the system for only short periods, preferably no more than 5 to 10 min, because of the profound impairment of the animals resulting from longer periods. The role of the liver will be assessable from the differences in metabolite pattern of blood or urine during the period of liver occlusion.

Another useful method is the cannulation of a side branch of the portal vein in order to both administer compounds and to sample directly at the portal circulation.[15] This can be achieved for freely moving rats when the cannula is brought to the skull of the animal; blood may be collected outside the cage, or an infusion may be given directly through the cannula. The technique allowing a 100% first passage in the liver and excluding first-passage metabolism by the gut mucosa that occurs after oral dosage has been described in detail.[15]

Bile Salt Supplementation and Choleresis in Bile Duct-Ligated Rats

When bile is sampled frequently or continuously for several hours, the pool size of bile salts decreases as does bile flow. The rat usually responds by increasing its rate of synthesis of bile salts. Nonetheless, bile salt depletion occurs and may influence the rate of excretion of compounds into bile and hence change the pattern of elimination. Therefore, it is usually desirable to support bile production by the intravenous infusion of sodium taurocholate, the main physiological bile salt component of the rat. Sodium taurocholate (4 mM in a saline solution) may be infused into the external jugular vein at a rate of approximately 12 ml/hr/kg body weight (45 to 50 μmol/hr/kg); this should ensure continuous bile flow for the period of infusion. Higher rates of infusion will result in choleresis, during which bile flow may easily be doubled. Alternatively, with a freely moving rat that has been surgically prepared as described in the section "Freely Moving Rats" the taurocholate infusion may be delivered directly into the gut by means of the returning cannula; a rapid response of bile secretion follows since the bile salts are rapidly and completely absorbed.

Urine Collection

Metabolism Cage

A simple way to collect the metabolites of a compound excreted in urine (and feces) is to place the rats in metabolism cages and collect urine and feces separately. Various types of metabolism cages are commercially

[13] K. S. Pang and J. R. Gillette, *Drug Metab. Dispos.* **6**, 567 (1978).
[14] S. Agoston, M. C. Houwertjes, and P. J. Salt, *Br. J. Pharmacol.* **68**, 637 (1980).
[15] T. Suzuki, Y. Saitoh, S. Isozaki, and R. Ishida, *J. Pharm. Sci.* **62**, 345 (1973).

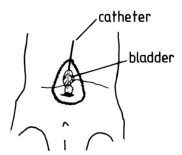

FIG. 5. Position of a catheter in the bladder of the rat.

available. It is obviously essential to prevent as much as possible contact between urine and feces. Therefore, the feces should be cleaned out very regularly, at least once a day. Diarrhea that may develop as a result of the drug treatment requires special attention. It is necessary that the cages be very thoroughly cleaned because of the possibility of contamination with *Escherichia coli* and other microorganisms, which contain enzymes that may chemically change the metabolites as excreted in the urine. For instance, β-glucuronidase activity may be present that would result in hydrolysis of glucuronides. Addition of an antimicrobial agent, e.g., sodium azide, to the urine collection vessel will inhibit bacterial growth.

Rather labile metabolites excreted in urine may break down during the collection period due to, for instance, air oxidation or exposure to light. In such instances urine must be collected over dry ice in a dark box. Finally, it is advisable to have food and water containers outside the cage but within reach of the rats, thereby preventing spillover into the cage. For the same reason, a powdered diet can be given instead of a pelleted diet.

All-glass metabolism cages have been described in which ^{14}C-labeled CO_2 and other gaseous compounds expired by rats can be completely recovered.[16-18]

Operative Techniques

BLADDER CANNULATION

The use of male rather than female rats is preferred because urine loss from the bladder can be blocked easily in males by tying off the penis under anesthesia. The rat should be under anesthesia, e.g., pentobarbital. A midline incision of about 2 cm is made along the linea alba in the lower abdomen of the rat and the apex of the bladder is pulled above the plane of

[16] L. J. Roth, *Nucleonics* **14**, April issue, 104 (1956).
[17] R. W. Wannemacher, A. S. Klainer, R. E. Dinterman, and W. R. Beisel, *Am. J. Clin. Nutr.* **29**, 997 (1976).
[18] This volume, Article [1].

the cut. The bladder wall in incised at its apex for about 3 mm and a catheter is inserted through this opening into the bladder. The catheter is fastened in place with a ligature (see Fig. 5) and is extended outside the animal toward its head. The skin and abdominal wall are subsequently closed with sutures. Because the catheter is put "deep" into the bladder, and the ligature is applied very low, only a very small volume of the bladder is left intact, i.e. there is little "dead space." If diuresis is not stimulated with diuretic agents, e.g., mannitol, the rate at which urine can be collected is very low and irregular.

URETHRA CANNULATION (FEMALE RATS)

In female rats, cannulation of the bladder through the urethra can easily be performed under anesthesia. The length of the urethra is very constant in rats of the same strain and bodyweight: between 2.5 and 3.5 cm for 200-g rats. After this length is determined in different sizes of rats, a mark may be made on the catheter to indicate the length of urethra so as to assure its insertion into the bladder. The catheter is fixed by a ligature and urine is collected. If a diuretic agent is not used, urine production rate is very low, and a reliable time course of the urinary excretion is not possible. This is attributed to the void volume of the bladder where urine is stored.

Although it is possible to cannulate a urethra in both male and female rats, the procedure is cumbersome in males, and the preparation is easily disturbed.

Mannitol Diuresis for Continuous Collection of Urine

Because normal production of urine in 200- to 300-g rats is low under anesthesia, less than 200 μl per hour, the use of a diuretic is required when the collection of urine samples is required during a short period, e.g., 5 min. In the anesthetized rat, an infusion of mannitol is normally used to ensure a constant flow of urine; its long-term use causes kidney damage but, for a few hours, this is a convenient drug. Mannitol (75 mg/ml in saline) is infused in the external jugular vein at a rate of 2 to 4 ml per hr in a 200- to 300-g rat (800 to 1600 μmol/hr). The resulting rate of urine production will be about 100 to 200 μl/min/kg body weight.[19] Although a production rate as high as 2 ml/min/kg can be reached at extremely high infusion rates of mannitol (7.5 g mannitol in 100 ml saline/hr/kg), the result, of course, is a very nonphysiological system.

Acknowledgments

We are very much indebted to Dr. K. Sandy Pang, Dept. of Pharmaceutics, University of Houston, TX, for expert advice during the preparation of this chapter.

[19] J. G. Weitering, K. R. Krijgsheld, and G. J. Mulder, *Biochem. Pharmacol.* **28**, 757 (1979).

[5] Collection of Feces for Studies of Xenobiotic Metabolism and Disposition

By HAZEL B. MATHEWS

Feces are an obvious, but often neglected route of excretion for drugs and other xenobiotics because feces are less desirable and more difficult to work with than is urine. This is an unfortunate oversight because quantitation of excretion in feces can provide valuable data for any study of xenobiotic metabolism and disposition, and in some studies, particularly those involving the more complex aromatics or halogenated aromatics, feces may contain greater than 90% of the metabolites excreted by exposed animals. Furthermore, feces are the only route for excretion of an unabsorbed parent compound administered orally; a study designed to quantitate excretion in feces following an oral versus an intravenous or intraperitoneal dose will provide an accurate estimate of gastrointestinal absorption and bioavailability of the compound administered. Therefore, accurate quantitation of excretion in feces should be an integral part of any quantitative study of xenobiotic disposition and metabolism.

The importance of feces as a route of excretion may vary with the species and the molecular weight of the excreted metabolites. For reasons that are not clear, a molecular weight threshold determines whether a compound will be excreted in urine or in bile and, subsequently, in feces. Compounds with molecular weights below the threshold are excreted primarily in urine, whereas larger molecules are excreted primarily in bile. The molecular weight threshold for excretion in bile is approximately 325 ±50 for the rat and varies with species up to 475 ±50 for the rabbit.[1] The molecular weight referred to is the total molecular weight of the compound excreted; therefore, metabolic conjugation and alkylation increase excretion in feces, and hydrolytic or reductive metabolism to smaller molecules increases excretion in urine. When compounds with molecular weights of 500 or greater are involved, excretion in feces is likely to be the major route of elimination in all species.

Methods of collecting feces from laboratory animals vary from very simple to quite elaborate cages. The complexity of cages required for collection of excrement will be dependent upon the volatility and chemical stability of the anticipated metabolites and the interest in the three major routes of excretion. If interest is restricted to feces, animals may be housed on dry bedding to absorb urine and the feces may be collected

[1] R. L. Smith, "The Excretory Function of Bile." Chapman & Hall, London, 1973.

METHODS IN ENZYMOLOGY, VOL. 77

directly from the bedding at the desired time intervals. When it is desirable to collect both feces and urine, specially designed metabolism cages are required. Metabolism cages for small laboratory animals are commercially available in designs that vary from the relatively inexpensive individual plastic cages[2] to batteries of a dozen or more metal cages,[3] to relatively complex and expensive all-glass cages[4] designed to collect urine, feces, and exhaled air. Here it should be noted that metabolism cages must meet the specifications outlined in the Animal Welfare Act[5] for all animal cages and should be constructed or purchased with these specifications in mind. Other factors to consider in the collection of excreta are the volatility and stability of the compounds of interest. For example, volatile metabolites must be collected in a closed system such as glass cages designed to collect exhaled air,[4,6] and some unstable metabolites may be collected in urine by refrigeration of the urine collector. However, the action of intestinal microbes, intestinal reabsorption, and the time required for intestinal transport make feces an unlikely place to look for unstable or volatile metabolites. If such a metabolite is anticipated in feces, it can best be studied by collecting bile from a cannulated bile duct.[1,7]

As a general rule the best system for studying excretion is the simplest system that will satisfy the needs of the study, because the best results from studies of excretion are obtained when the animal chamber can be disassembled and cleaned daily. In studies of excretion, total feces should be collected at each time point, at least daily, mixed well and accurately weighed and sampled. Because excretion in feces requires a time lag of 12 hr to 1 day, and sometimes longer for intestinal contents to be cleared, little is gained from collection of feces at shorter time periods. Mixing of rodent feces can be achieved by allowing them to dry and grinding them into a powder with a mortar and pestle or other suitable grinding device. Feces of larger animals dry more slowly and mixing is best achieved by adding a sufficient volume of water to permit homogenization. Homogenization of feces is not too unpleasant a task if they are kept frozen from the time they are collected until immediately prior to homogenization.

Assays of excretion in feces may utilize either analytical chemical or radiochemical methods or combinations of the two. Analytical chemical methods offer the advantage of greater specificity for the parent compound or its metabolites, but they are usually more time consuming and

[2] Maryland Plastics, 461 8th Ave., New York, 10001.

[3] Nalge Company, 75 Panorama Creek Dr., Rochester, NY 14602.

[4] Jencons Scientific Ltd. Mark Road, Hemel Hemstead Herts HP27DE, England.

[5] Animal Welfare Act or Public Health Law, 91-579.

[6] See also this volume, Articles [1] and [2].

[7] See this volume, Article [4].

yield less quantitative results for total excretion. Radiochemical assays are generally more rapid, sensitive, and quantitative assays of total excretion than are analytical chemical techniques, but radiochemical methods have the disadvantage of being nonspecific. That is, radiochemical assays measure only radiolabel without regard to the nature of the chemical containing the label. The best results are obtained from studies that utilize a radiolabel to quantitate total excretion and analytical chemical methods to isolate and identify the material excreted. The use of analytical chemical methods on every sample of feces collected is usually too time consuming to be practical and, in most cases, satisfactory results are obtained when they are restricted to a few, carefully selected time points: the initial sample collected; one or more intermediate samples; and the last sample collected if excretion in feces persists at a level sufficient for analytical chemical assay.

When selected samples of feces are chosen for more complete analytical chemical assays, it is usually sufficient to determine and report only the total radioactivity in the remainder of the samples. The more appropriate methods of reporting this data are as percentage of total dose excreted for each period during which feces were collected or as cumulative percentage of total dose excreted. This data is most rapidly and accurately obtained by using ^{14}C- or tritium-labeled compounds, and a sample oxidizer[8,9] that converts all organic material to CO_2 and H_2O. Sample oxidizers are designed to trap the CO_2 and H_2O separately. Thus, any ^{14}C or tritium in the sample is trapped as $^{14}CO_2$ or ^3H_2O and may be quantitated by liquid scintillation counting. Therefore, by keeping accurate records of the amount of feces excreted, oxidizing weighed samples of feces to CO_2 and H_2O in a sample oxidizer and liquid scintillation counting of the trapped CO_2 or H_2O, one is able to accurately determine the percentage of the dose excreted in feces during any time period in which feces were collected (see following example).

Example

Animal No. 14. Total dose = 193 μg [^{14}C] 2,3,6,2',3',6'-hexachlorobiphenyl (2.68×10^6 dpm); collection period = 0 hr to 1 day; feces collected = 4.4 g; sample oxidized = 0.10 g; radioactivity in sample = 48,199 dpm.

$$\text{Hexachlorobiphenyl-derived radioactivity excreted} = \frac{\text{total wt. of feces}}{\text{sample wt.}} \times \text{sample dpm}$$

$$= \frac{4.4 \text{ g}}{0.10 \text{ g}} \times 48,199 \text{ dpm} = 2.12 \times 10^6 \text{ dpm}$$

[8] R. J. Harvey Instrument Corp., Hillsdale, New Jersey.
[9] Packard Instrument Company, Inc., Downers Grove, Illinois.

$$\frac{\text{Percentage of total dose excreted}}{\text{in feces 0 hr to 1 day}} = \frac{\text{dpm in sample} \times 100}{\text{dpm in total dose}}$$

$$= \frac{2.1 \times 10^6 \text{ dpm}}{2.7 \times 10^6 \text{ dpm}} \times 100 = 79\%$$

The excretion of xenobiotics has been reviewed.[10]

[10] H. B. Matthews, in "Introduction to Biochemical Toxicology" (E. Hodgson and F. E. Guthrie, eds.), p. 162. Elsevier, Amsterdam, 1980.

[6] Germfree Rats

By MARTHA A. McLAFFERTY and PETER GOLDMAN

Evidence is accumulating that suggests that the bacterial flora has a pervasive influence on the physiology,[1,2] immune status,[3] and metabolism and nutrition[2,4] of the animal host. The characterization of this influence is frustrated by the enormous complexity of the animal flora, which consists of at least 400 different bacterial species.[5] Even if all of these organisms were classified and quantified, the information gained would not be useful in defining the function of this complex ecosystem.[6] The effect of the flora can best be studied by making observations on animals whose flora has either been simplified or eliminated entirely. Such studies require gnotobiotic animals.

A gnotobiotic animal is one that is derived either by aseptic cesarian section (or hatching of an egg) and reared and maintained under conditions that prevent the introduction of any additional associated forms of life, e.g., bacteria, viruses, fungi, and parasites. The term "germfree" has its limitations because it refers only to *demonstrable* forms of life. Thus the value of the statement that an animal is germfree or gnotobiotic depends on the sophistication of the diagnostic techniques used. A fuller description of these and other aspects of gnotobiotics as well as an extensive bibliography have been compiled.[7]

[1] H. A. Gordon and L. Pesti, *Bacteriol. Rev.* **35**, 390 (1971).

[2] B. S. Wostman, *World Rev. Nutr. Diet.* **22**, 40 (1975).

[3] M. Pollard and A. Nordin, in "Progress in Immunology" (B. Amos, ed.), p. 1295. Academic Press, New York, 1971.

[4] S. M. Levenson and E. Seifter, in "Protein Nutrition" (H. Brown, ed.), p. 74. Thomas, Springfield, Illinois, 1974.

[5] W. E. C. Moore and L. V. Holdeman, *Cancer Res.* **35**, 3418 (1975).

[6] D. C. Savage, *Annu. Rev. Microbiol.* **31**, 107 (1977).

[7] J. R. Pleasants, in "CRC Handbook of Laboratory Animal Science" (E. C. Melby, Jr. and N. H. Altman, eds.), Vol. 1, p. 119. Chem. Rubber Publ., Cleveland, Ohio, 1974.

METHODS IN ENZYMOLOGY, VOL. 77

This discussion is limited to a description of methods that can be employed by laboratories primarily engaged in fields such as bacteriology, pharmacology, toxicology, or nutrition, where gnotobiotic experiments may occasionally be useful. Only a few hundred square feet of space are required for the use of these methods and the space can readily be converted to other purposes when an isolator is not in use.

Our description is limited to a method of performing gnotobiotic experiments that we have found particularly suitable for studying the metabolism of xenobiotic compounds and for certain nutritional experiments. We have used these gnotobiotic techniques to establish whether a metabolic reaction occurring in a conventional rat can be attributed to its flora.[8] Obviously, metabolic transformations found in the conventional rat may result from the action of either mammalian or bacterial enzymes, whereas those occurring in the germfree rat can be attributed exclusively to mammalian enzymes. The role of individual species of bacteria in animal physiology can also be defined by gnotobiotic methods that permit the addition of selected bacteria to the animal host maintained in the gnotobiotic environment.

Germfree rats may be purchased from a number of suppliers and maintained for months or even years in the gnotobiotic environment provided by a plastic isolator. The isolator can be equipped with facilities for performing studies comparable to those done with conventional rats. We describe how the animals, their food and water, and any experimental materials can be transferred into the isolator; how manipulations can be performed within the isolator; and how animal products such as excreta can be removed from the isolator for analysis.

Animals

Rats of either the Sprague-Dawley or the Fisher strain may be obtained from Charles River Breeding Laboratories Inc. (Wilmington, MA 01887) or from Harlan Sprague-Dawley Laboratory (Madison, WI 53711). The rats are shipped throughout the country in special containers that maintain their germfree status and from which the rats may be transferred to isolators such as those described in the following section.

Isolator

The germfree environment is established within the boundaries of a clear flexible vinyl isolator (Fig. 1). The isolator we have used (24 × 24 × 48 in., Standard Safety Equipment Co., Pallatine, IL) readily

---

[8] P. Goldman, *in* "Metabolic Basis of Detoxication" (W. B. Jakoby, J. R. Bend, and J. Caldwell, eds.). Academic Press, New York (in press).

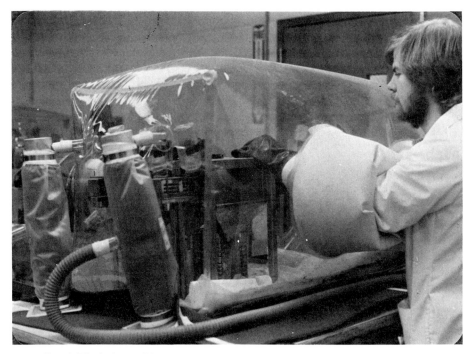

FIG. 1. The isolator with an operator performing tasks by means of gloves. At the left is the cylindrical filter for entering air (with hose attached) and a similar filter for exiting air.

accommodates six rat metabolism cages (#4-640-000, Acme Research Products, Cleveland, OH) and allows sufficient room for extra food, water, and ancillary equipment such as a balance or a container of an anesthetic agent. One of the isolator's long sides is fitted with a 12-in. circular stainless steel rim that provides the port through which material may be transferred between the isolator and the external environment (Fig. 2). Two parallel vinyl doors, separated by 6 in., span the port and provide a lock. In the outer door are two nipples that enable the lock to be sterilized by a peracetic acid spray.

The side of the isolator opposite the port is fitted with two heavy gloves that are attached to the side of the isolator by plastic sleeves (Gra-lite). Use of the gloves enables operations to be conducted within the isolator.

Two small nipples at one end of the isolator are connected to air filters. A small blower (Standard Safety Equipment Co.) provides a stream of circulating air at approximately 70 liters/min. The impedance of exit air through a second filter maintains the isolator in an inflated configuration.

FIG. 2. A view of the isolator from the direction opposite that of Fig. 1 shows the entry port at the side opposite the gloves. The nipples in the outer door can be seen.

Sterilization of the Isolator

Sterilization of the isolator prior to use and sterilization of the port prior to a transfer is done with a peracetic acid spray prepared on the day that it is used. The material for spraying is a mixture of 7.5 ml peracetic acid (40%, Pfaltz & Bauer, Stamford, CT) and 0.2 ml Conco AAS-60S triethanolamine dodecyl benzene sulfonate (Continentinal Chemical Co., Clifton, NJ) in 115 ml distilled water. CAUTION: This solution is a powerful irritant that must be used only in a well-ventilated room and by an operator equipped with a face mask containing an air filter (Northwest Safety Products, Auburn, MA). Rubber gloves are also recommended. The solution can be formulated in a plastic atomizer; attachment to a tank of compressed air provides the pressure for spraying.

For initial sterilization of a new isolator, approximately 100 ml of the peracetic acid solution is poured through the nipple of the door. By means of the gloves fitted into the isolator the isolator may be washed manually with paper towels that had been included when the isolator was assembled. Additional peracetic acid solution is sprayed through one nipple of the door (the other nipples on the isolator are stoppered) until the isolator becomes fully inflated. The inflated isolator is allowed to stand overnight to assure that no leaks are present.

Cages are autoclaved (121°, 20 psi) for 60 min and then passed through the port into the isolator. With the outer door replaced, the isolator is sprayed again with peracetic acid solution and then allowed to stand overnight to permit the mist to permeate all surfaces of the cages.

At this time, air filters are attached. A filter consists of a cylindrical metal frame, wrapped with fiberglass filter media (Standard Safety Supply Co.), held in place with Mylar tape. With the metal port for circulating air closed by means of a cotton plug, the filters are autoclaved for 1 hr. The sterile filters are then wrapped with a plastic cover that contains the nonsterile port of the filter. The plastic cover is held in place by tape and metal clamps. To attach the filter to the isolator, the cotton plug is removed from the sterile port and the isolator nipple and filter port are sprayed with peracetic acid. They are then rapidly joined and taped together.

Transfer

The stainless steel rim with its two parallel plastic doors provides a lock for sterilization of items being passed into the isolator; the isolator is protected as long as the inner door is in place. Small items may be inserted into the lock by temporarily removing the outer door; the lock is restored when the outer door is replaced. The lock and the surfaces of its contents may then be sterilized with the peracetic acid spray by inserting the atomizer nozzle into one of the nipples and generating a mist that permeates the entire lock. The stoppers are inserted into the outer door and the lock, restored, is allowed to remain undisturbed. After 1 hr the contents of the lock may be brought into the isolator after removal of the inner door by means of the gloves in the opposite side of the isolator.

For transfers of large amounts of material, the outer door of the port can be removed and a stainless steel cylinder, the "drum," can be connected to the port (Fig. 3). For this process, materials needed inside the isolator are loaded into the drum and the end of the drum covered with a sheet of mylar that is held in place by vinyl tape (#471, Standard Safety Supply Co.). The drum and its contents may then be autoclaved. We have found that sterilization can be achieved when the drum is autoclaved for 90 min at a temperature of 121°. However, it must be recognized that adequate sterilization may require that the autoclave be modified to assure steam circulation inside the drum.[7] After the drum is autoclaved it can be connected to the port by means of a transfer sleeve that is sprayed in the

FIG. 3. The sterilizing "drum" connected to the isolator port by means of the transfer sleeve. One nipple of the sleeve can be seen.

same manner as the lock. One hour after spraying, the inner door of the isolator is removed and the Mylar cover of the drum is torn open. The sleeve is then available for use as a conduit for transfer of the contents of the drum into the isolator.

Contents of the isolator that may be required in drug metabolism studies include the following (these items, except for food, which is sterilized separately, may be sterilized within a drum): food; water bottles and clamps; reserve water (900 ml volumes); urine collection tubes and caps; vials with caps to collect feces; spatula; forceps; pencil; Mylar bags for trash; intubation needles; syringes; scales; and mortar and pestle (for grinding food).

Food

Autoclavable diets and diets already sterilized (Agway RMH 3500) can be obtained from Charles River Breeding Laboratories (Wilmington, MA). Food, like the animals, can be delivered in small transfer units,

TABLE I
A DEFINED HIGH AND LOW FIBER DIET

Ingredient	Composition (%)	
	High fiber	Low fiber
Casein	23.0	23.0
Corn starch	46.6	64.6
Wheat bran[a] (coarse)	20.0	2.0
Salt mix[b]	4.7	4.7
Vitamin mix[c]	0.39	0.39
Corn oil mix[d]	5.0	5.0
Choline chloride	0.25	0.25
Total	99.94	99.94

[a] Vitamin, Inc., Chicago, IL. Other sources of fiber may be substituted for wheat bran.

[b] Salt mix, g per kg of diet: NaCl, 5.51; $MgSO_4$, 4; $Fe(C_6H_{11}O_7)_2$, 0.6; $MnCO_3$, 0.2; CuO, 0.025; ZnO, 0.025; $CoCl_2 \cdot 6H_2O$, 0.0005; NaF, 0.0001; MoO_3, 0.0005; KBr, 0.0001; Na_2SeO_3, 0.0001; $CaCO_3$, 17; K_2HPO_4, 10; Na_2HPO_4, 10; KI, 0.0002.

[c] Vitamin mix, g per kg of diet: thiamin-HCl, 0.06; riboflavin, 0.03; nicotinamide, 0.12; calcium pantothenate, 0.3; pyridoxine-HCl, 0.04; biotin, 0.01; folic acid, 0.01; vitamin B_{12} (0.1% in mannitol), 0.25; p-aminobenzoic acid, 0.05; inositol, 1; ascorbic acid, 2.

[d] Corn oil mix (5%) provides (g per kg of diet) 0.225 g (247 IU) DL-α-tocopherol; 0.016 g (16,000 IU) retinyl palmitate; 0.010 g vitamin K_1; 0.0001 g (1000 IU) cholecalciferol; 49.75 g corn oil.

whose sleeve may be sprayed with peracetic acid and attached to the isolator port.

Special Diets

Diets of invariant composition, e.g., L485,[9] can be purchased (Teklad Mills, Inc., Monmouth, IL 61462) or germfree diets for special purposes can be prepared. For example, the fully defined diet shown in Table I is a modification of that of Wostman and Kellogg,[10,11] which permits the fiber

[9] T. F. Kellogg and B. S. Wostmann, *Lab. Anim. Care* **19**, 812 (1969).

[10] B. S. Wostmann and T. F. Kellogg, *Lab. Anim. Care* **17**, 589 (1967).

[11] J. Sumi, J. Asai, M. Miyakawa, M. Arakawa, and M. Kanzaki, *in* "Germfree Research: Biological Effects of Gnotobiotic Environments" (J. B. Heneghan, ed.), p. 309. Academic Press, New York, 1973.

composition to be altered.[12] This diet supports weight gain in germfree rats that is comparable to that of conventional rats maintained on the AIN-76 Purified Test Diet.[13] To prepare this diet the bran is autoclaved for 90 min at 121° prior to its addition to the complete diet. The complete diet is then placed inside a drum and autoclaved for an additional 35 min at 121°.

Most of the diets used in germfree research provide vitamins in excess of requirements because the extent of deterioration of vitamins during the process of autoclaving has not been adequately examined. Other diets used in germfree research include one with a completely defined amino acid composition[14] and one with a low nitrate composition.[15]

Sterile diets may be prepared to meet other specifications. It must be recognized, however, that heat-labile components may have to be sterilized by filtration and added separately or that final mixing of the diet may have to be carried out within the isolator to avoid reactions that can occur in the autoclave.

Determination of Germfree Status

Assurance that animals are germfree requires that they be tested when obtained from the supplier, as well as periodically thereafter, for microbial contamination. Direct microscopic examination of fecal samples is the most rapid way of detecting the presence of bacteria but this method is only reliable for detection either of sudden increases in or of a relatively larger number of contaminants in the gastrointestinal tract. One must also be wary of this method because it can reveal bacteria that were present in sterilized food and were actually nonviable. Further assurance of the bacteriological sterility of the isolator is obtained when cultures from the isolator fail to show likely bacterial contaminants.

We examine feces microscopically by both gram stain and wet mount. Fresh samples of feces, bedding, water, etc. are transferred out of the isolator to be cultured at weekly intervals by the methods listed in Table II.[16] Cultures are retained for at least 1 week to detect possible slow-growing contaminants. Obviously, one cannot exclude the possibility that the "germfree" animal harbors organisms that escape detection by both

[12] M. N. Woods and P. Goldman, unpublished.
[13] Report of the American Institute of Nutrition Ad Hoc Committee on Standard for Nutritional Studies, *J. Nutr.* **107**, 1340 (1977).
[14] M. N. Woods and P. Goldman, *J. Nutr.* **109**, 738 (1979).
[15] L. Green, S. Tannenbaum, and P. Goldman, *Science* **212**, 56 (1981).
[16] L. V. Holdeman, E. P. Cato, and W. E. C. Moore, eds., "Anaerobe Laboratory Manual," 4th ed. Virginia Polytechnic Institute and State University Anaerobe Laboratory, Blacksburg, Virginia, 1977.

TABLE II
CULTURE CONDITIONS FOR DETECTION OF CONTAMINANTS

Media	Incubation conditions
Plates	
Brain-heart infusion[a] + 5% blood	Aerobic and anaerobic at 37°
Sabouraud dextrose[b] + 2% agar	Aerobic at room temperature
Tubes	
Tryptose-phosphate broth[c]	Aerobic at 37° and at room temperature
N_2C broth[a] or *Brucella* broth[c]	Anaerobic at 37°

[a] Made according to Ref. 16.
[b] BBL Microbiology Systems, Cockeysville, MD.
[c] Difco Laboratories, Detroit, MI.

microscopic examination and cultivation on specific media. Instances have been reported of viable bacterial contaminants that are visible microscopically but cannot be cultivated on standard media. Thus even more extensive routine cultivation methods[17] cannot totally exclude the possibility of a novel bacterial contaminant.

Rats Gnotobiotic with Certain Bacterial Strains

The association of germfree rats with known strains of bacteria provides a model for determining the *in vivo* role of these bacteria in the metabolism of the host. Bacteria are grown to a density of approximately 10^8 per ml in a suitable culture medium. The culture is placed in a screw-capped vial, the outside of which can be sterilized by spraying for its passage through the port and into the isolator. The animals may be associated with this culture either by gastric intubation or by wetting their diet with an aliquot of the bacterial culture medium. Association is assured when fecal samples, collected a few days later, show the presence of the desired bacteria.

Drug Metabolism Studies

A drug used in metabolism studies with germfree animals must be sterilized either by filtration or autoclaving before it is passed into the isolator. The desired dose may be administered to the animals either by gastric intubation[18] or by thorough admixture with an aliquot of the

[17] M. Wagner, *Ann. N. Y. Acad. Sci.* **78,** 89 (1959).
[18] Intubation needles are available from Popper & Sons, Inc., New Hyde Park, NY.

ground chow that would be consumed that day. The appropriate amount of drug may be added to the diet in solutions of ethanol or acetone but the animals will not consume the diet until the solvent is no longer detectable. Intubation is facilitated by treating the animals with a small dose of halothane anesthesia. Nonaqueous liquids, e.g., halothane, may be sterilized by passing them through a previously sterilized Swinny Filter holder fitted with a Fluoropore FG filter (Millipore Corp., Bedford, MA) into a sterile Vacutainer (Becton Dickinson, Rutherford, NJ). The Vacutainer can then be passed through the lock into the isolator.

Samples of urine and feces may be removed through the port by removing the inner door and inserting the samples into the lock.

Comments

It will be clear that each of the details of working with germfree rats is not included here. Rather, the object of our description is to present the broad outlines of the requirements for working with such animals, in the expectation that investigators will recognize that these methods are relatively simple and within the scope of their laboratories.

Acknowledgment

We thank Dr. Julian R. Pleasants for his helpful comments regarding this manuscript.

[7] Covalent Binding of Electrophilic Metabolites to Macromolecules

By LANCE R. POHL and RICHARD V. BRANCHFLOWER

During the past several years it has become evident that the cytotoxicity, mutagenicity, and carcinogenicity produced by many chemically inert xenobiotics is due to their metabolic conversion to reactive electrophiles. Although the mechanism by which the latter metabolites cause the observed tissue changes is generally not known, the irreversible interactions of the metabolites with critical nucleophilic sites of tissue molecules is believed to be involved. This idea is primarily based upon the observation, with several compounds, that parallel changes often occur between the incidence and severity of tissue alterations and the covalent binding of metabolites to such tissue macromolecules as protein, lipid, RNA, and

DNA in the target organs.[1-4] A detailed perspective on the role of covalent binding of chemically reactive metabolites in toxicity is available.[5,6]

Here we outline some of the more common procedures that have been employed to measure covalent binding of electrophilic metabolites to tissue macromolecules. In addition, we discuss precautions that should be considered in designing experiments and in interpreting the results.

General Considerations in Experimental Design

Position of Radiolabel in Parent Compound

Because the level of covalent binding of electrophilic metabolites to tissue macromolecules is usually in the range of picomoles to nanomoles per milligram of protein, lipid or nucleic acid, covalent binding is generally detected by measuring the binding of radiolabeled precursors. The position of the radiolabel in the parent compound must be chosen carefully. It should be in a position that is not expected to be lost either enzymatically or nonenzymatically. This is particularly true of ^3H labels, which can be eliminated as tritiated water during metabolic oxidations. Double-label experiments in which two regions of the parent compound are labeled with different isotopes, e.g., with ^3H and ^{14}C, can be particularly helpful in elucidating the structure of the chemically bound metabolite.[7,8]

Radiochemical Purity of Parent Compound

The radiochemical purity of the parent compound should be at least 99%, because the amount of covalent binding of metabolite is often less than 1% of the parent compound utilized. Therefore, if the sample compound contains more than 1% of radiolabeled impurities, there is the possibility that the measured covalent binding has resulted from an impurity or even from a metabolite of the impurity. Radiochemical purity can be readily checked by thin layer chromatography or high pressure liquid chromatography (HPLC). Similar checks should also be made of commercial products even if they are specified as being 99% radiochemically pure.

[1] E. C. Miller and J. A. Miller, *Pharmacol. Rev.* **18**, 805 (1966).
[2] J. H. Weisburger and E. K. Weisburger, *Pharmacol. Rev.* **25**, 1 (1973).
[3] J. R. Gillette, J. R. Mitchell, and B. B. Brodie, *Annu. Rev. Pharmacol.* **14**, 271 (1974).
[4] W. K. Lutz, *Mutat. Res.* **65**, 289 (1979).
[5] J. R. Gillette, *Biochem. Pharmacol.* **23**, 2785 (1974).
[6] J. R. Gillette, *Biochem. Pharmacol.* **23**, 2927 (1974).
[7] L. R. Pohl and G. Krishna, *Biochem. Pharmacol.* **27**, 335 (1978).
[8] L. R. Pohl, G. B. Reddy, and G. Krishna, *Biochem. Pharmacol.* **28**, 2433 (1979).

Vehicle for Radiolabeled Parent Compound

The solvent used for administering the radiolabeled parent compound should be chemically and pharmacologically inert. Although aqueous solutions are appropriate for water-soluble compounds, lipophilic compounds require carriers such as corn oil or sesame oil. For *in vitro* experiments, such solvents as ethanol and dimethylformamide have been used; concentrations of approximately 1% or less in the incubation mixture are safest. Whenever possible, the effect of the added solvent should be tested on the system being used.

Time Between Administration and Sacrifice

The amount of electrophilic metabolite bound to protein, lipid, DNA, or RNA *in vivo* usually rises steeply in the first few hours, levels off, and decreases slowly thereafter according to the rate of elimination of the parent compound, the chemical stability of the adduct, enzymatic excision of the bound adduct, and cell death.[4] Therefore the time dependence of the binding should be determined whenever possible.

Methods

Binding to Protein in Vitro and in Vivo

Although binding to protein is not always proportional to the toxicity of a chemical,[5,6,9] it does indicate that a reactive metabolite has been formed. Binding to protein is relatively easy to measure and, in most cases, is higher than binding to other types of tissue macromolecules, due to the relatively high intracellular protein concentration and the rapid reaction with nucleophilic groups on the proteins.

In Vitro. Radiolabeled substrate may be incubated with purified enzymes, microsomes, tissue homogenates, or isolated cells, employing the same conditions normally utilized for metabolic studies. Reactions are stopped by the addition of a protein precipitating agent, e.g., equal volume of 10% trichloroacetic acid or 3 volumes of absolute methanol, ethanol, or acetone. The reaction mixture is transferred to a conical centrifuge tube and centrifuged (clinical centrifuge) to pellet the precipitate. The supernatant solution is removed and the precipitate is washed by mixing with at least 5 volumes of organic solvents to remove noncovalently bound substrate and metabolites. The solvents are chosen on the basis of the solubility characteristics of the radioactive substrate and its metabolites. We have found that a mixture of methanol : ether (3 : 1) is

[9] J. Tyndal, *Drug Metab. Dispos.* 7, 451 (1979).

applicable to most experiments.[7,8] After each washing, the mixture is centrifuged and the solvent is removed by aspiration. The washing procedure is continued until the organic wash contains background levels of radioactivity, usually a matter of 5 to 10 washes. The pellet, which consists mainly of protein, is dissolved in 1 ml (or more if needed) of 1 N sodium hydroxide or a commercial tissue solubilizing agent such as NCS (Amersham/Searle); an aliquot of the clear solution is counted in scintillation fluid. Protein is determined on another aliquot of the dissolved precipitate by the method of Lowry et al.[10] with bovine serum albumin as a standard. Results are usually expressed as picomoles or nanomoles covalently bound per milligram of protein.

In Vivo. After the animals are killed, tissue samples are homogenized in 3 volumes of water.[8] Aliquots (0.5 ml) are transferred to conical centrifuge tubes and protein is precipitated with 10 ml of 10% trichloroacetic acid. The precipitate is washed and the extent of covalent binding is determined as outlined for *in vitro* studies.

Although the procedures presented do not remove nucleic acids from the protein precipitate, the error introduced by the presence of nucleic acids is usually negligible.[4]

Binding to Lipid in Vitro and in Vivo

Most studies dealing with covalent binding to lipid have involved volatile halogenated hydrocarbons of which carbon tetrachloride[11,12] and halothane[13,14] are examples.

In Vitro. Incubations are stopped by extracting total lipid with 19 volumes of chloroform : methanol (2 : 1, v/v).[15] The precipitated protein is removed by filtration and the chloroform : methanol extract is washed with 0.2 volumes of 0.1% NaCl (w/v). The aqueous layer is removed and the chloroform layer is repeatedly dried and redissolved in chloroform : methanol (2 : 1, v/v) until constant radioactivity and weight are attained. Volatile substrate and metabolites that are not bound are evaporated during this procedure, but nonvolatile lipid-soluble substrates and metabolites may not be easily removed (see later).

An aliquot of the evaporated lipid extract is solubilized in a commercial tissue solubilizer and the solution is counted in scintillation fluid.

[10] O. H. Lowry, N. J. Rosebrough, A. L. Farr, and R. J. Randall, *J. Biol. Chem.* **193**, 265 (1951).

[11] E. Gordis, *J. Clin. Invest.* **48**, 203 (1969).

[12] J. A. Castro and M. I. Diaz-Gomez, *Toxicol. Appl. Pharmacol.* **23**, 541 (1972).

[13] R. A. Van Dyke and C. L. Wood, *Drug. Metab. Dispos.* **3**, 51 (1975).

[14] A. J. Gandolfi, R. D. White, I. G. Sipes, and L. R. Pohl, *J. Pharmacol. Exp. Ther.* **214**, 721 (1980).

[15] J. Folch, M. Lees, and G. H. Sloane-Starley, *J. Biol. Chem.* **226**, 497 (1957).

Results are usually expressed as picomoles or nanomoles covalently bound per milligram of lipid.

In Vivo. Tissue samples (1 g is assumed to have a volume of 1 ml) are homogenized with 19 volumes of chloroform : methanol (2 : 1, v/v) after which the organic extract is processed as described for *in vitro* studies. Binding is expressed as the amount of covalently bound product per milligram lipid.

Binding of Nonvolatile Compounds

Because many exogenous biologically active compounds are lipophilic, the determination of the binding of the metabolites of nonvolatile compounds to lipid is complicated by the difficulty in separating unbound parent or metabolites from lipid by simple extraction procedures. This problem may be overcome by thin layer chromatography[16] or high-performance liquid chromatography,[17,18] using conditions that separate classes of lipid.

Binding to DNA or RNA in Vitro and in Vivo

Covalent binding of reactive metabolites to DNA or RNA is often measured after incubating radioactive substrate with cultured cells or with microsomal enzymes in the presence of exogenous DNA or RNA. The major difficulty in these studies, as with *in vivo* systems, is the separation of the classes of macromolecules from each other. Several suitable procedures have been reported,[19-27] most of which represent modifications of methods developed by Kirby,[28] Irving and Veazey,[29] and Okuhara[30] for

[16] V. P. Skipski and M. Barclay, this series, Vol. 14, p. 548.
[17] W. S. M. Geurts Van Kessel, W. M. A. Hax, R. A. Demel, and J. De Gier, *Biochim. Biophys. Acta* **486**, 524 (1977).
[18] B. J. Compton and W. C. Purdy, *J. Liq. Chromatogr.* **3**, 1183, (1980).
[19] L. Diamond, V. Defendi, and P. Brookes, *Cancer Res.* **27**, 890 (1967).
[20] N. H. Colburn and R. K. Boutwell, *Cancer Res.* **28**, 642 (1968).
[21] T. Kuroki, F. Huberman, H. Marquardt, J. K. Selkirk, C. Heidelberger, P. L. Grover, and P. Sims, *Chem. Biol. Interact.* **4**, 389 (1971).
[22] N. Kinoshita and H. V. Gelboin, *Proc. Natl. Acad. Sci. U.S.A.* **69**, 824 (1972).
[23] S. C. Maitra and J. V. Frei, *Chem.-Biol. Interact.* **10**, 285 (1975).
[24] C. M. King, N. R. Traub, R. A. Cardona, and R. B. Howard, *Cancer Res.* **36**, 2374 (1976).
[25] M. F. Wooder, A. S. Wright, and L. J. King, *Chem. Biol. Interact.* **19**, 25 (1977).
[26] J. W. Nicoll, P. F. Swann, and A. E. Pegg, *Chem.-Biol. Interact.* **16**, 301 (1977).
[27] P. G. Watanabe, J. A. Zempel, D. G. Pegg, and P. J. Gehring, *Toxicol. Appl. Pharmacol.* **44**, 571 (1978).
[28] K. S. Kirby and E. A. Cook, *Biochem. J.* **104**, 254 (1967).
[29] C. C. Irving and R. A. Veazey, *Biochim. Biophys. Acta* **166**, 246 (1968).
[30] E. Okuhara, *Anal. Biochem.* **37**, 175 (1970).

the isolation of DNA and RNA. The procedure presented here is based upon a combination of the several methods that have been reported.

Tissue samples are homogenized in a medium (5 ml/g tissue) that consists of sodium p-aminosalicylate : butan-2-ol : sodium triisopropyl-naphthalene sulfonate : water (6 : 6 : 1 : 87, w/w). An equal volume of phenol reagent (phenol : water : m-cresol : 8-hydroxyquinoline, (500 : 55 : 70 : 0.5, w/w) is added and the mixture is stirred vigorously at room temperature for 20 min. The resultant emulsion is broken by centrifugation at 10,000 g for 45 min at 4° and the phenol and aqueous phases are separated.

Isolation of DNA

The aqueous phase is adjusted to 1% with respect to sodium chloride (w/v), washed with 0.5 volume of phenol reagent and centrifuged. The aqueous phase is separated from the phenol phase and mixed with an equal volume of 2-ethoxyethanol to yield a fibrous precipitate of DNA. The DNA is wound onto a glass rod, washed with 75% aqueous ethanol, and dissolved in an aqueous solution containing 2% sodium acetate and 1.5% sodium chloride. Ribonuclease (1 mg/ml DNA solution), previously heated to 80° for 10 min to destroy deoxyribonuclease activity, is added. The mixture is incubated at 37° for 30 min. After treatment with phenol reagent, as previously described, the supernatant solution is centrifuged at 100,000 g for 1 hr at 4° to remove residual polysaccharide. DNA is precipitated with 2-ethoxyethanol and washed successively with ethanol, ethanol : ether (1 : 1, v/v), and ether until the washes contain no radioactivity above the background levels. The DNA is dried and dissolved in 1 mM potassium phosphate at pH 7.0. An aliquot is dissolved in 1 ml of tissue solubilizer and the radioactivity counted. Another aliquot is assayed for DNA by the diphenylamine reaction.[31] Results are usually expressed as micromoles covalently bound to DNA per mole of DNA phosphorus.

Isolation of RNA

The supernatant solution derived from the initial 2-ethoxyethanol precipitation of DNA is diluted with an equal volume of absolute ethanol and stored overnight at 0°–4°. The flocculent precipitate of RNA is collected by centrifugation (6000 g, 20 min, at 4°) and washed and dried as described for DNA. The RNA is dissolved in 1 mM potassium phosphate (pH 7.0) and an aliquot is measured for radioactivity. Another aliquot is assayed for RNA by the orcinol reaction.[31] Results are usually expressed as micromoles covalently bound to RNA per mole of RNA phosphorus.

---

[31] W. C. Schneider, this series, Vol. 3, p. 680.

The DNA isolated by the outlined procedure should not be contaminated by more than a few percent with protein or RNA. The isolated RNA should be approximately the same purity as DNA. However, even this small degree of contamination may cause inaccuracies.

Precautions in Interpreting Covalent Binding Data

Control Experiments

Artifacts in covalent binding can result from factors other than impurities in the radiolabeled parent compound. For example, during *in vitro* incubations, or during the covalent binding assay, it is possible that the parent compound may be transformed by chemical reactions such as air oxidation into products that bind covalently. This type of artifact can be corrected in *in vitro* studies by performing control incubations either with heat denatured preparations or in the absence of a necessary enzymatic cofactor, e.g., without NADPH in the case of cytochrome P-450-mediated reactions. The amount of covalent binding from such control studies is subtracted from the results of normal reactions, yielding a corrected value for covalent binding. It is not possible to control for similar artifacts in whole animal studies because the noted controls are not possible. Artifacts due to the covalent binding of radiochemical impurities present in the administered compound, however, can be reduced to a minimum by insuring that the compound is made as radiochemically pure as reasonably possible immediately before use.

Radioisotopically Labeled Macromolecules Resulting from Precursors of Endogenous Substrates

Potential Problems. When conducting experiments in cell culture or with living animals, the possibility exists that the measured radioisotopic labeling of macromolecules has resulted, at least in part, from the incorporation of radiolabeled precursors of endogenous substrates into macromolecules.[4,8,32,33] For example, tritiated water, formed from either the exchange or metabolic oxidation of a tritiated compound, can be incorporated into nucleotides.[4,33] This occurs, for example, in the reduction of ribose to 2-deoxyribose; a proton or tritium from the water pool replaces the 2-hydroxy group. Other metabolites, e.g., carbon dioxide, formaldehyde, formic acid, or glyoxylic acid, are precursors for the synthesis of amino acids and thus may be incorporated subsequently into protein.[8,32]

[32] J. R. Gillette and L. R. Pohl, *J. Toxicol. Environ. Health* **2**, 849 (1977).
[33] D. H. Phillips, P. L. Grover, and P. Sims, *Int. J. Cancer* **23**, 201 (1979).

Differentiating Incorporation from Covalent Binding

Distinguishing incorporation from covalent binding may be approached by studying the effects of protein synthesis inhibitors on the rate of covalent binding of radiolabeled drugs to tissue protein *in vivo*. However, in using this approach, the investigator assumes that the rates of metabolism and disposition of the compound and its metabolites are not altered by the inhibitor of protein synthesis.

In another approach, the compound is labeled with both ^3H and ^{14}C in different parts of the molecule and the relative abundance of the isotopes in the labeled macromolecules is determined periodically after administration of the drug. If the ratio remains constant, the two radiochemically labeled moieties of the drug are probably covalently bound to tissue macromolecules as a unit, and the labeling of the macromolecules is presumed to occur through the formation of a chemically reactive metabolite.

In order to show more convincingly that covalent binding of a reactive metabolite has occurred, rather than incorporation, it is necessary to hydrolyze the tissue molecules and subject the hydrolysates to a variety of separation procedures; to separate unmodified nucleosides from nucleoside adducts, Sephadex LH-20[33] and HPLC[34-36] have been used.

Absolute proof of adduct formation can only be obtained, however, when a covalent bond between the electrophilic metabolite and macromolecular substituent has been proven unequivocally by physical measurements such as mass spectra[35] or nuclear magnetic resonance analyses.[36]

[34] A. M. Jeffrey, S. H. Blobstein, B. Weinstein, and R. G. Harvey, *Anal. Biochem.* **73**, 378 (1976).
[35] K. M. Straub, T. Meehan, A. L. Burlingame, and M. Calvin, *Proc. Natl. Acad. Sci. U.S.A.* **74**, 5285 (1977).
[36] J. K. Lin, J. A. Miller, and E. C. Miller, *Cancer Res.* **37**, 4430 (1977).

[8] Chemical Depletion of Glutathione *in Vivo*

By JOHN L. PLUMMER, BRIAN R. SMITH,
HELMUT SIES, and JOHN R. BEND

The deliberate alteration of tissue glutathione (GSH) levels has proven to be a useful tool in detoxication and drug metabolism studies. GSH concentrations are not necessarily the same in different cell populations in a given organ and there may be discrete subcellular pools of GSH.[1,2] Nevertheless, it is clear that depletion of tissue GSH to less than about 30% of normal values, particularly in liver, can result in altered xenobiotic metabolism and increased toxicity of electrophilic metabolites. In rats, depletion of hepatic GSH resulted in increased toxicity of bromobenzene[3] and acetaminophen,[4] as well as altered metabolism of ethacrynic acid[5] and decreased tissue deposition of methylmercury.[6] Analogous toxicological effects were observed in mice,[4,7,8] goats,[9] hamsters,[10] and isolated rat hepatocytes.[11-14]

Several factors that can affect cellular or tissue GSH levels are summarized in Scheme 1. A widely used method for lowering tissue GSH

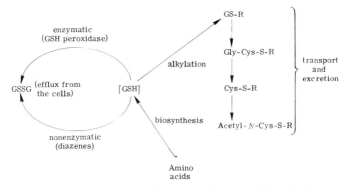

SCHEME 1. Factors that affect intracellular glutathione levels.

[1] J. E. Brehe, A. W. K. Chan, T. R. Alvey, and H. B. Burch, *Am. J. Physiol.* **231**, 1536 (1976).

[2] G. H. Mudge, M. W. Gemborys, and G. G. Duggin, *J. Pharmacol. Exp. Ther.* **206**, 218 (1978).

[3] D. J. Jollow, J. R. Mitchell, N. Zampaglione, and J. R. Gillette, *Pharmacology* **11**, 151 (1974).

[4] D. Jollow, W. Z. Potter, M. Hashimoto, D. C. Davis, and J. R. Mitchell, *Fed. Proc., Fed. Am. Soc. Exp. Biol.* **31**, 539 (1972).

[5] J. D. Wallin, G. Clifton, and N. Kaplowitz, *J. Pharmacol. Exp. Ther.* **205**, 471 (1978).

[6] R. J. Richardson and S. D. Murphy, *Toxicol. Appl. Pharmacol.* **31**, 505 (1975).

[7] R. Van Doorn, C.-M. Leijdekkers, and P. T. Henderson, *Toxicology* **11**, 225 (1978).

[8] J. R. Mitchell, D. J. Jollow, W. Z. Potter, J. R. Gillette, and B. B. Brodie, *J. Pharmacol. Exp. Ther.* **187**, 211 (1973).

[9] R. C. Hatch, J. D. Clark, A. V. Jain, and E. A. Mahaffey, *Am. J. Vet. Res.* **40**, 505 (1979).

[10] M. G. Parkki, *Scand. J. Work, Environ. Health* **4**, Suppl. 2, 53 (1978).

[11] I. Anumdi, J. Hogberg, and A. H. Stead, *Acta Pharmacol. Toxicol.* **45**, 45 (1979).

[12] H. Thor, P. Moldeus, R. Hermanson, J. Hogberg, D. J. Reed, and S. Orrenius, *Arch. Biochem. Biophys.* **188**, 122 (1978).

[13] T. D. Lindstrom, M. W. Anders, and H. Remmer, *Exp. Mol. Pathol.* **28**, 48 (1978).

[14] H. Thor, P. Moldeus, and S. Orrenius, *Arch. Biochem. Biophys.* **192**, 405 (1979).

levels is the administration of compounds that react enzymically with GSH to form conjugates. Inhibition of GSH synthesis can also result in depletion of this tripeptide in those organs with a sufficient turnover rate (see Article [9]). Conversion of GSH to its oxidized form, glutathione disulfide (GSSG), has also been used to deplete GSH, mainly in isolated cell preparations; the choice of the depleting agent will depend on the type of system under study. For example, compounds whose reaction with GSH is enzyme-catalyzed will be more effective in liver, where the glutathione transferase titer is high, and more effective in rodents with high transferase concentrations than in rabbits or monkeys.[15] Inhibitors of GSH synthesis are most effective in kidney, which has a high GSH turnover rate.[16] Depleting agents requiring activation by microsomal monooxygenases are most effective in those organs or cells with high concentrations of these enzymes.

Aside from the efficacy of an agent in depleting GSH, its unwanted effects must also be considered. Many depleting agents cause lipid peroxidation and cell lysis in isolated hepatocytes,[11] and increased rates of lipid peroxidation were observed with liver homogenates from treated animals,[17] probably as a consequence of GSH depletion. In addition, specific compounds may cause biochemical changes not directly related to GSH depletion.

Many electrophilic chemicals will lower tissue GSH levels by reaction with the sulfhydryl group of GSH. Some highly electrophilic compounds covalently bind nonselectively to nucleophiles including cellular macromolecules (see Article [6]), with consequent toxic effects. Hence, the use of moderately reactive compounds that require enzymic catalysis for GSH depletion is preferable. The advantage lies in the selectivity of the glutathione transferases in catalyzing the conjugation of many electrophiles with GSH, while minimizing the nonenzymic reaction rate with other nucleophilic macromolecules.

Types of GSH Depleting Agents

Substrates of Glutathione Transferases

α,β-*Unsaturated Carbonyl Compounds.* α,β-Unsaturated carbonyl compounds are typical weak electrophiles that react with GSH in the presence of the glutathione transferases.[18] The most widely used depleting agent in this class is diethyl maleate (DEM), introduced in 1969 by Boy-

---

[15] P. L. Grover and P. Sims, *Biochem. J.* **90**, 603 (1964).
[16] R. Sekura and A. Meister, *Proc. Natl. Acad. Sci. U.S.A.* **71**, 2969 (1974).
[17] M. Younes and C.-P. Siegers, *Res. Commun. Chem. Pathol. Pharmacol.* **27**, 119 (1980).
[18] E. Boyland and L. F. Chasseaud, *Biochem. Pharmacol.* **19**, 1526 (1970).

land and Chausseaud.[19] The commercially available product used by most investigators contains up to 20% diethyl fumarate, which is a poorer depleting agent than DEM.[18] Intraperitoneal administration of DEM (0.6 to 1.0 ml/kg) reduces the hepatic GSH levels of rats to 6–20% of control values in 30 min and for a period of 2 to 4 hr.[6,17,18,20,21] During the depleted state, the rate of GSH synthesis in the liver is increased.[22] The hepatic GSH content rises to a level of approximately twice that of control values in 24 hr and then returns to normal.[20] GSH in rat erythrocytes, kidney, lung, and brain is also depleted, but to a lesser extent than in liver.[6] DEM also decreased hepatic GSH levels in mice[17,23] and Syrian hamsters;[10] however, rabbits were resistant due to low hepatic glutathione transferase activity.[24] Repeated topical application of DEM to mice also depressed GSH levels in liver, kidney, and skin.[25] In addition to its use *in vivo*, DEM has been used to remove GSH from rat liver 9000-*g* supernatant fractions[3] and isolated rat hepatocytes.[13,26-28]

DEM treatment has also been reported to exert effects unrelated to GSH depletion, a consideration when selecting a depleting agent. Unfortunately, the role of impurities in the DEM in these undesirable effects has not been investigated. DEM stimulated bile flow in rats and dogs,[21] increased hepatic microsomal heme oxygenase activity after intraperitoneal administration to rats,[29] inhibited aryl hydrocarbon hydroxylase activity in mice *in vivo* and in rat liver microsomes,[25] and decreased *p*-nitroanisole demethylase activity and increased NADPH–cytochrome *c* reductase activity and cytochrome *P*-450 content in rat liver microsomes.[30] *In vitro*, benzphetamine *N*-demethylase and *p*-ethoxyacetanilide *O*-deethylase activities were inhibited and aniline and acetanilide hydroxylase activities were enhanced by DEM in rat liver microsomes.[31] Lipid peroxidation caused by DEM in isolated rat hepatocytes[11,26] and in liver homogenates

[19] E. Boyland and L. F. Chasseaud, *Biochem. J.* **104**, 95 (1967).
[20] P. J. Wirth and S. S. Thorgeirsson, *Cancer Res.* **38**, 2861 (1978).
[21] J. L. Barnhart and B. Combes, *J. Pharmacol. Exp. Ther.* **206**, 614 (1978).
[22] B. H. Lauterburg, Y. Vaishnav, W. G. Stillwell, and J. R. Mitchell, *J. Pharmacol. Exp. Ther.* **213**, 54 (1980).
[23] A. Wendel, S. Feuerstein, and K.-H. Konz, *in* "Functions of Glutathione in Liver and Kidney" (H. Sies and A. Wendel, eds.), p. 183. Springer-Verlag, Berlin and New York, 1978.
[24] G. L. Foureman, R. Roth, and J. R. Bend, unpublished data.
[25] A. H. L. Chaung, H. Mukhtar, and E. Bresnick, *JNCI, J. Natl. Cancer Inst.* **60**, 321 (1978).
[26] N. Stacey and B. G. Priestly, *Toxicol. Appl. Pharmacol.* **45**, 41 (1978).
[27] J. Hogberg and A. Kristoferson, *Eur. J. Biochem.* **74**, 77 (1977).
[28] J. Hogberg and A. Kristoferson, *Acta Pharmacol. Toxicol.* **42**, 271 (1978).
[29] R. F. Burk and M. A. Correia, *Res. Commun. Chem. Pathol. Pharmacol.* **24**, 205 (1979).
[30] R. A. Jordan, J. R. Gumbrecht, and M. R. Franklin, *Pharmacologist* **22**, 276 (1980).
[31] M. W. Anders, *Biochem. Pharmacol.* **27**, 1098 (1978).

of DEM-treated rats and mice[17] may be a consequence of GSH depletion rather than a separate effect, although DEM administration to mice did not result in exhalation of detectable amounts of ethane, a measure of lipid peroxidation.[23]

Phorone (diisopropylidene acetone) is another α,β-unsaturated carbonyl compound that may be useful as a GSH depleting agent.[7] As in the case of DEM, conjugation of phorone with GSH is catalyzed by the glutathione transferases. Phorone (250 mg/kg, ip) depressed hepatic GSH in rats to 9% of control levels, whereas in mice, 650 mg/kg decreased levels to 6% of control values in 2 hr. As observed after treatment with other depleting agents, GSH levels rose to above control values before returning to normal. Two days after administration of phorone (650 mg/kg) to mice, no hepatic lesions were detected by light microscopy, and no increases in serum glutamic oxaloacetic transaminase or serum glutamic pyruvic transaminase occurred. However, the number of mice used was insufficient to detect slight or occasional hepatotoxicity.

Other unsaturated compounds that deplete GSH by enzyme-catalyzed reactions include acrylonitrile,[19,32,33] acrylamide,[6] and esters of acrylic acid.[19,34,35] The organ distribution of GSH depletion with acrylamide suggests that prior metabolic activation is not necessary.[6]

Other Substrates of the Glutathione Transferases. Among compounds that deplete GSH by enzyme-catalyzed reactions are the aliphatic halo compounds, such as iodomethane,[36,37] chloroacetamide,[36] 2-chloroethanol,[36] and benzyl chloride[8]; aromatic halo compounds, such as 1-chloro-2,4-dinitrobenzene[38]; epoxides, such as styrene oxide[39] and trichloropropene oxide[12,25,40]; the organophosphates sumithion (dimethyl 3-methyl-4-nitrophenyl phosphorothionate) and sumioxon[41]; and N,N'-bis(2-chloroethyl)-N-nitrosourea (BCNU).[42]

Electrophiles Formed by Cytochrome P-450-Dependent Monooxygenase Activity. Some relatively inert compounds are metabolized to reactive electrophiles that, in turn, can deplete tissue GSH

[32] R. J. Jaeger, R. B. Conolly, and S. D. Murphy, *Toxicol. Appl. Pharmacol.* **29**, 81 (1974).
[33] H. Vainio and A. Makinen, *Res. Commun. Chem. Pathol. Pharmacol.* **17**, 115 (1977).
[34] E. H. Silver and S. D. Murphy, *Toxicol. Appl. Pharmacol.* **45**, 312 (1978).
[35] A. Suvarov and G. Kudin, *Farmakol. Toksikol. (Moscow)* **34**, 593 (1971).
[36] M. K. Johnson, *Biochem. Pharmacol.* **14**, 1383 (1965).
[37] B. G. Priestly and G. L. Plaa, *J. Pharmacol. Exp. Ther.* **174**, 221 (1970).
[38] A. Wahllander and H. Sies, *Eur. J. Biochem.* **96**, 441 (1979).
[39] J. Van Anda, B. R. Smith, J. R. Fouts, and J. R. Bend, *J. Pharmacol. Exp. Ther.* **211**, 207 (1979).
[40] R. B. Conolly and R. J. Jaeger, *Toxicol. Appl. Pharmacol.* **50**, 523 (1979).
[41] R. M. Hollingworth, *J. Agric. Food Chem.* **17**, 987 (1969).
[42] W. R. McConnell, P. Kari, and D. L. Hill, *Cancer Chemother. Pharmacol.* **2**, 221 (1979).

in vivo. Pretreatment of animals with inducers of the monooxygenase system such as phenobarbital or 3-methylcholanthrene increases the GSH-depleting effect, and usually the toxicity, of such agents.[2-4,8,43-45] Metabolically generated electrophiles are frequently more reactive than α,β-unsaturated carbonyl compounds and may result in hepatotoxicity (or toxicity in other organs) if excess electrophile remains after GSH levels have been substantially diminished. Compounds of this type include acetaminophen,[2,46-49] bromobenzene,[5,6,50,51] fluoroxene,[44,45] aniline and *p*-chloroaniline,[43] furans[52,53] (GSH depletion in erythrocytes), thiophene,[6] vinylidene chloride,[17,54] doxorubicin,[55] aromatic hydrocarbons,[56] and styrene.[10,33]

Thiol Oxidants

Oxidants that convert GSH to GSSG have been used mainly in isolated cell preparations to deplete GSH. The diazenecarboxylic acid derivatives developed by Kosower and colleagues[57] are the most important agents in this class.

At a ratio of about 1.5 moles per mole of erythrocyte GSH, azo ester (methylphenyldiazenecarboxylate) lowered GSH levels in washed human erythrocytes to about 10% of initial values in 1 to 2 min.[58] Use of excess azo ester resulted in irreversible functional damage and hemolysis.[59,60]

Diamide [diazenedicarboxylic acid bis(*N*,*N*-dimethylamide)] has been

[43] K. Aikawa, T. Satoh, K. Kobayashi, and H. Kitagawa, *Jpn. J. Pharmacol.* **28**, 699 (1978).

[44] M. A. Zumbiel, V. Fiserova-Bergerova, T. I. Malinin, and D. A. Holaday, *Anesthesiology* **49**, 102 (1978).

[45] G. G. Harrison and V. Marca, *S. Afr. Med. J.* **55**, 555 (1979).

[46] D. C. Davis, W. Z. Potter, D. J. Jollow, and J. R. Mitchell, *Life Sci.* **14**, 2099 (1974).

[47] H. S. Buttar, A. Y. K. Chow, and R. H. Downie, *Clin. Exp. Pharmacol. Physiol.* **4**, 1 (1977).

[48] J. M. Hassing, H. Rosenberg, and S. J. Stohs, *Res. Commun. Chem. Pathol. Pharmacol.* **25**, 3 (1979).

[49] M. Davis, G. Ideo, N. G. Harrison, and R. Williams, *Clin. Sci. Mol. Med.* **49**, 495 (1975).

[50] C.-P. Siegers, *J. Pharm. Pharmacol.* **30**, 375 (1978).

[51] F. P. Corongiu, M. Dore, S. Vargiolu, C. Montaldo, G. M. Ledda, and L. Congiu, *Res. Commun. Chem. Pathol. Pharmacol.* **15**, 121 (1976).

[52] M. Dershwitz, and R. F. Novak, *Pharmacologist* **21**, 170 (1979).

[53] M. Dershwitz and R. F. Novak, *Biochem. Biophys. Res. Commun.* **92**, 1313 (1980).

[54] R. J. Jaeger, R. B. Conolly, and S. D. Murphy, *Exp. Mol. Pathol.* **20**, 187 (1974).

[55] J. H. Doroshow, G. Y. Locker, J. Baldinger, and C. E. Myers, *Res. Commun. Chem. Pathol. Pharmacol.* **26**, 285 (1979).

[56] T. Suga, I. Ohata, and M. Akagi, *J. Biochem. (Tokyo)* **59**, 209 (1966).

[57] E. M. Kosower and N. S. Kosower, *Nature (London)* **224**, 117 (1969).

[58] N. S. Kosower, K.-R. Song, and E. M. Kosower, *Biochim. Biophys. Acta* **192**, 1 (1969).

[59] N. S. Kosower, K.-R. Song, and E. M. Kosower, *Biochim. Biophys. Acta* **192**, 15 (1969).

[60] N. S. Kosower, K.-R. Song, and E. M. Kosower, *Biochim. Biophys. Acta* **192**, 23 (1969).

the most widely used of the GSH oxidants. With a stoichiometry of about 2 mol of diamide per mole of cellular nonprotein thiol, GSH levels were lowered to less than 10% of initial values in isolated fat cells,[61] Erhlich ascites tumor cells, Chinese hamster cells, Chinese hamster ovary cells, and P-388 leukemia cells.[62] Similar results were obtained using rat kidney cortex slices.[63] However, reduced lipoic acid, NADH, and NADPH were oxidized slowly, and reduced flavin nucleotides rapidly, by diamide.[64,65] In addition, diamide inhibited protein kinase, sodium–potassium ATPase, and glucose 6-phosphatase,[66,67] and caused membrane damage.[60,68] Increased levels of GSSG resulting from this treatment can cause cessation of protein synthesis.[69]

Sodium tetrathionate has been used to oxidize GSH *in vivo*. At a dose of 860 mg/kg, it diminished kidney, erythrocyte, and liver GSH levels in rats to 15, 13, and 60% of control levels, respectively, in 2 to 5 hr.[6] Tetrathionate oxidized GSH and cysteine more rapidly than it oxidized protein thiol groups, but produced the undesirable effect of nephrotoxicity.[70]

An enzymatic mechanism for GSH depletion by oxidation also exists. Organic hydroperoxides are substrates for glutathione peroxidase (EC 1.11.1.9), the reaction products being GSSG and an alcohol derived from the hydroperoxide. GSSG, but not GSH itself, is released from the cells, possibly by a transport mechanism.[71-73] Isolated perfused rat livers or rat hepatocytes given cumene hydroperoxide or *tert*-butyl hydroperoxide released GSSG, resulting in diminished cellular GSH levels.[74] GSSG produced by enzymatic or nonenzymatic means, however, results in elevated intracellular levels of GSSG, which brings glutathione reductase (EC 1.6.4.2) into play. This enzyme regenerates GSH at the expense of

[61] B. J. Goldstein and J. N. Livingston, *Biochim. Biophys. Acta* **513**, 99 (1978).

[62] J. W. Harris, N. P. Allen, and S. S. Teng, *Exp. Cell Res.* **68**, 1 (1971).

[63] J. Hewitt, D. Pillion, and F. H. Leibach, *Biochim. Biophys. Acta* **363**, 267 (1974).

[64] E. M. Kosower, W. Correa, B. J. Konon, and N. S. Kosower, *Biochim. Biophys. Acta* **264**, 39 (1972).

[65] R. W. O'Brien, P. D. Weitzman, and J. G. Morris, *FEBS Lett.* **10**, 343 (1970).

[66] D. Pillion, F. H. Leibach, and H. Rocha, *Eur. J. Biochem.* **79**, 73 (1977).

[67] D. J. Pillion, L. Moree, H. Rocha, D. H. Pashley, J. Mendicino, and F. H. Leibach, *Mol. Cell. Biochem.* **18**, 109 (1977).

[68] J. A. Power, J. W. Harris, and D. F. Bainton, *Exp. Cell Res.* **105**, 455 (1977).

[69] N. S. Kosower and F. M. Kosower, *in* "Glutathione: Metabolism and Function" (I. M. Arias and W. B. Jakoby, eds.), p. 159. Raven, New York.

[70] J. L. Webb, "Enzyme and Metabolic Inhibitors," Vol. 2, p. 696. Academic Press, New York, 1965.

[71] H. Manzel, Dissertation Fachbereich Chemie, University of Tübingen (1973).

[72] S. K. Srivastava and E. Beutler, *J. Biol. Chem.* **244**, 9 (1969).

[73] S. K. Srivastava and E. Beutler, *Biochem. J.* **112**, 421 (1969).

[74] H. Sies and K. H. Summer, *Eur. J. Biochem.* **57**, 503 (1975).

NADPH. The effects of this and other biochemical stresses induced by GSH oxidation must be carefully considered before this technique is used in xenobiotic metabolism and toxicology studies.

Miscellaneous Compounds

A number of other chemicals not fitting into the preceding categories or whose mechanism of action is unknown are also effective in decreasing GSH levels in cells or tissues.

Ethylmorphine (1 mM) caused depletion of GSH to 32% of initial values in 1 hr in isolated rat hepatocytes.[11,75] Other compounds that undergo oxidative demethylation, such as benzphetamine and aminopyrine, also exhibited this effect.[75] The decrease in intracellular GSH was initially accompanied by an increase in extracellular GSSH. Formaldehyde also decreased intracellular GSH, but did not increase extracellular GSSG, indicating that direct reaction of GSH with formaldehyde is not responsible for the fall in GSH levels during oxidative demethylation. Aspirin (1000 mg/kg, intraperitoneally), or an equimolar dose of sodium salicylate, depressed rat hepatic GSH from 60 to 70% of control levels in 4 hr.[76] The effects of these agents may be the result of increased leakage of GSH from hepatocytes, or they may be due to hydroperoxide formation associated with drug oxidation[77] followed by GSH oxidation.

Intraperitoneal administration of glycylglycine lowered GSH levels in kidney and liver of rats.[78] At a dose of 440 mg/kg, kidney GSH was reduced to about 50% and hepatic GSH to 88% of controls in 1 hr. This effect is presumably due to increased breakdown of GSH mediated by γ-glutamyltranspeptidase. Maleate also increased the hydrolysis of γ-glutamyl compounds by γ-glutamyltranspeptidase[79]; it is not clear whether or not this effect played a role in the depletion of tissue GSH in rats that were administered disodium maleate.[6] With maleate, depletion was most pronounced in kidney, as would be expected if increased activity of γ-glutamyltranspeptidase were responsible.[6]

Hepatic GSH may also be depleted by fasting. Rats fasted for 18 hr had significantly lower hepatic GSH levels than fed controls, whereas kidney, brain, and erythrocyte GSH levels were not affected significantly.[6]

[75] D. P. Jones, H. Thor, B. Anderson, and S. Orrenius, *J. Biol. Chem.* **253**, 6031 (1978).

[76] N. Kaplowitz, J. Kuhlenkamp, L. Goldstein, and J. Reeve, *J. Pharmacol. Exp. Ther.* **212**, 240 (1980).

[77] H. Sies, G. M. Bartoli, R. F. Burk, and C. Waydhas, *Eur. J. Biochem.* **89**, 113 (1978).

[78] A. G. Palekar, S. S. Tate, and A. Meister, *Biochem. Biophys. Res. Commun.* **62**, 651 (1975).

[79] G. A. Thompson and A. Meister, *J. Biol. Chem.* **254**, 2956 (1979).

AGENTS FOR DEPLETION OF HEPATIC AND RENAL GLUTATHIONE

Compound	Dose	Species	Organ	Depletion (%)	Time after administration (min)	References
Diethyl maleate	0.6–1.0 ml/kg	Rat	Liver	80–94	30–120	6,17,18,20,21
			Kidney	40	30	6
		Mouse	Liver	80	60	17,23
Phorone	250 mg/kg	Rat	Liver	91	120	7
	650 mg/kg	Mouse	Liver	94	120	7

Comments

The table presents a summary of two compounds that are known to deplete hepatic GSH and that should be considered when selecting a depleting agent.

Despite its effects on monooxygenase activity and its possible stimulation of hepatic heme catabolism, DEM remains one of the most useful compounds for depleting hepatic GSH *in vivo*. Although other α,β-unsaturated carbonyl compounds share the depleting effect, little is known of their undesirable biochemical consequences. The development, principally by Meister and co-workers, of effective inhibitors of GSH synthesis has provided an important class of compounds useful for depressing tissue GSH levels *in vivo,* and this topic is discussed separately in this volume (Article [9]). Although oxidizing agents have been used successfully to deplete GSH for some *in vitro* preparations, their numerous biochemical effects, including increased GSSG levels, make their use *in vivo* unattractive for many experiments. Obviously, the choice of a chemical for depleting tissue GSH in metabolic and toxicological experiments must be made with considerable care and with the realization that parameters other than GSH content are also being altered.

[9] Depletion of Glutathione by Inhibition of Biosynthesis[1]

By OWEN W. GRIFFITH

Introduction

The intracellular glutathione concentration of intact organisms is normally maintained within certain fixed limits that are characteristic of each cell or tissue type. In this homeostatic condition the rate of the tripeptide's biosynthesis is precisely balanced by the rate of its utilization. If the rate of glutathione utilization is increased, e.g., by the presence of glutathione S-transferase substrates[2] or by a high extracellular amino acid concentration,[3] or if glutathione biosynthesis is slowed, the intracellular concentration of glutathione will decrease. Glutathione biosynthesis is most effec-

[1] These studies were supported in part by the National Institutes of Health, U.S. Public Health Services and the March of Dimes—Birth Defects Foundation. The technical assistance of Ernest B. Campbell is gratefully acknowledged.

[2] M. M. Barnes, S. P. James, and P. B. Wood, *Biochem. J.* **71,** 680 (1959).

[3] O. W. Griffith, R. J. Bridges, and A. Meister, *Proc. Natl. Acad. Sci. U.S.A.* **75,** 5405 (1979).

tively inhibited by buthionine sulfoximine, a tightly bound inhibitor of γ-glutamylcysteine synthetase, the enzyme catalyzing the first step of glutathione biosynthesis.[4] The synthesis of buthionine sulfoximine, its mode of action and the procedures for its use *in vitro* and *in vivo* are described in the following section.

Synthesis of DL-Buthionine-S,R-Sulfoximine

Preparation of DL-Buthionine[5]

A 2-liter 3-neck round bottom flask is fitted with a nitrogen gas bubbler, a water-cooled condenser, and a magnetic stirring bar. To the flask is added a cooled solution of 66 g (1.65 mol) of NaOH in 800 ml of water, and the solution is vigorously gassed with nitrogen. To that stirred solution is added 76.8 g (0.50 mol) or DL-homocysteine thiolactone hydrochloride[6]; a gentle flow of nitrogen is maintained throughout the procedure. After 30 min, 400 ml of ethanol and 101 g (0.55 mol, 62.6 ml) of 1-iodobutane are added, and the mixture is stirred overnight at room temperature. The solution is poured into a large beaker containing 600 ml of water, and the mixture is neutralized to pH 7 (pH paper) with concentrated HCl. Buthionine precipitates at once. After storing the mixture at 4° for a few hours, the product is collected by filtration, washed with a little water, and pressed as dry as possible on the filter. After drying overnight in a vacuum desiccator, the yield is 80 to 90 g of crude DL-buthionine. The crude material is dissolved in boiling water (about 10 g per liter), the hot solution is filtered, and the buthionine is allowed to crystallize at 4°. Crystals of pure DL-buthionine are collected as described earlier and dried; the yield is 70 to 80 g (74 to 84% overall).

Preparation of DL-Buthionine-S,R-Sulfoximine[5]

A 500-ml 3-neck round bottom flask is fitted with a magnetic stirring bar and a water-cooled reflux condenser protected by a $CaSO_4$ drying tube; it is placed in an oil bath maintained at 55°. To the flask are added 38.4 ml of sulfuric acid, 135 ml of dry chloroform, and 14.4 g (75 mmol) of DL-buthionine. The mixture is vigorously stirred, and 19.5 g (300 mmol) of sodium azide are added in portions of 0.5 to 1 g over a period of 5 to 8 hr;

---

[4] O. W. Griffith and A. Meister, *J. Biol. Chem.* **254**, 7558 (1979).

[5] The syntheses should be carried out in an exhaust hood to prevent exposure to 1-iodobutane and hydrazoic acid.

[6] DL-Homocysteine thiolactone hydrochloride was obtained from Sigma; 1-iodobutane was obtained from Aldrich. All other chemicals were reagent grade. The chloroform was dried over anhydrous Na_2SO_4.

the mixture is stirred at 55° overnight. During the addition of sodium azide, the sulfuric acid layer becomes increasingly viscous and foamy; a good quality magnetic stirrer is necessary to maintain agitation. The next day, the reaction mixture is allowed to cool and is poured into 800 ml of cold water. The aqueous layer is removed and the chloroform layer extracted with 200 ml of water. The combined aqueous layers are applied to a column (2.5 × 40 cm) of Dowex 50 (H⁺) and the column is washed with 1 to 2 liters of water. The buthionine sulfoximine is eluted with 1.2 liters of 3 M NH₄OH. Flash evaporation of the eluate and crystallization of the residue from aqueous ethanol yields 10 to 12 g of DL-buthionine-S,R-sulfoximine (60–70%).

Buthionine and buthionine sulfoximine may be characterized by their melting points [254–255° (dec) and 214–215.5° (dec), respectively] and by their elution times on an amino acid analyzer (66 min, 35 sec and 53 min, 39 sec, respectively, on a Durrum model 500 analyzer; for reference, phenylalanine elutes at 57 min, 45 sec).[4] The preparation of buthionine has been repeated several times and has never given any trouble providing that oxygen-mediated formation of homocystine is prevented by an adequate flow of nitrogen. The reaction may be carried out on a larger or smaller scale as desired. Successful conversion of buthionine to buthionine sulfoximine is dependent on the use of dry reagents and the maintenance of adequate stirring. Inappropriate reaction conditions will lead to a product that is contaminated with buthionine, buthionine sulfoxide, or buthionine sulfone (the latter two elute near aspartic acid on the amino acid analyzer). Recrystallization of buthionine sulfoximine from aqueous ethanol or water will remove small amounts (<10%) of the contaminants. It should be noted that sodium azide and hydrazoic acid (generated *in situ*) are toxic and potentially explosive. Although the preparation has been carried out several dozen times without incident, it should not be scaled up without adequate shielding. The procedure may be carried out on a smaller scale without difficulty. The flow-through of the Dowex 50 column may contain hydrazoic acid and should be promptly disposed of by flushing down a sink with a large volume of water.

Mechanism of Action of Buthionine Sulfoximine

Buthionine sulfoximine is a potent and apparently specific inhibitor of γ-glutamylcysteine synthetase. As shown in Fig. 1, the inhibitor is structurally similar to γ-glutamyl-α-aminobutyrate, the enzymatic product formed from glutamate and α-aminobutyrate.[7] In analogy to the more

---

[7] α-Aminobutyrate is a good alternative substrate that eliminates the oxidation problem of cysteine.

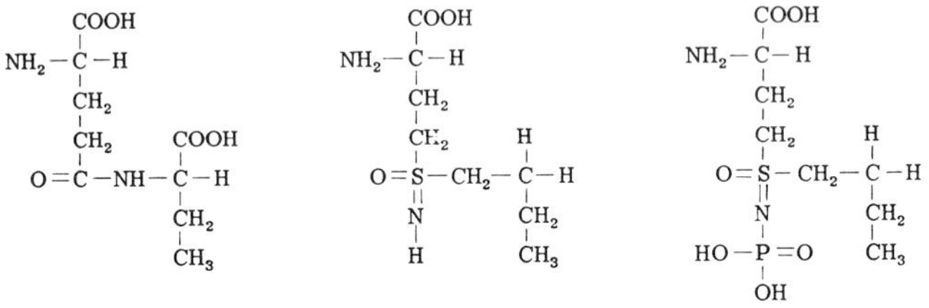

γ-Glutamyl-α-   Buthionine sulfoximine   Buthionine sulfoximine
aminobutyrate   phosphate

FIG. 1. Structures of γ-glutamylcysteine synthetase product and inhibitors.

thoroughly studied inhibition by methionine sulfoximine,[8-10] a less potent and less specific inhibitor, it is assumed that buthionine sulfoximine is initially bound to the enzyme as a γ-glutamyl-α-aminobutyrate analog. The compound is then phosphorylated enzymatically by ATP to yield enzyme-bound buthionine sulfoximine phosphate, the actual inhibitor. Buthionine sulfoximine phosphate resembles the transition state formed between γ-glutamylphosphate, the natural enzymatic intermediate, and α-aminobutyrate; it is very tightly but reversibly bound to the active site. As anticipated from the mechanism described, the rate and extent of inhibition by buthionine sulfoximine is dependent on the concentration of glutamate, which competes for the same binding site. In the presence of 5 mM glutamate and 10 mM ATP, 1 and 10 μM buthionine sulfoximine inhibit rat kidney γ-glutamylcysteine synthetase 52 and 100%, respectively, in 10 min.[4]

Buthionine sulfoximine has two centers of asymmetry (at the α-C and S), and consequently the synthetic product is a mixture of four isomers. Of the four possible isomers of methionine sulfoximine only the L-S isomer is phosphorylated by the enzyme[8-11]; a similar selectivity probably occurs with buthionine sulfoximine as well. In the present discussion the concentrations given are for DL-buthionine-S,R-sulfoximine; the concentration of active isomer would be about one-quarter of the values listed.

Use of Buthionine Sulfoximine to Deplete Glutathione in Isolated Cells

Incorporation of buthionine sulfoximine into the culture medium of isolated cells may produce a dramatic drop in the total intracellular glutathione content. The effect observed is dependent on the rate at which buthionine sulfoximine penetrates the cells and the rate at which

[8] P. G. Richman, M. Orlowski, and A. Meister, *J. Biol. Chem.* **248**, 6684 (1973).
[9] R. A. Ronzio, W. B. Rowe, and A. Meister, *Biochemistry* **8**, 1066 (1969).
[10] W. B. Rowe, R. A. Ronzio, and A. Meister, *Biochemistry* **8**, 2674 (1969).
[11] W. B. Rowe and A. Meister, *Proc. Natl. Acad. Sci. U.S.A* **66**, 500 (1970).

glutathione is consumed or released by the cells. With human red blood cells suspended in 4 mM buthionine sulfoximine, an intracellular inhibitor concentration of about 0.2 mM is obtained in 120 min, and GSH biosynthesis is inhibited more than 90% as judged by the decreased incorporation of [^{35}S]cysteine.[12] Measurable glutathione depletion does not occur for at least 6 hr, however, because the normal rate of glutathione turnover in erythrocytes is very slow ($t_{1/2}$ = 4 to 6 days[12,13]). With other cell types the depletion can be dramatic. For example, murine mastocytoma and lymphoma cells cultured in 0.2 mM buthionine sulfoximine show >90% depletion of glutathione in 12–15 hr.[14] Similarly, mouse resident peritoneal macrophages cultured in 0.2 mM buthionine sulfoximine lose 95% of their glutathione content in 7.5 hr.[15] Isolated kidney tubule cells or hepatocytes, both of which exhibit an active glutathione metabolism, would also be expected to show a rapid depletion of glutathione; experiments to verify this prediction have not yet been reported.

Use of Buthionine Sulfoximine to Deplete Glutathione in Intact Animals

Buthionine sulfoximine has been administered to mice or rats by subcutaneous or intraperitoneal injection; neutral solutions containing 50 to 200 mM buthionine sulfoximine are generally appropriate. Single doses of 32 mmol/kg[4] and multiple doses totaling 72 mmol/kg in 27 hr[16] have been given to mice without apparent effect other than inhibition of glutathione biosynthesis. The inhibitor is not significantly metabolized[12] and is excreted in the urine. Tissue levels of buthionine sulfoximine may be maintained at inhibitory levels by administration of 4 mmol/kg every 1.5 hr[16] or 8 mmol/kg every 4 hr. Two hours after a single injection of 4 mmol/kg, mouse kidney, liver, and pancreas concentrations of total glutathione are decreased to 18, 35, and 46% of control, respectively; both liver and kidney levels are held to 20% of control or less for several hours.[4,17] Similar results were obtained with rats.[17] Buthionine sulfoximine may also be administered orally by incorporation into the drinking water of mice. Mice will drink about 5 ml of water containing 20 mM inhibitor (about 5 mmol/kg) per day. After 15 days of treatment, glutathione concentrations of skeletal muscle, kidney, liver, and pancreas were 2, 4, 56, and 8% of control, respectively.[17]

[12] O. W. Griffith, unpublished observation (1980).
[13] E. Dimant, E. Landsberg, and I. M. London, *J. Biol. Chem.* **213**, 769 (1954).
[14] B. A. Arrick, C. F. Nathans, O. W. Griffith, and Z. A. Cohn, unpublished (1980).
[15] C. A. Rouzer, W. A. Scott, O. W. Griffith, A. L. Hamill, and Z. A. Cohn, *Proc. Natl. Acad. Sci. U.S.A.* **78**, 2532 (1981).
[16] B. A. Arrick, O. W. Griffith, and A. Cerami, *J. Expl. Med.* **153**, 720 (1981).
[17] O. W. Griffith and A. Meister, *Proc. Natl. Acad. Sci. U.S.A.* **76**, 5606 (1979).

[10] Whole-Body Autoradiography

By SVEN ULLBERG and BENGT LARSSON

Autoradiography is generally carried out in such a manner that a radioactively labeled chemical substance is allowed to distribute itself in an experimental animal. A section is taken and the distribution of the radioactivity within the specimen is recorded directly on a photographic film.

If, instead, a pulse counter is used, this will provide a very accurate quantitative measure of the radioactivity in the sample, but no localization within it.

In autoradiography, the registering instrument is the single silver halide grain in the emulsion. There is a slow and cumulative recording of the charged particles in the vicinity of the radiation source. Latent images are formed that are made permanent upon development. The autoradiographic exposure makes the procedure time consuming but the photographic emulsions can continue to register under favorable conditions for weeks and months without attention.

The final distribution picture is a document that can be examined, stored, and rechecked. If one is interested in the general body distribution of a substance, it can be studied in sections through the entire animal— whole-body autoradiography. Compared to a technique based on quantitative assay of removed organ pieces, whole-body autoradiography offers information that is not only more detailed but also more comprehensive. Practically all tissues may be scanned without preselection. A specific substance or its metabolite may be found to accumulate in an unforeseen site, which may shed new light on its function. When studying pregnant animals, the concentration in the fetal and maternal tissues can be compared (Fig. 1).[1]

The whole-body autoradiographic technique may be briefly summarized as follows: Commonly, a series of animals, e.g., mice, are injected intravenously with the same dose of a labeled compound. At intervals, the animals are rapidly frozen and sagittal sections are taken at the different levels of interest with a cryostat microtome. The sections are freeze-dried and pressed against a photographic film. After suitable exposure, section and film are separated and the film developed. The distribution pattern of the substance and its metabolites appear on the films—the autoradiograms. The sections may be stained and mounted under a cover

[1] L. Dencker, unpublished.

METHODS IN ENZYMOLOGY, VOL. 77

Brain   Liver   Kidney

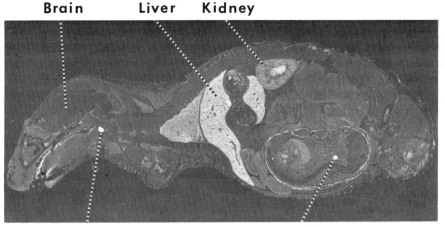

Thyroid   Fetal thyroid

FIG. 1. Thiouracil (^{14}C-labeled), which blocks the formation of iodinated thyroid hormones, is accumulated in the main site of action (thyroid gland) in both the mother and the fetus. The accumulation in the thyroid is more selective in the fetus because the radioactive substance is only slightly taken up in the excretory sites.[1]

glass, or they may be used in their unstained state as references for the interpretation of the autoradiograms. Quantitative data concerning the radioactivity in different tissues can be obtained by impulse counting of pieces taken from the sections collected on tape or by densitometry using an isotope scale as a reference source. Tissue pieces may also be punched out from the sections for microseparation of metabolites. By the use of different survival times, the variation in distribution can be followed as a function of time.

An important technical problem concerns the solubility of the substance under investigation in water or other solvents used in customary histologic processes. When applying the whole-body autoradiographic technique, the localization of a soluble compound is preserved by the freezing of the animals and by preventing liquids from coming into contact with the tissues until after the autoradiographic exposure.

Large cryosections can be used not only for autoradiography but for many other purposes. One possibility is to inject a fluorescent compound into an animal and record the fluorescence emitted by the whole-body section when exposed to ultraviolet light. The sections of noninjected animals may be used for various *in vitro* techniques, such as enzyme histochemistry (Fig. 2)[2] or the study of tissue affinity of labeled com-

[2] L.-E. Appelgren, *J. Reprod. Fertil.* **19,** 185 (1969).

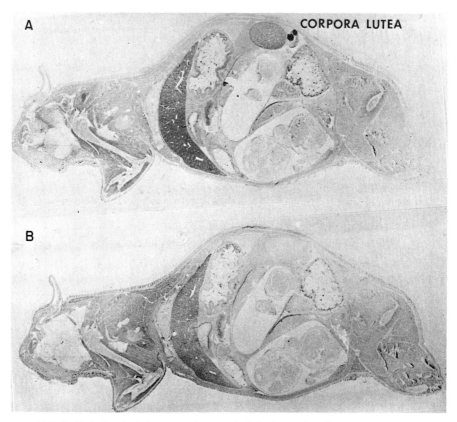

FIG. 2. Histochemical demonstration on whole-body sections from pregnant mouse of enzyme converting pregnenolone to progesterone (Δ^5-3β-hydroxysteroid dehydrogenase). (A) Shows enzyme activity confined to corpora lutea. (B) Shows the inhibitory effect on enzyme activity of a synthetic, nonsteroid compound, bis(p-acetoxyphenyl)-2-methyl-cyclohexylidenemethane.[2]

pounds (Fig. 7).[3] The affinity of fluorescent or radioiodine-labeled antibodies may reveal the tissue localization, for example, of a hormone or a virus. The *in vitro* technique also extends the studies to slaughterhouse material or human autopsy material. Finally, large cryosections may be of value for detailed anatomical or pathological inspection (Fig. 4).[3a]

Radiolabeled Compounds

Many radionuclides have been used successfully in whole-body autoradiography. The selection of a suitable isotope is restricted by such

[3] N. G. Lindquist and S. Ullberg, *Acta Pharmacol. Toxicol.* 31, Suppl. 2 (1972).
[3a] L. Dencker and S. Ullberg, in "Advances in the Detection of Congenital Malformation" (E. B. van Julsingha, J. M. Tesh, and G. M. Fara, eds.), p. 249. 1976.

Brain   **Lung**

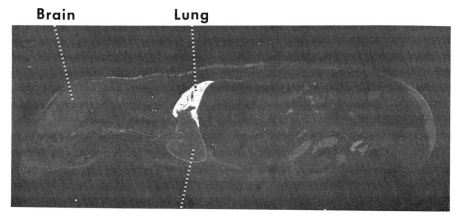

Heart blood

FIG. 3. Autoradiogram of a mouse 24 hr after intravenous injection of nickel carbonyl labeled with nickel-63. There is a very selective accumulation in the lung with activity localized in the alveolar and not the bronchial cells. Note that the substance was not inhaled but injected intravenously.[4]

factors as the radiation energy, the half-life, and the chemical requirements for labeling. The most frequently used radionuclides seem to be carbon-14 (^{14}C), tritium (^{3}H), iodine-125 (^{125}I), and sulfur-35 (^{35}S). Others are iodine-131, calcium-45, phosphorus-32, phosphorus-33, sodium-22, iron-59, cobalt-57, cobalt-58, cobalt-60, mercury-203, cadmium-109, fluorine-18, bromine-80m, bromine-82, nickel-63, lead-203, cesium-137, selenium-75, zinc-65, chromium-51, thallium-204, gold-198, vanadium-48, strontium-89, and strontium-90.

The nuclides that are most frequently used in autoradiography are ^{14}C and ^{3}H because most compounds of biological interest are organic. In the choice between ^{14}C and ^{3}H, the following must be considered: Tritium has the advantage of giving better resolution due to its weak beta-energies (see table) with a consequent short range in the section and photographic emulsion and a rather small spread of the autoradiographic image. Another advantage of tritium is the high specific activity that may be obtained due to its relatively short half-life of 12.3 years. The half-life of ^{14}C (5730 years) is almost 500 times longer; it is interesting that tritiated substances are about 500 times as "hot" as the ^{14}C-labeled ones, broadly speaking.

On the other hand, tritiated compounds may be less reliable with respect to their radiochemical purity. There may also be risks of exchange reactions of the label and of isotope effects, i.e., chemical reactions are slowed if the tritium occurs in reactive positions within the molecule.

[4] A. Oskarsson and H. Tjälve, *Brit. J. Ind. Med.* **36**, 326 (1979).

Eye   Brain   Palate

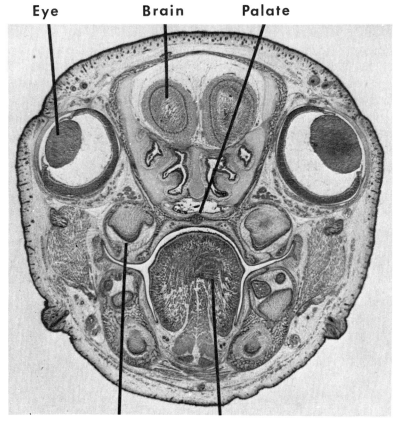

Tooth germ   Tongue

FIG. 4. For teratology: Examination of fetal defects such as cleft palate can be made on a transverse section through the head of a newborn rat. A 10-μm frozen section was immediately fixed in a cold ($-20°$) formalin–ethanol solution. Goldner's collagen stain.[3a]

Carbon-14 has dominated until recently because of its higher efficiency. Its radiation (see table) is easily registered by conventional X-ray films whereas most of the tritium beta-particles are too weak to be recorded. This limitation, however, has been greatly reduced by the availability of the LKB Ultrofilm ^{3}H, which is especially designed for detection of the radioactive emission of tritium and other weak beta emitters.

In our studies we inject experimental animals with 0.25 to 0.50 μCi per gram body weight when ^{14}C and relatively fine-grained common X-ray films are used. The corresponding dose for ^{3}H is 20 times higher (5–10 μCi) provided that the LKB Ultrofilm ^{3}H is used. With increasing section thickness, the difference in doses is greater. Because ^{14}C-labeled substances, grossly estimated, seem to be about 10 times more expensive

RADIATION PROPERTIES OF THE MOST FREQUENTLY USED
RADIONUCLIDES IN WHOLE-BODY AUTORADIOGRAPHY

Radionuclide	Half-life	Particle energy in keV (percentage particle intensity)[a]
Carbon-14	5760 years	156(100)
Tritium	12.3 years	18.6(100)
Iodine-125[b]	60 days	2.9(29)
		3.6(80)
		22.6(14)
		30.9(11)
		34.6(2)
Sulfur-35	87.2 days	167(100)

[a] The particle intensity as a percentage of the total number of nuclear transformations of the radioisotope.

[b] X rays and γ rays are present in addition to electron emission.

than tritiated substances, the choice of ¹⁴C generally involves slightly lower costs. This circumstance, however, may be of minor importance when related to the obvious advantages of the much higher specific activity of the tritiated compounds: It is possible to increase the radioactive dose of ³H substantially without increasing the rate of excretion and without reaching toxic or unphysiological levels of concentration.

Another radioisotope frequently used is ¹²⁵I. It emits extranuclear electrons of different discrete energies (see table); the more energetic ones, those of 30 keV, are dominant in whole-body autoradiography. Iodine-125 is extensively used in the labeling of peptides and proteins by simple iodination of their tyrosine units, but also as a replacement for chlorine in organic molecules such as DDT and chloroquine.

A radioisotope that was previously used extensively is ³⁵S. Currently, its relative importance seems to have decreased because of the limited possibilities of storing labeled products. It has radiation properties similar to those of ¹⁴C except for its shorter half-life of 87 days. In work with larger experimental animals, radiosulfur-labeled preparations are still preferable. By combining different radioisotope techniques, it may be possible to carry out very detailed metabolic studies. For example, autoradiography may be combined with a microseparation method, e.g., thin-layer chromatography, and the isolated radioactive material may then be located by using a photographic film or a pulse counter, preferably a scanning device. The autoradiograms are of great help in selecting the proper tissue pieces for the separation work—these may actually be cut or punched out from thick whole-body sections.

An interesting possibility for obtaining additional chemical information is the treatment of whole-body sections with a solvent before autoradiographic exposure. The technique is useful if either the unchanged substance or a major metabolite is selectively soluble in a specific solvent. Adjacent sections can be used as controls without treatment. The identity of the substance that is extracted can be checked by radiochromatography.

Another approach is to use two differently labeled preparations of the same substance in which the radionuclide is located at different sites in the molecule.

The radioactive compound or its metabolites are frequently bound and retained for a long time at their target tissues, and remain there after other tissues are free of them.

It may also be possible to map the sites of biotransformation of toxic or carcinogenic compounds. Examples are N-nitrosoamines and organic solvents. The metabolites of these substances can frequently be observed to be covalently bound close to the site of their formation. They will appear especially clear after the original, unmetabolized substance has been excreted or after selective extraction or evaporation.

Experimental Animals and Administration

In our investigations we mainly use adult mice. Old animals should be avoided because their use results in poorer section quality and greater damage to the microtome knives. To follow changes with time of the distribution and excretion pattern of a labeled-compound, a series of pregnant mice sacrificed after different postinjection intervals may be used. The special requirements for working with pregnant animals have been discussed by Dencker.[5]

Other animals that have been sectioned with good results are rats, cats, monkeys, newborn dogs, pigs, reptiles (lizard), amphibians (frog), birds (quail), and fishes (pike and trout). The largest specimen from which good quality sections have been obtained was a pregnant monkey (*Macaca irus*) weighing 5 kg.[6]

Our most common route of administration of a labeled substance has been by intravenous injection administered in a tail vein. Water-soluble compounds are generally injected in 0.2 ml of saline. Other routes include intraperitoneal, intramuscular, and subcutaneous injections and oral and topical cutaneous administration. The intravenous injection is preferred because it allows short-term studies to be carried out. It also facilitates

[5] L. Dencker, *Acta Pharmacol. Toxicol.* 39, Suppl. 1 (1976).
[6] L. Dencker, N. G. Lindquist, and S. Ullberg, *Toxicology* 5, 225 (1975).

quantitative estimations of the autoradiograms because variations in the rate of absorption from the injection site or other depot of the labeled substance is avoided.

Short-term studies of the distribution of lipophilic substances are difficult because solvents other than water are toxic and can be given only in very low doses. However, by using microsyringes (e.g., Hamilton), small amounts (20 μl for an adult mouse) of dimethyl sulfoxide (DMSO) have been used successfully as a solvent; propylene glycol and ethanol are other possible solvents. Ethanol, however, may cause local coagulation when it enters the circulation; this can result in emboli of the labeled blood in the lung. The use of a suspension is another alternative; the fat emulsion, Intralipid (Vitrum, Stockholm), marketed as an agent for intravenous fat nutrition, is an example. Topical application to the skin is another method whereby the rate of absorption may be increased if application is in a solution of DMSO.

The Mounting of Specimens

After the administration of the labeled substance, the animals are killed in an atmosphere of CO_2 and quickly frozen by immersion in a cold liquid. Various postinjection intervals are used to follow the distribution as a function of time. As a freezing liquid, hexane cooled with dry ice (about $-75°$) is generally used. If a very rapid freezing is desired, isopentane cooled with liquid nitrogen is effective. The mice, embedded in (carboxymethyl)cellulose (CMC gel), are normally frozen directly on the microtome stage. The CMC gel forms a firm support of reinforced ice around the animal on the stage. During freezing, the stage is surrounded by a metal frame. Animals investigated after very short postinjection periods (1–5 min) are generally prepared for the sectioning in two steps: (1) they are frozen without the CMC gel and then (2) mounted on the microtome stage as described earlier.

Equipment for Sectioning

Whole-body sectioning requires specially constructed microtomes and cryostats. Several different types have been made, but only a few of these are marketed. Two advanced models that have been relatively recently developed will be briefly described; they differ mainly in size. The larger one (LKB 2250, LKB Sweden AB, Bromma) allows the sectioning of large specimens such as a 5-kg pregnant monkey (45 × 15 cm).[6] The smaller model (LKB 2258) can be used for mice and rats (16 × 15 cm). Both can also be used for small specimens, e.g., a rat fetus. Even such hard tissues as teeth may be sectioned because of the good stability that the microtomes maintain.

Each cryostat is supplied with a control panel with which the microtome can be programmed to trim the specimen automatically down to the desired level. The operator then collects the number of sections needed at that particular level of the specimen. Section thickness can be varied in a continuous range from 1 to 999 μm. The microtome can then be reprogrammed to section down to the next level of interest.

The cutting stroke length may be adapted to the size of the specimen block by a separate control. The knife retracts about 30 μm upward at the end of each stroke while the specimen returns to its starting position. In this way, smearing of radioactively labeled material back over the surface of the block is avoided.

Sectioning Procedure

After the freezing and mounting of the specimen, usually carried out at $-75°$, the frozen block is generally allowed to reach the temperature of the cryostat ($-20°$). The frozen animal is then sectioned("planed") sagittally (discarding the sections), until a section surface of interest appears. Figure 5 shows, in principle, the process of a whole-body sectioning. A piece of transparent tape (Scotch tape) is fastened with a brush or a cotton wad onto the section surface of the block. The microtome knife then cuts

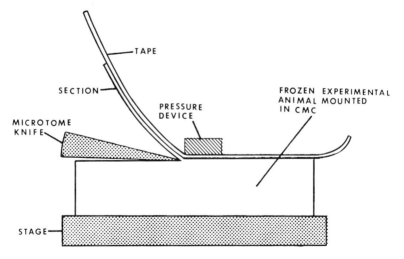

Fig. 5. Method for obtaining intact whole-body sections. A tape is fastened on the surface of the block. The section comes off, attached to the tape. CMC, (Carboxymethyl) cellulose.

under the tape, and the section will, from the beginning, be attached to the tape, which supports it and keeps it intact (Fig. 6). The tape has a tendency to loosen from the block just in front of the knife edge: this will reduce the quality of the section. This effect can be avoided by gently pressing down with a plastic, matchbox-shaped, device on the tape in front of the knife edge while the section is being cut.

Sections are generally cut 20 μm thick, but some thicker sections, e.g., 100 μm, may also be taken to decrease the exposure time and get preliminary results although at the expense of detail. The total useful range of section thickness is 5–200 μm. Thick sections may be used for impulse counting, thus simplifying quantitative measurements.

Whole 5-μm-thick sections provide autoradiograms with particularly good resolution. For contact autoradiography, these thin sections are pressed against thin fine-grained emulsions. Even better resolution, however, can be achieved if thin, whole-body sections or portions cut from such sections are attached permanently by dry mounting to nuclear plates.[7,8] The resulting autoradiograms can then be examined under the microscope by focusing alternately on the section and on the developed grains of the emulsion. In this way, a labeled substance may be localized to single cells.

Dehydration of Sections

Before the sections are apposed to film for exposure, they must be freeze-dried. This is attained by storing them at $-20°$ for 1–2 days, attached to frames of Plexiglas or wood. A vacuum is not needed. Before being removed from the freezer, the dried sections are placed in an air-tight box to prevent condensation of atmospheric moisture onto the sections.

Film

Ordinary X-ray film is the most sensitive commercial photographic material for all beta emitters except tritium and similar low-energy radionuclides. X-ray films are relatively coarse grained and are generally double coated. The radiation from strong beta emitters, such as ^{32}P passes through the backing of the film and will cause blackening in both emulsion layers. However, the radiation from ^{14}C and most beta emitters passes through only one of the two gelatin layers. The other layer does not add to the sensitivity but only to background fog and it is advantageous to remove it after the photographic processing.

[7] L.-E. Appelgren, *Acta Physiol. Scand., Suppl.* 301 (1967).
[8] S. Ullberg and L.-E. Appelgren, *in* "Autoradiography of Diffusible Substances" (L. J. Roth and W. Stumpf, eds.), p. 279. Academic Press, New York, 1969.

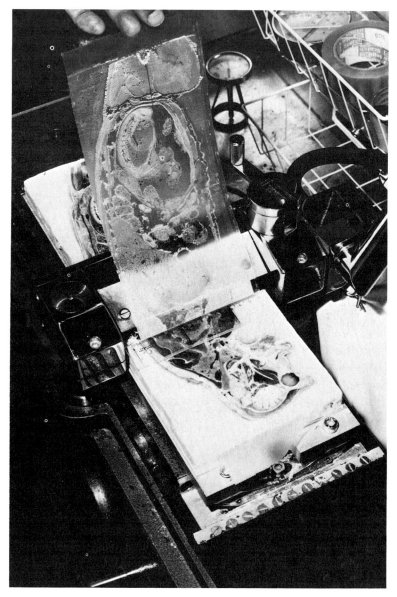

FIG. 6. Sectioning of pregnant monkey on microtome in cryostat ($-20°$). A section has been cut through the rear half of the monkey. The tape is lifted with the section attached to the tape. Note fetus on section. The head part of the specimen is covered with tape. A plastic ruler is placed on the block just in front of the knife edge.

The most sensitive X-ray films are the so-called no-screen X-ray films (intended for use in X-ray work without amplifying screens). Generally speaking, the more coarse grained the X-ray film, the more sensitive it is, and the higher is the background fog.

Kodak Ltd., Great Britain, produces two very useful X-ray films, No-Screen X-ray and Industrex C, the latter being finer grained and about one-quarter as sensitive as the former but giving better resolution. Gevaert's Structurix film is comparable to Industrex C.

The registration of tritium requires special film. All conventional X-ray films tested in our laboratory have shown a low sensitivity to tritium. These films are covered with a protective (antiscratch) gelatin layer, and this coating absorbs the majority of the tritium radiation. Recently a tritium-sensitive film has been marketed under the name "LKB Ul-trofilm ^{3}H." It has no antiscratch layer and is single-coated with a thin layer of relatively large-sized silver bromide crystals. The silver/gelation ratio is high to further reduce absorption of the tritium radiation. The film base lacks the blue stain that is nearly always present in the usual X-ray films. This tritium-sensitive film may also be suitable for use in au-toradiography of chromatograms, as with thin-layer plates employed for separation of tritium-labeled compounds.

Exposure

After freeze-drying, the sections are pressed directly against photo-graphic films for exposure. Control exposure of a section without radioac-tivity has not resulted in any blackening of the film. For quantification, a suitable radioactive scale may be placed on the film. The blackening caused by this reference scale on the film will later allow a quantitative estimation of the optical density in the various regions of the autoradio-gram (see the Quantitation section). A sandwich of photographic film and section, interposed between paper layers, is placed in a simple press con-sisting of two aluminum bricks pressed together with two paper (bulldog) clamps. The press is then placed in a light-tight box, which is usually stored at a low temperature ($-10°$ to $-20°$) during exposure; the sections are preserved better at lower temperature.

Factors influencing autoradiographic exposure time are (1) radiation properties of the isotope (energy and half-life) (2) radioactive dose, (3) rate of excretion from the animal, (4) section thickness, and (5) sensitivity of the film.

In work with long-lived isotopes, we have generally tried to choose conditions that will yield satisfactory results within a month's exposure. Average exposure times after 10 μCi of ^{14}C are 2–4 weeks for 20 μm sections on relatively fine-grained X-ray films [Industrex C (Kodak) and Structurix (Agfa Gevaert)] and 4–6 days for 100 μm sections on No Screen X-ray Film (Kodak). Thus the use of thick sections and coarse-grained emulsions can shorten the exposure time considerably (in work with most isotopes except tritium) and can also be used to obtain preliminary biological information and help in estimating the exposure time for the autoradiograms from the thinner sections.

When using 200 μCi of the tritiated substances combined with Ultrofilm ^{3}H (LKB), the exposure time is also 2–4 weeks. In cases in which the labeled compound is rapidly excreted, a considerably longer exposure time may be needed for animals studied a long time after administration.

Photographic Processing

After completion of the exposure, the section and the film are separated in the dark and the film is developed and fixed in accordance with the recommendations of the manufacturer. The film should be thoroughly rinsed before drying. To standardize the photographic processing, autoradiograms from the same series of animals may be placed on the same rack before being passed through the photographic solutions for development. The optical density of the developed autoradiograms may be compared with the optical density under an isotope scale. The sections may be stained and mounted under a coverslip (see following section).

We generally present our autoradiograms as positive prints, which means that a white area in the published picture corresponds with high activity in the section. The prints are made by copying the original autoradiograms on photographic paper in an enlarger. The main reason for publishing our autoradiograms as positives is that only one photographic step is necessary for obtaining a photopaper copy. In addition, the autoradiograms are in this way easier to distinguish from the black-and-white photos of stained sections.

Sections

With a certain degree of experience, the interpretation of autoradiograms may be done without comparison with the corresponding histological section, although histological identification of specific structures is often helpful. Occasionally, the unstained section will do but a stained one is preferable. Therefore, a few sections from each animal may be selected for staining. Some types of Scotch tape (among them 3M Type 800) allow the section to be stained while still attached to the tape. The section does not come off, and the tape is not stained to any significant extent. Any staining method can be tried. One that has been used successfully is Goldner's collagen stain.[9] The staining method used most frequently is hematoxylin-eosin.

The stained and dehydrated sections are mounted under a cover slip without being removed from the tape. As a mounting medium, the ethanol-soluble substance Euparal (CHROMA-Gesellschaft, Schmid, GmbH & Co., Stuttgart, W. Germany) is used. Farebrother and Woods[10] have suggested a convenient way of preparing stained whole-body sections for storage by spraying the section surface with Trycolac (Aerosol Marketing Co., Ltd., 30 Nottinghill Gate, London W11 3HX).

Quantitation

The amount of radioactivity in the various tissues of whole-body sections is generally determined by one of the following methods.

Densitometry

Densitometer readings of the autoradiogram are made, using an isotope scale (radioactive staircase, or step-wedge) as a source of reference. The isotope staircase, which contains known concentrations of the particular radionuclide (^{14}C, ^3H, or ^{203}Hg), is placed beside the whole-body section on the film during the exposure. The isotope staircase consists, for example, of a sequence of gelatin layers containing different concentrations of the labeled compound arranged in a geometric series in which the radioactivity of adjacent steps is related in the ratio 2 : 1, thus forming a gradually changing scale.[11] After the autoradiogram has been developed, the blackness of the various tissues can be compared with the blackness of the scale by the use of densitometer readings. The densitometer we use for this purpose is a Schnell-photometer Type GII (Jena). Even rather small areas can be examined because we are able to enlarge the autoradiogram with the densitometer.

Cross et al.[12] have described in great detail a method for quantitation by densitometry including the construction of a very convenient, small aperture (0.5 mm) densitometer. Liss and Kensler[13] have used a slightly different method and estimated the half-life of drugs in tissues. The con-

[9] B. Romeis, "Mikroskopische Technik." Oldenbourg, München, 1968.
[10] D. E. Farebrother and N. C. Woods, *J. Microsc. (Oxford)* **97**, 373 (1973).
[11] M. Berlin and S. Ullberg, *Arch. Environ. Health* **6**, 589 (1963).
[12] S. A. M. Cross, A. D. Groves, and T. A. Hesselbo, *Int. J. Appl. Radiat. Isot.* **25**, 381 (1974).
[13] R. H. Liss and C. J. Kensler, *in* "Advances in Modern Toxicology: New Concepts in Safety Evaluation" (M. A. Mehlman, ed.), p. 273. Hemisphere Publ., Washington, D.C., 1976.

centration in very small areas may also be estimated on photographic enlargements on photopaper made from autoradiograms that include an isotope scale. The scale is then cut out and simply placed on the print for visual comparison with the tissue areas to be estimated.

An advantage of densitometry is that small areas can be measured. It is also a rapid method. Its main disadvantage is that the quantitative accuracy obtained is not very high.

Direct Analysis

This method employs impulse counting or other direct analysis of the concentration of a radioactive substance in pieces cut or punched out from thick sections. The advantages of this technique as compared to taking samples during an ordinary dissection are the greater precision with which specimens can be taken and the reduced or eliminated risk of artifacts from diffusion, postmortem changes, and contamination. Sometimes it is also of interest to use the tissue pieces for separation of metabolites.[14]

Volatile Substances

When autoradiography with volatile labeled substances is performed on dried, thin whole-body sections, the nonmetabolized volatile substance evaporates and only nonvolatile metabolites are registered.

However, registration of the total radioactivity, including volatile substances, has been made possible by low-temperature techniques. After embedding the animal and freezing in liquid nitrogen, the frozen block can be divided sagittally with a saw in one or more planes. X-Ray film is pressed against the flat surface, and the block is transferred to a freezer at $-80°$. After a first analysis at this low temperature, the block may be sectioned at $-20°$ in the normal manner. The thin sections are then dried and autoradiographed. This technique has been applied to inhalation anesthetics,[15] ethylene oxide, several organic solvents,[16] and nitrosamines.[17]

In Vitro Autoradiography

Gross cryostat sections may be used in several *in vitro* techniques. A common method is to incubate sections, which are usually fixed, in a medium (generally aqueous) containing a radiolabeled substance. The sec-

[14] R. d'Argy, *Acta Pharmacol. Toxicol.* **41**, Suppl. 1, 16 (1977).
[15] E. N. Cohen and N. Hood, *Anesthesiology* **30**, 306 (1969).
[16] K. Bergman, *Scand. J. Work Environ. Health* **5**, Suppl. 1 (1979).
[17] E. Johansson-Brittebo and H. Tjälve, *Chem.-Biol. Interact.* **25**, 243 (1979).

Placenta   **Eye**

Amniotic fluid

FIG. 7. Detail of an autoradiogram of a section through a pregnant mouse showing a fetus. The section was incubated in a water solution of [14C]chloroquine at room temperature. This illustrates the affinity of chloroquine for melanin in the fetal eye.[3]

tions are rinsed and dried before being apposed to film for autoradiographic exposure. One purpose of such *in vitro* investigations is to reveal the receptor sites, the "target tissues," for a physiological substance or for a drug that may imitate the physiologic substance or interfere with its action by blocking the receptor sites. Another possibility is to study the binding of ^{125}I-labeled antibodies to their corresponding antigens. The affinity of many polycyclic amines to melanin[3,18] (Fig. 7) and the binding of bisquaternary ammonium compounds to cartilage (chondroitin sulfate)[19] have been investigated.

The advantages of the *in vitro* autoradiography compared to *in vivo* techniques seem to be that the former is less affected by the rapid degradation of the labeled substance; the high concentrations of radioactivity in blood and accumulations in excretory sites are avoided. The *in vitro* method is not limited to the use of experimental animals; organs and tissues also may be used.

An obvious disadvantage of autoradiography *in vitro*, on the other hand, is that nonspecific tissue absorption, which is not easily prevented, may yield misleading results unless controlled. A technical problem arises from the necessity for use of different experimental conditions with different substances. Therefore, much more experience is needed before we can evaluate the potential of *in vitro* autoradiography.

However, in our laboratory, we observed *in vitro* selective binding of two ^{125}I-labeled gonadotropins (LH and HCG) to their known main target, the corpora lutea, using whole-body sections from mice in the fifteenth day of pregnancy. Immediately after sectioning, 20 μm sections were fixed in $-20°$ acetone or methanol. They were then incubated in a buffer solution containing labeled hormone at room temperature for 0.5 to 2 hr, rinsed for 2 hr, and dried before autoradiographic exposure. The uptake in the corpora lutea was totally selective, with the exception of some absorption into the gastrointestinal contents.[20]

Immunotechniques

An interesting but little explored field is the use of whole-body sections for immunofluorescence. This approach should be useful for the localization of the sites of formation of peptide hormones, for example, or the sites of virus infection. A problem results from the rather large amount of antiserum needed for large sections. Smaller amounts would, however, be required if the peroxidase method were used instead, and still smaller amounts with autoradiography or radioiodine-labeled antibodies.

[18] N. G. Lindquist, *Acta Radiol., Suppl.* 325 (1973).

[19] H. Shindo, *Ann. Sankyo Res. Lab.* 24, 1 (1972).

[20] S. Ullberg and R. d'Argy, *Acta Pharmacol. Toxicol.* 41, Suppl. 1, 84 (1977).

[11] Isolated Perfused Rat Liver Technique

By DIRK K. F. MEIJER, KATJA KEULEMANS, and GERARD J. MULDER

Introduction

The function of the liver in the metabolism of xenobiotics can be studied with a spectrum of liver preparations ranging from the intact organ *in vivo*, through perfusion systems, liver slices, isolated hepatocytes, homogenates, and membrane fractions, to purified enzymes. Each preparation has its special advantages: some of the *in vitro* techniques are helpful in elucidating processes that cannot be studied well *in vivo* but, on the other hand, the use of less integrated liver preparations has the potential of introducing artifacts that may cause misunderstanding of what is observed.

However, many factors cannot be controlled *in vivo*, including the hepatic blood supply (portal and arterial), as well as neuronal and hormonal influences. The use of the isolated perfused liver is attractive because of the intact architecture of the organ, the possibilities of controlling blood and bile flow, and the ease of manipulation of the composition of the perfusion medium. Advantages include the large number of perfusate samples that may be collected and, compared with the intact animal, the smaller number of interactions with endogenous compounds. Furthermore, it is possible to use concentrations of substrates and metabolic inhibitors that would not be tolerated *in vivo* because of their toxicity.

Compared with isolated hepatocyte preparations, the perfusion system does not require damaging treatment with Ca^{2+}-free solutions and digesting enzymes and the normal functional polarity of the cells and their localization in the liver lobule is maintained. In contrast to other types of preparation such as liver homogenates or organelles isolated from them, i.e., microsomes, it has the advantage of allowing the study of sequential metabolic pathways in an environment that includes presumably physiological cosubstrate levels.

There are, of course, disadvantages to the use of isolated perfused liver. In contrast to isolated hepatocyte preparations, it is not possible to obtain many identical liver samples at the same time. With the perfusion technique only relatively short-term experiments are feasible, generally of no more than 6 to 8 hr duration. The different functions of the various cell types within an organ may also be expected to complicate the interpretation of results. It is advisable to use the isolated perfused liver technique in combination with other approaches and other tissue preparations.

Here we present a technique for perfusion of the isolated liver of the rat for the purpose of drug conjugation studies. The investigator can find additional detailed information on the technique for the rat as well as other species in the proceedings of two international symposia[1,2] and in two monographs on the subject.[3,4] The current interest in the technique is also reflected by more recent reviews on the application of liver perfusion in the study of lipogenesis,[5] hormone action,[6] and microsomal electron transport in cytochrome P-450 systems[7] and general metabolic aspects.[8]

Perfusion Apparatus

General Introduction

Many modifications of the original perfusion arrangement described by Bauer and Miller have been described.[1,2] Both circulating and noncirculating techniques are used. It is possible to use a liver that is completely isolated from the animal or that is left *in situ* in the killed animal. A complete liver perfusion apparatus can be purchased.[9] Most investigators use a modification of the apparatus first described by Brauer and Miller in 1951 (see Refs. 1,2) that was elaborated by Schimassek, Northrop and Parks, Hems *et al.*, and Staib *et al.* (see Refs. 1,2,4,8,10) and also used in our laboratory.[10-13]

The essential parts of any liver perfusion apparatus are a thermostatically controlled cabinet (at 38°); a pump providing direct or hydrostatic pressure for perfusion of the organ; an artificial lung allowing adequate oxygenation of the perfusion medium; a perfusate flow meter; a perfusion-medium reservoir and silicon tubing to contain the perfusate;

[1] W. Staib and R. Scholz, eds. "Stoffwechsel der isoliert perfundierten Leber." Springer-Verlag, Berlin and New York, 1968.

[2] I. Bartosek, A. Guaitani, and L. L. Miller, eds., "Isolated Liver Perfusion and Its Application" Raven, New York, 1973.

[3] J. C. Norman, "Organ Perfusion and Preservation." Appleton, New York, 1968.

[4] B. D. Ross, "Perfusion Techniques in Biochemistry." Oxford Univ. Press (Clarendon), London and New York, 1972.

[5] H. Brunengraber, M. Boutry, Y. Daikuhara, L. Kopelovick, and J. M. Lowenstein, this series, Vol. 35, p. 597.

[6] J. H. Exton, this series, Vol. 30, p. 25.

[7] H. Sies, this series, Vol. 52, p. 48.

[8] J. F. O'Donnell and L. Schiff, *Prog. Liver Dis.* 2, 41 (1965).

[9] L. Sestoff and L. O. Kristensen, *Am. J. Physiol.* 236(5), 1 (1979).

[10] D. K. F. Meijer and J. G. Weitering, *Eur. J. Pharmacol.* 10, 283 (1970).

[11] G. J. Mulder, K. Keulemans, and N. E. Sluiter, *Biochem. Pharmacol.* 24, 103 (1975).

[12] D. K. F. Meijer, R. J. Vonk, K. Keulemans, and J. G. Weitering, *J. Pharmacol. Exp. Ther.* 202, 8 (1977).

[13] G. J. Mulder and K. Keulemans, *Biochem. J.* 176, 959 (1978).

and a liver chamber to house as well as support the isolated liver, including its cannulas. In addition, pH, O_2, and CO_2 electrodes, temperature probes, and constant infusion pumps are often included.

Perfusion Cabinet

The perfusion cabinet can be designed to perform two perfusions simultaneously or to perfuse one liver sequentially with different media. It is advantageous to use a spacious perfusion chamber to facilitate rapid assembly and dismantling of the apparatus. All parts should be easily accessible from the outside for rapid adjustments or replacements. Some investigators include automatic valves and fraction collectors for sampling perfusate and bile throughout the perfusion period. The perfusion chamber can be heated by a thermostatically controlled heating fan mounted in the top of the chamber. The heated air should be evenly distributed over the cabinet (for instance by a second fan on the bottom, see Fig. 1). A Plexiglas front in the cabinet is advisable for a good overview of the entire apparatus. Side doors can be provided with some panels to allow sampling of perfusion medium and bile without appreciably altering the temperature.

Perfusion Pump

A roller pump[14] or a peristaltic pump acting on silicon rubber tubing is used to obtain a flow of about 80 ml/min up to a hydrostatic reservoir; about 40 ml/min (cell-free media) is used for perfusion of the organ and 40 ml/min serves as an overflow to the main reservoir, which is important for adequate mixing of the oxygenated medium with the effluent perfusate. The rate of pumping can be temporarily increased for rapid mixing of added substrates.

Control of Perfusate Flow

The height of the hydrostatic overflow reservoir above the liver chamber determines the perfusion pressure and can be used for taking portal perfusate samples. If it is necessary to change the pressure, the reservoir can be moved along one of the vertical stainless steel rods mounted in the chamber as a frame for the perfusion apparatus. For perfusion with oxygen carrier-containing media, the normal portal pressure of 13–14 cm (water) is sufficient. For perfusion with media without oxygen carriers, a pressure of about 20–25 cm (water) is often required for a perfusate flow of 4 ml per gram of liver per minute. The perfusion medium should pass through a nylon filter (for instance, those used in

[14] Watson Marlow, type MHRE. Falmouth, Cornwall, U.K.

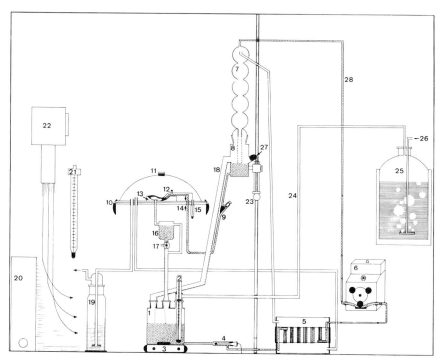

FIG. 1. Isolated rat liver perfusion apparatus. (1) Main reservoir with perfusion medium; (2) thermometer; (3) magnetic stirrer; (4) flow cuvette with pH electrode; (5) silastic tube oxygenator with outlets and inlets for perfusion medium and gases; (6) peristaltic roller pump; (7) 6-bulb glass column film-oxygenator; (8) hydrostatic reservoir with nylon filter; (9) silicon rubber tube including a "flow control" tap for portal "blood" supply; (10) Perspex liver chamber with (11) airtight cover; (12) portal cannula with air bubble trap; (13) removable liver platform with a hole for the hooked caval cannula and grooves for portal and biliary cannulas; (14) "pull through" inlet for rubber tubing connected to the portal vein cannula; (15) bile collecting tube with biliary cannula; (16) perfusate-flow meter reservoir; (17) three-way tap for sampling of effluent perfusate; (18) overflow tube for perfusion medium; (19) gas trap; (20) fan for air circulation; (21) thermostated thermometer connected to (22) heating fan; (23) steel rod of perfusion apparatus frame; (24) tube for passing carbogen gas; (25) flask with saline and bubble device for humidification of O_2–CO_2 gas mixture; (26) inlet for O_2–CO_2 from external cylinders; (27) stopcock for taking "portal" perfusion medium samples; (28) silicon rubber tubing for perfusion medium. Dotted reservoirs and tubes indicate circulating perfusion medium; open tubes are for gas flow. Flow of gas occurs from (25) through (24) to (1), up to (8) and (7) via (18), from (7) into (5), from which it goes to (10) and finally leaves the apparatus via (19).

disposable blood transfusion systems or a nylon gauze of mesh 100) before entering the liver, in order to remove particulate material. In the inlet tube near the portal cannula, an air bubble trap device is essential for prevention of obstruction of flow in the liver. Perfusion, solely through the portal vein with an oxygen-rich medium, is sufficient for most studies. Simultaneous perfusion via the hepatic artery has been reported for rat liver[15] but may only be important for perfusion of the plexus of the biliary tree in studies of biliary reabsorption.

Oxygenation of the Perfusion Medium

The perfusion medium is gassed with O_2–CO_2, which is humidified and warmed to 38° before it enters the apparatus by bubbling it through a 2-liter flask of saline (maintained at 38° in a water bath). O_2 and O_2–CO_2 (95:5) are supplied from cylinders outside the cabinet. A mixing device constructed from flow meters is used for preparing the desired mixture of CO_2 and O_2 that enables adjustment of the pH of the perfusion medium.

Gas exchange with the medium can be achieved with film oxygenators or, alternatively, by passing the perfusion medium through silastic tubing (0.5 mm wall thickness, 1.5 mm i.d.[16]) that is permeable to O_2 and CO_2 and is coiled around a Perspex cylinder jacketed to expose the tubing to the O_2–CO_2 gas mixture. For hemoglobin-free perfusion, we use a 6-bulb glass column film oxygenator that is placed vertically on the hydrostatic reservoir; a countercurrent of O_2–CO_2 is provided (length 26 cm, maximal i.d. 2.8 cm) with a silicon rubber tube oxygenator containing three parallel silastic tubes, each with a length of 200 cm (Fig. 1). The silastic tubing should be replaced after about 20 perfusion experiments of 4 hr each.

Liver Support and Cannulas

The isolated liver is placed in a chamber manufactured of a Perspex bottom with an airtight cover. After passing through the oxygenators, the gas is introduced into the liver chamber, providing the proper humidity for the liver surface. Excess gas escapes through a gas trap that maintains a slightly positive pressure in the apparatus and enables a visual control of gas flow. Because the apparatus is air tight, CO_2 from the system can be collected in the gas trap, a useful system for following oxidation of, for instance, ^{14}C-labeled methyl groups in the liver. The liver itself is in a chamber on a removable platform that fits into the Perspex bottom of the chamber. The chamber has a hole in the bottom for the caval cannula; grooves at the edge fix the portal and biliary cannulae (Fig. 1). The organ is placed in an inverted position against a hemispherical support, thereby resembling the liver's position against the diaphragm in the intact animal. This guarantees a normal position and blood supply to the liver lobes. Caval perfusate is drained from a hooked cannula (polyethylene tubing,

[15] R. Abraham, W. Dawson, P. Grasso, and L. Goldberg, *Exp. Mol. Pathol.* **8**, 370 (1968).
[16] Silastic Medical-Grade Tubing, Dow Corning, Midland, USA.

i.d. 1.5 mm) inserted into the inferior vena cava. The outlet tip of the cannula is 1 cm below the liver thereby providing a small negative pressure. Before entering the main reservoir (200 ml capacity), the perfusate passes a small reservoir of defined volume under the liver chamber; a three-way tap in that chamber can be used for sampling the effluent perfusate. By measuring the time necessary for filling this reservoir, the flow rate of the perfusate can be determined.

Control of pH and Bile Flow

Between the flat-bottomed, glass main reservoir and the pump, a flow cuvette is included that can be used as a site for pH, CO_2, or O_2 electrodes.[17] The pH of the perfusion medium (initially 7.40) is determined by the bicarbonate content in the medium and the CO_2 concentration in the gas mixture. The release of nonvolatile acids due to an initial period of ischemia, the surgical trauma, and subsequent perfusion of the liver, may produce a slight metabolic acidosis, which can be compensated for by lowering the CO_2 concentration. However the concentration of CO_2 in the gas mixture should not be less than 2% (v/v) because a lower value would disturb normal metabolic pathways. Alternatively, addition of sodium bicarbonate by infusion with a pH-stat can be used throughout the experiment. The bulk of the perfusion medium in the main reservoir, coming from the liver and overflow reservoir, is stirred with a magnetic stirrer.

Bile production can be measured by attaching preweighed tubes in which the bile is drained from the biliary cannula to the liver chamber. Bile flow is well maintained by replacing the excreted bile salts with a constant infusion of 15 μmol/hr of taurocholate (sodium taurocholate dissolved in saline infused at 0.9 ml/hr).

Cleaning Procedures and Sterility

The apparatus can be cleaned by washing with tap water and storing in detergent solution. After washing with hot water and rinsing with distilled water, the Perspex parts are placed in 5% formaldehyde solution and the tubes and glassware are sterilized in an autoclave or a dry sterilizer. Care should be taken in using ethanol because small amounts, remaining after cleaning procedures, can have major effects on liver function. After sterilizing, the Perspex parts are thoroughly cleaned with sterile distilled water. To the perfusion medium, 10 μg/ml of ampicillin is added for prevention of gross bacterial contamination. We prefer to isolate the liver completely; if the dead animal is left in the perfusion cabinet for several

---

[17] Ingold Electronics, LOT 401-M7/NS/10, pH electrode.

hours at 38°, autolysis will occur and toxins and bacteria from the abdominal cavity will contaminate the liver, the perfusion medium, and the apparatus.

The Choice of Perfusion Medium

Media Containing Red Cells

The medium most closely resembling a rat's own blood, i.e., fresh blood from donor rats, requires the puncture with sterile equipment of the abdominal aorta of many rats under ether narcosis. The blood can either be defibrinated by mechanical stirring or be heparinized. In the latter case, siliconized glassware should be used in the perfusion apparatus to prevent clotting.[18] Usually, whole rat blood is diluted 1:1 or 1:2 with Krebs–bicarbonate solution to prevent obstruction of liver flow due to partial sludging of sinusoids with aggregated erythrocytes. The major advantage of this medium is that all the normal constituents are present—an important factor for the process under study. The immunological properties of the homologous blood are nearly identical and the oxygen-carrying capacity permits physiological flow rates. Disadvantages include the lack of control of the complex composition of the medium, the presence of vasoconstrictive substances originating from the donor rats, and the time-consuming preparation of the sterile medium. The glycolytic capacity of red cells in the relatively large volume of perfusion medium may also influence the perfusion conditions, while hemolysis of the fragile erythrocytes cannot be prevented even under optimal pump conditions. Such destruction of red cells results in abnormal exposure of the liver to high concentrations of hemoglobin and red cell ghosts. Aged red cells, which may have the advantage of a slower glycolytic rate, hemolyze easily. Bovine, sheep, or human red cells may be used in semisynthetic media. After extensive washings (5 to 6 times with five volumes of Ringer's lactate at 4°) and separation from white cells, such erythrocytes may be suspended in Krebs–bicarbonate buffer to a hematocrit of 15 to 20%.

Red Cell-Free Media

The alternative is the use of a Krebs–bicarbonate buffer solution with glucose added. It is taken for granted in this discussion that if the flow of perfusate is high enough, adequate oxygenation of the liver is attained. Oxygen consumption, gluconeogenesis, and urogenesis in either hemoglobin-free rat liver perfusion or media containing red cells are re-

---

[18] G. Northrop and R. E. Parks, *J. Pharmacol. Exp. Ther.* **145**, 135 (1964).

ported to be identical[7] (O_2 consumption is about 2.0 μmol per minute per gram of liver at 37°). Even the addition of albumin for osmotic purposes is claimed not to be mandatory[7]; however, fatty acid synthesis appears to be severely impaired in albumin-free perfusion media.[5] Moreover, complete omission of albumin results in the release of fibrinous material from the liver, which finally obstructs the flow.[19] Addition of 1 to 4% serum albumin as a physiological "drug carrier" in the blood is advisable. Dialysis or gel filtration of all commercial preparations of bovine serum albumin is recommended, especially for liver perfusion that last longer than 3 hr.[5,20] The use of plasma expanders such as polyvinylpyrrolidone, dextran, gelatin hydrolysates, or sucrose may be suitable for special purposes, e.g., in assessing the influence of protein binding of drug on hepatic disposition, but is not routinely advantageous. Indexes of various liver functions, including bile production, have been reported to deteriorate when using such substances,[1] although sucrose may be relatively safe.[21] The use of a fluorocarbon–polyol emulsion as oxygen carrier in the perfusion medium is not recommended because of the release of inorganic fluoride that could affect metabolic processes[7]; the preparation of such media is usually cumbersome.

The bicarbonate-buffered "Krebs" solution used by us contains 118 mM NaCl, 5.0 mM KCl, 1.1 mM $MgSO_4$, 2.5 mM $CaCl_2$, 1.2 mM KH_2PO_4, 25 mM $NaHCO_3$, and 1.0% bovine serum albumin. After solution of the albumin, it may be necessary to adjust the pH with 0.1 N NaOH, depending on the batch and pretreatment of the albumin preparation that is used. Some reports suggest that the concentration of phosphate should probably be raised to 2 mM because that concentration is more physiological for the rat and improves the energy status of the isolated liver.[9] It is worth emphasizing that, in a recirculating liver perfusion system for periods as short as 2 hr, the composition of the perfusion fluid may gradually change because of accumulation of metabolites.[22–24] Such metabolites can influence hepatic uptake and change the rate of disposition of drugs as well as of bile flow. This problem is overcome by using a single-pass system. It has been reported that inclusion of a dialysis unit in a recirculation setup is helpful in improving the constancy of the preparation with regard to such parameters as bile production, ATP/ADP ratio in the liver, and pH of the perfusate.[22,23]

[19] S. Duca, *Drug Res.* **26**, 858 (1976).

[20] D. Lee and R. K. Holland, *Transplantation* **27**, 384 (1979).

[21] A. R. Beaubien, L. Tryphonas, A. P. Pakuts, M. MacConaill, and H. A. Combley, *J. Pharmacol. Methods* **2**, 213 (1979).

[22] I. Bartosek, A. Guaitani, and S. Garrattini, *Pharmacology* **8**, 244 (1972).

[23] J. Graf, R. Kaschnitz, and M. Peterlik, *Res. Exp. Med.* **157**, 12 (1972).

[24] D. L. Schmucker and J. C. Curtes, *Lab. Invest.* **30**, 201 (1974).

Single-Pass Perfusion

The perfusion apparatus and the perfusion media discussed here can, in principle, also be used for single-pass studies. For a 2-hr perfusion experiment, a volume of about 5 liters of hemoglobin-free medium is required; if red blood cells are added, the perfusate flow can be reduced to 15 ml/min (hematocrit 20%) and about 2 liters of medium suffice. The use of inexpensive media without erythrocytes and albumin should be considered in this case. The advantage of the single-pass procedure, in comparison with the recirculating system, is that accumulation of endogenous products or metabolites that could influence liver function are minimized. Drugs can be infused at a constant concentration and metabolic processes in the liver will be reflected in a steady-state output of metabolites or removal of the unchanged drug in perfusate and bile. It should be stressed, however, that with this technique permeable cofactors or substrates, e.g., taurine, will be constantly washed out of the liver, leading to a decrease in any reaction in which they or their products participate. In principle, only a roller-pump for providing a constant flow through the liver, a perfusate reservoir and a calibrated flask to collect the effluent perfusate are necessary. If one leaves the liver *in situ,* one can even do without a special liver chamber. Although the entire perfusion medium can be gassed and brought to temperature before starting the perfusion, inclusion of a silastic tube oxygenator is recommended for producing a standardized oxygenation of the medium. If the perfusion medium is sterilized beforehand by autoclaving or bacterial filtration, antibiotics may be omitted.

General Points

Before starting a major project, it may be worthwhile to perform pilot experiments investigating the influence of the composition of the perfusion medium on the process to be studied, e.g., the effect of diluted homologous blood versus simple buffer solution. One should also consider the advantages and disadvantages of single-pass and recirculating experiments as well as the length of the perfusion period necessary for obtaining relevant data. An antibiotic may interaction at the levels of protein binding, membrane transport, or biotransformation. Obviously, age, sex, and feeding status of the liver donors may influence the data. Finally, comparison with data obtained *in vivo* may be essential in interpreting the results obtained with the perfusion technique.

Surgical Procedure and Preparation of the Isolated Liver

Careful surgical technique is essential in obtaining a viable rat liver for purposes of perfusion. The procedure should be directed to very careful manipulation of the liver during removal from the animal, avoidance of a lengthy period of hepatic anoxia, and adoption of a regimen that minimizes microbial contamination.

Surgical Methods

The fed donor rat is anesthetized with sodium pentobarbital (Nembutal, 50 mg/kg intraperitoneally) and placed on a mobile operating table. The entire operative field of the rat is disinfected by vigorous swabbing with polyvinylpyrrolidone iodine. An abdominal midline incision is made to expose the liver without damaging the diaphragm. The liver is freed from ligaments and surrounding tissues as much as possible, and double, loose ligatures (surgical silk) are placed around the common bile duct and the portal vein. One ligature is placed around the inferior vena cava, proximal to the insertion of the right renal vein. The common bile duct is cannulated (see Article [4]). The bile cannula is secured by a ligature that, at the same time, ties off the hepatic artery. The portal vein is subsequently tied off with the distal ligature and is gently pulled upward using this ligature. With a fine pair of scissors, an incision is made in the ventral surface close to the distal ligature and the first flush of blood is permitted to flow away. A polyethylene cannula is introduced with the tip about 3 mm away from the first branches of the portal vein; the ligature is then closed. Alternatively, cannulation of the portal vein can be carried out with an intravenous catheter placement unit. The polyethylene cannula (1.3 mm i.d.) fits tightly into a small connecting tube attached to the silicon rubber tubing (85-cm length) coming from the hydrostatic reservoir. This tube can be pulled through the liver table (Fig. 1), enabling initial perfusion of the liver *in situ* outside the cabinet and a permanent connection of the liver to the perfusion system. Immediately after connecting the liver, the organ is perfused for only 10 sec to remove blood and to supply oxygen. Swelling of the liver should be prevented. The abdominal aorta is now severed, the diaphragm is rapidly cut, and the thorax is opened to expose the right atrium; the rat dies during these procedures. A double ligature is placed around the thoracic inferior vena cava and, through an incision in the right atrium, a hooked polyethylene cannula (1.5 mm i.d.; 2.4 mm o.d.) is brought in and shifted until its tip is close to the hepatic veins. The use of smaller vena cava cannulas can result in liver congestion. After securing the cannula and tying off the vena cava ligature close to the right kidney, the liver is again perfused with about 20 ml of perfusion medium to remove the remaining blood. The perfusate should easily drain from the caval cannula without swelling of the liver. Subsequently the esophagus is severed and adhesions and ligaments of the stomach and the left anterior and posterior small liver lobes are carefully cut. Next, the thoracic vena cava inferior is cut near the heart.

After incising the right part of the diaphragm, the liver is carefully lifted up a bit and the rest of the diaphragm and adhesions to the gut are severed.

The liver, with the connected cannulae, is transferred to the Perspex platform that, in turn, rests on a glass beaker (400 ml), and is placed against a hemispherical support. The arrangement of the lobes is checked; the portal and biliary cannulas should be clearly visible and positioned exactly opposite to the caval cannula. The residual diaphragm is spread out under the liver to prevent obstruction of caval flow. The caval cannula is put through a hole in the liver platform and the portal vein cannula is fixed to the edge. Remaining blood clots are removed and the surface of the liver is rinsed with 10 ml of saline at 37°. At this stage, a few milliliters of perfusion medium are run through and should drain without producing swelling of the liver lobes.

The operative procedure, from cannulating the portal vein to this point, should not take longer than 4 min; because of the intermediate perfusions, the liver is without oxygen for no longer than 2 min. Obviously, some practice in the procedure will be necessary to meet these time restrictions.

The small liver platform is transferred to the liver chamber by lifting it up and pulling the silicone tubing connected to the portal cannula through a hole in the bottom of the liver chamber. The position of the portal and caval cannula is checked again and the tip of the biliary cannula is placed in the collection tube under the liver table.

Control of Perfusion Conditions

The airtight cover of the liver table is replaced and the flow of O_2–CO_2 gas is controlled. After closing the cabinet, the temperature of the perfusion medium should reach a constant value of 38° within 5 min. Due to the short period of ischemia and manipulation of the liver, the pH of the perfusion medium may drop up to 0.02 units initially but should spontaneously recover within 15 min. Perfusate flow through the liver should be 4 ml per min per g of liver by the time that constant temperature is reached. Thereafter, 30 min is allowed for equilibration before constant flow and bile production is attained. At this stage, the amount of perfusate is brought to the desired volume, i.e., 100 ml in our apparatus. Determination of flow rate and pH is performed at 10-min intervals, prior to sampling of perfusate and bile and inspection of the gross appearance of the liver. Small liver lobes or parts of a liver lobe can be sampled during perfusion without leakage by careful ligation and excision of the tissue. Liver weight after the perfusion is determined by removing nonhepatic tissue, blotting with filter paper, and weighing.

Aseptic Conditions

Although surgery is not performed under entirely sterile conditions, bacterial contamination can largely be prevented by disinfecting the operating field of the rat, by touching the liver only with disinfected hands, and by avoiding use of contaminated instruments. An initial concentration of 10 μg/ml ampicillin in the perfusion fluid is usually sufficient to prevent gross sepsis during the 2- to 3-hr period of perfusion. Sterilization of glassware, rubber tubing, and disinfection of Perspex parts will prevent gross bacterial contamination. These steps are necessary because even this short period of perfusion can result in sufficient bacterial growth to cause obstruction of flow and an abnormal drop of pH in the perfusion medium.

Function Control, Viability Tests, and Performance of the Isolated
    Perfused Liver

Function Control During Perfusion

Numerous tests of liver function have been applied to control performance of the perfused liver. Most well-defined tests of function may be used; often the investigator chooses tests that are to some extent related to the matter under study. Among the classical tests of liver function *in situ* are the measurement of oxygen consumption by inclusion of Clark-type platinum electrodes and the monitoring of pH and potassium concentration in the perfusion medium with ion-selective electrodes. Bile production and hepatic clearance of cholephilic dyes such as bromosulfophthalein,[25] dibromosulfophthalein,[12] and indocyanine green may be used to evaluate the viability of the isolated liver during the experiment. Organ absorbance spectrophotometry of liver lobes (cytochrome *P*-450 and reduced nicotinamide nucleotides) and fluorescence intensity emitted from the surface of the organ (reduced nicotinamide nucleotides) can be used to monitor the metabolic status during perfusion, but these methods require expensive instrumentation.[7] The combination of control of the acid–base balance, oxygen consumption, bile flow, perfusate flow, and outer appearance of the liver will, in practice, provide adequate information to the investigator. The liver should be uniformly pinkish brown; red

[25] J. Jam, M. Reeves, and R. J. Roberts, *J. Lab. Clin. Med.* 87, 373 (1976).

spots caused by congestion and white spots caused by air bubbles should be absent.

Retrospective Estimates of Function

Tissue levels of certain critical normal metabolites (glycogen; ADP/ATP) and the wet and dry weight of liver can be determined. Lactate–pyruvate, glucose, and urea concentrations, as well as the concentration of enzymes (lactate dehydrogenase, transaminases, alkaline phosphatase) in the perfusion medium can be used as parameters of viability. Histology and histochemistry may provide very useful information on the integrity of the perfused livers[10,24,26]: vacuolated hepatocytes, distended sinusoids, and endothelial cell injury are indicative of an inadequate oxygen supply and excessive hepatic pressure. Abnormally high activity of lysosomal enzymes, e.g., acid phosphatase, may point to hyperoxia.[26] After 2 hr of perfusion, the dry-wet mass ratio should not be lower than 0.28.

Performance of the Perfused Liver

The pH of the perfusate, without corrections by infusion of bicarbonate or lowering of the pCO_2, will normally decrease from 7.40 initially to 7.35 after 2.5 hr of perfusion. Bile production should be about 12 μl per min per 10 g of liver at the start and will drop by about 15% in two hours if a taurocholate infusion is used; without replacement of bile acids, bile flow may gradually drop 30 to 40% after 2 hr. The flow of perfusate through the liver should be constant during this period; increases in perfusion pressure are only necessary if added substrates or metabolic inhibitors increase the resistance of the vascular bed or induce distension of endothelial and parenchymal cells. For a biochemical comparison of performance of isolated rat liver perfused with hemoglobin-containing and hemoglobin-free media, the reader is referred to a review by Sies.[7]

Examples of Use of the Method

The isolated rat liver perfusion technique has been used extensively in studies of drug conjugation, hepatic uptake, and biliary excretion of drugs. Conjugation has been examined with this technique for lidocaine,[27] 1-naphthol,[28,29] harmol,[11,13] 4-dimethylaminophenol,[30] 4-nitrophenol,[31]

[26] T. Sugano, K. Suda, M. Shimada, and N. Oshuvo, *J. Biochem.* (*Tokyo*) **83**, 995 (1978).
[27] K. S. Pang and M. Rowland, *J. Pharmacokinet. Biopharm.* **5**, 655 (1977).
[28] R. Scholz, W. Hansen, and R. G. Thurman, *Eur. J. Biochem.* **38**, 64 (1973).
[29] K. W. Bock, E. Huber, and W. Scholte, *Naunyn-Schmiedeberg's Arch. Pharmacol.* **296**, 199 (1977).
[30] P. Eyer and H. G. Kampffmeijer, *Biochem. Pharmacol.* **27**, 2223 (1978).
[31] N. Hamada and T. Gessner, *Drug. Metab. Dispos.* **5**, 407 (1975).

paracetamol and phenacetin[32,33] (glucuronidation and sulfation), sulfanilamide,[3,34] isoniazide[35] (acetylation), paracetamol,[32] bromosulfophthalein[25] (glutathione conjugation). For the kinetic analysis of xenobiotic metabolism in isolated perfused livers, useful information is also available.[36–39]

[32] F. Grafström, K. Ormstad, P. Moldeus, and S. Orrenius, *Biochem. Pharmacol.* **28**, 3573 (1979).

[33] K. S. Pang and J. R. Gilette, *J. Pharmacokinet. Biopharm.* **7**, 275 (1979).

[34] A. Bettschart, R. Kok, and D. Bovet, *Helv. Physiol. Pharmacol. Acta* **15**, 241 (1957).

[35] D. Notter, G. Catau, and S. Besson, *in* "Drug Measurement and Drug Effects in Laboratory Health Science," p. 135. Karger, Basel, 1980.

[36] M. Rowland, *Eur. J. Pharmacol.* **17**, 352 (1972).

[37] M. S. Anwer and R. Gronwall, *Can. J. Physiol. Pharmacol.* **54**, 277 (1976).

[38] A. Rane, G. R. Wilkinson, and D. G. Shand, *J. Pharmacol. Exp. Ther.* **200**, 420 (1977).

[39] K. S. Pang, *Trends Pharmacol. Sci.* **2**, 247 (1980).

[12] Isolated Perfused Rat Kidney

By John F. Newton, Jr. and Jerry B. Hook

Historically, the role of the kidney in elimination of drugs and other foreign chemicals was believed to be restricted to excretion of the parent compound or metabolites of hepatic origin. Consequently, the ability of the kidney to metabolize drugs and xenobiotics has been frequently underestimated. Although the kidney has measurable capacity for metabolizing xenobiotics, the *in vitro* metabolism of model substrates is often one-tenth to one-hundredth that of the liver.[1,2] However, there appear to be several factors that may enhance the ability of the intact kidney to metabolize foreign compounds. Xenobiotics may reach high concentrations in cells of the kidney as a result of preferential reabsorption and/or secretion. In addition, certain compounds may be removed from plasma binding sites and actively transported into cells. Therefore, a compound that is unavailable to other organs as a result of extensive protein binding may be accessible to the renal cells. Consequently, the renal contribution to *in vivo* metabolism of compounds may not be predicted from reports of *in vitro* renal drug metabolism. For instance, using a model substrate for

[1] P. Moldeus, *Biochem. Pharmacol.* **27**, 2859 (1978).

[2] D. P. Jones, G. Sundby, K. Ormstad, and S. Orrenius, *Biochem. Pharmacol.* **28**, 929 (1979).

study of conjugation reactions, Diamond *et al.*[3,4] observed that the kidney can account for at least 41% of the urinary conjugates of *p*-nitrophenol in the chicken and 22% in the rat.

To determine the contribution of the kidney to total drug metabolism in the intact animal, it was essential to develop methods for these studies that may reflect the transport processes and concentrating ability of the kidney *in vivo*. Therefore, a technique was developed in which the cellular structure is maintained and, as well, the basic unit of renal structure, the nephron, remains intact as a single unit.

The isolated perfused kidney (IPK) has been utilized effectively to study the transport and/or metabolism of a number of endogenous compounds. The interrelationship of choline transport and metabolism in the kidney have been described by Trimble *et al.*[5] The renal activation of vitamin D has also been quantified in the isolated perfused kidney.[6]

Only recently has the isolated perfused kidney been used for the study of renal drug metabolism. Mitchell *et al.*[7] demonstrated the "consumption" of gentamicin, cephaloridine, and neomycin by the isolated perfused rat kidney; whether this represented sequestration of the drugs in renal tissue or metabolism to inactive metabolites was not clear. Szefler and Acara[8] studied the transport of isoproterenol as well as its metabolism to 3-*O*-methylisoproterenol in the IPK. Bekersky *et al.*[9] evaluated the renal clearance and metabolism of salicylate and salicyluric acid in the IPK; renal clearance of salicylate administered directly was different than clearance of salicylate derived from renal metabolism of salicyluric acid.

Several laboratories are now using the IPK as a technique to determine the role of drug metabolism in the nephrotoxicity of xenobiotics. Tange *et al.*[10] used the IPK to determine initial biochemical effects of the nephrotoxicant *p*-aminophenol. Hart *et al.*[11] demonstrated metabolism of acetaminophen to a reactive intermediate that conjugated with renal glutathione. We have quantified the conjugation of acetaminophen and/or its reactive intermediate with sulfate, glucuronide, and glutathione with

[3] G. L. Diamond, M. J. Cooper, L. M. Tremaine, and A. J. Quebbemann, *Pharmacologist* **22**, 244 (1980).

[4] G. L. Diamond and A. J. Quebbemann, *Pharmacologist* **21**, 193 (1979).

[5] M. E. Trimble, M. Acara, and B. Rennick, *J. Pharmacol. Exp. Ther.* **189**, 570 (1974).

[6] A. M. Rosenthal, G. Jones, S. W. Kooh, and D. Fraser, *Am. J. Physiol.* **239**, E12 (1980).

[7] C. J. Mitchell, S. Bullock, and B. D. Ross, *J. Antimicrob. Chemother.* **3**, 593 (1977).

[8] S. J. Szefler and M. Acara, *J. Pharmacol. Exp. Ther.* **210**, 295 (1979).

[9] I. Bekersky, L. Fishman, S. A. Kaplan, and W. A. Colburn, *J. Pharmacol. Exp. Ther.* **212**, 309 (1980).

[10] J. D. Tange, B. D. Ross, and J. G. G. Ledingham, *Clin. Sci. Mol. Med.* **53**, 485 (1977).

[11] S. Hart, I. Calder, B. D. Ross, and J. D. Tange, *Abstr. Int. Congr. Nephrol., 7th, 1978* p. W-2 (1978).

this system. In addition, we have determined the effect of various modulators of drug metabolizing activity on the metabolism and excretion of acetaminophen and its metabolites in the isolated perfused kidney.[12,13]

Several excellent reviews have appeared describing the use of IPK.[14–16] Herein we describe how the preparation has been modified in our laboratory for studies on renal drug metabolism.

General Considerations for Studies Involving Drug Metabolism in the IPK

Rapid turnover of perfusate in the system is essential. Dead volumes in the perfusion system should be minimized. This may be accomplished by using small volumes of perfusate; however, the substrates necessary for normal kidney function may become depleted, possibly sacrificing kidney viability. We have used a system in which perfusate diverted from a separate, rapidly mixing system, is infused into the kidney. The effluent perfusate is then allowed to reenter and equilibrate with the recirculating system. With this technique, larger volumes of perfusate can be used without sacrificing complete mixing. However, the method requires that points of large pressure drop, e.g., filters and oxygenators, be minimized. To this end we utilize parallel systems with double oxygenators and filters. In addition, filters of large particle-holding capacity are used, thereby reducing the chance of a large pressure drop due to filter occlusion.

Extreme care should be taken to minimize the effect of perfusate on renal drug metabolism. Addition of antibiotics to the perfusate for reduction of bacterial growth may have an effect on drug transport and/or metabolism. Use of oxygen-emulsifying fluorocarbons as oncotic agents may inhibit the cytochromes involved in drug metabolism.[17] In addition, the perfusate itself may have drug metabolizing capabilities. Bovine serum albumin may be contaminated with monoamine oxidase activity.[18] Therefore, it is critical to document the stability of a drug during perfusion without a kidney in the circuit. Any metabolism of the compound by the perfusate can be subtracted to quantify the actual extent of renal metabolism. This technique should also be extended to identify and quantify the stability of possible metabolites in the perfusate.

The stability of a drug and its metabolites in the urine is also a factor.

[12] J. F. Newton, W. M. Kluwe, and J. B. Hook, *Toxicol. Appl. Pharmacol.* **48**, A19 (1979).
[13] J. F. Newton, W. M. Kluwe, C.-H. Kuo, W. E. Braselton, Jr., and J. B. Hook, *Abstr. Soc. Toxicol Annu. Meet.*, p. A68 (1980).
[14] R. H. Bowman, this series, Vol. 39, p. 3.
[15] B. D. Ross, *Clin. Sci. Mol. Med.* **55**, 513 (1978).
[16] B. D. Ross and A. Nizet, *Proc. Int. Congr. Nephrol., 7th, 1978* p. 653 (1978).
[17] M. N. Goodman, R. Parrilla, and C. J. Toews, *Am. J. Physiol.* **225**, 1384 (1973).
[18] R. A. Roth and R. Alper, personal communication (1980).

For many compounds, collecting the urine into a vial in dry ice will suffice. For others, stability may be enhanced by collecting urine directly into acidic or alkaline solutions. For highly reactive compounds, it is a common technique to react the urine with a "trap." The metabolite reacts chemically with the "trap" in such a way as to allow identification of the metabolite's original structure. Gemborys and Mudge[19] allowed the unstable metabolite of acetaminophen (N-hydroxyacetaminophen) to react with cysteamine, thereby reducing the amount of reactive intermediate converting back to the parent molecule. With the IPK, efforts to stabilize metabolites are often a compromise, especially when several metabolites are involved. For example, a solution of low pH will stabilize N-hydroxyacetaminophen but may degrade some of the other metabolites of acetaminophen. Therefore, several collection techniques must be utilized to total evaluation of renal drug metabolism.

Apparatus

The perfusion system for our studies is shown in Fig. 1. It consists of a 3.5-cm arterial cannula made from a 19-gauge thin-wall needle, specially beveled and smoothed; a pressure transducer connected by a Y-connector preceding the cannula; an infusion pump (Harvard Apparatus Model 1203 or Masterflex Model 7557 with 7014 pump head) that can be adjusted to keep a constant pressure of 120 mmHg at the cannula tip; an arterial perfusate reservoir; two membrane oxygenators connected in parallel, each containing 3.8 meters of silastic tubing (Dow Corning, 602-235, i.d. 1.47 mm by o.d. 1.96 mm); 95:5 $O_2:CO_2$ is passed through the oxygenators into a beaker of water (rate of bubbling reflects gas flow rate); two 47-mm Millipore Swinnex filter holders containing 47-mm diameter prefilters (Millipore type AP25 04700), in parallel; a recirculating pump (Masterflex pump 7545 with pump head 7016); a two-way sampling valve; a venous perfusate reservoir; and a venous effluent collector and flow rate monitoring device that is the cylinder of a 20-ml syringe onto which a stopcock had been mounted. All components except the filters, pumps, and sampling valve are housed in a $27 \times 27 \times 90$ cm thermostatically controlled Plexiglas box. Except where required for connections to glassware and other parts of the perfusion apparatus, all components are connected by Tygon tubing (i.d. 3/16 in. by o.d. 5/16 in. chemical resistant, formulation R-3603). The two perfusate reservoirs are connected with a short piece of tubing that allows perfusate to be pumped through the shaded tubing (Fig. 1) at a rate approximately four times the rate of flow into the kidney. This provides increased oxygenation and more complete

[19] M. Gemborys and G. H. Mudge, personal communication (1980).

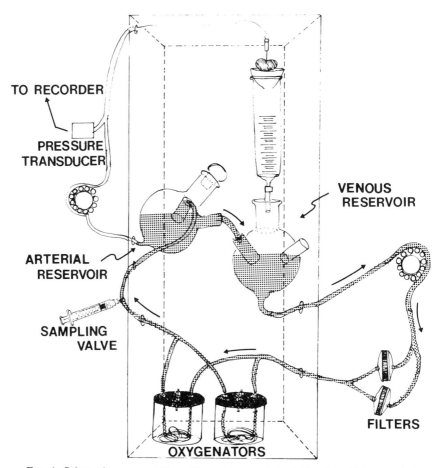

FIG. 1. Schematic representation of apparatus used for isolated rat kidney perfusion. Crosshatched areas indicate the perfusate that is undergoing rapid recirculation. Arrows indicate the direction of flow through the rapidly recirculating system (not drawn to scale).

filtering. The system also produces very rapid and complete mixing of the perfusate.

The high rate of perfusate flow requires the use of multiple filters and oxygenators in parallel. When single pieces or multiple pieces in series were used, the high resistance often produced leaks in the system. An additional advantage of the parallel arrangement is that oxygen exchange and filtration areas are doubled. Even with a dual filter system, the 8-μm filters originally used in our preparations frequently became clogged, causing a large pressure drop distally and leakage proximally. Recently, prefilters (Millipore AP25 04700) instead of filters have been employed because of the larger particle-holding capacity of the former; prefilters have produced excellent results.

The 19-gauge thin-wall cannula is easier to manipulate during cannulation than the 18-gauge cannula preferred by many investigators. Cannulae with smaller radii do not support a viable kidney due to the high flow velocities at the cannula tip. The straight cannula was found to be more effective than those with a 90° angle. The recirculating pump is set to run at a constant speed from experiment to experiment, unlike the infusion pump whose speed is varied to maintain constant pressure. Operating the recirculating pump at a constant speed is of particular advantage; the Masterflex pump 7545 can accommodate several pump heads. Thus, a single pump can service several kidney perfusion systems, thereby reducing cost.

The oxygenators are basically modifications of those reported by Hamilton et al.[20] Briefly, the oxygenator is 10.1 cm high and 8.8 cm in diameter with a plastic screw-on top. Four holes are drilled into the top, where custom-lathed Teflon fittings are inserted. Two fittings in the oxygenator are connected to silastic tubing for inlet and outlet of the perfusate and the other two are for inlet and outlet of the oxygen–carbon dioxide mixture. A stub adaptor (16-gauge, Intramedic Luer Stub Adaptor) is used to connect the Teflon fitting to the silastic tubing inside the oxygenator. Tygon tubing is connected directly to the fitting outside.

Also note that all flow toward the kidney comes from the bottom of the arterial reservoir, negating the need for a bubble trap. Bubble traps may be a disadvantage in drug metabolism studies because they increase circulation time through the system.

The air in the Plexiglas box is circulated by a fan and the perfusate temperature is maintained at 37° by heating elements attached to the side and back walls.

Perfusate Preparation

The perfusion medium is a modified Krebs–Ringer bicarbonate buffer containing lactic acid, 10 mM; glucose, 6 mM; bovine serum albumin (Fraction V, Pentax), 7%; inulin, 2 mg %; and urea, 7 mM, adjusted to pH 7.4. The final electrolyte concentrations in the perfusing medium are Na^+, 122 mM; K^+, 4 mM; Ca^{2+}, 2.5 mM; Mg^{2+}, 1 mM; Cl^-, 104 mM; HCO_3^-, 25 mM; SO_4^{2-}, 1 mM; and phosphate, 2 mM. All water used in the perfusate is glass distilled.

[20] R. L. Hamilton, M. N. Berry, M. C. Williams, and E. M. Severinghaus, J. Lipid Res. 15, 182 (1974).

Many different types and concentrations of oncotic agents may be used in this preparation. Haemaccel,[21,22] dextran,[23] hydroxyethyl starch,[22] fluorocarbons,[22] and bovine serum albumin[24] have all been used successfully for this purpose for isolated kidney perfusion. The merits of each will vary with the investigation. For instance, although relatively inexpensive, Haemaccel and dextran produce excessive potassium excretion.[15] The oxygen-emulsifying fluorocarbons, although allowing excellent renal function, may not be applicable to drug metabolism studies because of the high solubility of lipid soluble compounds in fluorocarbons. Bovine serum albumin binds drugs to a large extent and may contain enzymes capable of metabolizing certain compounds.[18] Regardless of the oncotic agent, it is important to quantify fractional drug binding, at the concentrations to be studied, in order to determine the fraction of total drug available to the renal cells.

Bovine serum albumin, Fraction V, is probably the most widely used oncotic agent for isolated kidney perfusion. However, kidney function may vary drastically depending on the source of the albumin. Our best kidney function studies (Fig. 2) have been with Miles Biochemicals Pentax Bovine Serum Albumin Fraction V (No. 81-003-4). The concentration of albumin required for optimum function, i.e., for high sodium reabsorption and inulin clearance, varies considerably among the products of different manufacturers. In our perfusion system, the optimal albumin concentration with Sigma albumin was 6.0% whereas that for Miles Pentax albumin was 7.0%. Pentax albumin at 7% supports a physiological inulin clearance, good fractional sodium reabsorption, and high urine flow rates (Fig. 2).

In our laboratory, albumin is dialyzed against a 10-volume excess of perfusate buffer without $CaCl_2$ for 72 hr at 4°. Albumin is prepared for dialysis as a 30% solution in buffer without $CaCl_2$. Some investigators use antibiotics during dialysis to prevent bacterial growth[14] although this may have an effect on transport, metabolism, or excretion. Following dialysis, the concentration of the bovine serum albumin in solution is usually between 13–17% as determined by the method of Lowry et al.[25] It is diluted with buffer without $CaCl_2$ to 7% prior to perfusion. Substrates are added and the pH adjusted to 7.45 with 0.1 N NaOH. Aliquots for individual experiments may be prepared and frozen. On the day of perfusion, the solution is thawed and $CaCl_2$ added to a concentration of 2.5 mM. The

[21] H. J. Shurek, J. P. Brecht, H. Lohfert, and K. Hierholzer, *Pfluegers Arch.* **354**, 349 (1975).
[22] H. Franke and C. Weiss, *Adv. Exp. Med. Biol.* **75**, 425 (1976).
[23] F. J. Koschier and M. Acara, *J. Pharmacol. Exp. Ther.* **208**, 287 (1979).
[24] J. R. Little and J. J. Cohen, *Am. J. Physiol.* **226**, 512 (1974).
[25] O. H. Lowry, N. J. Rosebrough, A. L. Farr, and R. J. Randall, *J. Biol. Chem.* **193**, 265 (1951).

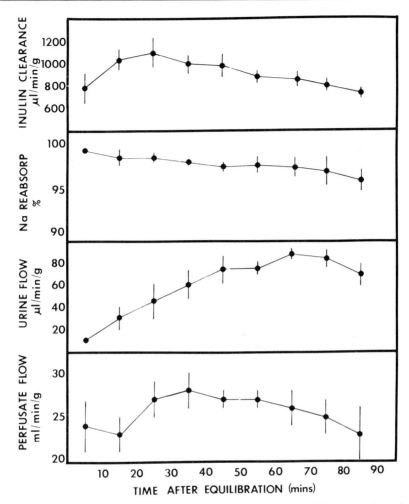

FIG. 2. Functional characteristics of the isolated perfused rat kidney at various times after equilibration. Each point represents mean for 12 kidneys and vertical lines indicate ± SEM.

perfusate is filtered with a Millipore disc filter holder (No. YY30-142-30) through a prefilter and filters with 8 μm, 0.45 μm and 0.22 μm pore sizes; small particulate matter and a large percentage of bacterial contamination that may be present are thereby removed. Because filtration will not remove bacterial toxins, preparation of perfusate under sterile conditions is suggested. A measured volume of perfusate is placed in the perfusion apparatus and allowed to warm and equilibrate with 95% O_2 : 5% CO_2 for 30 min prior to cannulation.

Operative Technique

The right renal artery is cannulated by the method described by Nishiitsutsuji-Uwo et al.[26] with a few modifications. In most cases the right renal artery of the rat arises directly across the aorta from the superior mesenteric artery. This anatomical arrangement allows for insertion of a cannula into the superior mesenteric artery for retrograde cannulation across the aorta into the renal artery. This is accomplished without interruption of perfusate flow to the kidney. We use male, Fisher-344 rats (200–225 g) as kidney donors because of their low incidence of aberrant anatomy, i.e., the renal artery arising from the aorta at a place other than directly opposite the mesenteric artery. When first attempting the procedure, it is best to use large rats (350–450 g) for ease in cannulation. However, heavy rats usually have excessive amounts of fat surrounding the area around the origin of the right renal artery. Prior to cannulation of the large animals, the area must be cleared of fat, which often results in tactile stimulation of the renal artery, possibly leading to impaired functional performance of the isolated perfused kidney. Thus, it is advantageous to use a smaller rat.

The rat is anesthetized with sodium pentobarbital (50 mg/kg). The left jugular vein is cannulated with PE50 tubing (i.d. 0.58 mm, o.d. 0.965 mm) and 100 mg of mannitol and 100 IU of heparin in 1.4 ml of saline are injected. Ten minutes after the injection, a midline incision is made from the bladder to the sternum with transverse cuts at the bladder and sternum. The intestines are laid off to the left side of the animal, thereby exposing the right kidney. A small, saline-soaked sponge (3.8 × 3.8 cm) is placed over the right kidney and a larger sponge (3.8 × 7.6 cm) is forced under the sternum to keep the liver away from the area of cannulation.

The right ureter is exposed and isolated approximately 4 cm from the renal pelvis. It is raised over a wooden Q-Tip where it can easily be cleaned of connective tissue and cannulated. A loose ligature is placed around the ureter cranial to the area of intended cannulation. We have had the most success with 000 silk; 00 silk tended to loosen easily. The right ureter is cut with microdissecting scissors about two-thirds of the way through at an angle of about 45°. An 8-cm length of PE10 tubing (i.d. 0.28 mm, o.d. 0.61 mm) is then inserted up the ureter, as close to the renal pelvis as possible; this step will minimize twisting of the ureter after cannulation. The ureter below the point of cannulation is severed.

The fat around the area of the origin of the right renal artery should be gently dissected away with extreme care to avoid tactile stimulation of the renal artery. The depression of flow resulting from tactile stimulation can be prevented by α-receptor blocking agents such as phentolamine or

[26] J. M. Nishiitsutsuji-Uwo, B. D. Ross, and H. A. Krebs, *Biochem. J.* **103**, 852 (1967).

phenoxybenzamine.[21] However, the use of such agents may be detrimental in studies of drug metabolism.

After removal of fat, the adrenal artery should be located and ligated. Ligation may not be necessary in larger rats in which the cannula can be positioned in the renal artery beyond the point of origin of the adrenal artery; in smaller animals, movement of the cannula beyond the origin of the adrenal artery is extremely difficult. Using curved forceps, the vena cava is retracted toward the right kidney, exposing the right adrenal artery. In most instances the adrenal artery will arise from the cranial side of the renal artery; it can arise from the caudal side or, in some cases, directly off the aorta.

Forceps are placed underneath the right renal artery, as close to the aorta as possible, and a 7-cm length of 000 surgical silk is drawn back under the vessel. The same procedure is followed with the mesenteric artery, as close to the aorta as possible, and both ligatures are loosely tied. Only three ligatures (adrenal, renal, and mesenteric arteries) are used; a ligature on the mesenteric artery distal to the cut does not appear to be essential if there is flow from the cannula during cannulation; a single ligature on the right renal artery is adequate for holding the cannula in place if 000 silk is used. Minimal use of ligatures reduces manipulation of the renal artery and the time necessary for cannulation.

Cannulation Technique

Due to the anatomical arrangement of the mesenteric and renal arteries of the rat, retrograde cannulation of the mesenteric artery allows for perfusion without interruption of blood flow. Several investigators have chosen to perfuse the aorta rather than the renal artery; we find this method less effective. This may be due to several small arteries that arise dorsally from the aorta, each of which must be ligated for precise measurement of perfusion pressure and flow. Extensive manipulation of the area surrounding the renal artery would be required.

The area of the mesenteric artery just proximal to its bifurcation into the intestinal mesentery is an appropriate area for cannulation. The area is gently exposed in preparation for cannulation. The cannula is taken from the perfusion chamber and placed next to the rat in preparation for perfusion. A flow rate of 2.5 ml/min/100 g body weight of rat is set. The following procedure must be performed extremely quickly and delicately. Cannulation of the renal artery must be accomplished before breathing ceases because this may cause a dramatic yet transient increase in perfusion pressure.[21]

Small curved serrafine forceps are used to carefully occlude flow to the mesenteric artery and by gently lifting, to prepare it for cannulation. The mesenteric artery is cut, with microdissecting scissors, at a 45° angle at a point just proximal to the bifurcation. The cannula is inserted into the mesenteric artery using an abbreviated twisting motion, always with pressure toward the right kidney. Placement of the flat side of the bevel against the dorsal side of the artery is a convenient approach for successful entry into the vessel. Once the cannula is securely in the mesenteric artery, flow rate is increased to 6 ml per min per 100 g body weight. This will allow easier movement across the aorta and into the renal artery. The cannula is moved as far as possible into the renal artery. The beveled part of the cannula must be entirely beyond the point of origin of the renal artery. While holding the cannula securely in place, the ligature around the renal artery is tightened, followed by the ligature around the mesenteric artery. A single knot is sufficient; we find that double knotting increases the chance of a loose ligature. Once the cannula is secured, it is held at an angle of approximately 25° from the body while the kidney is excised. Excision of the kidney involves basically two large cuts after the kidney is freed from perirenal fat. The first severs the vena cava and aorta caudal to the renal artery as well as the mesenteric artery distal to the cannulation point. The cannula and kidney can then be raised and a second cut used to sever the aorta and vena cava above the renal artery. The kidney should then be free from the body of the rat. It is moved into the perfusion chamber and the first 30 ml of perfusion medium passing through the kidney are diverted from the recirculating system. From the time the kidney is placed in the chamber, pressure is maintained at 120 mmHg, mean, by adjusting the flow rate. The kidney will dilate dramatically for the first 15 min of perfusion, after which the flow rate will stabilize. The ureteral cannula is positioned for optimal urine flow being very careful not to twist the ureter. Once the ureter is positioned, the cannula is cut as short as possible and the urine allowed to fall into a preweighed 2-ml vial. The short cannula minimizes resistance to urine flow.

Experimental Procedure

Once the kidney is placed in the perfusion chamber, it is imperative that perfusion pressure be maintained at 120 mmHg, mean, above the pressure needed to overcome the cannular resistance. Because the pressure due to cannula resistance varies with perfusion flow, it is important to determine pressure values for cannula resistance at different flow rates before the operative procedure is begun. During the 15-min equilibration period, the desired volume of perfusate in the system is established.

This may require addition of more perfusate to the system due to perfusate loss during cannulation. Not more than 10% of the desired final volume should be added in less than 5 min. We routinely use an initial volume of 200 ml perfusate.

After equilibration, nine 9-min clearance samples are collected. Between each period we delay 1 min to change preweighed vials. Therefore, the complete experiment runs for 90 min after equilibration. At the midpoint of each clearance period, a 1-ml sample of perfusate is taken from the sampling port. This volume allows sufficient sample to monitor kidney functional and biochemical status while limiting the volume of perfusate lost from perfusate sampling and urine formation to less than 10% of initial volume. Perfusate flow rate is also determined during each clearance period. Urine flow is determined gravimetrically. Most physiological and biochemical functions of the isolated perfused kidney are reported as values per gram of kidney tissue before perfusion. These data are obtained by determining the weight of the left kidney and assuming that the right (perfused) kidney weight is equal before perfusion.

Acknowledgments

We thank Dr. J. J. Cohen for his helpful advice during the development of our perfusion system. We also acknowledge L. Brody for her excellent art work, D. Hummel and T. Palucki for manuscript preparation, and B. Hook, W. Telford, and M. Bailie for their technical assistance. The authors' research is supported by U.S.P.H.S. grant ES 00560.

[13] Lung Perfusion Techniques for Xenobiotic Metabolism and Toxicity Studies

By BRIAN R. SMITH and JOHN R. BEND

The isolated perfused lung preparation is a versatile technique for toxicological, pharmacological, or biochemical studies. The utility of this system arises from the simplicity and flexibility of an organ available in a living and breathing state in the absence of other tissues that would mask the pulmonary contribution to xenobiotic metabolism, and from the ease in which experimental parameters can be manipulated. These parameters include blood (or perfusate) flow rate, temperature, oxygen partial pressure, and concentration of test material in the circulation. Some complications associated with whole animal studies, such as substrate or metabolite elimination, metabolism by other organs, or redistribution of test materials to other organs, are avoided in perfused lung experiments. Furthermore, several advantages are realized by use of the perfused lung technique. Blood samples can be readily obtained for analysis, and cell-free perfusion media can be employed that allow facile recovery of test compounds, metabolites, or biogenic endogenous substances that may be synthesized by lung tissue and released into the circulation. Another advantage includes the ease with which test materials may be administered directly into the pulmonary artery or into the trachea. The system can also be adapted to allow administration of test substances into the trachea by inhalation. While avoiding several complexities inherent to whole animal studies, the perfused organ technique allows biotransformation studies to be performed in lungs in which tissue structure and cellular organization remain intact. These are important factors when one considers the cellular heterogeneity of lung tissue and the fact that the metabolism of a substance often involves the sequential interaction of several enzymatic systems that may reside in different subcellular compartments.

The principal limitation of the perfused lung technique is that experiments are restricted to about 4 hr in duration; use of cell-free perfusion media further decreases the duration of lung viability to about 1.5–2 hr. Although this is an adequate time period for most metabolism or toxicity studies, short-term organ cultures have been successfully exploited to examine reactions that proceed slowly in lung tissue; these include covalent binding of metabolites to DNA[1,2] and sulfation[3,4] or glucuronidation conjugation.[4-8]

While perfusion of lungs from horses,[9] dogs,[10-13] and cats[14] has been

[1] C. C. Harris, A. L. Frank, C. van Haaften, D. G. Kaufman, R. Connor, F. Jackson, L. A. Barrett, E. M. McDowell, and B. F. Trump, *Cancer Res.* **36**, 1011 (1976).
[2] G. M. Cohen, P. Uotila, J. Hartiala, E.-M. Suolinna, N. Simberg, and O. Pelkonen, *Cancer Res.* **37**, 2147 (1977).
[3] G. M. Cohen, S. M. Haws, B. P. Moore, and J. W. Bridges, *Biochem. Pharmacol.* **25**, 2561 (1976).
[4] R. Mehta and G. M. Cohen, *Biochem. Pharmacol.* **28**, 2479 (1979).
[5] G. M. Cohen and B. P. Moore, *Biochem. Pharmacol.* **26**, 1481 (1977).
[6] B. P. Moore and G. M. Cohen, *Cancer Res.* **38**, 3066 (1978).
[7] G. M. Cohen, A. C. Marchok, P. Nettesheim, V. E. Steele, F. Nelson, S. Huang, and J. K. Selkirk, *Cances Res.* **39**, 1980 (1979).
[8] M. J. Mass and D. G. Kaufman, *Biochem. Biophys. Res. Commun.* **89**, 885 (1979).
[9] A. C. Hammond, R. G. Breeze, and J. R. Carlson, *Fed. Proc., Fed. Am. Soc. Exp. Biol.* **39**, 306 (1980).
[10] D. E. Donald, *J. Appl. Physiol.* **14**, 1053 (1959).
[11] B. Eiseman, L. Bryant, and T. Waltuch, *J. Thorac. Cardiovasc. Surg.* **48**, 798 (1964).
[12] K. C. Weber and M. B. Visscher, *Am. J. Physiol.* **217**, 1044 (1969).
[13] D. W. Jirsch, R. L. Fisk, G. Boehme, D. Modry, and C. M. Couves, *Chest* **60**, 44 (1971).
[14] M. Kiese and H. Uehleke, *Naunyn-Schmiedebergs Arch. Exp. Pathol. Pharmakol.* **242**, 117 (1961).

reported, this article is concerned only with perfusion of lungs from common laboratory animals, i.e., rats, guinea pigs, and rabbits. Perfusion of lungs from these species can be accomplished with the same basic equipment and surgical procedures although rat and guinea pig lungs require a smaller artificial thorax than that employed for rabbit perfusion. Niemeier and Bingham[15] devised a controlled and physiologically defined method and apparatus for the perfusion of rabbit lungs. This system was further adapted for more convenient use in the study of xenobiotic metabolism.[16]

Perfusion of guinea pig[17] and rat[18–22] lungs has also been previously described. An elaborate apparatus for perfusing guinea pig lungs should work equally well with rats.[17] It is our objective here to present a practical perfusion apparatus and to describe the surgical procedures in sufficient detail to allow investigators not experienced in animal surgery to successfully utilize the technique.

Perfusion Apparatus

The apparatus employed for lung perfusion consists of several discrete parts: a peristaltic (roller) pump for circulating perfusion fluid, a pH meter, an animal ventilator, an artificial thorax with upper reservoir, and a recirculating water bath. A meter for measurement of blood flow rate may be desirable but the need for this device is negated by the use of a perfusion pump with a calibrated output. Constant pressure perfusion is carried out more conveniently than constant flow perfusion and is generally satisfactory; the flow rate usually changes little during a 90-min experiment, and animal-to-animal flow rate differences are small. Flow rates are normally about 25 ml per min per g of lung. If constant flow perfusion is desired, a pressure monitor for the pulmonary artery is also required.

A versatile perfusion pump is the Cole-Parmer Masterflex 7565-10. This device has a calibrated pumping rate so that an approximation of the flow rate can be monitored on an analog meter mounted on the front panel. This pump accommodates several pump heads, allowing the same instrument to be used for perfusing lungs or other organs from several species

[15] R. W. Niemeier and E. Bingham, *Life Sci.* **11**, 807 (1972).

[16] T. C. Orton, M. W. Anderson, R. D. Pickett, T. E. Eling, and J. R. Fouts, *J. Pharmacol. Exp. Ther.* **180**, 482 (1973).

[17] A. L. Delaunois, *Arch. Int. Pharmacodyn. Ther.* **148**, 3 (1964).

[18] A. Hauge, *Acta Physiol. Scand.* **72**, 33 (1968).

[19] J. Hughes, C. N. Gillis, and F. E. Bloom, *J. Pharmacol. Exp. Ther.* **169**, 237 (1969).

[20] A. F. Junod, *J. Pharmacol. Exp. Ther.* **183**, 341 (1972).

[21] J. Hartiala, P. Uotil, and W. Nienstedt, *J. Steroid Biochem.* **7**, 527 (1976).

[22] K. Vahakangas, K. Nevasaari, O. Pelkonen, and N. T. Karki, *Acta Pharmacol. Toxicol.* **41**, 129 (1977).

that may require widely different flow rates. The flow rate delivered by this pump can be adjusted with such accuracy that flow controlling devices[15] are unnecessary. If it is desirable to monitor the perfusion rate, a Biotronex (Silver Springs, MD) BL-615 flow meter coupled with a Biotronex BLX-2024-F09 extracorporeal flow transducer gives excellent results. The flow transducer is incorporated into the lid of the artificial thorax, as shown in Fig. 5.

A pH meter is desirable, although cell-free media can be periodically monitored with pH paper if a meter is not available. Five to six percent CO_2 in the ventilating gas, however, usually maintains the pH of the perfusion medium near 7.4. If CO_2 is not mixed with the ventilating gas, the lungs rapidly hyperventilate (in the absence of the CO_2-generating capacity of the intact animal) and the ensuing alkalosis leads to constriction of the pulmonary vasculature and markedly decreased circulation.

A mixture of 5% CO_2 and 95% O_2 is often used as a ventilating gas in lung perfusion experiments. The use of this mixture has been questioned, however, on the basis that oxygen in high concentrations has toxic effects on pulmonary tissue.[23-25] Therefore, it is recommended that the isolated lungs respire warm, humidified air to which sufficient CO_2 has been added to control the pH of the perfusion medium.[11,12,15,18,26] Humidification is accomplished by bubbling air and CO_2 through gas washing bottles containing water at 37°. The humidified air and CO_2 are combined using a Y-connector and added to an uncapped 500-ml Nalgene wash bottle, which serves as a gas ballast reservoir. The wash bottle spout is shortened and serves as a connector for the tubing that attaches to the tracheal cannula. An alternative gas system that uses a rubber breathing bag with solinoid operated valves and that presents a stream of gas to the lungs at a constant pressure (0.25 cm H_2O) has been reported.[15] Although this apparatus performs very well, it is much more complex than the ballast system described here.

Excellent animal ventilators are available from Harvard Apparatus (Mills, MA). These instruments have variable stroke volume, breathing rate, and inspiration : expiration phase ratios. The Model 666 is recommended for rabbits, whereas the Model 680, which has a smaller stroke volume, is suitable for rats and guinea pigs. The "positive stroke" and "negative stroke" outputs of these ventilators must be combined with a Y-

[23] H. Witschi, Fed. Proc., Fed. Am. Soc. Exp. Biol. **36**, 1631 (1977).

[24] E. R. Block and J. K. Cannon, Lung **155**, 287 (1978).

[25] P. R. B. Caldwell and E. R. Weibel, in "Pulmonary Diseases and Disorders" (A. P. Fishman, ed.), p. 800. McGraw-Hill, New York, 1980.

[26] F. C. P. Law, T. E. Eling, J. R. Bend, and J. R. Fouts, Drug. Metab. Dispos. **2**, 433 (1974).

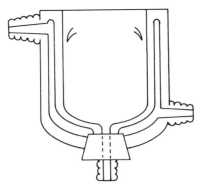

FIG. 1. The artificial thorax is jacketed for temperature control. The projections on the outside are hooks to which the springs or rubber bands are fastened for holding the lid in place. The inside volume is 600 ml for rats and guinea pigs or 2000 ml for rabbits.

connector; the total output produces the alternating positive pressure changes that are applied to the artificial thorax. By use of a vacuum source attached to the chamber, the pressure within the artificial thorax is never allowed to reach atmospheric pressure. The size of the orifice leading from the vacuum source to the chamber must be minimized to prevent air movement through this opening from interfering with the operation of the animal ventilator. Control of the vacuum source is discussed in the "Surgical Procedures and Perfusion" section.

A water-jacketed chamber serves as an artificial thorax[27] (Fig. 1). Its dimensions must be sufficient to accommodate the fully inflated lungs from the animal species under investigation. The artificial thorax has an outlet at the bottom from which the effluent from the pulmonary vein is pumped or through which the pulmonary vein cannula passes if a single-pass perfusion is carried out.

Affixed to the top of the artificial thorax is a 0.5-in Plexiglas disc that seals the chamber and also contains the connectors that interface with the animal ventilator, the vacuum pump, the manometer, the ventilating gas, and the circulating liquid (Fig. 2). The disc is cut to match the outer diameter of the artificial thorax. The outer, lower surface is rabbeted 0.25 in. deep and 0.5 in. wide to accommodate a circular washer (cut from 0.25 in. latex stock) that forms a gas-tight seal between the Plexiglas plate and the top of the artificial thorax (Fig. 2). The Plexiglas disc has 2 compression fittings through which the tubing that attaches to the tracheal and arterial (centrally located) cannulas are passed. The fittings are machined from 0.75-in. round Plexiglas stock and are threaded (20 per in.). The

[27] The artificial thorax is available in 600 ml (for rats or guinea pigs) and 2000 ml (for rabbits) sizes from Research Triangle Glass Blowing, P.O. Box 149, Morrisville, NC 27560.

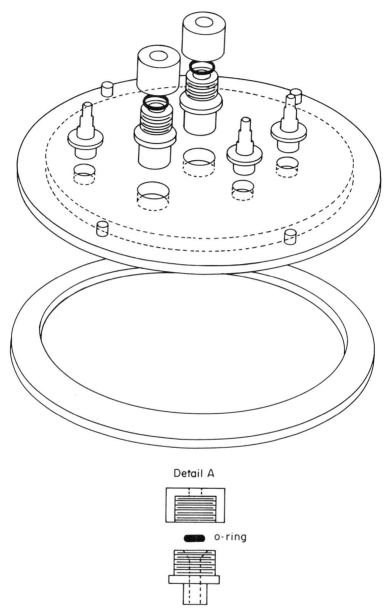

Detail A

o-ring

FIG. 2. The artificial thorax is sealed by a Plexiglas disc that has fittings for all attachments, including gas and vacuum sources. The centrally located fitting is for the perfusion medium. The large washer is cut from 0.25-in. latex stock. The projections near the edges are for the springs or rubber bands that fasten this disc to the artificial thorax. Detail A shows the construction of the compression fittings. These are 0.5-in. in diameter and have 20 threads per inch.

TABLE I

TUBING DIMENSIONS USED FOR CANNULAS AND TO TRANSPORT GAS OR
PERFUSION MEDIUM INTO THE APPARATUS

Animal	Tracheal cannula[a]	Apparatus tubing	Pulmonary artery cannula[a]	Apparatus tubing
Rat or	0.062 × 0.082[b]	0.070 × 0.110[c]	0.066 × 0.095	0.085 × 0.128[c]
guinea pig	(PE 205)	(PE 260)	(PE 240)	(PE 280)
Rabbit	0.085 × 0.128	0.115 × 0.147	0.115 × 0.147	0.115 × 0.147[c]
	(PE 280)	(PE 330)	(PE 330)	(PE 330)

[a] All dimensions are in inches. The cannulas are approximately 1 in. long and are cut at a 45° bevel on one end to facilitate coupling to the apparatus. To retain the cannulas in the tissue, they are slightly flared on the nonbeveled end by holding them near a flame.

[b] If young rats or guinea pigs are used, the tracheas are smaller and must be cannulated with 0.45 × 0.062 (PE 160) tubing. These can be coupled to the 0.070 × 0.110 (PE 260) tubing in the apparatus using an adapter made from 0.062 × 0.082 (PE 205) tubing.

[c] This tubing is sufficiently enlarged to accept the cannula by using the tips of a small pair of hemostats.

fittings are center-bored (5/32 in. for both fittings in the rabbit apparatus; in the rat or guinea pig apparatus, 1/8 in. for the tracheal fitting and 9/64 in. for the pulmonary artery fitting) to accept the appropriate diameter tubing (Table I), and their tops are countersunk (80° bevel) to a depth of 1/8 in. (Detail A to Fig. 2). The fittings are glued into holes that have been drilled in the Plexiglas disc (Fig. 2). The caps to the fittings are also constructed from 0.75-in. round Plexiglas stock and are center-bored to accommodate the tubing that passes through the fitting (Detail A to Fig. 2). An O-ring is placed around the tubing and between the fitting and the cap to form a compression seal when the cap is tightened. Transverse sections of latex tubing (1/8 in.) can be used for the compression seals if O-rings are not available. Tubing that has the correct internal diameter to accommodate the arterial and tracheal cannulas is passed through these fittings (Table I). This allows the cannulated lungs to be connected to and disconnected from the perfusion apparatus with minimum difficulty, and without introducing air bubbles into the tubing. If the desired diameter tubing is not available, the appropriate fit may be attained by expanding tubing of the closest available size with the tips of a small hemostat. Polyethylene tubing is convenient for the apparatus because it is available in a variety of exact sizes, but has the disadvantage that it may adsorb or absorb lipophilic substances.[28] Teflon tubing is less apt to do this, but is not manufactured to the close tolerances available in polyethylene, and fitting the cannulas to it is more difficult.

The disc also contains three gas fittings for the animal ventilator, the

[28] B. W. Blase and T. A. Loomis, *Toxicol. Appl. Pharmacol.* 37, 481 (1976).

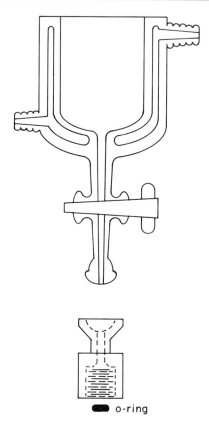

FIG. 3. Fifty-milliliter upper reservoirs are available from Research Triangle Glass Blowing (see Ref. 27). The overall length is 4.75 in. for rat and guinea pigs or 5.5 in. for rabbits. The stem must be 3 in. longer if the flow transducer is not used. The adapter is milled from 0.75-in. round Teflon stock and fastens to the top of the flow transducer or to the center fitting on the disc shown in Fig. 2.

water-column manometer, and the vacuum pump (Fig. 2). These fittings can be constructed by gluing, with 24-hr epoxy glue, tubing connectors into holes that have been drilled in the Plexiglas disc. The disc is fastened to the top of the artificial thorax with rubber bands or springs extending from projections on the top of the disc to glass projections on the sides of the artificial thorax (Figs. 1, 2, and 5).

The upper reservoir (Fig. 3) ranges from 30 to 200 ml in capacity depending on the animal species under investigation and whether single-pass or recirculating perfusion is desired.[29] The reservoir, like the artificial

[29] A 50-ml upper reservoir is available with a long or short stem from Research Triangle Glass Blowing (see Ref. 27).

TABLE II
COMPOSITION OF THE FORTIFIED KREBS–RINGER
BICARBONATE BUFFER[a] USED FOR
PERFUSION MEDIUM

Reagent	Concentration	
	g per liter	mM
NaCl	6.9	118
KCl	0.35	4.75
CaCl$_2$ · 6H$_2$O	0.55	2.54
KH$_2$PO$_4$	0.16	1.19
MgSO$_4$ · 7H$_2$O	0.29	1.19
NaHCO$_3$	2.1	25
Glucose	0.97	5
Bovine serum albumin	45	—

[a] The pH is adjusted to 7.4 with NaOH.

thorax, is jacketed for temperature regulation. The reservoir stem contains a stopcock that allows the perfusion rate to be adjusted or stopped. The reservoir also serves as a mixing chamber for test materials, and a small stirrer may be situated above the reservoir to facilitate complete mixing. If a pH meter is used, the electrode can also be placed in this chamber. The stem of the reservoir is of appropriate length to result in the desired total pressure head at the pulmonary artery (25 cm for rabbits or 22 cm for rats and guinea pigs) and is terminated with a 1/2 in. male ball joint. The ball joint is clamped to its Teflon female counterpart, which is threaded on the other end and attaches to the center fitting in the Plexiglas disc. If a flow transducer is desired, a holder for this device (see Fig. 4) can be incorporated onto the top of the Plexiglas disc (Fig. 5). In this case the Teflon female ball joint adapter screws onto the top of the transducer, and the bottom of the transducer is connected to the Plexiglas disc with tubing and compression fittings. Use of the flow transducer requires a short stem on the upper reservoir as illustrated in Fig. 3 in order to maintain the proper height of liquid. An abbreviated drawing of the assembled apparatus is shown in Fig. 5.

Perfusion Media

Several media have been used for perfusion fluid. Krebs–Ringer bicarbonate buffer fortified with 4.5% bovine serum albumin and 5 mM glucose was introduced by Junod[20] and seems to be the most widely used medium (Table II). Albumin helps in solubilizing test materials that have low

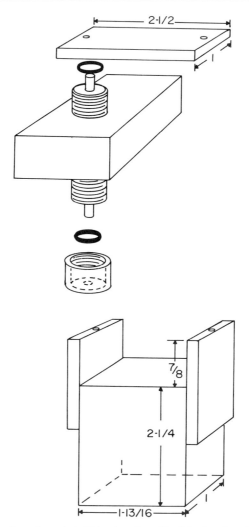

FIG. 4. The flow transducer is held in place by a Plexiglas bracket that is fastened with glue or machine screws to the upper disc for the artificial thorax (Fig. 2). The fittings on the flow transducer are the same size as those used on the disc.

water solubility and allows perfusion periods of 1.5 hr without development of edema.[30] This is sufficient time for most drug uptake and metabolism studies. This medium has the advantage of being extractable with many organic solvents. Its major disadvantage is that highly nonpolar solvents such as hexane will form stable emulsions in the presence of

[30] G. Nicolaysen, *Acta Physiol. Scand.* **83**, 563 (1971).

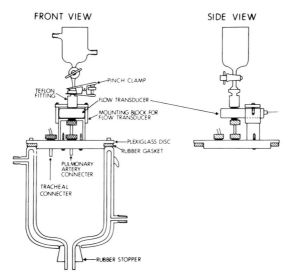

FIG. 5. This abbreviated drawing depicts the appearance of the assembled apparatus (to improve clarity, the three gas fittings shown in Fig. 3 have been omitted). If the flow transducer and its holder are deleted, the stem of the upper reservoir must be extended an additional three inches. If the flow transducer is not used, the Teflon ball joint fitting is connected directly to the center fitting of the Plexiglas disc.

albumin. This problem can be overcome by adding about 1/4 volume of methanol to the aqueous samples prior to extraction. Krebs–Ringer buffer can also be fortified with red blood cells, which will allow lung viability to be maintained for up to 4 hr.[30]

Autologous heparinized whole blood is the best medium for long-term organ perfusion but is difficult to handle in terms of preparation, perfusion (due to clotting), and sample extraction. Details for the use of autologous blood in lung perfusion experiments have been presented for rabbits[15] and for rats.[18] Plasma and defibrinated blood[31] have also been used, but these media offer little advantage over albumin-fortified Krebs–Ringer buffer in perfused lung experiments.[30]

Surgical Procedures and Perfusion

Nembutal is the anesthetic of choice for rats (50 mg/kg) and guinea pigs (45 mg/kg). Ether or halothane are also appropriate for these species, but none of the aforementioned agents work well with rabbits. The best approach is to anesthetize rabbits with CO_2 until respiration has ceased

[31] I. De B. Daly, Q. J. Exp. Physiol. Cogn. Med. Sci. **28**, 357 (1938).

(about 2 min) and then to perform the surgery quickly. The short duration of the operation avoids thrombus formation, which is important for successful perfusion.

In all species, the surgical procedures are the same. The only differences are in the cannula sizes (see Table I for cannula sizes and construction). After the animal has been anethetized, it is strapped to a dissection board, or it is held in place in a shallow tray by an assistant. From this point, the completion of the surgical procedure will require about 25 min. The ventral side of the animal is wetted with 70% ethanol, and the integument is removed from just below the rib cage to the mid-throat area and laterally to the forelegs. The trachea is dissected free from surrounding tissues (including the esophagus) and cannulated. A transverse incision is made immediately below the bottom of the rib cage. Before the diaphragm is cut away, a syringe filled with air (50 ml for rabbits, 10 ml for guinea pigs and rats) is connected to the tracheal cannula to prevent lung collapse. The cartilaginous, ventral area of the rib cage is cut away to expose the thoracic cavity, and the pericardium is removed. Excessive fatty tissue, usually present in rabbit, is removed to allow easy access to the anterior area of the heart. A loose ligature (000 surgical silk) is passed around the pulmonary artery (also around the aorta in rats and guinea pigs). Some teasing is required to get the tips of a small pair of curved, blunt forceps between the pulmonary artery and the aorta. An incision is made in the right ventrical and a cannula is inserted through the semilunar valve into the pulmonary artery. It is best to have the cannula attached to an infusion vessel containing perfusion medium so that the blood is immediately flushed from the pulmonary vascular bed, the outflow being through an incision in the left ventrical. This immediate *in situ* perfusion negates the need for anticoagulants and also minimizes the likelihood of introducing air bubbles into the vasculature. During this time, the lungs can be gently inflated and deflated with the air-containing syringe to promote circulation through the lung capillary bed, but care must be taken not to overinflate the lungs as young animals (particularly rats) are susceptible to the induction of air emboli by lung overinflation. It is imperative that a negative pressure is never applied to the trachea.

When single-pass studies are desired, the pulmonary vein must also be cannulated. This allows the perfusion fluid effluent from the lungs to leave the apparatus via a tube that exits through the bottom of the artificial thorax. Otherwise, air would be drawn into the chamber during negative pressure cycles and inhalation would not occur. This requires an airtight seal between the tubing and the chamber. A cannula (the same size as that used for the pulmonary artery) is inserted through the incision in the left ventricle, passed through the mitral valve, and secured within the pulmonary vein by a ligature around the left atrium. If the pulmonary vein is not cannulated, an incision in the left atrium will allow the perfusion medium to escape with minimum resistance.

The lungs are partially inflated with air before they are removed from the animal, the tracheal and arterial cannulas are clamped with small hemostats, and the infusion vessel and syringe disconnected from them. It is convenient to support the lungs by the esophagus rather than by the trachea during surgical removal from the animal, as the trachea is easily torn. While lifting gently on the esophagus with a hemostat, trim away the tissue supporting the trachea, and with blunt scissors pointed toward the head of the animal, cut the hilar ligaments along the anterior sides of the thoracic cavity while minimizing contact with the lung itself. Next, while lifting gently on the esophagus, trim and tease the fascia that connects the trachea, esophagus, lungs, and aorta to the dorsal area of the thoracic cavity. At this time, the lungs should be above the animal, but still attached to the diaphragm and to the structures mentioned in the previous sentence. While holding onto the ligature around the tracheal cannula, gently pull the esophagus downward to separate it from the trachea. Cut the dorsal aorta near the heart, and carefully sever the minute basal ligaments that tie the lung lobes to the diaphragm (close attention is needed here to avoid cutting small extensions of the lungs that are pulled downward by the ligaments). At this point the lungs are freed from the animal and transferred to a shallow dish filled with warm perfusion medium (37°). Remaining fatty tissue is dissected away from the heart and trachea. Remnants of the aorta are removed, but with care to avoid perforating the pulmonary artery that lies immediately underneath the dorsal aorta. The cannula in the pulmonary artery is dissected free from the heart so that the lung can hang freely from the cannula. This is accomplished by inserting blunt scissor points through the space followed by the ligature that fastens the pulmonary artery cannula in place, and then cutting the cardiac tissue below. The ventricles are cut away and the lungs attached to the perfusion apparatus, which has been previously filled and primed. A precaution here will prevent air emboli: Krebs–Ringer buffer tends to effervesce slightly on standing, and small bubbles that can lead to air emboli will collect in the tubing. These bubbles can be flushed into the upper reservoir using a 25-ml syringe containing perfusion medium that is connected to the outlet tubing for the pulmonary artery cannula.

The tubing that carries the ventilating gas to the trachea is clamped to prevent the lungs from collapsing before the apparatus is closed and breathing initiated. After the lung cannulas are attached to the appropriate connectors, the Plexiglas disc is lowered to the artificial thorax and fastened down. A negative pressure of about −3 cm (water) is applied to the artificial thorax before the clamp on the ventilating gas is released. By following this procedure, the lungs are never collapsed during the entire operation. After the animal ventilator is activated, a negative pressure of −25 cm (water) is momentarily applied to the artificial thorax to assure that the lungs are fully ventilated (this is the pressure equivalent to a yawn). The stroke volume of the ventilator is adjusted to produce a pressure change from −1 to −11 cm (water) for all species. The negative pressure can be controlled by dividing the vacuum source into two parallel lines using Y-connectors and adjusting the vacuum with pinchcocks. One line is permanently adjusted to maintain the desired negative pressure in the thorax, and the second can be manipulated to generate the −25-cm pressure excursions without affecting the "fine tuning" of the first.

If a recirculating system is used, flow from the perfusion pump is initiated, and the pumping rate adjusted to produce a constant liquid level in the upper reservoir. The stirrer and the pH electrode are placed in the upper reservoir. Before the addition of any test material, a 5- to 10-min period is usually allowed for the circulation to stabilize to a constant flow rate and to allow any damaged areas of the lungs to become visible.

Viability

Although a variety of biochemical[15] and stain exclusion techniques[32] have been reported to assess lung viability, experience should allow the investigator to accurately determine the lung status by more empirical methods. Collapsed areas do not follow lung expansions or contractions as breathing occurs; such areas usually become edematous after a few minutes. Edematous areas can be identified by their translucent appearance and lack of ventilation. Areas that are not perfused because of thrombi or air emboli retain the pink color imparted to lung tissue by blood pigments and are clearly visible. Loss of lung compliance, which accompanies atelectasis and edema, is evidenced by an increased negative pressure in the artificial thorax induced by the constant stroke volume of the animal ventilator. Blood flow rate is not a good index of lung status but will usually remain constant even after edema develops.

Specialized Methods

Several highly specialized lung perfusion techniques have also been developed, but will only be mentioned here. A preparation was reported by Gillis and Iwasawa[33] in which right and left lungs of a complete pair

---

[32] T. E. Eling, R. D. Pickett, T. C. Orton, and M. W. Anderson, *Drug. Metab. Dispos.* **3**, 389 (1975).

[33] C. N. Gillis and Y. Iwasawa, *J. Appl. Physiol.* **33**, 404 (1972).

were perfused independently. This is a valuable approach when comparing the effects of one or more test materials on lung because considerable animal-to-animal variability in test parameters is frequently seen.

A simplification of this technique was reported by Plummer et al.[34] in which one lung of a complete pair is removed for cytochrome P-450 determination after perfusion for several minutes, but before any test materials were added. Using this procedure, subtle decreases in pulmonary cytochrome P-450 levels that accompany exposure to some aromatic solvents were detected.[34] It was also demonstrated that microsomal cytochrome P-450 content in microsomes from left and right lungs from the same rabbit were not statistically different, and that these levels were not affected by 60-min perfusion periods.

It is sometimes desirable to determine the rate at which a substance that has been accumulated by perfused lungs is released from the tissue back into the circulation. This procedure usually entails single-pass perfusion with or without the test material in the perfusion medium. These experiments are greatly facilitated by use of a dual upper reservoir that attaches to the apparatus through a three-way stopcock. With this control system, drug-containing or drug-free media can be directed to the lungs.[35]

Because perfusion through the pulmonary artery generally does not supply the bronchial tree with circulation, a method has been developed for independently perfusing this aspect of the lung vasculature in guinea pigs.[31] This method could feasibly be adapted to rats and rabbits.

Another specialized technique has been developed to measure changes in bronchiolar constriction induced by test materials. Bronchiolar perfusion has been utilized to accomplish this measurement.[36]

The perfused lung system described here is easily adapted for inhalation studies by replacing the gas ballast reservoir with a small chamber where ventilating gases and test materials can be mixed. It is also necessary to install a one-way valve assembly just above the tracheal cannula to prevent the exhaled gas mixture from re-entering the mixing chamber.[37]

Comments

The perfused lung apparatus and the appropriate surgical details have been described with adequate detail such that a dexterous investigator can use this technique. The aid of a machine shop in constructing certain components will, of course, be helpful. This system has been instrumental

[34] J. L. Plummer, C. R. Wolf, R. M. Philpot, J. R. Bend, and B. R. Smith, unpublished data.
[35] A. G. E. Wilson, F. C. P. Law, T. E. Eling, and M. W. Anderson, J. Pharmacol. Exp. Ther. 199, 360 (1976).
[36] T. Sollmann and W. F. von Oettingen, Proc. Soc. Exp. Biol. Med., p. 692 (1927).
[37] R. W. Niemeier, Environ. Health Perspect. 16, 67 (1976).

in several biochemical, toxicological, and pharmacological studies of lung and will likely become even more widely used as increasing emphasis is placed on nonrespiratory functions of the lung, including its role in xenobiotic biotransformation.

[14] Vascular Autoperfusion of Rat Small Intestine *in Situ*

By HERBERT G. WINDMUELLER and ALBERT E. SPAETH

There is a growing awareness that the small intestine may be an important site for the metabolism of drugs and other foreign compounds, particularly orally administered ones because these must traverse mucosal epithelial cells of the intestine before entering the body. A wide variety of xenobiotic metabolizing activities have been demonstrated *in vivo* in intestinal cells or subcellular fractions. These include *O*- and *N*-dealkylation, hydroxylation, esterification, deesterification, glucuronidation, reduction, oxidation, desulfuration, alkylation, and sulfate conjugation.[1-5] Oral drug administration frequently induces higher levels of activity. For lack of adequate methods, however, the quantitative importance of the intestine's role in whole body drug metabolism *in vivo* has been more difficult to assess.[6] Vascular perfusion techniques are well suited for providing such information. They permit study of the intact organ under nearly normal physiological conditions while allowing access to the lumen, the arterial and venous circulation, and the lymphatic drainage.

In one technique, the entire small intestine of a rat can be removed *en bloc* and vascularly perfused in a recycling system with heparinized or defibrinated blood.[7] This method requires extensive equipment and surgical practice. Furthermore, to achieve adequate intestinal function, a glucocorticoid and norepinephrine must be added to the perfusate, apparently to compensate for denervation.[8,9]

[1] K. Hartiala, *Physiol. Rev.* 53, 496 (1973).
[2] R. S. Chhabra, R. J. Pohl, and J. R. Fouts, *Drug. Metab. Dispos.* 2, 443 (1974).
[3] H. Hoensch, C. H. Woo, S. B. Raffin, and R. Schmid, *Gastroenterology* 70, 1063 (1976).
[4] M. J. Rance and J. S. Shillingford, *Biochem. Pharmacol.* 25, 735 (1976).
[5] R. S. Shirkey, J. Chakraborty, and J. W. Bridges, *Biochem. Pharmacol.* 28, 2835 (1979).
[6] M. Gibaldi and D. Perrier, *Drug. Metab. Rev.* 3, 185 (1974).
[7] H. G. Windmueller and A. E. Spaeth, *in* "Intestinal Permeation" (T. Z. Csaky, ed.), Hand. Exp. Pharmacol. Springer-Verlag, Berlin and New York (submitted for publication).
[8] H. G. Windmueller, A. E. Spaeth, and C. E. Ganote, *Am. J. Physiol.* 218, 197 (1970).
[9] H. G. Windmueller and A. E. Spaeth, *J. Lipid Res.* 13, 92 (1972).

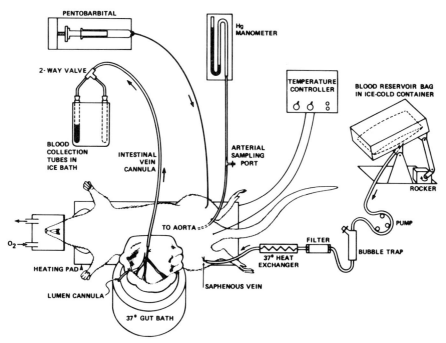

FIG. 1. Schematic diagram of apparatus for autoperfused rat intestinal segment *in situ* with pentobarbital anesthesia. (Reprinted from Windmueller and Spaeth.[7])

A second approach, simpler and more normal physiologically, is vascular autoperfusion of a segment of small intestine *in situ*.[10] Here the arterial, neural, and lymphatic connections of the segment remain intact while the venous effluent is totally collected (Fig. 1). The collected blood is replaced by continuous transfusion, into a saphenous vein, of heparinized blood obtained from other donor rats. With appropriate adaptations, this technique, described below, allows for nearly the entire range of experimentation possible with the isolated preparation.

Setting Up and Maintaining the Preparation

Apparatus

The equipment is shown schematically in Fig. 1 and the surgical platform and associated apparatus are seen in Fig. 2. The Plexiglas platform has a metal surface that accepts magnets, useful for positioning and securing various cannulae. The gut bath is fabricated using a plastic cup base,

[10] H. G. Windmueller and A. E. Spaeth, *Arch. Biochem. Biophys.* **171**, 662 (1975).

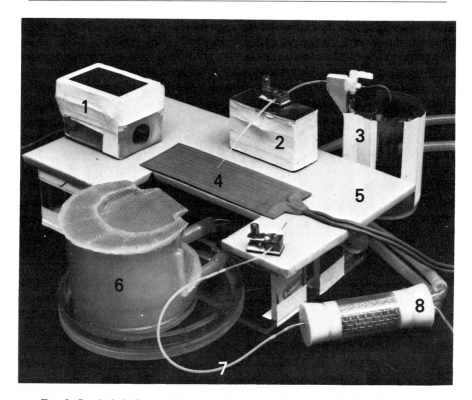

FIG. 2. Surgical platform and associated apparatus for autoperfused rat intestinal segment *in situ*. (1) Box for rat's head; (2) block supporting intestinal vein cannula—the top surface of the block is metal, the bottom surface is a magnet, and the cannula is held in a modified chromatography clip lined with silicone rubber sponge tape (The Connecticut Hard Rubber Co., New Haven, CT) and also attached to a magnet; (3) ice bath with two blood collection tubes and two-way valve; (4) heating pad; (5) Plexiglas surgical platform, 18×30 cm, with surface covered by a thin metal sheet; (6) gut bath thermoregulated with $37°$ H_2O; (7) blood infusion cannula held in clip attached to a magnet and positioned for saphenous vein cannulation; (8) $37°$ heat exchanger, containing a coiled length of tubing. (Reprinted from Windmueller and Spaeth.[7])

with the upper portion formed from silicone rubber sheeting glued over stainless steel wire mesh. Water at $37°$ is circulated through the body of the bath to warm the bottom and sides of the shallow intestinal compartment. Water from the same source circulates through the heat exchanger to warm the replacement blood. Replacement blood is drawn by aortic puncture from anesthetized donor rats, usually 500–700 g ex-breeder males, and collected in ice in a silicone rubber envelope that serves as the blood reservoir. The envelope is placed within a flat metal box surrounded, in turn, by a styrofoam box. Ice water is circulated by pump through the space between the two boxes to maintain the blood temperature near 0°. To prevent the blood cells from settling out, the unit is rocked mechanically at 30 cycles/min. Photographs of this unit and other apparatus as well as additional details will be published.[7] The replacement blood is pumped with a peristaltic pump (Model RL-175 Holter pump, Extracorporeal Medical Specialties, King of Prussia, PA) through a bubble trap, a glass-wool filter and a heat exchanger—a coil of tubing in a water jacket. To maintain body temperature at 37°, the rat lies on a heating pad thermoregulated with a proportional temperature controller and a rectal thermistor. A small amount of additional heat is supplied by a heat lamp. The rat's head is inserted through a snug hole in a rubber diaphragm covering one end of a plastic box, which is continuously flushed with gas at 1 liter/min. When pentobarbital anesthesia is used (50 mg/kg body weight, i.p.) the gas is 100% O_2; anesthesia is maintained by a continuous infusion of sodium pentobarbital into the left saphenous vein (3.0–3.5 mg/hr delivered from a continuously variable infusion pump, e.g., Model 906, Harvard Instrument Co., S. Natick, MA) (Fig. 1). Alternatively, when ether anesthesia is used, the flushing gas is a mixture of ethyl ether and O_2 regulated as shown in Fig. 3. The same anesthetic is used for intestine donor and blood donor rats. Cannulae and tubing are made of silicone rubber. For blood vessel cannulation, done by direct puncture, the cannulae are fitted with a 7–10 mm piece of sharpened, thin-wall 23-gauge stainless steel needle tubing, except for the infusion of pentobarbital where a 25-gauge needle is used.[7] These needles are carefully sharpened under 40× magnification with hard Arkansas stones (rotating wheel and hand-held). After an experiment, all equipment is rinsed thoroughly with tap and distilled water and air-dried.

Setting Up the Basic Preparation

The anesthetized intestine-donor rat (350–400 g) is laid in a supine position on the heating pad, the legs are taped to the surgical platform, and the hair is clipped from the abdomen, both thighs, and from around the base of the tail. Through a midline abdominal incision, the small intestine and cecum are exteriorized onto a gauze pad moistened with warm gut bath solution (Earle's balanced salt solution[11] plus 5.6 mM glucose and 0.6 mM L-glutamine and equilibrated with 95% O_2 : 5% CO_2). Then the gut bath is positioned alongside the rat with the lip of the bath extending into the abdominal cavity. The contact surface between the front of the bath and the rat's abdomen is coated liberally with Vaseline, thereby prevent-

[11] W. R. Earle, *J. Natl. Cancer Inst.* **4**, 165 (1943).

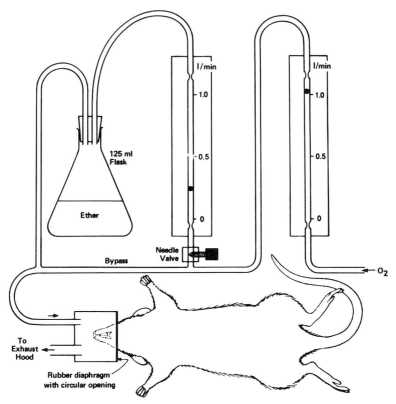

FIG. 3. Schematic diagram of apparatus for maintaining ether anesthesia. Total O_2 flow is 1 liter per minute. The ether content of the gas flow to the rat is regulated with the needle valve. Light anesthesia can be easily maintained for several hours when about 25% of the O_2 passes through the flask of ether, as shown. (Reprinted from Windmueller and Spaeth.[7])

ing loss of fluid by siphoning. The intestine is placed into the bath, which is filled with 10 ml of gut bath solution. Using moistened swabs to handle the intestine gently, a 2- to 5-cm segment, drained by a single vein, is selected and positioned as shown in Fig. 1. Suitable segments are generally available in the distal two-thirds of the jejunum and in the ileum. The remaining intestine and cecum are returned into the abdomen and the segment and bath are covered with a thin, transparent plastic sheet. Care is taken to maintain the moistness of the intestine.

When pentobarbital anesthesia is used, the left saphenous vein is exposed and cannulated and the continuous pentobarbital infusion begun. The cannula is secured by taping it to the surgical platform. A dissecting microscope with 4–8× magnification and fiber-optic light are used for this and subsequent surgery. Next, working through a small opened flap in the plastic cover, the selected intestinal vein is fully exposed near its junction with the larger superior mesenteric vein by carefully removing the overlying mesentery with ultra-micro scissors and forceps (see Fig. 4A). A 6-0 silk ligature is passed under the intestinal vein. These steps in the surgery

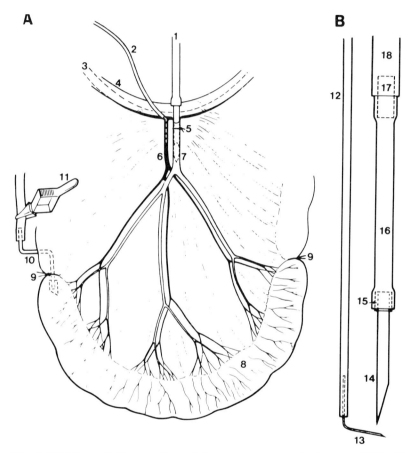

FIG. 4. (A) Schematic diagram of intestinal segment with cannulae in place. (B) Details of intestinal artery cannula (left) and intestinal vein cannula (right). (1) Intestinal vein cannula; (2) intestinal artery cannula (optional, see text); (3) superior mesenteric artery; (4) superior mesenteric vein; (5) 6-0 silk ligature; (6) intestinal artery; (7) intestinal vein; (8) jejunal or ileal segment; (9) 4-0 silk ligature; (10) lumen infusion cannula (20-gauge needle tubing with silicone rubber tubing at each end); (11) clamp; (12) silicone rubber tubing, 0.25 mm i.d., 0.51 mm o.d.; (13) 36-gauge needle, 5.4 mm long (overall); (14) 23-gauge thin-wall needle, 8 mm long; (15) polyethylene collar, PE-50, 2 mm long; (16) silicone rubber tubing, 0.64 mm i.d., 1.19 mm o.d., 3 cm long; (17) Teflon connector, 0.76 mm i.d., 1.27 mm o.d., 5 mm long; (18) silicone rubber tubing, 1.02 mm i.d., 2.16 mm o.d., 10 cm long.

are completed early to allow time for any constriction in the intestinal vein caused by the surgery to disappear before cannulation. Then the right saphenous vein is exposed and cannulated with the primed blood infusion cannula (Fig. 2), which is tied into the vein with a 6-0 ligature. Next, 4-0 ligatures are secured around each end of the selected segment, forming a closed loop with an infusion cannula inserted into the proximal end (Fig. 4A). Now the edges of the abdominal incision are retracted and, by wiping with dry swabs, a 10–15 mm section of the aorta is exposed near the iliac bifurcation. To position the aortic cannula, used to monitor blood pressure and obtain arterial blood samples, the tip of a blunt trocar is tunneled from the iliac bifurcation out through a small skin incision near the base of the tail. Through this hollow trocar, the aortic cannula, attached to a mercury manometer, is positioned with its needle tip lying over the cleared portion of aorta. Just before aortic cannulation, the rat is injected with 3 mg of sodium heparin in 0.3 ml of 0.15 M NaCl by means of the blood infusion cannula in the saphenous vein. After cannulation, the aortic cannula is secured by tape to the surgical board. The blood pressure will be 130–150 mmHg with pentobarbital anesthesia and 100–110 mmHg with ether anesthesia.

Finally, the intestinal vein is cannulated, after positioning the cannula carefully with the magnetic holder (Fig. 2). Cannulation is aided by lifting and pushing slightly with a metal probe on the junction of the intestinal and superior mesenteric veins. The cannula needle is securely tied in place with the previously placed ligature (Fig. 4A). The venous blood is collected through a two-way valve into tared tubes in ice (Figs. 1 and 2). Coincident with the flow of blood, infusion of replacement blood is begun, initially at 1.5–2.0 ml/min, depending upon the length of the segment. The time required for the entire surgical preparation is about 40 min after familiarity with the method is attained.

Maintaining the Preparation

The rate of blood infusion is adjusted to equal the rate of blood collection, determined gravimetrically on samples collected over 4- to 8-min intervals, plus an additional amount to compensate for the slow, continuous surgical loss, typically 0.3 ml/min. Normal arterial blood pressure is thereby maintained. Mean blood flow rates for jejunal segments from rats fasted 18 hr is 3 ml per minute per gram of intestine. For nonfasted rats, it is 30% higher. The wet weight of tissue in the experimental segments averages 0.4 g. Body temperature and blood pressure remain normal, as does the pale pink color and gross appearance of the segments, for at least 4 hr—the longest experiments done to date. It is important that the needle and tubing used to cannulate the intestinal vein (Fig. 4B) be sufficiently large to prevent excessive resistance to flow, which would produce an engorgement and tortuosity of the mesenteric veins and darkening in color of the tissue. Ether or pentobarbital delivery is adjusted to maintain a light level of anesthesia: costal respiration and a tail-pinch reflex should be preserved.

Applications and Specialized Techniques

Permeation and Metabolism of Luminal Compounds

Test compounds, unlabeled or labeled with radioactivity, can be administered as a single dose into the closed intestinal segment in 0.3 ml or less of Earle's balanced salt solution, or an outflow cannula may be tied into the distal end of the segment allowing test solutions to be continuously perfused through the loop.[7,10] The permeation rates and extent of intestinal metabolism during permeation are determined from analyses of the collected venous blood samples. The metabolism of a variety of luminally administered amino acids,[10,12] sugars,[13] and drugs[14–20] have been studied in this way. By using an intestinal segment preparation from a rat with a previously established mesenteric lymph cannula,[21,22] one can determine whether the administered compound or its metabolic products pass from the intestine directly into blood capillaries or whether they reach the circulation associated with proteins or lipoproteins by way of the lymph. Mesenteric lymph cannulae are implanted[23,24] 1 day prior to setting up the segment preparation and the rats are maintained in restraining cages, to allow for recovery and to ensure a normal lymph flow. Intestinal segments can also be prepared in germ-free rats to evaluate the role of the intestinal microflora in the metabolism of absorbed compounds.[12]

It may become important to chill the collected intestinal venous blood very rapidly, to stabilize the absorbed compound or its metabolites. A

[12] H. G. Windmueller and A. E. Spaeth, *Arch. Biochem. Biophys.* **175**, 670 (1976).
[13] H. G. Windmueller and A. E. Spaeth, *J. Biol. Chem.* **255**, 107 (1980).
[14] W. H. Barr and S. Riegelman, *J. Pharm. Sci.* **59**, 154 (1970).
[15] W. H. Barr and S. Riegelman, *J. Pharm. Sci.* **59**, 164 (1970).
[16] H. Ochsenfahrt and D. Winne, *Naunyn-Schmiedeberg's Arch. Pharmacol.* **264**, 55 (1969).
[17] D. Winne and J. Remischovsky, *Naunyn-Schmiedeberg's Arch. Pharmacol.* **268**, 392 (1971).
[18] K. W. Bock and D. Winne, *Biochem. Pharmacol.* **24**, 859 (1975).
[19] D. Josting, D. Winne, and K. W. Bock, *Biochem. Pharmacol.* **25**, 613 (1976).
[20] U. Breyer and D. Winne, *Biochem. Pharmacol.* **26**, 1275 (1977).
[21] A.-L. Wu and H. G. Windmueller, *J. Biol. Chem.* **253**, 2525 (1978).
[22] A.-L. Wu and H. G. Windmueller, *J. Biol. Chem.* **254**, 7316 (1979).
[23] J. L. Bollman, J. C. Cain, and J. H. Grindlay, *J. Lab. Clin. Med.* **33**, 1349 (1948).
[24] A. L. Warshaw, *Gut* **13**, 66 (1972).

simple 0° heat exchanger that can be placed around the intestinal vein cannula has been designed for this purpose.[7] If chemical change of the compound is catalyzed by enzymes in the blood, appropriate inhibitors can be infused continuously into this same cannula.[12,25]

Net Uptake and Metabolism of Vascular Compounds

The net uptake rate from the blood and the net release rate of metabolic products into the blood can be determined from arteriovenous concentration difference measurements across the segment and from the measured blood flow.[13,25] Arterial blood for such measurements is obtained from the aortic cannula, by opening the sampling port after temporarily clamping the tubing between the port and the manometer (Fig. 1). The first 10 drops of blood flush the cannula and are discarded. Arterial and venous samples should ideally be collected simultaneously. This can be approximated by sampling arterial blood near the mid-time for collection of the corresponding venous sample. For completely simultaneous sampling, important when A/V differences are small and the arterial concentration is variable, blood from the intestinal vein and aortic cannulae can be withdrawn at identical rates into paired syringes mounted in parallel on a withdrawal pump.[13] Such A/V measurements can be made for normal blood constituents, to assess the metabolic activity of the intestinal segment,[13,25] or for drugs or other foreign or labeled compounds administered intravenously.

The test compound can also be infused directly into the blood stream of the single artery supplying the experimental segment, thereby avoiding exposure of other tissues to it.[7,25] Because all the venous blood is collected, the compound will not reach the general circulation and will gain access only to the experimental segment. The exchange of substances between the blood and lumen can also be studied in this way. In order to permit arterial infusion, an intestinal segment must be selected that is supplied by a single artery and a single vein, as illustrated in Fig. 4A. The artery is carefully cleared by microdissection adjacent to the site selected for intestinal vein cannulation. Arterial cannulation immediately precedes intestinal vein cannulation in the surgical routine. The artery is entered with a 36-gauge needle that is bent, sharpened, and attached to the infusion tubing (Fig. 4). The small size of this needle does not interfere with normal blood flow through the artery. An infusion rate of 0.01 ml/min is typical.

[25] H. G. Windmueller and A. E. Spaeth, *J. Biol. Chem.* **253**, 69 (1978).

Choice of the route of delivery of a foreign compound provides some control over which cell population in the intestinal segment becomes exposed to it. Compounds administered luminally reach only mucosal cells, whereas all the cells in the preparation become exposed if the compound is administered into the arterial blood supply to the segment.[26]

Biosynthesis of Specific Lymph and Plasma Proteins

The ability of an intestinal segment to synthesize a specific protein *in vivo* can be determined by luminal or arterial administration of a labeled precursor, e.g., an amino acid, with subsequent isolation of the labeled product. Because, with either route of administration, all nonincorporated label will leave the tissue in the intestinal venous blood (all of which is collected), no label will reach the general circulation, and labeled proteins appearing in the collected venous blood or in mesenteric lymph from the rat will be, unequivocally, products of the experimental intestinal segment. This technique has been used to identify lymph and plasma apolipoproteins of intestinal origin.[21,22]

General Considerations

Several features make the autoperfused intestinal segment a particularly attractive experimental model. The tissue remains in a viable condition because, unlike intestinal preparations *in vitro*, it continues to receive nutrients and dispose of waste products via the normal vascular channels. The composition of the perfusing blood remains constant and physiologically normal, except for the presence of heparin and an anesthetic. Single-pass perfusion of both vasculature and lumen for long periods with access to the arterial and venous blood, the lymph, and the luminal effluent offers a wide range of experimental possibilities. With tissue weight and blood flow rate both known, absolute rates for the uptake and metabolic conversion of foreign compounds from both the vascular and luminal compartment can be determined. For comparative purposes, specific regions along the small intestine may be studied independently or, with modifications in technique,[27] even simultaneously.

[26] L. M. Pinkus and H. G. Windmueller, *Arch. Biochem. Biophys.* **182,** 506 (1977).
[27] M. Galluser, A. Pousse, G. Ferard, and J. F. Grenier, *Biomedicine* **25,** 127 (1976).

[15] Preparation of Mammalian Hepatocytes

By Jeffrey R. Fry

Isolated hepatocytes are being increasingly used in a wide range of biochemical and pharmacological investigations, including studies into the pathways of conjugation of xenobiotics and their regulation. The reasons behind this increasing use of isolated hepatocytes are not difficult to imagine. Isolated hepatocytes provide a very effective cellular "half-way house" between tissue homogenates/fractions on the one hand and the intact animal on the other; they also possess certain advantages over other cell-based *in vitro* systems such as perfused liver and liver slices. This topic is discussed more fully in recent reviews[1,2] and will only be briefly discussed at the end of this chapter.

The interest in the use of isolated hepatocytes can be traced to the work of Howard and Pesch,[3] who in 1968 described a technique for the reproducible isolation of reasonable numbers of viable rat hepatocytes by treatment of liver slices with a mixture of collagenase and hyaluronidase. Prior to that time numerous chemical and mechanical techniques for the isolation of rat hepatocytes had been described, but the cells so obtained were of doubtful viability.[1] The basic enzymic digestion procedure of Howard and Pesch was subsequently adapted to a perfusion system by Berry and Friend,[4] with the aim of increasing cell yield. This enzyme-perfusion technique has since been modified by various workers including Seglen,[5] who demonstrated that initial removal of calcium ions with a Ca^{2+}-chelator, followed by addition of Ca^{2+} back to the enzyme mixture (to activate collagenase), improved the technique still more. An excellent account of the enzyme-perfusion technique for the isolation of liver cells is given in a recent volume of this series.[6]

Much valuable information on the conjugation of xenobiotics can be gained from species comparison studies. The enzyme-perfusion dissociation technique was originally developed for use with rat liver but it has proved difficult to readily adapt this technique to other species and cannot be used where only pieces of liver are available, e.g., human liver biopsy material. Furthermore, the technique requires skill in operation and fre-

[1] J. R. Fry and J. W. Bridges, *Prog. Drug Metab.* **2,** 71 (1977).
[2] J. R. Fry and J. W. Bridges, *Rev. Biochem. Toxicol.* **1,** 201 (1979).
[3] R. B. Howard and L. A. Pesch, *J. Biol. Chem.* **243,** 3105 (1968).
[4] M. N. Berry and D. S. Friend, *J. Cell Biol.* **43,** 506 (1969).
[5] P. O. Seglen, *Exp. Cell Res.* **74,** 450 (1972).
[6] P. Moldéus, J. Högberg, and S. Orrenius, this series, Vol. 52, Part C, p. 60.

METHODS IN ENZYMOLOGY, VOL. 77

quently involves two workers. Because of these limitations we have developed a method for the isolation of viable hepatocytes using collagenase and hyaluronidase that does not require perfusion at any stage and is therefore readily adaptable to a wide range of different species and samples.[7] This chapter outlines our modified isolation technique and the conditions of cell incubation and discusses some of the uses of such isolated cells in studies on xenobiotic conjugation reactions.

Hepatocyte Isolation Technique

The technique is based on incubation of finely cut liver slices with a collagenase–hyaluronidase mixture in a Ca^{2+}-containing medium after prior removal of Ca^{2+} from the slices by incubation with a suitable chelator. By this means, moderate yields of hepatocytes with good viability are obtained and the cells possess a smooth, spherical appearance. The viability is comparable to that obtained using the perfusion method but the yield is somewhat lower.

Solutions

Wash Solution. This is based on Dulbecco's phosphate buffered saline, pH 7.4 (NaCl, 8.0 g; KCl, 0.20 g; KH_2PO_4, 0.20 g; $Na_2HPO_4 \cdot 12H_2O$, 2.89 g; phenol red, 0.01 g; all dissolved in 1 liter of distilled water), to which is added glucose (from a freshly prepared stock solution of 1 g in 10 ml distilled water) to a final concentration of 1 mg/ml.

Chelating Solution. Wash solution (see preceding solution) containing 0.5 mM ethylene glycol-bis(β-aminoethyl ether) N,N^1-tetraacetic acid (EGTA). This may be conveniently made up by preparing a stock solution of 25 mM EGTA in water, adjusting pH to 7.4 with sodium hydroxide, and diluting 1 : 50 with wash solution on the day of the cell isolation.

Enzyme Solution. Collagenase (5 mg) and hyaluronidase (10 mg) are dissolved in 10 ml Hank's buffer (NaCl, 8.0 g; KCl, 0.4 g; KH_2PO_4, 0.06 g; $Na_2HPO_4 \cdot 12H_2O$, 0.125 g; phenol red 0.01 g; all dissolved in 1 liter of distilled water) to which is added 0.2 ml of 250 mM $CaCl_2$, 0.1 ml of 1 M $NaHCO_3$, and 0.1 ml of a freshly prepared glucose solution (100 mg/ml). The pH of this enzyme solution is 7.0 and is not readjusted before use.

The glucose-free wash solution, stock EGTA, $CaCl_2$, and $NaHCO_3$ solutions and the Hank's buffer may be stored for up to a month at 4°. The stock glucose solution (100 mg/ml) should be made up fresh.

Cell Incubation Medium. This is based on a HEPES-buffered salt solution, pH 7.45 (NaCl, 8.00 g; KCl, 0.40 g; KH_2PO_4, 0.06 g; $Na_2HPO_4 \cdot$

[7] J. R. Fry, C. A. Jones, P. Wiebkin, P. Bellemann, and J. W. Bridges, *Anal. Biochem.* **71**, 341 (1976).

$12H_2O$, 0.125 g; $CaCl_2 \cdot 2H_2O$, 0.185 g; $MgSO_4 \cdot 7H_2O$, 0.246 g; HEPES (free acid), 4.76 g; glucose, 1.0 g; 5 N NaOH, 2 ml; all dissolved in 1 liter of distilled water). The solution may be stored in 90-ml lots at $-20°$. To 90 ml of this salt solution is added calf serum (10 ml), $100\times$ MEM vitamins (1 ml), 200 mM L-glutamine (1 ml), and $50\times$ MEM amino acids (2 ml), and the pH is readjusted to 7.45 with approximately 2 ml 0.5 N NaOH. If desired, 1 ml of a solution of phenol red (1 mg/ml) may also be added to monitor the pH during incubation.

Collagenase (Type IV), hyaluronidase (Type II), EGTA, and HEPES (free acid) are available from Sigma Chemical Co., whereas the calf serum, vitamins, glutamine, and amino acid concentrates have been obtained from Gibco-Biocult, Paisley, Scotland.

Procedure

The procedure adopted for rat liver will be described; procedures for other liver samples are essentially identical from the slicing stage onward.

Male Wistar albino rats are used (weight range 90–120 g). The animals are killed by cervical dislocation and the liver lobes rapidly removed into 10 ml of wash solution maintained at room temperature. The lobes are weighed, quickly blotted dry, and placed on a filter paper resting on a glass or Perspex plate. The filter paper acts as an anchor for the tissue and the plate provides a flat cutting surface. The lobes are cut into thin slices approximately 0.5 mm thick. This thin slicing is a crucial part of the technique because it provides intimate contact of tissue with dissociating enzymes. With experience it is possible to achieve satisfactory slices by hand slicing using single-edged blades such as those manufactured by the British American Optical Co., Slough, Bucks. (disposable microtome blades); it should be noted that hand slicing requires some practice. It is possible, however, to devise a simple piece of apparatus with which liver slices of reproducible thickness can easily be obtained, even by inexperienced workers. It is important to degrease the blades with alcohol prior to use.

The slices obtained from approximately 3 g liver are transferred to a 250-ml Erlenmeyer flask containing 10 ml wash solution, and the flask is shaken at 37° for 10 min (90 oscillations per min). The solution is removed by means of a Pasteur pipette attached to a water-driven vacuum pump. This washing stage is repeated twice and serves to remove debris, red blood cells, and some connective tissue from the slices. The slices are next subjected to two 10-min cycles with the chelating solution; this serves to remove the calcium ions from the slices and thereby aids cell dissociation. The solution present after the second cycle is carefully removed and replaced with 10 ml of the enzyme solution. The flask and its contents are then shaken at 37° for up to 1 hr. The solution initially becomes more alkaline in the presence of the liver slices; by adjusting the pH of the enzyme solution to 7.0 prior to mixing with the slices, subsequent extreme fluctuations in pH are minimized.

When the slices appear to be well digested (usually after 45–60 min of incubation) the resulting cell suspension is filtered through a single layer of surgical gauze (held in a small mesh tea strainer) into a beaker containing 20 ml of wash solution at room temperature. Any material remaining in the flask is washed into the beaker by addition of 10 ml wash solution. It is important not to triturate any undigested material retained on the filter in an effort to improve cell yield as this simply decreases cell viability with minimal ($<10\%$) increase in viable cell yield. The filtered cell suspension is gently mixed by swirling and allowed to stand for 5 min. This allows the viable cells to settle as a layer at the bottom of the beaker; the debris remains in suspension. The supernatant liquid is carefully removed by aspiration, and the settled layer of cells is gently suspended in 10 ml of wash solution. The suspension is centrifuged at 50 g for 90 sec, the supernatant fluid is removed, and the washing step is repeated twice more. Finally, the cell pellet is gently resuspended in 10 ml incubation medium and a sample of this suspension is taken for viability assessment prior to use of the cells.

Viability Criteria

A number of widely different criteria have been proposed as indexes of viability of isolated hepatocytes.[8] However, it is readily apparent that no one criterion may be regarded as ideal in terms of reproducibility, sensitivity, and speed. In my opinion, the most rational use of the several criteria of viability would entail use of more than one criterion when a technique is being established and only one or two when the technique has become routine.

The viability criterion most commonly used is that of trypan blue exclusion and is based on the premise that viable cells exclude the dye whereas nonviable cells take up the dye, which then stains the nucleus blue. The test is carried out as follows. A sample (0.25 ml) of the cell suspension is mixed with 0.1 ml of 0.4% trypan blue in isotonic saline and the numbers of viable and nonviable cells are counted in an Improved Neubauer counting chamber. From these figures, a percentage of viability value can readily be calculated as can the viable cell yield; these two values are useful quality control checks for each hepatocyte isolation.

[8] H. Baur, S. Kasperek, and E. Pfaff, *Hoppe-Seyler's Z. Physiol. Chem.* **356**, 827 (1975).

An alternative viability test is that of leakage of cytosolic enzymes such as lactate dehydrogenase. This test appears to detect the same irreversible damage to the plasma membrane as does the trypan blue exclusion test[8] and is the test of choice when assessing cell viability during an incubation. Details of this test are given by Moldéus *et al.*[6]

By the procedure outlined in the preceding section, cell yields of 15–30 × 10[6] cells per g liver are routinely produced with a viability of greater than 90%. Further evidence of the good viability of hepatocytes isolated by this method is given by the retention of a number of different vital functions, including retention of hormone and immunoglobulin receptor sites,[7,9] maintenance of normal cofactor levels (e.g., ATP, NADPH, glutathione),[10,11] retention of integrated xenobiotic metabolism,[12–14] and the ability to culture the differentiated cells *in vitro.*[15–17]

Adaptation of Dissociation Technique to Other Species and Tissues

The major advantage of this method is that it can be readily adapted to the isolation of hepatocytes from species other than the rat, including gerbil, hamster, guinea pig, rabbit, ferret, dog, sheep, and monkey.[18,19] It has also been successfully used in the isolation of viable fetal human hepatocytes.[20] The variables that may have to be altered on changing the species are the concentration of the enzymes (particularly of collagenase) and the length of time to which the liver slices are exposed to the enzymes. Otherwise, the protocol for cell isolation is identical to that outlined for rat hepatocytes. This technique may also be used for the isolation of viable rat hepatoma cells[18] and, with some modification, the isolation of viable renal cortex cells from a number of species.[13,19]

[9] E. Orlans, J. Peppard, J. R. Fry, R. H. Hinton, and B. M. Mullock, *J. Exp. Med.* **150**, 1577 (1979).
[10] P. Wiebkin, G. L. Parker, J. R. Fry, and J. W. Bridges, *Biochem. Pharmacol.* **28**, 3315 (1979).
[11] J. Gwynn and J. R. Fry, unpublished observations.
[12] P. Wiebkin, J. R. Fry, C. A. Jones, R. K. Lowing, and J. W. Bridges, *Biochem. Pharmacol.* **27**, 1899 (1978).
[13] J. R. Fry, P. Wiebkin, J. Kao, C. A. Jones, J. Gwynn, and J. W. Bridges, *Xenobiotica* **8**, 113 (1978).
[14] R. J. Shirkey, J. Kao, J. R. Fry, and J. W. Bridges, *Biochem. Pharmacol.* **28**, 1461 (1979).
[15] R. K. Lowing, J. R. Fry, L. J. King, and J. W. Bridges, *Chem.-Biol. Interact.* **25**, 303 (1979).
[16] J. R. Fry, P. Wiebkin, and J. W. Bridges, *Biochem. Pharmacol.* **29**, 577 (1980).
[17] J. R. Fry and J. W. Bridges, *Naunyn-Schmiedeberg's Arch. Pharmacol.* **311**, 85 (1980).
[18] C. A. Jones, Ph.D. Thesis, University of Surrey (1979).
[19] J. Kao, C. A. Jones, J. R. Fry, and J. W. Bridges, *Life Sci.* **23**, 1221 (1978).
[20] J. R. Fry, unpublished observations (1978).

Adaptation to the Sterile Isolation of Hepatocytes and Their Subsequent Culture

For the isolation of sterile hepatocytes, all glassware, solutions, and instruments are sterilized by suitable means and the sterile wash solution is supplemented with gentamycin at 70 $\mu g/ml$ final concentration. The protocol for the isolation is otherwise identical to that described earlier with the exception that the cells are finally resuspended in a complete tissue culture medium (Leibowitz L-15 medium supplemented with 10% tryptose phosphate broth, 10% fetal calf serum, and antibiotics). The cells are cultured on either tissue culture-treated polystyrene or collagen-coated coverslips.

General Considerations on Incubation of Hepatocytes with Substrates and Assay of Conjugates

For reasonably short-term studies (up to 4 hr) cells may be incubated in the incubation medium in Erlenmeyer flasks under an atmosphere of air with minimal loss of viability. Ideally, the volume of the cell suspension should occupy no more than 10% of the stated volume of the flask in order to maintain adequate aeration of the medium. The use of a complex incubation medium (as described in the preceding section) is essential for maintenance of cell viabilities at a high level during incubation.[21] The use of simple salt solutions is not recommended unless very short incubation periods, i.e., less than 10 min, are contemplated.

Substrates that are readily water-soluble can be added to the cell suspension in a small volume of isotonic saline. However, these water-soluble substrates are often acidic salts and it is worthwhile checking the pH of the cell suspension after such additions; the presence of phenol red as a pH indicator is useful in this respect. Most substrates are not very water-soluble, and it is our practice to prepare solutions of them in dimethylformamide at a concentration 500-fold greater than required. This solution is added to the cell suspension at 2 μl per ml of cell suspension so as to achieve a final DMF concentration of 0.2% (v/v), a concentration that has no effect on the viability or metabolic activity of the liver cells.

Finally, it is important to check the toxicity of the substrate to the cells so as to ensure subsequent work at nontoxic substrate levels. If this check is not carried out, conjugation reactions under study may be measured against a changing ability of the cells to metabolize the substrate due to substrate-mediated cytotoxicity. Cytotoxicity may be assessed by trypan blue exclusion or lactate dehydrogenase leakage as outlined; if the leakage

[21] A. J. Dickson and C. I. Pogson, *FEBS Lett.* **83**, 27 (1977).

of dehydrogenase is assessed, it is obviously important to check that the substrate does not interfere with the enzyme assay.

The assays are usually performed at a cell concentration of $2-5 \times 10^6$ viable cells per ml medium. This corresponds to approximately 15–40 mg wet weight of liver per ml medium. Because of the small amount of tissue and the probable production of more than one type of conjugate, analysis of the conjugates can be difficult. Two general approaches may be used: (1) hydrolysis of the conjugates with selective enzymes (particularly for glucuronides and sulfates) followed by appropriate measurement of the aglycones, or (2) chromatographic separation of the conjugates. The first approach has been the one most widely adopted, probably because of its general applicability to a wide range of substrates, but the second approach may possess some advantages particularly if a radiolabeled substrate is available. If hydrolysis of the conjugates with selective enzymes is the desired method, the choice of enzyme may be important. For example, some sulfate conjugates, e.g., 4-methylumbelliferone (4-MU) sulfate, are acid labile and thus assay of 4-MU glucuronide in the presence of 4-MU sulfate requires the use of a β-glucuronidase with a pH optimum of 7.0 (obtained from *Escherichia coli*) rather than an acid β-glucuronidase (obtained from *Helix pomatia*) so as to eliminate any interference from spontaneous 4-MU sulfate hydrolysis. A crude β-glucuronidase/sulfatase preparation, e.g., Sigma Type H-1, is useful for analysis of total glucuronide and sulfate conjugates,[22] whereas a pure β-glucuronidase preparation, e.g., Sigma Glucurase or Type $\bar{V}$-A is useful for analysis of glucuronide conjugates.[23] Thus, with both crude and purified β-glucuronidase preparations, it is possible to analyze total and glucuronide conjugates and, by subtraction, sulfate conjugates.

The Value of Isolated Hepatocytes in Conjugation Studies

The main attractions in the use of isolated viable hepatocytes for conjugation studies are that (1) the levels of cofactors used are those generated within the cells and (2) the full range of competing conjugation reactions are present without distortion of their cellular topography. For these reasons, use of isolated hepatocytes can yield valuable information on the events that most likely occur in hepatocytes *in vivo*.

---

[22] The crude preparation is used at a final concentration of 1 mg per milliliter and incubated with substrates overnight at 37°.

[23] The purified enzyme, at a final concentration of 100 to 500 units per milliliter is incubated with substrates overnight at 37°.

Acknowledgment

The author wishes to thank Professor J. W. Bridges for having provided the stimulus to most of the work presented in this chapter.

[16] Preparation and Characteristics of Isolated Kidney Cells

By Kari Ormstad, Sten Orrenius, and Dean P. Jones

Suspensions of intact, fully functional cells from adult mammalian tissues have widespread utility in biochemical research. The following method for preparation of cells from rat kidney[1] was derived from our prior experience with preparation of rat liver cells.[2] It involves perfusion with a medium containing EGTA (ethyleneglycol-bis(aminoethyl ether) tetraacetic acid) to weaken cell adhesions by removal of Ca^{2+}, followed by perfusion with a medium containing collagenase to hydrolyze the connective material. The resultant preparation contains mainly spherical cells of uniform size, which are predominantly derived from the tubular epithelium. The preparation procedure is rapid and the cells retain viability *in vitro* for at least 4 hr. The cells are comparable to intact kidney tissue with regard to amino acid transport, glutathione synthesis, drug metabolism, and respiratory characteristics. The preparation is therefore suitable for a wide variety of metabolic studies.

Because it is of great importance to ensure the metabolic integrity of isolated cells, we summarize the characteristics of this preparation, methods for assessment of cell viability, and incubation conditions found most suitable for short-term (less than 4 hr) experiments.

Isolation of Renal Cells

Materials

The modified Hanks' buffer, pH 7.4, that is used for cell preparation contains 8.0 NaCl, 0.4 g KCl, 0.2 g $MgSO_4 \cdot 7H_2O$, 0.06 g $Na_2H\,PO_4$, 0.06 g $KH_2\,PO_4$, 2.19 g $NaHCO_3$, and 3.0 g HEPES (N-2-hydroxyethyl-piperazine-N'-2-ethanesulfonic acid) per liter of distilled water. The inclusion of HEPES is not essential; however, its addition improves the quality of the cell preparation by increasing the buffering capacity as well

---

[1] D. P. Jones, G.-B. Sundby, K. Ormstad, and S. Orrenius, *Biochem. Pharmacol.* **28,** 929 (1979).

[2] P. Moldéus, J. Högberg, and S. Orrenius, this series, Vol. 52, p. 60.

as making the solution slightly hypertonic. The solution is bubbled with 95% O_2 : 5% CO_2 for several minutes and the pH is adjusted to 7.4 with 1 M NaOH. Typically, 250 ml of this solution are needed per cell preparation; 150 ml containing 0.5 mM EGTA and 1% (w/v) bovine serum albumin (Fraction V) are used for the first perfusion and 100 ml containing 0.1% (w/v) collagenase and 4 mM $CaCl_2$ are used for the second perfusion.

Modified Krebs–Henseleit buffer, pH 7.4, (6.9 g NaCl, 0.3 g KCl, 0.13 g KH_2PO_4, 0.292 g $MgSO_4 \cdot 7H_2O$, 0.374 g $CaCl_2 \cdot 2H_2O$, 2.0 g $NaHCO_3$ in 1 liter) containing 25 mM HEPES and equilibrated with 95% O_2 : 5% CO_2 is used for the final washing of cells and for incubations.

Collagenase (Grade II) from Boehringer Mannheim GmbH, Germany, is typically used, although enzyme from other sources is also suitable. Other chemicals are at least of reagent grade and purchased locally. Distilled, deionized water is used for all solutions.

Preparation of Kidney Cells

Male Sprague-Dawley rats (200–250 g) are anesthetized with ether, and heparin (1000 IE) is injected into a tail vein. The peritoneal cavity is opened and the aorta is freed below and above the renal arteries. A ligature is placed just below the renal arteries, and a second one is placed as high up in the abdomen as possible. The coeliac and superior mesenteric arteries are ligated to avoid leakage of perfusion fluid. An oblique incision is made in the aorta below the upper ligature and a conical-tipped probe cannula (1.5 × 80 mm) is immediately inserted and secured with a new ligature (Fig. 1, inset). The perfusion is started with the kidneys *in situ* using modified Hanks' buffer containing (0.5 mM) EGTA and 1% bovine serum albumin (w/v). The buffer is oxygenated (95% O_2 : 5% CO_2) and maintained at 37°. To avoid gas embolism, the tubes and cannula are filled with perfusion medium, and liquid is allowed to drip out of the cannula during insertion. The flow rate is adjusted to a pump pressure of about 1 m H_2O, which allows a flow of 5–10 ml per kidney per minute. Within about 1 min, both kidneys become pale.

The kidneys are excised from the posterior abdominal wall by cutting from the lateral aspect to the median line. The renal capsule and some connective tissue are left in place. The kidneys are transferred to the beaker from which the perfusate is withdrawn to provide a recirculating system (Fig. 1).

Cells are usually prepared from two rats. Kidneys from the first rat are perfused during the second cannulation (usually about 6 min). After perfusion of the second pair of kidneys for 5 min, the perfusion assembly containing the four kidneys is transferred to another beaker with modified Hanks' buffer containing 4 mM $CaCl_2$ and 0.10% collagenase (w/v). Be-

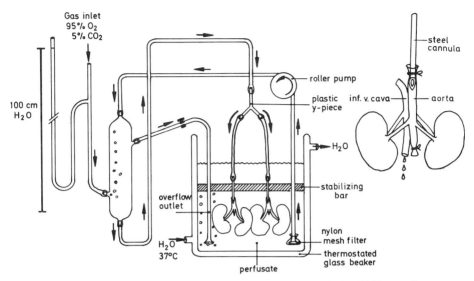

FIG. 1. Schematic drawing of apparatus for preparation of isolated kidney cells.

fore starting perfusion with this fluid, the tubing and the oxygenator are emptied as completely as possible of EGTA-containing perfusate. Then the collagenase solution is circulated through the kidneys under constant pressure for 12–20 min. Kidneys that are adequately perfused become slightly swollen and pale. The kidneys are removed from the perfusion apparatus, fat and connective tissue are gently removed, and the kidney cells are dispersed in modified Krebs–Henseleit buffer.

The dispersed cells are filtered through nylon mesh to remove connective tissue, glomeruli, and larger tissue fragments. The cells are allowed to settle for 2–3 min at ambient temperature and the excess medium is aspirated. The yield of cells is increased 50–100% by dispersing the tissue retained in the nylon mesh sieve in the medium containing collagenase and incubating at 37° for 4–5 min before transfer of cells to the Krebs–Henseleit buffer.

Characterization of Renal Cells

Cell Yield and Viability

Cell yield is estimated by counting a suitable aliquot of cells in a Bürker chamber or by sedimenting cells in hematocrit tubes centrifuged at 150 g for 3 min (Table I). Smaller cells, of endothelial and reticuloendothelial origin, constitute a small fraction of the total cell yield and are not included in the cell count. Large fragments, recognizable as tubular

TABLE I
Yield and Viability Parameters for Isolated
Rat Kidney Cells

Yield	10^7 cells/kidney 2×10^7 cells/kidney[a]
Trypan blue exclusion	90–95%
NADH penetration	10–15%
Packed cell volume[b]	14 μl
Protein content[c]	2 mg/10^6 cells
GSH content	30 nmol/10^6 cells

[a] Incubated with collagenase-containing medium for an additional 5 min after dispersal.
[b] Obtained by centrifugation of cells for 3 min at 150 g.
[c] Determined by the method of Bradford[3] relative to bovine serum albumin.

and glomerular fragments, are present when perfusion with collagenase is performed for less than 10 min. Consequently, perfusion time is normally longer than 10 min.

Viability is estimated by trypan blue exclusion and by the NADH penetration assay.[2] For trypan blue exclusion, a suitable aliquot of cells (final concentration of $0.5–1.0 \times 10^6$/ml) is pipetted into modified Krebs–Henseleit buffer containing 0.18% trypan blue, and the percentage of cells that do not stain are counted. The NADH penetration assay is performed by suspending $0.2–0.6 \times 10^6$ cells in 1 ml modified Krebs–Henseleit buffer containing 1.3 mM pyruvate and 0.2 mM NADH. Absorbance decrease at 340 nm measures the activity of lactate dehydrogenase to which NADH is accessible. Addition of 100 μl of 5% Triton X-100 results in cell lysis and makes NADH accessible to 100% of the lactate dehydrogenase. The ratio of the rate of NADH oxidation before lysis to that after lysis is expressed as the percentage NADH penetration. Yield and viability parameters are summarized in Table I. In some cases, it is more convenient to use packed-cell volume or protein content[3] for quantitation (Table I).

Isolated kidney cells respire at a rate comparable to perfused kidney[4] (Table II). The cells are responsive to exogenous substrates, such as succinate, but not to added ADP. Inclusion of bovine serum albumin provides an enhanced respiratory rate, presumably due to supply of respiratory substrates (e.g., adsorbed fatty acids) because this enhancement does not

[3] M. Bradford, Anal. Biochem. 72, 248 (1976).
[4] J. S. Stoff, F. H. Epstein, R. Narins, and A. S. Relman, Annu. Rev. Physiol. 38, 46 (1976).

TABLE II
RESPIRATORY CHARACTERISTICS OF ISOLATED KIDNEY CELLS

Conditions	O_2 consumption (nmol/10^6 cells per min)
Krebs–Henseleit buffer	24.8
+ 0.5 mM ADP	24.8
+ 1.0 mM Succinate	48.6
+ 1.0 mM Succinate + 0.5 mM ADP	49.3
+ 1% Bovine serum albumin	45.2
+ 0.5 μM Antimycin A	5.4

occur in the presence of succinate (Table II).[1] Most of the oxygen consumption under these conditions can be attributed to cytochrome oxidase because over 80% of the O_2 consumption is inhibited by antimycin A. The calculated turnover number for endogenous cellular respiration relative to cytochrome oxidase is 2.5 nmol/sec per nanomole cytochrome oxidase, which is comparable to that calculated for isolated liver cells and intestinal epithelial cells.[5]

Spectroscopy of Isolated Cells

Techniques for direct optical spectroscopy of isolated cells[6,7] can be applied directly to study of hemoproteins in kidney cells. Cells are suspended in modified Krebs–Henseleit buffer to a concentration of about 0.5×10^6 cells/ml (10^6 cells/ml is the maximum cell concentration in which absorbance changes obey Beer's law). Unlike liver cells, in which cytochrome b_5 and a significant fraction of mitochondrial cytochrome exist in the reduced form, most cytochromes are nearly completely oxidized in the kidney cells. Consequently, mitochondrial cytochromes can be quantitated (Table III) from the dithionite-reduced minus methyl viologen-oxidized difference spectrum from 500 to 650 nm. Using this approach, it is clear that mitochondrial cytochromes account for most of the cellular hemoproteins.

Cytochrome P-450 represents a smaller fraction of the total hemoprotein in kidney cells than in liver cells. It can be quantitated by a differential reduction method that corrects for interference by cytochrome oxidase. Cells are suspended to a final concentration of 10^6 cells/ml in modified Krebs–Henseleit buffer containing 20% glycerol (v/v) and 5 mg sodium cholate per ml. Sodium succinate is added to a final concentration of 5

[5] D. P. Jones, R. Grafström, and S. Orrenius, *J. Biol. Chem.* **255**, 2383 (1980).

[6] D. P. Jones and H. S. Mason, *J. Biol. Chem.* **253**, 4874 (1978).

[7] D. P. Jones, S. Orrenius, and H. S. Mason, *Biochim. Biophys. Acta* **576**, 17 (1979).

TABLE III
HEMOPROTEIN CONTENT OF ISOLATED KIDNEY CELLS

Hemoprotein	Wavelength	$mM^{-1}\,cm^{-1}$	$nmol/10^6$ cells
Catalase	600–630	5.4	0.043
Cytochrome $a + a_3$	605–630	13.1	0.385
Cytochrome $c + c_1$	550–540	19	0.785
Cytochrome $b_{561} + b_{566}$	561–545	23	0.270
Cytochrome P-450	450–490	102	0.068

mM, and the mixture is bubbled with CO for 1 min. The cytochrome oxidase becomes reduced under these conditions. After 5 min the suspension is divided into two cuvettes and a baseline is recorded between 400 and 500 nm. A few grains of powdered sodium dithionite are added to the sample cuvette to reduce cytochrome P-450. The spectrum from 400 to 500 nm represents the difference spectrum of the reduced cytochrome P-450 · CO complex minus oxidized cytochrome P-450, as well as other cytochromes that are not completely reduced by succinate in the reference cuvette.[1] The extinction coefficient used to calculate the cytochrome P-450 from this difference spectrum is 102 mM^{-1} cm^{-1}.[8]

Catalase is estimated from the absorbance change at 660 nm minus 630 nm following addition of 1 mM KCN to cells that have been depleted of Compound I by addition of 25 mM methanol.[9] The extinction coefficient used is 5.4 mM^{-1} cm^{-1}.[10]

Spectra are recorded on samples at room temperature. The cytochrome content is constant during short-term experiments, but cells should not be maintained without oxygenation because this leads to cell damage. Cells may be stored on ice for 4–5 hr without effect on the cytochrome content. Prolonged storage of cells, however, even on ice, leads to excessive cell aggregation and complicates optical spectroscopy. Therefore, studies should be performed as rapidly as possible after preparation.

Metabolic Incubations

The cell yield and viability characteristics make this preparation suitable for a wide variety of metabolic studies. Short-term incubations can be satisfactorily performed in Erlenmeyer flasks at 37° using modified

[8] D. P. Jones, S. Orrenius, and S. W. Jakobsson, in "Extrahepatic Metabolism of Drugs and Other Foreign Compounds" (T. E. Gram, ed.), p. 123. Spectrum Publ. Jamaica, New York, 1980.

[9] D. P. Jones, H. Thor, B. Andersson, and S. Orrenius, J. Biol. Chem. 253, 6031 (1978).

[10] N. Oshino, R. Oshino, and B. Chance, Biochem. J. 131, 555 (1973).

Krebs–Henseleit buffer. Cell concentration usually should not exceed 10^6 cells/ml for incubations longer than 30 min unless controls are run to ensure adequate oxygenation. This is usually done by performing incubations at two different cell concentrations to confirm that metabolism is linearly dependent upon cell concentration. A more convenient method for routine use is to calculate the oxygen availability in solution using a measured oxygen-transfer constant for the incubation flask.[11]

The oxygen-transfer constant relates the rate of oxygen entry (v) into solution to the difference between the oxygen concentration in solution (expressed as partial pressure, $P_{O_2}^e$) and the partial pressure in the incubation atmosphere ($P_{O_2}^g$),

$$v = 1.4K(P_{O_2}^g - P_{O_2}^e) \tag{1}$$

where 1.4 μM/torr is the factor for converting torr to μM O_2. The units for each term are v, μM/min; K, min^{-1}; P_{O_2}, torr. K is dependent upon temperature, mixing rate, and surface area to volume ratio. K can be readily calculated by measuring the rate of O_2 transfer into a deoxygenated buffer with a gas mixture of known O_2 content. This is done by measuring O_2 concentration with an O_2 electrode as a function of incubation time. If mixing rate and temperature are constant, then K can be calculated for different incubation volumes and used to determine whether oxygen supply is sufficient for a specific experiment. The calculation also can be performed using Eq. (1) for steady-state conditions because the rate of oxygen entry into solution is equivalent to the rate of oxygen consumption. Thus, using data from Table II, the determined rate constant, and a gas mixture of known O_2 content, O_2 concentration of the solution can be estimated. For study of systems dependent upon cytochrome P-450, an oxygen concentration higher than 30 μM should be maintained,[6] although other systems may require a considerably higher or lower oxygen concentration for optimal function.

Because of the shape of Erlenmeyer flasks, increased volume decreases the surface area to volume ratio and thus decreases K; to compensate, the shaking speed must be increased. Because cells are subject to physical damage, the shaking should not exceed about 100–120 cycles per min. An improved incubation system is offered by roundbottom flasks (50 ml) to optimize the surface area to volume ratio. Flasks are placed on a rotatory evaporator (Büchi RSB/40-5-50), equipped with an "udder" attachment, and adjusted to rotate through a water bath at 38°. Oxygen is supplied through the central vacuum exit of the evaporator, and rotation speed is normally set to give 30 rpm. Using these conditions, drug metabolism is linear for 2 hr without loss of cell viability.

[11] D. P. Jones and H. S. Mason, *Anal. Biochem.* **90,** 155 (1978).

Because kidney cells adhere to glassware, particularly after acid washing, we routinely siliconize glassware to eliminate this problem.

Drug Oxidation and Conjugation

The kidneys are active in metabolism and excretion of drugs and drug metabolites. Metabolism of a variety of substrates has been associated with function of the cytochrome P-450 system in kidney microsomes from rat, mouse, rabbit, hamster, guinea pig, and man. These activities currently have been studied to a lesser extent in isolated cells. The most complete data are available for the metabolism of acetaminophen,[1] but Phase I metabolism has also been measured for benzo[a]pyrene, 7-ethoxycoumarin, and biphenyl, and Phase II metabolism has been measured for 4-methylumbelliferone and benzoic acid.[12] Activities are typically lower in kidney than in liver when expressed either as a function of cell count, weight, milligrams microsomal protein, or nanomoles total cellular cytochrome P-450. In addition, cytochrome P-450-dependent activities in kidney typically show different patterns of induction than liver following pretreatment with compounds such as phenobarbital or 3-methylcholanthrene.

In isolated kidney cell suspensions, glutathione conjugates do not accumulate in the incubation medium as they do in suspensions of isolated liver cells. These derivatives are further metabolized by γ-glutamyltransferase and cysteinylglycinase to the cysteine derivatives.[13] These products then appear to be taken up by the cells and acetylated to the N-acetyl derivatives. With acetaminophen as substrate, the total production of the cysteine and N-acetylcysteine derivatives is linear with time; however, the formation of the N-acetylcysteine derivative occurs at only about half of the rate of the γ-glutamyltransferase and cysteinylglycinase activities.

Turnover of the glutathione pool in kidney is very rapid, and because of the lower rate of glutathione conjugate formation, the kidney cells are more resistant to GSH depletion than are isolated liver cells. However, because the preparation represents a mixture of cell types, it is unclear whether the observed decrease in GSH occurs to a small degree in the whole cell population or to a much larger degree in only a small fraction of the cells. Whereas 3-methylcholanthrene pretreatment causes only a small increase in total cellular cytochrome P-450, it results in an increased rate

[12] J. R. Fry, P. Wiebkin, J. Kao, C. A. Jones, J. Gwynn, and J. W. Bridges, *Xenobiotica* **8**, 113 (1978).

[13] D. P. Jones, P. Moldéus, A. H. Stead, K. Ormstad, H. Jörnvall, and S. Orrenius, *J. Biol. Chem.* **254**, 2787 (1979).

TABLE IV
AMINO ACID UPTAKE AND GLUTATHIONE METABOLISM
BY ISOLATED KIDNEY CELLS

Function	nmol/10⁶ cells per min
Cystine uptake	1.9
Cysteine uptake	5.2
Methionine uptake	2.8
Glutamate uptake	3.3
Glycine uptake	5.6
Glutathione synthesis rate[a]	1.5
GSH oxidation rate[b]	35
GSSG metabolism rate[c]	20

[a] In the presence of 1 mM glutamate, 1 mM glycine, and 0.2 mM cystine.
[b] Initial concentration of GSH was 1 mM.
[c] Initial concentration of GSSG was 1 mM.

of GSH depletion in the presence of acetaminophen.[14] Enhanced GSH depletion by acetaminophen has also been seen in isolated cells from animals that had undergone chronic exposure to ethanol.

Amino Acid Transport and GSH Metabolism

Isolated cells offer an ideal model for study of the relationship of cellular transport to metabolic processes because many intracellular processes are rate-limited by the uptake of necessary precursors. Isolated kidney cells are active in amino acid uptake (Table IV) and under suitable conditions can be used to study GSH synthesis.[15] Rates of uptake of the precursor amino acids for GSH synthesis are high but, after depletion of intracellular GSH by pretreatment with diethyl maleate,[16] it is possible to observe recovery of the GSH pool at essentially the same rate as for uptake of sulfur-containing amino acid, i.e., cystine.[15]

Isolated cells also allow direct testing of mechanisms of amino acid uptake, such as that proposed by Meister and co-workers[17] involving γ-glutamyltransferase. Incubations with serine · borate, an inhibitor of this enzyme, are consistent with this pathway but suggest that it may represent only a small fraction of total amino acid transport.[15] Incubations

[14] P. Moldéus, B. Andersson, A. Norling, and K. Ormstad, *Biochem. Pharmacol.* **29**, 1741 (1980).
[15] K. Ormstad, D. P. Jones, and S. Orrenius, *J. Biol. Chem.* **255**, 175 (1980).
[16] This volume, Article [8].
[17] O. W. Griffith, R. J. Bridges, and A. Meister, *Proc. Natl. Acad. Sci. U.S.A.* **75**, 5405 (1978).

with methionine sulfoxime, an inhibitor of γ-glutamylcysteine synthetase, in the presence of either GSH or the precursor amino acids suggest that to a certain extent, GSH can be taken up by the cells without prior hydrolysis to the constituent amino acids[18,19]; serine · borate inhibits such transport.

Several methods are available for study of amino acid uptake. We find that the most convenient and reliable method is simply to sediment the cells by centrifugation, aspirate the supernatant liquid, wash the cells once by resuspending in buffer and resedimenting, and measure the component in the cell pellet. This process takes about 5 min and therefore limits the earliest time points, but it is very reliable for 5- to 30-min incubations. Earlier time points are obtained by rapid filtration of the cells on a Millipore (100 mesh) filter. The latter method is most convenient for measurement of uptake of radiolabeled amino acid. The combination of these two techniques provides a reliable means for measuring uptake of added constituents, equilibration of these components with the intracellular pool, and incorporation into other cellular components.

Comments

Studies of metabolic function using cell suspensions require that cellular integrity be maintained. While the NADH penetration assay provides a convenient assay of cell viability, additional characterization, such as trypan blue exclusion, sensitivity of respiratory rate to added ADP, and ability to maintain a normal GSH content, should be examined. It is also important to ensure that oxygenation is adequate and that excessive changes in pH do not occur.

A major advantage of this preparation is that cells are uniformly exposed to exogenously added materials. In addition, control incubations can be readily performed for most metabolic manipulations. Additional refinement of the preparation, such as fractionation of cell types,[18,20] can be expected ultimately to yield a more complete understanding of renal function.

Acknowledgment

The work presented in this chapter was supported by grants from the Swedish Medical Research Council and Karolinska Institutet.

[18] T. G. Pretlow, J. Jones, and S. Dow, *Am. J. Pathol.* **74**, 275 (1974).
[19] K. Ormstad, T. Låstbom, and S. Orrenius, *FEBS Lett.* **112**, 55 (1980).
[20] J. I. Kreisberg, A. M. Pitts, and T. G. Pretlow, *Am. J. Pathol.* **86**, 591 (1977).

[17] Isolation of Pulmonary Cells and Use in Studies of Xenobiotic Metabolism

By THEODORA R. DEVEREUX and JAMES R. FOUTS

The development of unique isolated cell populations from lung provides an opportunity to localize pulmonary xenobiotic biotransformation to specific cell types and study metabolism at the cellular level.[1-3] Numerous reports[4] have described the pulmonary cytochome P-450-dependent monooxygenase/mixed-function oxidase enzymes, which can metabolically activate or detoxify a variety of toxic chemicals entering the lung via inhalation or circulation. However, most of these studies have not distinguished the cellular localization of the pulmonary mixed-function oxidase system because the lung is a complex organ and may contain some 40 or more varied cell types.[5] This article describes our techniques for isolation of alveolar type II cells and nonciliated bronchiolar epithelial (Clara) cells and briefly discusses how these freshly isolated cells can be used to study xenobiotic biotransformation.

Animal Models

The rabbit was chosen as a model for pulmonary cell separation because of the relatively large size of its lungs (compared to other laboratory animals) and because of its high pulmonary mixed-function oxidase activity.[6,7] The rabbit P-450-dependent monooxygenase system, however, is unresponsive to the classic phenobarbital- or 3-methylcholanthrene-type inducers.[4] Therefore, techniques have also been developed for separation of certain pulmonary cell types from rat in order to investigate inducible systems of xenobiotic metabolism in pulmonary cells.[8,9]

[1] T. R. Devereux and J. R. Fouts, in "Microsomes, Drug Oxidations and Chemical Carcinogenesis" (M. J. Coon, A. H. Conney, R. W. Estabrook, H. V. Gelboin, J. R. Gillette, and P. J. O'Brien, eds.), Vol. 2, p. 825. Academic Press, New York, 1980.

[2] T. R. Devereux and J. R. Fouts, In Vitro 16, 958 (1980).

[3] T. R. Devereux and J. R. Fouts, Biochem. Pharmacol. (in press).

[4] Reviewed in R. M. Philpot, M. W. Anderson, and T. E. Eling, in "Metabolic Functions of the Lung" (Y. S. Bakhle and J. R. Vane, eds.), p. 123. Dekker, New York, 1977.

[5] S. P. Sorokin, AEC Symp. Ser. 21, 3 (1970).

[6] G. E. R. Hook, J. R. Bend, D. Hoel, J. R. Fouts, and T. E. Gram, J. Pharmacol. Exp. Ther. 182, 474 (1972).

[7] J. R. Bend, G. E. R. Hook, R. E. Easterling, T. E. Gram, and J. R. Fouts, J. Pharmacol. Exp. Ther. 183, 206 (1972).

[8] K. G. Jones and J. R. Fouts, Pharmacologist 22, 277 (1980).

[9] K. G. Jones, T. R. Devereux, B. R. Smith, J. R. Fouts, and J. R. Bend, Toxicology (in press).

METHODS IN ENZYMOLOGY, VOL. 77

TABLE I
BUFFER SOLUTIONS[a]

1. Lung perfusion medium	Krebs–Ringer bicarbonate solution
	Bovine serum albumin, 4.5%
	Glucose, 5 mM
2. Balanced salt solution	NaCl, 0.15 M
	KCl, 6 mM
	KH$_2$PO$_4$, 5 mM
	Glucose, 5 mM
	HEPES, 25 mM[b]
3. Cell isolation buffer	Balanced salt solution, 3 parts
	F12K growth medium, 1 part[c]
	BSA, 0.5%
	DNase, 0.05%[d]

[a] All buffers titrated to pH 7.4.
[b] N-3-Hydroxyethylpiperazine-N'-2-ethanesulfonic acid.
[c] Grand Island Biological Co., Grand Island, NY.
[d] Deoxyribonuclease I-Sigma Chemical Co., St. Louis, MO.

Preparation of Macrophages, Alveolar Type II Cells, and Clara Cells from Rabbit Lung

Male New Zealand White rabbits weighing 2–3 kg are anesthetized by injecting 3 ml sodium pentobarbital (50 mg/ml) into a marginal ear vein. The chest cavity is opened and the lungs are perfused by gravity (pressure head about 26 cm H$_2$O) with a cannula inserted into the pulmonary artery via the right ventricle. The left auricle is cut to allow the blood to wash out of the lungs. The perfusion medium (Table I) is warmed to 37° prior to use. During the perfusion, the lungs are ventilated manually with air through a cannula inserted into the trachea.

The lungs are perfused until the exiting perfusion medium appears to be free of blood and the lungs turn white; about 200 ml of perfusion medium needed. The lungs are removed intact from the thorax and cold HEPES-buffered balanced salt solution (Table I) at pH 7.4 is instilled into the trachea and allowed to drain out (repeated five times) to remove macrophages. Alveolar macrophages (used for metabolic studies) are isolated by low speed centrifugation (500 g for 10 min) of this lavage fluid.[10] Following the lavages, a solution of 0.1% Protease I (Sigma Chemical Co., St. Louis, MO) and 0.5 mM EGTA [ethylene glycol-bis(β-aminoethyl ether) N,N'-tetraacetic acid] in balanced salt solution is instilled into the trachea. The trachea is closed off and the lungs are incubated in a beaker of the balanced salt solution at 37° for 15 min. The trachea and bronchi are

[10] G. E. R. Hook, J. R. Bend, and J. R. Fouts, *Biochem. Pharmacol.* **21**, 3267 (1972).

removed and the lungs placed in a small beaker in ice and minced into 1–2 mm^3 pieces with scissors. The mince is suspended in a total of 200 ml of cold cell isolation buffer (Table I). The remainder of the cell separation procedure utilizes this buffer and is performed at 4°.

The cell suspension is poured through two layers of cheese cloth and consecutively filtered through 160-μm and 40-μm mesh nylon filters to achieve a suspension of single cells. These cells are centrifuged for 8 min at 500 g, resuspended in cell isolation buffer, washed again, and resuspended to about 10 ml. At this point, the cells from two rabbits are combined and counted.

Initial separation of alveolar type II and Clara cells is achieved by centrifugal elutriation. Cell fractionation by this method is based on cell size and is performed by four stepwise increases in flow rate or decreases in elutriator rotor speed, the smallest cells and particles eluting first. A 15–20 ml suspension of cells (5×10^8 to 1×10^9 cells, total) is loaded into the Beckman JE-6 elutriator rotor (Beckman Instruments, Palo Alto, CA). It is important that the elutriator buffer (same as cell isolation buffer) contains sufficient BSA and DNase to prevent clumping. BSA protects the cells during centrifugation and DNase disrupts sticky DNA, which would otherwise cause clumping in the elutriator chamber. Elutriator fraction 1 (150 ml) is collected at a rotor speed of 2000 rpm and a flow rate of 9 ml/min. This fraction consists of cell debris and small cells and is discarded. Elutriator fraction 2 (100 ml) is collected at a flow rate of 15 ml/min and a rotor speed of 2000 rpm and contains 50% alveolar type II cells. Elutriator fraction 3 (100 ml) is collected at a rotor speed of 2000 rpm and a flow rate of 28 ml/min and is discarded because it contains a mixture of cells. Elutriator fraction 4 (100 ml) contains about 30% Clara cells and is collected at a rotor speed of 1200 rpm and flow rate of 30 ml/min.

The alveolar type II cells found in elutriator fraction 2 can be further purified by metrizamide (Accurate Chemical and Scientific Corp., Hicksville, NY) or Percoll (Pharmacia Fine Chemicals, Piscataway, NJ) density gradient centrifugation.[1,2] A 5-ml suspension containing about 8×10^7 cells harvested from elutriator fraction 2 (by low speed centrifugation at 500 g for 10 min) is layered onto a 25-ml solution of 28% Percoll in balanced salt solution (v/v). These gradients are centrifuged in a Beckman TH-4 swinging bucket rotor at 800 g for 30 min at 4°. The fraction of cells enriched at the top of the 1.05 density Percoll contains an average of 80% type II cells. A yield of about 1×10^8 cells is obtained from four rabbits. These cells are washed and resuspended in HEPES-balanced salt solution for use in metabolism studies.

Further purification of Clara cells, which are found principally in elutriator fraction 4, is achieved by Percoll discontinuous density gradient centrifugation. Five-milliliter suspensions (each containing about 2×10^7 cells) of elutriator fraction 4 are layered onto discontinuous gradients that consist of 15 ml 28% Percoll in HEPES-balanced salt solution on top of 10 ml 36% Percoll in HEPES-balanced salt solution. These gradients are centrifuged in the same manner as the alveolar type II cells (see preceding paragraph). The Clara cells are found scattered between the upper 1.05 density band and the bottom of the tube. The cells in the upper band and the ones in the pellet at the bottom of the tube are discarded. The Clara cell fraction is washed and resuspended in the cell isolation buffer. Purification of this fraction is performed by another elutriator centrifugation step. Elutriator fraction $E_2 1$ is collected at a flow rate of 25 ml/min and a rotor speed of 2000 rpm. This fraction is discarded. Elutriator fraction $E_2 2$ is collected at a rotor speed of 1200 rpm and a flow rate of 30 ml/min and contains 70% Clara cells. About $1-2 \times 10^7$ cells are obtained in this fraction from four rabbits. These cells are washed and resuspended in balanced salt solution for enzyme assays.

A two-polymer aqueous phase system composed of 5.0% dextran T500 and 3.8% polyethylene glycol 6000 in 0.15 M sodium phosphate buffer, pH 7.2, has also been used to partition Clara cells from elutriator fraction 4.[2] This system is technically easy to use and a fraction can be obtained from the polyethylene glycol layer, which contains 70% Clara cells. However, the yield of cells is very low, about 5×10^6 cells from 4 rabbits.

Preparation of Rat Lung Cells

Alveolar type II cells are isolated from rat lung in a manner similar to rabbit except that the perfused lungs are incubated with 0.5% Protease type I for 15 min at 37°.[8] Following metrizamide density gradient centrifugation of the type II cell-enriched sample from the elutriator, a fraction that contains 90% alveolar type II cells ($2-3 \times 10^6$ cells/rat lung) is obtained. About 5×10^7 cells from elutriator fraction 2 are layered on 25 ml 13.25% metrizamide in balanced salt solution and centrifuged in a Beckman TH-4 swinging bucket rotor at 800 g for 10 min at 4°. The type II cells remain at the top of the gradient. The cell digest prepared from rats contains only about 1% Clara cells and these cells have not yet been extensively purified.

Other Cell Types

It would be useful to isolate other pulmonary cells such as tracheal epithelial, alveolar type I, endothelial, and mesothelial cells for investigations in xenobiotic metabolism. These cell types may be important as targets for damage by inhalation of chemicals or for disease derived from accumulation of chemicals in the lung via the capillaries. Methods for isolation of tracheal[11] and type I[12] cells have been described. Techniques based on cell size and density such as those described earlier may be useful for isolation of some of these cell types.

Cell Identification and Viability

Alveolar type II cells are identified and enumerated by the modified Papanicolou stain without acid alcohol,[13] which stains the lamellar bodies dark purple. Clara cells are identified under the light microscope by a nitroblue tetrazolium (NBT) staining method developed in our laboratory[4]: Air-dried smears of cell fractions are fixed in 10% formalin for 30 sec. Slides are rinsed in balanced salt solution and incubated in a solution of 0.1% NBT and 0.1% NADPH in balanced salt solution at 37° for 10 min. The slides are drained and counterstained in a 1% aqueous methylene green solution for 4 min. Excess stain is rinsed away with water and the slides are mounted in balanced salt solution : glycerin, 1 : 3. Clara cells are stained purple whereas other cells appear blue or light pink.

Cell viability is measured by exclusion of 0.04% trypan blue dye. Because NADPH does not cross intact cell membranes, NADPH-dependent mixed-function oxidase enzyme activities in freshly isolated cells are compared in the presence or absence of added NADPH as another indicator of viability. Intact viable cells usually contain sufficient NADPH to support xenobiotic metabolism.

Xenobiotic Metabolism in Isolated Cells

Although the freshly isolated pulmonary cells can be used for a variety of research objectives, the discussion that follows focuses on one area of interest, pulmonary metabolism of xenobiotics. Pulmonary cells can be isolated under sterile conditions in order to initiate cell cultures. In fact, alveolar type II cells have been cultured in several laboratories.[13-15] However, most cell culture systems fail to maintain levels of cytochrome P-450 and monooxygenase activities near what is found in the freshly isolated cells.[16] The freshly isolated alveolar type II and Clara cells can be

---

[11] W. E. Goldman and J. B. Baseman, *In Vitro* **16**, 306 (1980).
[12] P. E. Piciano and R. M. Rosenbaum, *Am. J. Pathol.* **90**, 99 (1978).
[13] Y. Kikkawa and K. Yoneda, *Lab. Invest.* **30**, 76 (1974).
[14] W. H. J. Douglas and M. E. Kaighn, *In Vitro* **10**, 230 (1974).
[15] R. J. Mason, M. C. Williams, R. D. Greenleaf, and J. A. Clements, *Am. Rev. Respir. Dis.* **115**, 1015 (1977).
[16] P. S. Guzelian and J. L. Barwick, *Biochem. J.* **180**, 621 (1979).

TABLE II
ENZYME ACTIVITIES IN CELL FRACTIONS AND LUNG HOMOGENATE

	7-Ethoxy-coumarin-deethylase[a]	Coumarin hydroxy-lase[a]	Benzpyrene hydroxy-lase[b]	NADPH cyto-chrome c reductase[c]	Microsomal cytochrome P-450 content[d]
Cell digest	50	4	16	20	0.152
Type II cells (80% purity)	30	1	13	44	0.074
Clara cell-enriched elutriator fraction (30% Clara cells)	85	20	44	60	—
Clara cells (70% purity)	120	42	60	60	—
Macrophages	3	ND	9	7	ND
Lung homogenate	130	4	—	28	0.146

[a] Picomoles umbelliferone per milligram protein per minute. All samples sonicated, 1 mM NADPH added, reference in footnote 3.

[b] Picomoles benzo[a]pyrene-OH per milligram mg protein per minute. All samples sonicated, 1 mM NADPH added, reference to method in footnote 3.

[c] Nanomoles cytochrome c reduced per milligram protein per minute. Reference to method in footnote 3.

[d] Nanomoles P-450 per milligram microsomal protein. Reference in footnote 3.

ND, Not detectable.

maintained at 4° overnight with only about 10% loss in mixed-function oxidase activity. Gentamycin, 5 μg per ml, is added to these cells.

Because about 5×10^6 cells are required for each mixed-function oxidase incubation, cell separation techniques have been developed to produce a high yield of cells. The elutriator has the advantage over other cell isolation techniques of being able to separate 5×10^8 to 1×10^9 cells into several distinct fractions in a short period of time without overloading the system.

Clara cells isolated from rabbit lung exhibit higher activities for all mixed-function oxidase enzymes measured than are observed for the alveolar type II cells (Table II); the Clara cell of the bronchiolar epithelium was the first pulmonary cell shown to be a concentrated site of cytochrome P-450 monooxygenase activity.[3] Studies utilizing the freshly isolated pulmonary cells show that the alveolar type II cell also exhibits some mixed-function oxidase activity (Table II).[1,3] The 7-ethoxycoumarin deethylase activity measured in the isolated type II cells cannot be accounted for by the 1–2% contamination of these cells by Clara cells. However, these studies with isolated cells indicate that coumarin hydroxylase is specifically enriched in Clara cells. The trace amounts of this activity observed in the type II cell fraction are probably due to Clara cell contamination.

Pulmonary cells from rats are being isolated in order to examine the effects of mixed-function oxidase inducers on these enzymes in different cell populations. Aryl hydrocarbon hydroxylase activity has been measured in alveolar type II cells isolated from control rats and β-naphthoflavone-treated rats.[8] A 90-fold increase in activity is seen in this enzyme activity in type II cells prepared from β-naphthoflavone-treated animals as compared to untreated animals. These experiments support the role of the type II cell in pulmonary xenobiotic metabolism as well as emphasizing the need to study these enzyme systems in other pulmonary cells.

Comments

Our methods for isolation of pulmonary cells should be applicable to other animal species. The comparatively uniform size and low density (due to the presence of a high content of phospholipid) of the alveolar type II cell make this cell type relatively easy to separate from others; this has been done in various species in several laboratories.[13–15] In contrast, because the Clara cell has variable densities and exists in small numbers per lung, it is more difficult to isolate, especially in species other than rabbit. The large size of the Clara cell has proved to be the most useful characteristic in its separation. One drawback to elutriator separation of Clara cells is that clumped cells (doublets and triplets of small cells) may fractionate with the larger sized Clara cells. Use of DNase and BSA in the elutriator is, therefore, essential in minimizing clumping.

While developing the purification technique, it was found that different proteolytic enzymes yielded different proportions of pulmonary cell populations.[2] Protease I (Sigma) released a cell digest containing 30% type II cells and 5% Clara cells, the best yield of any protease tested. Because these proportions may not represent what is found in the whole lung, enzyme activities are presented as specific activities. Total activities by cell populations can only be speculated on at this time until more precise morphometric studies on cell numbers in lung have been made.

A possible limitation of our methods is enzyme alteration caused by or occurring during the cell isolation procedures. Our studies have shown that Protease I drastically inhibits the mixed function oxidase if it enters the cells, although it does not alter the enzyme activities of cells with intact membranes. Much of the cell digest is composed of cell debris and, therefore, exposed to the destructive action of protease. This may explain why lung homogenate exhibits higher activities (e.g., 7-ethoxycoumarin deethylase activity) than the cell digest (Table II). Loss of nutrients, cell–cell communication and regulation, or other factors and changes occurring during cell separation may also cause alterations in the enzymes that are being measured. All these factors must be considered when comparing enzyme activities in the isolated cells to activities observed in other study models.

[18] Separation and Use of Enterocytes

By LAWRENCE M. PINKUS

The epithelial cells of the small bowel, here referred to as "enterocytes," consist principally of the mature absorptive (villous) cells (about 75%) and their precursors, the rapidly proliferating (crypt) cells (about 25%); small numbers of other specialized cells, e.g., goblet, enteroendocrine, and Paneth, are also present.[1-3] Enterocytes have been isolated by vibrational,[4] enzymatic,[5-7] and chelation–elution methods[8-12]; excellent reviews are available.[13,14] Many metabolic characteristics of the isolated rat[15-21] and guinea pig[22-24] cells have been described and their capacity for

[1] C. P. Leblond and C. G. Stevens, *Anat. Rec.* **100**, 357 (1948).

[2] H. Cheng and C. P. Leblond, *Am. J. Anat.* **141**, 537 (1972).

[3] H. Moe, *Int. Rev. Gen. Exp. Zool.* **3**, 241 (1968).

[4] D. D. Harrison and H. L. Webster, *Exp. Cell Res.* **55**, 257 (1969).

[5] A. D. Perris, *Can. J. Biochem.* **44**, 687 (1966).

[6] P. J. A. O'Doherty and A. Kuksis, *Can. J. Biochem.* **53**, 1010 (1973).

[7] G. A. Kimmich, *Biochemistry* **9**, 3659 (1970).

[8] B. K. Stern, *Gastroenterology* **51**, 855 (1966).

[9] S. Reiser and P. A. Christiansen, *Biochim. Biophys. Acta* **225**, 123 (1973).

[10] E. M. Evans, J. M. Wrigglesworth, J. M. Burdett, and W. F. R. Pover, *J. Cell Biol.* **51**, 452 (1971).

[11] M. M. Weiser, *J. Biol. Chem.* **248**, 2536 (1973).

[12] C. A. Ziomek, S. Schulman, and M. Edidin, *J. Cell Biol.* **86**, 849 (1980).

[13] W. C. Hülsmann, J. W. O. Van den Berg, and H. R. De Jonge, this series, Vol. 35 [67].

[14] G. A. Kimmich, *Methods Membr. Biol.* **5**, 151 (1975).

[15] G. I. Leslie and P. B. Rowe, *Biochemistry* **11**, 1696 (1972).

[16] M. M. Weiser, *J. Biol. Chem.* **248**, 2552 (1973).

[17] F. A. Wilson and L. L. Treanor, *Biochim. Biophys. Acta* **406**, 280 (1975).

[18] C. M. Towler, G. P. Pugh-Humphreys, and J. W. Porteous, *J. Cell Sci.* **29**, 53 (1978).

[19] M. Wadford, P. Lund, and H. A. Krebs, *Biochem. J.* **178**, 589 (1979).

[20] H. R. De Jonge, *Biochem. Pharmacol.* **22**, 2659 (1973).

[21] F. Raul, M. Kedinger, P. Simon, J. Grenier, and K. Haffen, *Biol. Cell.* **33**, 163 (1978).

[22] E. M. Evans and K. Burdett, *Gut* **14**, 98 (1973).

drug metabolism has been investigated (see table). For *in vitro* studies, suspensions of enterocytes offer certain advantages compared with everted gut sacs or rings because they contain fewer cell types and their cell surfaces are better exposed to oxygen and substrates. Negative aspects are a loss of polarity (as with brush border or basolateral transport) and the inherent possibility of damage to plasma membranes during isolation. This chapter describes chelation–elution methods that subsequently release villous and crypt cells or yield mixed suspensions of enterocytes from the intestinal epithelia of rats and guinea pigs. The utility of such preparations and some criteria for their viability are discussed.

Rat

Weiser's improvement[11] of Stern's method[8] allows rapid separation of villous and crypt cells from the jejunum, ileum, or whole intestine. Attachment of cells to the basement membrane is loosened by chelation and the loosened cells are released in a villous-to-crypt gradient by repeated expansion and contraction of the intestinal wall. The method described here incorporates modifications to satisfy the nutritional preferences[25] and to help maintain the membrane integrity of the isolated cells.

Reagents and Supplies

Phosphate Buffered Saline (PBS): NaCl, 137 mM; Na$_2$HPO$_4$, 8.16 mM; KH$_2$PO$_4$, 1.47 mM; KCl, 3.22 mM; pH 7.4; store at 25-fold this concentration

Solution A: NaCl, 96 mM; sodium citrate, 27 mM; KH$_2$PO$_4$, 5.6 mM; KCl, 1.5 mM; pH 7.3

EDTA(Na$_2$), 0.25 M, pH 7.4

CaCl$_2$, 1 M

Bovine serum albumin (BSA), fatty acid-free, 30% w/v

Solution B: Phosphate-buffered saline (1×) containing EDTA (1.5 mM), BSA (0.1%), and dithiothreitol (DTT) (1 mM); prepare immediately before use

Krebs–Ringer bicarbonate (KRB): NaCl, 122 mM; NaHCO$_3$, 25 mM; KCl, 4.72 mM; MgCl$_2$, 2.56 mM; KH$_2$PO$_4$, 1.2 mM; pH 7.4

KRB-DTT, this solution is prepared just prior to use by adding CaCl$_2$ (2.5 mM), DTT (1 mM), and BSA (1%)

Animals. Rats, fed or fasted overnight (80–250 g). Immediately before use, supplement PBS, Solution A, and Solution B with D-glucose (5.5

[23] H. Binder, G. F. Lemp, and J. Gardner, *Am. J. Physiol.* **238**, G190 (1980).

[24] L. M. Pinkus, submitted for publication in *Arch. Biochem. Biophys.*

[25] H. G. Windmueller and A. E. Spaeth, *J. Biol. Chem.* **253**, 69 (1978).

SYSTEMS OF DRUG METABOLISM IN THE INTESTINAL EPITHELIUM

System(s)	Assay substrate(s)	Primary location	Reference[a]
Cytochrome P-450 content	—	Microsomes, mitochondria	a,b,c,e,f,g,h,t
Mixed-function oxidase(s)	Aniline, benzopyrene, biphenyl	Microsomes	a,b,c,*d*,e,f,h,r,t,z
N-Demethylase	Ethylmorphine, benzphetamine	Microsomes	c,h,r
O-Deethylase	7-Ethoxycoumarin, phenacetin	Microsomes	h,*j*,*k*,r,s
O-Demethylase	Harmine, p-nitroanisole	Microsomes	*d*
Cytochrome b_5	—	Microsomes	b,e,f,g
NADPH-cytochrome reductase	—	Microsomes	a,b,c,e
Indoleamine dioxygenase	D-Tryptophan	Cytosol	x,y
Amine oxidase	Histamine, putrescine	Cytosol	q,w
N-Acetyltransferase	Aniline, sulfanilamide, isoniazid	Cytosol	e,i,*k*,p
Aryl sulfotransferase	7-Hydroxycoumarin, phenol, 1-naphthol		*d*,*j*,*k*,l
UDPglucuronosyltransferase	N-Acetyl-p-aminophenol, morphine, phenol, 2- or 4-hydroxybiphenyl, 7-hydroxycoumarin, 1- or 2-naphthol		*d*,i,*j*,*k*,l,r,u,v
Glutathione S-transferase	1-Chloro-2,4-dinitrobenzene, N-acetyl-p-aminophenol	Cytosol	m,*d*
N-Methyltransferase	Histamine	Cytosol	q
Thiol S-methyltransferase	Hydrogen sulfide, 2-thioacetanalide	Cytosol	n

Glycine N-acyltransferase	Benzoic acid	—	o
Arylesterase	4-Nitrophenyl acetate, indoxyl acetate	Microsomes	e,r
β-D-Glucuronidase	β-D-Glucuronyl-p-nitrophenol	Lysosomes, microsomes	v

[a] References: *italics*, assayed in intact cells; **boldface**, assayed in cell extracts; Roman, assayed in other preparations.
(a) S. J. Stohs, R. C. Grafström, M. D. Burke, P. W. Moldéus, and S. G. Orrenius, *Arch. Biochem. Biophys.* **177**, 105 (1976). (b) H. Hoensch, C. H. Woo, S. B. Raffin, and R. Schmid, *Gastroenterology* **70**, 1063 (1976). (c) R. S. Chhabra, R. J. Pohl, and J. R. Fouts, *Drug Metab. Dispos.* **2**, 443 (1974). (d) R. Grafström, P. Moldéus, B. Andersson, and S. Orrenius, *Med. Biol.* **57**, 287 (1979). (e) R. J. Shirkey, J. Chakraborty, and J. W. Bridges, *Anal. Biochem.* **93**, 73 (1979). (f) D. P. Jones, R. Grafström, and S. Orrenius, *J. Biol. Chem.* **255**, 2383 (1980). (g) Y. Takesue and R. Sato, *J. Biochem. (Tokyo)* **64**, 873 (1968). (h) C. L. Miranda and R. S. Chhabra, *Biochem. Pharmacol.* **29**, 1161 (1980). (i) K. Hartiala and T. Tehro, *Nature (London)* **201**, 1036 (1965). (j) J. R. Dawson and J. W. Bridges, *Biochem. Pharmacol.* **28**, 3299 (1979). (k) R. J. Shirkey, J. Kao, J. F. Fry, and J. W. Bridges, *Biochem. Pharmacol.* **28**, 1461 (1979). (l) J. R. Dawson and J. W. Bridges, *Biochem. Pharmacol.* **26**, 2359 (1977). (m) L. M. Pinkus, J. N. Ketley, and W. B. Jakoby, *Biochem. Pharmacol.* **29**, 2885 (1980). (o) N. R. Strahl and W. H. Barr, *J. Pharm. Sci.* **60**, 278 (1971). (p) J. W. Jenne, *J. Clin. Invest.* **44**, 1992 (1965). (q) S. L. Taylor and G. R. Lieber, *Comp. Biochem. Physiol.* **63C**, 21 (1979). (r) R. J. Shirkey, J. Chakraborty, and J. W. Bridges, *Biochem. Pharmacol.* **28**, 2835 (1979). (s) E. J. Pantuck, K. C. Hsiao, S. A. Kaplan, R. Kuntzman, and A. H. Conney, *J. Pharmacol. Exp. Ther.* **191**, 45 (1974). (t) K. Ichihara, I. Kasaoka, E. Kusunose, and M. Kusunose, *J. Biochem.* **87**, 671 (1980). (u) K. W. Bock and D. Winne, *Biochem. Pharmacol.* **24**, 859 (1975). (v) G. W. Lucier, B. R. Sonawone, and O. S. McDaniel, *Drug Metab. Dispos.* **5**, 279 (1977). (w) J. Kuscher, H. Richter, J. Schmidt, R. Hesterberg, C. Specht, and W. Lorenz, *Agents Actions* **3**, 182 (1973). (x) T. Shimizu, S. Nomiyama, F. Hirata, and O. Hagaishi, *J. Biol. Chem.* **253**, 4700 (1978). (y) J. S. Cook, C. I. Pogson, and S. A. Smith, *Biochem. J.* **189**, 461 (1980). (z) L. W. Wattenberg, J. L. Leong, and P. J. Strand, *Cancer Res.* **22**, 1120 (1962).

mM) and L-glutamine (2 mM); gas with 100% O_2 at 37°. Place KRB-DTT on ice; gas with 95% O_2–5% CO_2 and dispense 30 ml to each of 20 centrifuge tubes (50 ml) to be used in the collection of cells.

Procedure

STEP 1. Anesthetize a rat of about 160 g with water-saturated ether or sacrifice by a blow on the head. (CAUTION: Significant changes in cellular concentrations of metabolites result from anoxia.)[26,27] Quickly excise the small intestine (5 cm proximal from the ileocecal valve to ligament of Treitz) by applying mild traction and cutting the adherent mesentery.

STEP 2. Gently place on a moist diaper. Cut in half and force out fecal matter with warm PBS dispensed from a 50-ml syringe. (NOTE: The whole intestine can be used but is more difficult to manipulate.)

STEP 3. Clamp off the distal end of each segment, i.e., the jejunum and ileum. Fill with Solution A (about 0.25 ml per cm; 4 ml/g wet weight), maintaining mild pressure with the syringe, and clamp the proximal end.

STEP 4. Incubate each segment with gentle agitation (75 cycles per min) at 37° in oxygen-saturated PBS for 15 to 17 min. Discard Solution A.

STEP 5. Rinse and fill the segment with Solution B as in Step 3. Incubate each segment for ten intervals as follows: (1) 4 min; (2) 2 min; (3) 2 min; (4) 3 min; (5) 4 min; (6) 5 min; (7) 7 min; (8) 10 min; (9) 10 min; (10) 10 min. After each interval, place the segment in a funnel lined with Nitex 300-μm mesh,[28] remove one clamp, lift, allowing the intestine to collapse, and drain the suspension of released cells into a centrifuge tube containing KRB–DTT at 4°. Refill with Solution B. Note that the segments stretch during the procedure, sometimes requiring the injection of 10 to 15% additional Solution B at the end. However, too much pressure can cause rupture of the intestinal wall.

STEP 6. Keep cells on ice. Collect by centrifugation (300 g, 5 min). Resuspend in KRB–DTT, transfer to 15-ml graduated centrifuge tubes and centrifuge (300 g, 5 min). Record the packed cell volume of each fraction. (NOTE: Fractions 8 to 10 are largely crypt cells and should contain no more than 25% of the total volume of cells.)

STEP 7. Before use, wash the cells with KRB (minus DTT) or with the buffer appropriate to the experiment.

Comments. With a little practice the method is effective for all common strains of laboratory rats. One separation is illustrated in Fig. 1 in which alkaline phosphatase and thymidine kinase are used as marker enzymes for villous and crypt cells, respectively.[11]

[26] J. R. Bronk and H. J. Leese, *J. Physiol. (London)* **235**, 183 (1973).
[27] J. T. Brosnan, H. A. Krebs, and D. H. Williamson, *Biochem. J.* **117**, 91 (1970).
[28] Tetko, Inc., Valhalla, New York.

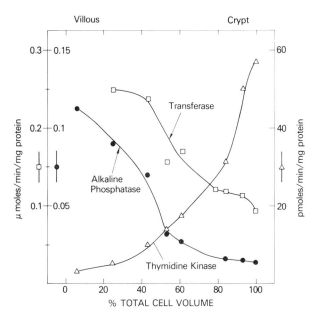

FIG. 1. Distribution of glutathione S-transferase activity in rat enterocytes. Alkaline phosphatase and thymidine kinase are used as marker enzymes for villous and crypt cells, respectively. Taken from L. M. Pinkus, J. N. Ketley, and W. B. Jakoby, *Biochem. Pharmacol.* **26**, 2359 (1977).

Very large amounts of intestinal mucus (possibly caused by infection or diet) or improper handling (twisting) of the intestine can result in a poor separation. Dithiothreitol helps reduce mucus and promotes firmer packing of cells upon centrifugation.

Suspensions of mixed enterocytes, primarily villous cells, can be prepared by several methods.[5,6,8,9,18,19] In KRB–DTT buffer, they exhibit little change in viability for at least 1 hr at 0°. Upon incubation at 37°, they usually remain viable for only 30 to 45 min and then progressively deteriorate.[18,19] However, a method for maintaining rat enterocytes in tissue culture has been described.[29]

Isolated rat enterocytes synthesize macromolecules[16,18–20] and lipids,[30–32] oxidize glutamine, ketone bodies, glucose, and other metabolic fuels,[13,18–20,33] respond to hormones with increases in cAMP[34,35]; and carry

[29] A. Quaroni and R. J. May, *Methods Cell Biol.* **21B**, 403 (1980).

[30] K. M. M. Shakir, S. G. Sundaram, and S. Margolis, *J. Lipid Res.* **19**, 433 (1978).

[31] P. J. A. O'Doherty, *Can. J. Biochem.* **58**, 527 (1980).

[32] H. Muroya, H. S. Sodhi, and R. G. Gould, *J. Lipid Res.* **18**, 301 (1977).

[33] L. M. Pinkus, *Abstr., Int. Congr. Biochem., 11th, 1979* p. 427 (1979).

[34] J. Besson, D. H. Hoa, and G. Rosselin, *C.R. Hebd. Seances Acad. Sci.* **284**, 2139 (1978).

out many characteristic reactions of drug metabolism (see table). They also serve as a convenient starting material for assays of cytochrome P-450 and for isolation of membranes.[36,37]

Guinea Pig

Reagent. Villous and crypt cells can be obtained from a fasted (18–24 hr) guinea pig by a method described by Porteous *et al.*[38]

Prepare a modified Solution A containing NaCl, 96 mM; KH$_2$PO$_4$, 8 mM; Na$_2$HPO$_4$, 5.6 mM; KCl, 1.5 mM; EDTA, 5 mM; EGTA, 5 mM; DTT, 0.5 mM; mannitol, 185 mM; and BSA, 0.25%; pH 6.8. Add glucose and glutamine and bubble with 100% O$_2$ during the isolation procedure, as above.

Procedure. Excise the intestine, cut it in half and rinse and fill it with modified Solution A. Incubate in 225 mM NaCl at 37° for 5 min. Drain, fill with cell isolation buffer (200 mM sucrose, 19 mM KH$_2$PO$_4$, 76 mM NaHPO$_4$, pH 7.4), and incubate for five intervals of 5 min each. After each interval, place the intestine on a flat surface, gently press along its entire length, drain to collect the released cells (as in Step 5 above), and refill. Early fractions contain only villous cells. After the final interval, press firmly and *roll* the intestine to release additional crypt cells; this requires some practice. The degree of enrichment in various fractions can be estimated by intraperitoneal injection of 50 μCi of [^{3}H]thymidine, 2 to 4 hr prior to sacrifice, and measuring the specific activity of alkaline phosphatase (villous cells) and [^{3}H]thymidine uptake (crypt cells).

For the isolation of mixed enterocytes, incubate with Solution A for 10 min to loosen cells and follow with Solution B (supplemented with 1% albumin) for two 5-min periods. Cells are released by pressing and rolling the intestine. Collect as in Step 5. Centrifuge, resuspend in KRB–DTT, and filter thru Nitex 125-μm mesh.[28] Wash with the appropriate buffer prior to use.

Isolated guinea pig enterocytes are virtually free of mucus and, upon incubation, remain viable much longer than rat cells (over 90 min[24,39]). Like rat cells, they catalyze many reactions of drug metabolism (see table). They have been used as starting material in the isolation of nuclei[38] and mitochondria.[13,24]

---

[35] B. Simon, H. Seitz, and H. Kather, *Biochem. Pharmacol.* **29**, 673 (1980).

[36] V. Scalera, C. Storelli, C. Storelli-Joss, W. Haase, and H. Murer, *Biochem. J.* **186**, 177 (1980).

[37] R. J. Shirkey, J. Chakraborty, and J. W. Bridges, *Anal. Biochem.* **93**, 73 (1979).

[38] J. W. Porteous, H. M. Furneaux, C. K. Pearson, C. M. Lake, and A. Morrison, *Biochem. J.* **180**, 455 (1979).

[39] J. R. Dawson and J. W. Bridges, *Biochem. Pharmacol.* **28**, 3299 (1979).

Other Species

Chick enterocytes can be isolated by the methods of Kimmich[7] or Topping and Visek[40]; mouse enterocytes by the method of Ziomek *et al.*[12]; and hamster enterocytes by a procedure of Gaginella *et al.*[41] Considerable metabolic data are available for chick enterocytes.[14,42]

The Criteria of Enterocyte Viability

The preceding procedures yield sheets or groups of columnar cells as well as more rounded individual cells and small numbers of lymphocytes and bare nuclei. Damaged cells may appear swollen or have blebs in plasma membranes.[10] As determined by dye exclusion, viability is usually between 80 to 90%. More sensitive functional criteria[43] include the capacity to maintain ion gradients for Na^+/K^+, oxygen uptake, energy charge, active sodium-dependent transport,[14] and [5-^{3}H]uridine incorporation into RNA. A rapid rate of oxidation of 2 mM [U-^{14}C]glutamine to $^{14}CO_2$, virtually unchanged in the presence of glucose, is a characteristic of rat and guinea pig enterocytes. The apparent $K_{1/2}$ for this process is 0.2–0.5 mM,[24] which is in the range of physiological blood glutamine concentration.

The following values have been confirmed in our laboratory for enterocytes of rat and guinea pig: 16–24 mg protein/mg DNA[18,20]; intracellular Na^+/K^+ ratio is 0.5–2, plus ouabain, 20–25 (in KRB); oxygen uptake, endogenous substrates, 12–15 nmol O_2/min/mg protein; maximum exogenous substrates, 25–30 nmol O_2/min/mg protein[18,19]; 2 mM [U-^{14}C]-glutamine oxidation to $^{14}CO_2$, 2–4 nmol/min/mg protein; energy charge, 0.6–0.7; cAMP (total formed in 30 min at 25°), basal, 1–10 pmol/mg protein, and hormone-stimulated, 10–300 pmol/mg protein[18,23,24]; glutamine synthetase activity (nmol/min/mg protein), rat, 0.5–1.5, and guinea pig, 8–12.[24]

Significance

Metabolism in the gut influences both the absorption and bioavailability of a variety of drugs. Although the overall level of drug-metabolizing activity is low, most of the systems of drug metabolism found in liver have also been identified in the gut epithelium[44] (see table). Furthermore, many of these systems can be induced by nutritional components, phenobarbi-

[40] D. C. Topping and W. J. Visek, *Am. J. Physiol.* **233**, E341 (1977).
[41] T. S. Gaginella, A. C. Haddad, V. L. W. Go, and S. F. Phillips, *J. Pharmacol. Exp. Ther.* **201**, 259 (1977).
[42] J. W. Porteous, *Biochem. J.* **188**, 619 (1980).
[43] H. Baur, S. Kasperek, and E. Pfatt, *Hoppe Seyler's Z. Physiol. Chem.* **356**, 827 (1975).
[44] K. Hartiala, *Physiol. Rev.* **53**, 496 (1973).

tal, aryl hydrocarbons, and ingested toxins.[45-47] Therefore techniques for the isolation of villous and crypt cells may be of value in studying not only the development of drug metabolizing systems but also their mechanism(s) of induction in the gut.

[45] E. Heitanen, M. Laitinan, and U. Koivusaari, *Enzyme* **25**, 153 (1980).
[46] C. M. Schiller, *Environ. Health Perspect.* **33**, 91 (1979).
[47] Symposium on "Target Organ Toxicity: Intestine" (C. M. Schiller, guest ed.), in *Environ. Health Perspect.* **33**, 1–126 (1979).

[19] Mammalian Nerve Cell Culture

By W. Dimpfel

Tissue culture offers a unique possibility for combining the advantages of experimentally well-controlled *in vitro* conditions with the advantage of working with living cells. Nevertheless, a great number of methodological difficulties had to be overcome with neuronal tissue before the technique could advance from a mere morphological description of living tissue *in vitro*[1] to a model system that allows meaningful statements about the function of the nervous system.[2] The technical procedure for obtaining reproducible primary nerve cell cultures as described herein has been successfully applied in several laboratories for more than 6 years. It can be regarded as the further development of the successfully introduced, and already widely used, explant cultures.[3-5]

Standard Procedure

The close interrelations between different cell types within the central nervous system (CNS) and the extensive arborization of neurites in the adult account for the difficulty in dissociating tissue into a single cell suspension that could be plated in culture dishes.

Therefore, embryonic tissue is best suited for culturing purposes. The embryos (12–14 days *in utero* for mice; 14–16 days for rats) are obtained under sterile conditions by cesarean section and placed under a stereo microscope at low magnification into prewarmed buffered saline, preferably Ca^{2+} free.

[1] R. G. Harrison, *Anat. Rec.* **1**, 116 (1906).
[2] W. Dimpfel, *Arch. Toxicol.* **44**, 55 (1980).
[3] W. Hild and I. Tasaki, *J. Neurophysiol.* **25**, 277 (1962).
[4] S. M. Crain and E. R. Peterson, *J. Cell. Comp. Physiol.* **64**, 1 (1964).
[5] R. P. Bunge, M. B. Bunge, and E. R. Peterson, *J. Cell Biol.* **24**, 163 (1965).

The brain or portions of the CNS are excised by means of iridectomy scissors and watchmaker forceps, placed into a small separate dish together with a drop of saline, and minced until the tissue pieces are smaller than 1 mm^3.[6] Warm saline, 2 ml, containing 0.6 units per ml of Dispase (Boehringer, Mannheim, FRG) is added and the dish kept at 37° for 10 min. The tissue suspension is triturated cautiously at least 5 to 6 times by means of a Pasteur pipet and diluted further to give 8–10 ml of complete medium. The suspension is gently forced through a nylon mesh (gauze: 135 μm) by means of a glass rod. At this stage, a single cell suspension should have been attained, enabling a cell count. After further dilution of the single cell suspension, approximately 10^6 cells/ml are plated onto collagen-coated plastic dishes. Careful handling, without major shaking, prevents aggregation of the cells. The dishes are placed in an automatically controlled CO_2 incubator. Either 5 or 10% CO_2 is used, depending on the amount of bicarbonate buffer in the growth medium (usually Dulbecco's modification of Eagle medium). The cultures are fed three times per week by aspirating two-thirds of the fluid and replacing it with fresh medium.

There are two reasons for retaining one-third of the fluid in the dish: the cells are conditioning the medium according to their needs; and, drying out of the cell layer is prevented, especially if 30 or 40 dishes are changed together—the usual practice needed to save time. The daily observation of one or two cultures from each dissection by phase-contrast microscopy is recommended in order to detect microbial contamination. Such contamination is always a possibility, especially if antibiotics are not used in the culture medium. Although working without antibiotics is practicable, it is difficult to achieve without contamination. On the other hand, because neurons are extremely sensitive to antibiotics, one should at least strive to use very low concentrations of antibiotics if they cannot be avoided, i.e., less than 10 units per ml of a penicillin—preferably carbenicillin because of its lower neurotoxicity.

If glial cells tend to overgrow the neurons, a mitotic inhibitor such as fluorodeoxyuridine (2 μg/ml) together with uridine (50 μg/ml) may be given for a short period (2 days) after the initial week of culture. The cultures are generally grown for 2 to 3 weeks before experiments are performed with them; this interval appears to be needed for this system of cells to reach a certain degree of "maturity" in terms of synaptogenesis. The cultures are harvested for biochemical purposes by scraping with a rubber policeman following a washing procedure, which depends on the specific purpose for which the cells are to be used.

[6] E. W. Godfrey, P. G. Nelson, B. K. Schrier, A. C. Breuer, and B. R. Ransom, *Brain Res.* **90**, 1 (1975).

By use of Petriperm dishes (Haereus, Göttingen, FRG), which have a semipermeable membrane as a bottom, cultures may be harvested simply by cutting out the bottom for further processing, i.e., for extraction or measurement of radioactivity.

Variations

There have been a number of different suggestions for improving the standard procedure for obtaining increased neuron survival. Coating the dishes with collagen[7] might be replaced by coating with polylysine.[8] Enzymatic dissociation of the cells has also been attained by using trypsin instead of Dispase.[9] The pore size of the nylon mesh used for filtration may be smaller than 135 μm, but at the expense of excluding certain larger cell types. The amount and type of serum has been a matter of debate: 10% fetal calf serum and 10% horse serum have been used in most cases, but replacement of the horse serum by neonatal calf serum or pig serum works equally well. The newest trend consists of leaving out the serum completely, in order to obtain a chemically defined growth medium.[10] Much hope is placed on the effect of insulin in combination with high glucose concentrations. Recent findings provide some optimism.[11] Addition of gangliosides to the medium is currently under trial, because these glycosphingolipids have been shown to induce neurite outgrowth *in vivo* and in tumor cell cultures.[12]

Applications

There are no major problems in the experimental design of morphological techniques with these dissociated cell cultures. Feasibility and interpretation of the results depend largely on the identification of particular cell types. Two major lines of approach can be observed with regard to methodological aspects. The first consists of improvement of "mixed" cultures that contain neurons and the variety of glial cells, ependymal cells, and fibroblasts. The objective is to retain, as nearly as possible, the normal physiological interactions between the different cell types. Such studies aim at the use of a physiological system as a tool for the study of drug activity at the cellular level. The close similarity of these types of

[7] M. B. Bornstein, *Lab. Invest.* 7, 134 (1958).
[8] E. Yavin and Z. Yavin, *J. Cell Biol.* 62, 540 (1974).
[9] E. L. Giller, Jr., B. K. Schrier, A. Shainberg, H. R. Fisk, and P. G. Nelson, *Science* 182, 588 (1973).
[10] P. Honnegger, D. Lenoir, and P. Favrod, *Nature (London)* 282, 305 (1979).
[11] E. Y. Snyder and S. U. Kim, *Brain Res.* 196, 565 (1980).
[12] W. Dimpfel, W. Möller, and U. Mengs, *Arch. Pharmacol.* 308, Suppl., K46 (1979).

cultures to *in vivo* features has been described in a series of papers dealing with morphological, biochemical, and electrophysiological characteristics,[13-16] and the successful use of this approach in studying toxin or drug actions has emerged as a completely new area of pharmacology.[17] The second approach attempts to separate the different cell types before culturing in order to achieve pure (mostly neural or glial) cultures. An alternative method of achieving this goal is to kill specific cell types within a mixed culture (by use of specific toxins or antibodies). These studies may allow new insights into cell-to-cell interactions and into the trophic requirements. Interactions between nerve and muscle cell have already received attention,[9] particularly with respect to receptor sensitivity.[18] In addition, trophic or nutritional factors in the CNS have been studied with regard to myelination.[19]

Biochemical analysis is inherently difficult because of the low amount of tissue that can be obtained from such cultures. The usual 60-mm-diameter dish contains only about 1.5 mg of protein. A partial solution to the problem is the use of isotopes in order to increase sensitivity. Another serious limitation, which applies only to the mixed cultures, is cell heterogeneity. The only way of approaching this goal is to quantitate the particular cell types and to correlate the observed measurements to the relative amount of a specific cell type under study. The discovery of tetanus toxin as a highly selective marker for neurons is currently simplifying research in this direction.[20] Other cell types have been quantitatively identified by immunological methods, for which radioimmunoassays must be developed before quantitation can be achieved.[21,22] The choice of the culture system thus depends on the type of question asked.

The "window to the brain"—as the dissociated cultures have been called—also offer advantages to electrophysiologists. Impalement of morphologically distinguishable cells under complete visual control by mi-

[13] B. R. Ransom, E. Neale, M. Henkart, P. N. Bullock, and P. G. Nelson, *J. Neurophysiol.* 40, 1132 (1977).
[14] B. R. Ransom, C. N. Christian, P. N. Bullock, and P. G. Nelson, *J. Neurophysiol.* 40, 1157 (1977).
[15] B. R. Ransom, P. N. Bullock, and P. G. Nelson, *J. Neurophysiol.* 40, 1163 (1977).
[16] P. G. Nelson, B. R. Ransom, M. Henkart, and P. N. Bullock, *J. Neurophysiol.* 40, 1178 (1977).
[17] W. Dimpfel, *Exp. Neurol.* 65, 53 (1979).
[18] G. D. Fischbach and S. A. Cohen, *Dev. Biol.* 31, 147 (1973).
[19] S. U. Kim and D. Pleasure, *Brain Res.* 145, 15 (1978).
[20] W. Dimpfel, J. H. Neale, and E. Habermann, *Naunyn-Schmiedeberg's Arch. Pharmacol.* 290, 329 (1975).
[21] M. Schachner and M. Willinger, *Prog. Brain Res.* 51, 23 (1979).
[22] M. C. Raff, J. P. Brockes, K. L. Fields, and R. Mirsky, *Prog. Brain Res.* 51, 17 (1979).

croelectrodes with tip diameters far less than 1 μm are reported by a number of laboratories. For these purposes, the culture dish is mounted on the fixed stage of an inverted phase contrast microscope and kept at 37°. Microelectrodes are held by micromanipulators and are guided into the cells either by free hand movement or by the aid of automatic microdrives. The electronic equipment is conventional and has been used widely.[23] Because neurons usually sit on top of a monolayer consisting of several other cell types, they are easily accessible although extremely fragile. This means that electrodes of at least 30 to 50 megohms (3 M KCl) have to be drawn in order to give a very fine tip. With experience it is even possible to insert two electrodes into the same cell. As neurons in these cultures mature to the stage at which synaptic connections are formed, it is also possible to record synaptic potentials, either occurring spontaneously or by stimulation of a different cell by a second electrode. One of the most exciting techniques currently applied is the use of iontophoresis of drugs or transmitters in order to determine the sensitivity of a specific neuron.[23] The possibilities provided by these cultures should facilitate examination of drugs acting at the neuronal membrane or at synapses.

[23] J. L. Barker and B. R. Ransom, *J. Physiol. (London)* **280**, 331 (1978).

Section II

Enzyme Preparations

[20] 4-Nitrophenol UDPglucuronyltransferase (Rat Liver)

By BRIAN BURCHELL and PHILIP WEATHERILL

4-Nitrophenol + UDPglucuronate $\rightleftharpoons$ 4-nitrophenyl-*O*-glucuronide + UDP

Liver microsomal UDPglucuronyltransferases catalyze the glucuronidation of a wide variety of endogenous and xenobiotic compounds.[1] For the present, our knowledge of the number of multiple forms and their substrate specificity is inadequate to justify a strict nomenclature based on the function of these versatile enzymes. However, multiple forms of the transferase probably have evolved specifically to metabolize endogenous compounds in various tissues and each form of transferase so far isolated might be tentatively classified as specifically glucuronidating a single endogenous substrate.[2]

A method is described here for the purification of nitrophenol UDPglucuronyltransferase to apparent homogeneity and is based on that previously described.[3-5] Other procedures for preparation of this enzyme have been reported.[6,7]

Assay Method

Principle. 4-Nitrophenol exhibits an absorption maximum (405 nm, alkaline pH) that disappears on formation of the colorless ether glucuronide. The assay procedure is based on this decrease of absorbance at 405 nm; the original spectrophotometric method[8] has been modified.[9]

Reagents

UDPglucuronate, triammonium salt (Sigma Chemical Co.), 20 mM, adjusted to pH 7.4 with sodium hydroxide

4-Nitrophenol, 1 mM, in 0.5 M Tris-maleate, pH 7.4, containing, 10 mM MgCl$_2$

Trichloroacetic acid, 0.5 M

Sodium hydroxide, 2 M

[1] G. J. Dutton and B. Burchell, *Prog. Drug. Metab.* **2**, 1 (1977).

[2] B. Burchell, *Rev. Biochem. Toxicol.* **3**, 1 (1981).

[3] B. Burchell, *Biochem. J.* **161**, 543 (1977).

[4] B. Burchell, *FEBS Lett.* **78**, 101 (1977).

[5] B. Burchell, *Biochem. J.* **173**, 749 (1978).

[6] J. P. Gorski and C. B. Kasper, *J. Biol. Chem.* **252**, 1336 (1977).

[7] K. W. Bock, D. Josting, W. Lilienblum, and H. Pfeil, *Eur. J. Biochem.* **98**, 19 (1979).

[8] K. J. Isselbacher, M. F. Chrabas, and R. C. Quinn, *J. Biol. Chem.* **237**, 3033 (1962).

[9] A. Winsnes, *Biochim. Biophys. Acta* **191**, 279 (1969).

METHODS IN ENZYMOLOGY, VOL. 77

Procedure. The standard incubation mixture containing 125 μl Tris-maleate and 1 mM 4-nitrophenol at pH 7.4, along with enzyme and water in a final volume of 200 μl, is warmed to 37°. The reaction is initiated by the addition of 50 μl of 20 mM UDPglucuronic acid.

After incubation at 37° for 10 min, 0.5 ml of ice-cold (0.5 M) trichloroacetic acid is added and the mixture is rapidly vortexed and kept in ice for 5 min before removal of the protein pellet by centrifugation at 2000 g for 10 min. The supernatant solution, 400 μl, is mixed with 400 μl of 2 M sodium hydroxide and then diluted by addition of 2.2 ml of distilled water. Absorbance is measured at 405 nm to assess the reduction in color caused by the formation of 4-nitrophenylglucuronide. The extinction coefficient of 4-nitrophenol at pH >10 is 18.1 $\times$ 10^3 cm^2 mol^{-1}. One unit of transferase activity represents the formation of 1 nmol 4-nitrophenylglucuronide per min.

Purification Procedure

The table summarizes the results obtained in this laboratory over a 3-year period.[3,5] Protein concentrations are determined by the method of Bradford.[10]

Step 1. Preparation of the Microsomal Fraction. Male Wistar rats (200–250 g) are given 2 g sodium phenobarbital per liter in their drinking water before the animals are killed. All subsequent operations are performed at 4°. Livers from five rats are chopped with scissors and homogenized in 3 volumes of ice-cold 0.25 M sucrose by three strokes of a Teflon–glass homogenizer. This homogenate is centrifuged for 10 min at 10,000 g and the resulting supernatant fluid is centrifuged at 105,000 g for 60 min. The microsomal pellets are collected and immediately used for further purification.

Step 2. Solubilization. The microsomal fraction is suspended in 1% (w/v) Lubrol 12A9 (ICI Organics Division) and 0.2 M potassium phosphate at pH 7.0 by gentle homogenization and mixed for 10 min. One milliliter of detergent solution is used per 20 mg of microsomal protein. The suspension is centrifuged at 105,000 g for 60 min. The supernatant fluid contains about 85–90% of the UDPglucuronyltransferase activity toward six aglycone substrates.[3,5,11]

Step 3. Salt Fractionation. Finely ground solid ammonium sulfate is slowly added to the Lubrol-soluble supernatant liquid to 25% saturation (13.4 g/100 ml). The mixture is stirred for 20 min and centrifuged for 15 min at 15,000 g giving three phases: a floating viscous pink layer; a clear

[10] M. M. Bradford, *Anal. Biochem.* **72**, 248 (1976).
[11] P. J. Weatherill, Ph.D. Thesis, University of Dundee (1980).

PURIFICATION OF LIVER MICROSOMAL 4-NITROPHENOL UDPGLUCURONYLTRANSFERASE

Fraction	Volume (ml)	Protein (mg)	Activity (nmol/min)	Specific activity[a] (units/mg protein)
1 Microsomes	—	2880	5760	2
2 Lubrol extract[b]	143	2304	89,856	39 (32–54)
3 Salt fractionation	94	1026	55,423	54 (34–64)
4 DEAE-cellulose	15	397	32,256	81 (50–94)
5 CM-cellulose	5	162	14,465	89 (45–110)
6 DEAE-Sephadex	23	18	4822	262 (217–289)
7 UDP-hexanolamine-Sepharose	6	0.5	1441	2928 (2261–3336)

[a] The value in parentheses is the range of specific activity obtained in a series of purification experiments.

[b] Note that Lubrol extraction of microsomes causes a large activation rather than purification of the transferase.

red intermediate phase; and a very small pink pellet. The red intermediate phase is collected using a peristaltic pump; the other precipitates are discarded. Solid ammonium sulfate is added to the red solution to 60% saturation (22.7 g/100 ml) and the resultant mixture stirred for 20 min. The precipitate obtained by centrifugation at $15,000\,g$ for 20 min is dissolved in Buffer A (0.05% Lubrol,[12] 25 mM potassium phosphate, 2 mM EDTA, and 5 mM 2-mercaptoethanol, adjusted to pH 7.5) and dialyzed against 50 volumes of the same buffer overnight. This fraction is stable for several days when stored on ice.

Step 4. DEAE-Cellulose. The clear red ammonium sulfate fraction is applied (60 ml/hr) to a column (6 × 20 cm) of DEAE-cellulose (Whatman DE-52) previously equilibrated with Buffer A. UDPglucuronyltransferase activity toward 4-nitriphenol appears in the first protein fractions eluted from the column by Buffer A and is separated from the majority of the red material (cytochromes b_5, P-450, and P-420). The fractions containing transferase activity are pooled, concentrated 5-fold by vacuum dialysis, and dialyzed overnight against 50 volumes of Buffer B (0.05% Lubrol, 5 mM potassium phosphate, 5 mM 2-mercaptoethanol, adjusted to pH 6.5).

Step 5. CM-Cellulose. The dialyzed solution applied (60 ml/hr) to a column (3 × 12 cm) of CM-cellulose (Whatman CM-52) previously equilibrated with Buffer B. The remaining red material in the DE-52 fraction binds in a narrow band to the top of this column, whereas the yellow solution containing transferase activity is eluted. The active fractions are

[12] Lubrol 12A9, 0.05% (w/v), does not exhibit a significant absorbance at 280 nm and thereby facilitates tracing of protein profiles during chromatography.

pooled and concentrated to 3 to 5 ml by vacuum dialysis and dialyzed overnight against 150 volumes of Buffer A.

Step 6. DEAE-Sephadex. The greenish-yellow protein solution is applied (40 ml/hr) to a column (3.5 × 25 cm) of DEAE-Sephadex (Pharmacia), previously equilibrated with Buffer A. The green/yellow material remains at the top of this column, whereas transferase activity is eluted by Buffer A. These colorless fractions are pooled and stored at 0° overnight.

Step 7. UDP-Hexanolamine-Sepharose. UDP-hexanolamine-Sepharose 4B is prepared by methods described in the next section.

The preparation from Step 5 (approximately 10 ml, containing 1 protein per ml) is routinely applied (5 ml/hr) simultaneously to two separate UDP-hexanolamine-Sepharose 4B columns (2 × 6 cm) previously equilibrated with Buffer A containing 5 mM magnesium chloride.

Approximately 90% of the protein sample routinely applied and about 30% of the transferase activity is eluted from the column with Buffer A. Approximately 50% of the bound UDPglucuronyltransferase is eluted with 3 ml of 5 mM UDPglucuronic acid in Buffer A.[13] The purified transferase exhibiting the highest purity is applied to a small DEAE-Sephadex column (1 × 10 cm) previously equilibrated with Buffer A for removal of the nucleotide. The enzyme is eluted on the void volume with Buffer A, whereas UDPglucuronic acid is considerably retarded.

Synthesis of UDP-Hexanolamine-Sepharose 4B

UDP-hexanolamine is the ligand required for the affinity chromatography step in the purification of UDPglucuronyltransferase. The synthesis is composed of several steps, each of which is the manufacture of an essential intermediate. The methods used for Steps A to F are as described by Barker *et al.*[14] and Sadler and Hill.[15]

When anhydrous solvents are used, these are dried over 4-Å molecular sieves for at least 48 hr before use. Crystalline phosphoric acid and 6-amino-1-hexanol are stored over dessicant at room temperature. N,N'-Carbonyldiimidazole is purchased before each synthesis and the melting point ascertained as being approximately 114° (not less than 110°) before use. Pyridinium Dowex is prepared by treating the acid form with 2.5 M pyridine for 48 hr and then washing thoroughly with distilled water.[16]

---

[13] The remaining protein and transferase may be eluted with 2 M KCl in Buffer A, although this nonspecific eluate is impure.

[14] R. Barker, K. W. Olsen, J. W. Shaper, and R. L. Hill, *J. Biol. Chem.* **249**, 7135 (1972).

[15] J. E. Sadler and R. L. Hill, personal communication.

[16] I. P. Trayer, H. R. Trayer, D. A. P. Small, and R. C. Bottomley, *Biochem. J.* **139**, 609 (1974).

Ascending paper chromatography, using isobutyric acid : concentrated ammonium hydroxide : water (66 : 1 : 33) as solvent, is performed to assess each stage of the synthesis.

Nucleotide-containing spots are visualized under ultraviolet light. Chromatograms are developed with ninhydrin to detect amines and the Hanes–Isherwood molybdate spray for the detection of phosphates[17] using a UV lamp to develop the spots that contain phosphate.

A. *Synthesis of O-Phosphoryl-6-Amino-1-Hexanol (Hexanolamine Phosphate).* 6-Amino-1-hexanol (23.4 g, 0.2 mol) and crystalline phosphoric acid are mixed in a round-bottom flask that is connected via a condenser and a vacuum adaptor to another round-bottom flask. The vacuum adaptor is linked via a dry ice–ethanol trap to a vacuum pump and the apparatus evacuated to 0.1 mm Hg.

The reaction vessel is heated in a sand bath to 150 ± 5° for 24 hr. The opaque, tan product is cooled to room temperature, dissolved in 400 ml distilled water, and the pH of the solution is adjusted to pH 10.5 with 4 M lithium hydroxide. Unreacted phosphoric acid precipitating as lithium phosphate is removed by filtration through a pad of Celite on a scintered glass filter. The clear solution is adjusted to pH 3.0 with glacial acetic acid and changed onto a 300-ml column of Dowex 50W-X8 (20–50 mesh) pyridinium cycle. Fractions that contain ninhydrin-positive material are concentrated to dryness under reduced pressure and the residue is heated in a vacuum oven at 100° for 2–3 hr to remove pyridinium acetate. The resulting brown solid is dissolved in 200 ml distilled water and filtered through a Whatman GF-C glass-fiber filter. This solution is stirred into 1 liter of absolute ethanol and left for 24 hr at 4°.

The precipitated hexanolamine phosphate is removed from the ethanol by filtration. After drying, the hexanolamine phosphate is a cream-colored, flaky solid with a melting point of 244°, in 41% yield (16.4 g). Paper chromatography gave a single spot of R_f 0.57 with both ninhydrin and Hanes–Isherwood phosphate reagent. This product is used in the next stage of the synthesis without further recrystallization.

B. *Synthesis of N - Trifluoroacetyl - O - Phosphoryl - 6 - Amino - 1 - Hexanol (TFA-Hexanolamine Phosphate).* This reaction is carried out in a hood. Hexanolamine phosphate (1.0 g, 5.08 mmol) is added to 10 ml distilled water and the mixture is cooled in an ice bath. The pH is adjusted to 9.5 with 5 M potassium hydroxide to dissolve the hexanolamine phosphate. Ethyltrifluorothiolacetate (0.5 ml, 3.9 mmol) is added to this solution, and finely dispersed by vigorous stirring. The pH is maintained between 9.0 and 9.5 by the addition of 5 M potassium hydroxide. Another 0.5 ml ethyltrifluorothiolacetate was added after 1 hr and the stirring is con-

[17] R. S. Bandurski and B. Axelrod, *J. Biol. Chem.* **193**, 405 (1951).

tinued. The reaction is judged as being complete when a barely percepti-
ble pink spot resulted from a ninhydrin test. If necessary a third 0.5 ml
ethyltrifluorothiolacetate is added. The reaction was usually complete
after 2–3 hr if the pH was maintained within the range 9.0 to 9.5.

On completion the reaction mixture was adjusted to pH 5.0 using
trifluroroacetic acid and concentration to dryness under reduced pressure
at 40–45°. The residue is dissolved in 10 ml of distilled water and concentrated to dryness twice more to remove residual reagents. The product is
dissolved in 10 ml distilled water, cooled to room temperature. The solution is adjusted to pH 1.5 with trifluoroacetic acid and passed over a 70-ml
Dowex 50W-X8 (20–50 mesh) H$^+$ cycle column to remove cations and
unreacted amine. Fractions with pH less than 2.5 are pooled, concentrated to a light yellow oil under reduced pressure at 40–45°, and dissolved
in dimethylformamide to a final volume of approximately 5.0 ml.

Paper chromatography of the product gave a single ninhydrin-
negative, Hanes–Isherwood-positive spot with R_f 0.72.

C. *Synthesis of N-Trifluoro-O-Phosphoryl-6-Amino-1-Hexanolimidazo-
lide (TFA-Hexanolamine Phosphate Imidazolide)*. The TFA-hexanolamine phosphate in dimethylformamide is concentrated to a minimum
volume at 40° to 45° under reduced pressure. The resulting viscous liquid is dissolved in 5 ml anhydrous dimethylformamide and concentrated
in the same manner. This procedure is repeated twice to remove remaining water. *N,N'*-Carbonyldiimidazole (1.6 g, 10 mmol) is added, followed
by 5 ml dimethylformamide and the mixture is stirred under a CaSO$_4$
drying tube at room temperature until the *N,N'*-carbonyldiimidazole is
dissolved. The reaction is left for 4–6 hr at room temperature. Anhydrous
methanol (0.2 ml, 5 mmol) is added and the reaction left at room temperature for 20 min to hydrolyze any remaining *N,N'*-carbonyldiimidazole.
The solution is concentrated at 40–45° under reduced pressure for 5 min to
remove excess methanol and carbon dioxide and then used immediately in
Stage E.

Paper chromatography shows complete conversion of TFA-hexanolamine phosphate to a faster moving, ninhydrin-negative, Hanes–Isherwood-positive, spot at R_f 0.81.

D. *Synthesis of Uridine-5'-Monophosphoric Acid, Tributylamine Salt
(UMP-TBA)*. Uridine-5'-monophosphoric acid, sodium salt (2 g, 2.84
mmol) is dissolved in 10 ml of distilled water, applied to a 2.0 ml Dowex
50W-X8 (20–50 mesh) pyridinium cycle column and eluted with distilled
water. Fractions with an absorbance at 280 nm of greater than 0.4 are
pooled and concentrated under reduced pressure to approximately 50 ml.
Tributylamine (1.4 ml, 5.86 mmol) is added along with 40 ml of ethanol to
aid the solution of tributylamine. The solution is concentrated to dryness

under reduced pressure to yield a clumped white powder. Dimethylformamide (10 ml) is added and the suspension taken to dryness. This procedure is repeated an additional three times to remove any remaining water. The fine, white powder is used immediately in Stage E.

E. *Synthesis of* P^1-*(N-Trifluoroacetyl-6-Amino-1-Hexyl)*-P^2-*(5'-Uridine)* *Pyrophosphate* *(TFA-Hexanolamine-UDP)*. The TFA- hexanolamine imidazolide is added to the dry UMP-TBA along with 5 ml of anhydrous dimethylformamide. Two 5-ml rinses of the TFA-hexanolamine imidazolide flask are added and the suspension stirred until the UMP-TBA is dissolved. The flask is covered with aluminum foil and left for 24 to 36 hr at room temperature under a $CaSO_4$ drying tube.

Therefore, the solution is diluted with an equal volume of distilled water and applied to a 50-ml column of Dowex 1-X2 Cl^- cycle equilibrated with 50% (v/v) aqueous methanol. Fractions are monitored for absorbance at 271 nm by diluting 10 μl into 1.0 ml of 30 mM sodium hydroxide. The column is washed with 50% (v/v) aqueous methanol until the absorbance of the diluted fraction at 271 nm is zero. The column is developed with a linear gradient of 250 ml 0.01 M HCl and 250 ml 0.01 M HCl/0.4 M lithium chloride.

TFA-hexanolamine-UDP is identified by alkaline hydrolysis and paper chromatography and is eluted by 0.17–0.3 M LiCl. A single UV-positive, ninhydrin-negative, Hanes–Isherwood-positive spot of R_f 0.43 is converted to a UV-positive, ninhydrin-positive, Hanes–Isherwood-positive spot with R_f 0.52 after hydrolysis.

F. *Synthesis of* P^1-*(6-Amino-1-Hexyl)*-P^2-*(5'-Uridine)* *Pyrophosphate* *(UDP-Hexanolamine)*. The TFA-hexanolamine-UDP is adjusted to pH 11.6 using 5 M KOH and incubated at 37° until a single ninhydrin-positive, Hanes–Isherwood-positive spot R_f 0.52 is observed after paper chromatography. The hydrolyzed material is concentrated to dryness under reduced pressure, dissolved in 10 ml of water, adjusted to pH 7.0, and taken to dryness under reduced pressure. The UDP-hexanolamine is stored in a vacuum desiccator at $-20°$. The final yield of UDP-hexanolamine is approximately 42% (2.15 mmol).

G. *Coupling of UDP-Hexanolamine to CNBr-Activated Sepharose.* CNBr-activated Sepharose CL-4B (Pharmacia), 8 g, is swollen and washed with 1 mM HCl. UDP-hexanolamine solution (30 mM) is prepared in 0.2 M sodium carbonate at pH 10.0; 8 ml are added to the moist gel. The pH of the slurry is between 9.4 and 9.7. The slurry is tumbled for 18 hr at 4°. The gel is recovered, washed with 400 ml 0.1 M sodium carbonate at pH 9.0, and the absorbance of the wash is measured at 262 nm. Assuming a molar extinction coefficient of 10,000, the amount of UDP-hexanolamine bound is determined by comparison of the amount present at the start of the incubation and the amount recovered in the wash. Approximately 90% of the UDP-hexanolamine (6–7 μmol UDP-hexanolamine/g of moist gel) is routinely bound to the Sepharose 4B. The gel is washed with 400 ml of a solution of 0.05% (w/v) Lubrol, 25 mM potassium phosphate, 2 mM EDTA, 5 mM 2-mercaptoethanol, 5 mM MgCl$_2$, adjusted to pH 7.4, before use.

Properties

Purity and Stability. SDS gel electrophoresis reveals a single polypeptide of 57,000 daltons.[4,5] Antiserum raised in rabbits against the transferase, when reacted with a crude ammonium sulfate fraction of solubilized hepatic microsomes by double diffusion analysis in agarose, formed a single sharp immunoprecipitin line, which suggests that the immunogen contained only one polypeptide species.[5]

The enzyme preparation has a half-life of 18 to 20 days when stored at 0°. Addition of small phospholipid vesicles[18] (1 mg egg lecithin/mg enzyme) considerably improves the stability of the transferase. Storage by freezing at −20° in 20% (w/v) glycerol results in a complete loss of activity. The enzyme is also thermolabile at 37° for longer than 15 min. Dialysis and vacuum dialysis result in a loss of up to 40% of enzyme activity.[5]

Aglycone Substrate Specificity

The true substrate specificity of this purified transferase is only revealed after the protein has been reconstituted with phosphatidylcholine. Addition of phospholipids do not increase the activity of freshly purified transferase toward 4-nitrophenol, but reveal previously undetectable activity toward testosterone.[2,11,19] The rates of glucuronidation of other substrates by this preparation, in the absence of phospholipids, are 1-naphthol (4512 units/mg), 2-aminophenol (38 units/mg), 2-aminobenzoate (20 units/mg), and morphine (158 units/mg).

Sugar Nucleotide Substrate Specificity

The purified transferase appears to exhibit absolute specificity toward UDPglucuronic acid as the sugar nucleotide. The enzyme does not catalyze the glucosidation or galacturonidation of 4-nitrophenol.[5]

---

[18] B. Burchell and T. Hallinan, *Biochem. J.* **171**, 821 (1978).
[19] P. J. Weatherill and B. Burchell, *Biochem. J.* **189**, 377 (1980).

Activators and Inhibitors

Transferase activity is increased approximately 30% by 2 to 20 mM magnesium ions whereas EDTA has no effect. Diethylnitrosamine and alkyl ketones, but *not* UDP-N-acetylglucosamine, activate the purified enzyme 2- to 3-fold toward 2-aminophenol as substrate. The transferase is not inhibited by 0.11 mM bilirubin, 2 mM N-ethylmaleimide, or 0.05% Lubrol 12A9.

[21] Estrone and 4-Nitrophenol UDPglucuronyltransferases (Rabbit Liver)

By Robert H. Tukey and Thomas R. Tephly

UDPglucuronic acid + R-OH → R-O-glucuronic acid + UDP

Assay Methods

Principle. Catalytic activity is measured by radiometric assays based on the conjugation of [4-^{14}C]estrone or 4-nitro[U-^{14}C]phenol with glucuronic acid. Unreacted substrate is removed by organic solvent extraction and the remaining water-soluble, radiolabeled glucuronides quantified by liquid scintillation spectrometry.

Reagents

Tris-HCl, 0.5 M, pH 8.0, containing 0.1 M MgCl$_2$
Tris-HCl, 0.5 M, pH 7.5, containing 0.1 M MgCl$_2$
Estrone, 5 mM, in dimethyl sulfoxide
[4-^{14}C]Estrone (from New England Nuclear Corp.), specific activity 60 mCi/mmol, diluted in propylene glycol to 800,000 cpm/ml.
4-Nitro[U-^{14}C]phenol (from California Bionuclear), specific activity 24 mCi/mmol, diluted to 800,000 cpm ml in 10 mM 4-nitrophenol
Tris-HCl, 2 M, pH 9.0, containing 4 mM 4-nitrophenol
UDPglucuronic acid (UDPGA), 10 mM and 30 mM

Procedure. [4-^{14}C]Estrone in benzene/ethanol (9 : 1) is added to propylene glycol and the solvent evaporated under a gentle stream of nitrogen. A sample of the propylene glycol mixture (1.25 ml) and 250 μl of the estrone solution in dimethyl sulfoxide are mixed with 12.5 ml of 0.5 M Tris-HCl at pH 8.0, and brought to a volume of 50 ml in a volumetric flask.

The standard incubation reactions are carried out in 15-ml glass-stoppered centrifuge tubes with 400 μl of the [4-^{14}C]estrone solution (final concentration; 50 mM Tris-HCl, pH 8.0, containing 10 mM MgCl$_2$, 10 μM estrone, 50,000 cpm [4-^{14}C]estrone), 100 μl of the 10 mM UDPglucuronic acid solution, and enzyme in a final volume of 1 ml. When assaying activity in fractions taken after column chromatography, 1.0 mM UDPglucuronic acid is used. For specific activity determinations of the various pooled fractions, 3 mM UDPglucuronic acid is used. Reaction mixtures are incubated at 37° for 5 min and terminated by addition of 10 ml of water-saturated dichloromethane. The mixtures are briefly shaken for 10 sec and centrifuged for 5 min at 1500 rpm with a clinical centrifuge to separate the phases. The aqueous phase (500 μl), which contains the [^{14}C]estrone glucuronide, is placed in a scintillation vial; 10 ml of Aquasol 2 (New England Nuclear) are added, and the radioactive glucuronide is determined with liquid-scintillation counting techniques.

4-Nitrophenol UDPglucuronyltransferase activity is also analyzed by a radiochemical method. Incubation mixtures contain 100 μl of Tris-HCl at pH 7.5, 100 μl of the 4-nitro[U-^{14}C]phenol solution and 100 μl UDPglucuronic acid (1 mM or 3 mM as described for estrone UDPglucuronyltransferase). Reactions are initiated by adding an enzyme solution to yield a final volume of 1.0 ml. Incubations are conducted for 10 min at 37° and terminated by adding 0.5 ml of 10% (w/v) trichloroacetic acid. The precipitated protein is removed by centrifugation and a 1-ml sample of the clear supernatant fluid is added to 500 μl of the 2 M Tris buffer. Unreacted 4-nitrophenol is quantitatively removed by extracting three times with 15 ml of diethyl ether. The aqueous phase (500 μl), which contains 4-nitro[U-^{14}C]phenol glucuronic acid, is quantitated with a liquid-scintillation counter.

When estrone and 4-nitrophenol UDPglucuronyltransferase activities are assayed from the fractions following chromatography on UDP-hexanolamine Sepharose 4B, 25 μg of bovine phosphatidylcholine (Sigma) are added to the incubation mixtures. The addition of phospholipid is important because both enzyme activities have been shown to be dependent on phospholipid.[1]

One unit of either estrone or 4-nitrophenol UDPglucuronyltransferase is defined as the amount catalyzing the formation of 1.0 nmol of product per minute. Protein determinations using microsomes are assayed by the method of Lowry et al.[2] Protein determinations on DEAE-cellulose and affinity chromatography column fractions are calculated using the Bio-

[1] R. H. Tukey and T. R. Tephly, Life Sci. **27**, 2471 (1980).
[2] O. H. Lowry, N. J. Rosebrough, A. L. Farr, and R. J. Randall, J. Biol. Chem. **193**, 265 (1951).

Rad protein assay, as developed by Bradford,[3] and standardized using microsomal protein.

Preparation of Microsomes

Female rabbits weighing approximately 3 kg are starved for 48 hr and killed by a sharp blow to the head. Their livers are perfused with ice-cold 1.15% KCl to remove blood. The tissue is homogenized with 5 parts (w/v) of 1.15% KCl in a hand homogenizer (loose fit) followed by homogenization with a motor-driven Potter-Elvejhem homogenizer. The homogenate is centrifuged for 15 min at 3000 rpm in a Sorval RC-2B centrifuge using the GSA rotor. The supernatant liquid is decanted, filtered through cheese cloth, and centrifuged for 15 min at 9000 rpm. The resulting supernatant suspension is filtered through cheese cloth and centrifuged for 2 hr at 105,000 g. The microsomal pellets are suspended in 1.15% KCl, gently homogenized with a hand homogenizer (loose fit), and centrifuged for 60 min at 105,000 g. The supernatant fluid is decanted, the pellets overlayed with 5 ml of 1.15% KCl, and stored frozen at $-70°$.

Purification

Buffers

Buffer A: 25 mM Tris-HCl, pH 8.0, containing 20% glycerol and 10^{-4} M dithiothreitol

Buffer B: 25 mM Tris-HCl, pH 8.0, containing 20% glycerol, $10^{-4} M$ dithiothreitol, and Emulgen 911 (0.05%, Kao Atlas Ltd.)

Buffer C: 15 mM Tris-HCl, pH 8.0, containing 20% glycerol, $10^{-4}M$ dithiothreitol, and Emulgen 911 (0.05%)

Buffer D: 20 mM Tris-HCl, pH 8.0, containing 20% glycerol, and $10^{-4}M$ dithiothreitol

Buffer E: 20 mM Tris-HCl, pH 8.0, containing 20% glycerol, $10^{-4} M$ dithiothreitol, and Emulgen 911 (0.05%)

PURIFICATION PROCEDURES

Two procedures have been developed to separate and purify enzymes that catalyze either the glucuronidation of estrone or 4-nitrophenol. Procedures I[4] and II[5] differ primarily in the manner in which microsomes are solubilized and in the quantities of protein applied to a column of DEAE-

[3] M. M. Bradford, *Anal. Biochem.* **72**, 248 (1976).

[4] R. H. Tukey and T. R. Tephly, *Arch. Biochem. Biophys.* **209**, 565 (1981).

[5] R. H. Tukey, R. Robinson, B. Holms, and T. R. Tephly, submitted for publication.

cellulose. The final purification step, affinity chromatography, is conducted under identical conditions for both transferases independent of the procedure chosen for the partial purification and separation of the enzyme activities. Procedure I provides a greater quantity of both enzymes. Procedure II is the more rapid procedure, allowing both enzyme activities to be purified in 2 days. Tables I and II show the results of each step of a typical purification by Procedures I and II, respectively.

Purification Procedure I. Microsomes, 4.5 g of protein, are suspended in Buffer A to a final concentration of 10 mg of protein/ml of buffer. Sodium cholate is slowly added to a final concentration of 1 mg of sodium cholate/mg protein, and the mixture stirred on ice for 30 min. The preparation is centrifuged at 105,000 g for 2 hr and the supernatant liquid is retained. Solid ammonium sulfate is added to 35% of saturation (19.9 g/100 ml), stirred on ice for 30 min, centrifuged at 16,000 rpm in a Sorval RC-5 centrifuge for 15 min, and the precipitate discarded.

Ammonium sulfate is slowly added to the supernatant to 70% of saturation (22.2 g/100 ml), stirred on ice for 30 min, and centrifuged at 16,000 rpm for 25 min. The precipitate is suspended in Buffer A to a concentration of 10 mg protein/ml, immediately passed through a column (2.5 × 50 cm) of Sephadex G-25, and dialyzed against 2 liters of Buffer A with at least three changes over 24 hr.

After centrifugation of the suspension at 30,000 g for 1 hr, Emulgen 911 is added to the suspension to give a concentration of 50 μg of detergent per mg of protein. This preparation is slowly placed on a column (2.5 × 35 cm) of DEAE-cellulose (DE-52) previously equilibrated with Buffer B. (It is important to assure that the DE-52 has been properly activated by washing the resin in 500 mM Tris-HCl, pH 8.0, overnight. The ionic strength of the equilibration buffer must be 25 mM with respect to Tris-HCl at pH 8.0. If the ionic strength is 20 mM, for example, separation of the enzyme activities will not occur using this procedure.) The column is washed with Buffer B, which results in elution of 93% of the recovered estrone UDPglucuronyltransferase activity (Fraction A) and 5 to 10% of the 4-nitrophenol UDPglucuronyltransferase activity. With a 0–0.25 M linear KCl gradient, the remaining 4-nitrophenol UDPglucuronyltransferase activity and a small amount of estrone UDPglucuronyltransferase activity can be eluted at approximately 0.05–0.07 M KCl (Fraction B). The 4-nitrophenol UDPglucuronyltransferase activity present in Fraction A can now be removed by diluting the preparation 1:2 with 20% glycerol, 100 μM dithiothreitol, and 0.05% Emulgen 911, and passing the fraction through another DEAE-cellulose column equilibrated with Buffer C. The estrone UDPglucuronyltransferase activity passes through the column while the 4-nitrophenol UDPglucuronyltransferase

TABLE I

PURIFICATION OF ESTRONE AND 4-NITROPHENOL UDPGLUCURONYLTRANSFERASES BY PROCEDURE I

Purification step	Volume (ml)	Protein (mg)	Activity (units)		Specific activity (units/mg)	
			Estrone	4-Nitro-phenol	Estrone	4-Nitro-phenol
Microsomes	450	4500	38,300	405,000	8.5	90
Ammonium sulfate	200	1990	25,800	278,000	13	140
First DEAE-cellulose						
Fraction A	800	210	9,450	11,600	45	55
Fraction B	150	375	750	150,000	2.0	400
Second DEAE-cellulose						
Fraction A	320	115	8,050	NA[b]	70	NA
Affinity chromatography						
Fraction A[a]	100	9.5	3,280	NA	345	NA
Fraction B[a]	75	7.5	180	37,100	24.0	4950

[a] All fractions assayed in the presence of 25 µg phosphatidylcholine.
[b] NA, No detectable activity.

TABLE II

PURIFICATION OF ESTRONE AND 4-NITROPHENOL UDPGLUCURONYLTRANSFERASES BY PROCEDURE II

Purification step	Volume (ml)	Protein (mg)	Activity (units)		Specific activity (units/mg)	
			Estrone	4-Nitro-phenol	Estrone	4-Nitro-phenol
Microsomes	40	400	6720	54,400	16.8	136
DEAE-cellulose						
Fraction 1	75	19.1	470	NA[b]	24.6	NA
Fraction 2	400	29.7	2620	NA	88.1	NA
Fraction 3	90	54.4	95	21,800	1.7	400
Affinity chromatography						
Fraction 1[a]	60	1.3	201	NA	160	NA
Fraction 2[a]	80	2.4	806	NA	340	NA
Fraction 3[a]	60	2.1	28	6600	13.3	3220

[a] All fractions assayed in the presence of 25 μg phosphatidylcholine.
[b] NA, No detectable activity.

activity is retarded. Although the usual amount of 4-nitrophenol UDPglucuronyltransferase activity can be removed with KCl, this fraction has not been studied.

Purification Procedure II. Microsomes, 500 mg, are suspended to a final concentration of 10 mg/ml in Buffer D and solubilized with Emulgen 911 (0.5 mg detergent/mg protein). After stirring the mixture on ice for 30 min, the suspension is diluted 1 : 2 with Buffer D and directly applied to a DEAE-cellulose column (2.5 × 50 cm) previously equilibrated with Buffer E. If the column is washed with 200–400 ml of Buffer E, a sharp peak of estrone UDPglucuronyltransferase is eluted that does not contain any detectable 4-nitrophenol UDPglucuronyltransferase activity (Fraction 1). The column is washed with Buffer E containing 15 mM KCl, and a second peak of estrone UDPglucuronyltransferase activity is eluted (Fraction 2). Buffer E is passed through the column until no additional estrone UDPglucuronyltransferase activity is removed. Neither Fraction I nor Fraction 2 contain detectable 4-nitrophenol UDPglucuronyltransferase. Removal of 4-nitrophenol UDPglucuronyltransferase from the DEAE-cellulose column is accomplished by elution with Buffer E containing 50 mM KCl (Fraction 3). As was observed with the fraction that contained 4-nitrophenol UDPglucuronyltransferase activity isolated by Procedure I (Fraction B), this third fraction also contains a small amount of activity toward estrone.

Chromatography of 4-Nitrophenol and Estrone UDPglucuronyltransferase Activities on UDP-Hexanolamine-Sepharose 4B

Procedure I and Procedure II yield fractions that contain only estrone UDPglucuronyltransferase activity (Fraction A; Fractions 1 and 2) and a fraction that contains 4-nitrophenol UDPglucuronyltransferase activity with a trace of estrone UDPglucuronyltransferase activity (Fraction B and Fraction 3). Because purification of these respective enzyme activities by affinity chromatography requires different optimizing conditions for maximal yields, the purification of the different enzyme activities are presented separately. Preparation of the affinity matrix is described in detail at the end of this chapter.

Affinity Chromatography of 4-Nitrophenol UDPglucuronyltransferase Activity

Once Fraction B of Procedure I or Fraction 3 of Procedure II are eluted from DEAE-cellulose, MgCl$_2$ is added to a final concentration of 5 mM. Neither fraction is diluted or dialyzed. After the addition of MgCl$_2$, the preparations are applied directly to a bed (2.5 × 15 cm) of UDP-hexanolamine Sepharose 4B and washed extensively with Buffer E containing 50 mM KCl and 25 μg phosphatidylcholine per milliliter. After washing, no detectable 4-nitrophenol UDPglucuronyltransferase or protein should be eluted from the column. The specifically bound 4-nitrophenol UDPglucuronyltransferase can then be removed by washing the column with Buffer E containing 50 mM KCl, 25 μg per ml phosphatidylcholine, and 5 mM UDPGA. If the fractions are stored under nitrogen at 4° in the presence of phosphatidylcholine, enzyme activity does not start to decline significantly for about 1 week.

Affinity Chromatography of Estrone UDPglucuronyltransferase Activity

The fractions from the DEAE-cellulose column that contain only estrone UDPglucuronyltransferase activity are pooled and estrone (10 μM final concentration) in dimethyl sulfoxide (DMSO, 0.1% final volume) is added to the preparations; estrone appears to increase the binding of estrone UDPglucuronyltransferase to the affinity column. The enzyme is then slowly applied to a bed of UDP-hexanolamine Sepharose 4B (2.5 × 15 cm), previously equilibrated in Buffer E. The columns are washed extensively with Buffer E containing 25 mM KCl and 25 μg phosphatidylcholine per ml. Estrone UDPglucuronyltransferase activity is eluted by washing the column with the same buffer containing 5 mM UDPGA.

It is important that the salt be included in the buffer washes, because, as with the purification of 4-nitrophenol UDPglucuronyltransferase, the higher ionic strength removes nonspecifically bound proteins that otherwise would elute when the transferase is removed with 5 mM UDPGA. However, if the ionic strength of the buffer is increased by greater than 25 mM KCl, specifically bound estrone UDPglucuronyltransferase is gradually eluted. Because estrone UDPglucuronyltransferase has been found to be completely dependent upon the presence of phospholipid for maximal catalytic activity[6] and because affinity chromatography removes residual phospholipids,[7] lipid is included in the elution buffer along with UDPglucuronic acid.

Properties

Molecular Weight, Purity, and Stability. Polyacrylamide slab gel electrophoresis in the presence of sodium dodecyl sulfate (SDS) reveals that estrone and 4-nitrophenol UDPglucuronyltransferases, purified either

[6] R. H. Tukey and T. R. Tephly, *Life Sci.* **27**, 2471 (1980).
[7] B. Burchell and T. Hallinan, *Biochem. J.* **171**, 821 (1978).

from Procedure I (Fraction A, Fraction B) or Procedure II (Fraction 2, Fraction 3), contain a single polypeptide band with an approximate molecular weight of 57,000.

When both transferases are mixed and subjected to electrophoresis, only one protein band can be observed (by staining). The highly purified estrone UDPglucuronyltransferase activity in Fraction 1, isolated by Procedure II, also exhibits a prominent protein band similar to that of the enzyme in Fraction 2, although several minor higher- and lower-molecular-weight contaminants are present. However, if the protein in Fraction 1 is excised from the acrylamide, subjected to limited proteolysis by either α-chymotrypsin or *Staphylococcus aureus* V8 protease and subjected to electrophoresis in polyacrylamide in the presence of SDS,[8] the peptide fragments formed are identical to the fragments that are formed following limited proteolysis of the estrone UDPglucuronyltransferase activity in Fraction 2. It therefore appears that the estrone UDPglucuronyltransferase activities in Fraction 1 and Fraction 2 isolated by Procedure II are the same enzyme.

The purified enzyme preparations maintained at 4° in the presence of phospholipid, are stable for up to 1 week; activity is lost upon freezing. If estrone UDPglucuronyltransferase is purified in the absence of phospholipid and stored at 4°, irreversible loss of activity can result overnight.

Enzyme Activity, Recovery, and Substrate Specificity. The purification of estrone UDPglucuronyltransferase activity or 4-nitrophenol UDPglucuronyltransferase activity from either Procedure I or II routinely resulted in yields ranging from 8 to 14% for both enzymes. The estrone UDPglucuronyltransferase in Fraction 1 isolated by Procedure II results in 1 to 3% recovery although the specific activity of this preparation is only half that of the specific activity in Fraction 2.

For substrate specificity studies, all enzyme assays were performed in the presence of 25 μg phosphatidylcholine. The transferase catalyzing glucuronidation of estrone exhibited no activity when 4-nitrophenol was used as substrate. The purified 4-nitrophenol UDPglucuronyltransferase catalyzes the glucuronidation of estrone to only a slight extent. Both enzymes catalyze the glucuronidation of estradiol at rates that are identical to the glucuronidation of estrone, suggesting that estrone and estradiol are conjugated by the same enzymes. Testosterone and morphine do not serve as substrates for either enzyme.

Physical Properties. Although both enzymes exhibit similar subunit molecular weights in the presence of SDS, physical characterization studies reveal several distinct differences between the two enzymes. Limited proteolysis in the presence of SDS[8] resulted in clear differences in their

[8] E. F. Johnson, M. C. Zounes, and U. Müller-Eberhard, *Arch. Biochem. Biophys.* **192,** 282 (1979).

TABLE III

PROPERTIES OF PURIFIED RABBIT LIVER MICROSOMAL ESTRONE AND
4-NITROPHENOL UDPGLUCURONYLTRANSFERASES

Properties	Estrone glucuronyl-transferase	4-Nitrophenol glucuronyl-transferase
Molecular weight		
SDS-polyacrylamide	57,000	57,000
Ultragel A-34	230,000	230,000
pH optimum for catalytic activity	8.0	7.5
Isoelectric point	7.6	6.8
Percentage hydrophobic amino acids	60	53
Substrates conjugated	Estrone, estradiol	4-Nitrophenol, estrone, estradiol

peptide maps. Table III is a composite of the physical and functional characteristics of the highly purified transferases.

Preparation of UDP-Hexanolamine Sepharose 4B

Reagents and General Procedures. Methanol and dimethylformamide (DMF) are dried by storage over 4A molecular sieves. Phosphoric acid (K&K Laboratories) and 6-amino-1-hexanol (Aldrich Chemical Co.) are stored desiccated, whereas 1,1'-carbonyldiimidazole (CDI, Sigma Chemical Co.) is stored in a desiccator at −20°. Bottles containing CDI are discarded after opening twice. Aqueous solutions are concentrated with a Büchler flash evaporator using water aspiration. Solutions in DMF are concentrated under reduced pressure with a mechanical pump attached to an acetone–dry ice trap. Ascending paper chromatography and the development of the paper for either amines or phosphates was conducted as described by Barker *et al.*[9]

Synthesis of UDP-Hexanolamine. UDP-hexanolamine was synthesized by the method of Barker *et al.*,[9] as modified by Sadler *et al.*[10] for the synthesis of CDP-hexanolamine. However, to obtain good yields of the ligand, and to ensure that preparations are free of unbound UDP, i.e., hexanolamine phosphate, minor modifications of these procedures have been developed. A brief description of the procedure is presented.

N-Trifluroacetyl-*O*-phosphoryl-6-amino-1-hexanol   (TFA-hexanol-amine-phosphate, 80 mmol) is synthesized as described by Barker *et al.*[9]

[9] R. Barker, K. W. Olsen, J. W. Shaper, and R. L. Hill, *J. Biol. Chem.* **247**, 7135 (1972); see also this volume, Article [20].
[10] E. J. Sadler, J. I. Rearick, J. C. Paulson, and R. L. Hill, *J. Biol. Chem.* **254**, 4434 (1979).

and stored in 50 ml of anhydrous dimethylformamide. Carbonyldiimidazole, 100 mmol, in 50 ml DMF, is added and the solution stirred under a $CaSO_4$ drying tube until the CDI is completely dissolved (usually 15 min); the reaction is kept at room temperature overnight. When the imidazolide reaction is completed, excess CDI is hydrolyzed by adding 2 ml of anhydrous methanol. After 20 min the solution is evaporated for 5 min to remove excess methanol and CO_2.

Sodium uridine monophosphoric acid, 15 g, is applied to a Dowex 50W-X8 column (2.5 × 50 cm, pyridine cycle) and developed with water. Fractions are monitored at 280 nm and those with an A_{280} greater than 0.2 are pooled and concentrated to about 200 ml under reduced pressure. Three equivalents of tributylamine and 200 ml of absolute ethanol are added, and the solution taken to dryness under reduced pressure. Dimethylformamide (50 ml) is added to the dried preparation and the solution taken to dryness.

TFA-hexanolamine-phosphate-imidazolide in DMF is added to the 40 mmol of UMP-tributylamine. The contents are covered with aluminum foil maintained under a $CaSO_4$ drying tube for 24 to 36 hr. At the end of this period, 200 ml of water are added and the solution applied to a Dowex 1-X2 column (2.5 × 50 cm; 200–400 mesh, Cl^- cycle) that has been equilibrated with 50% methanol/10 mM pyridine-HCl or pH 5.4. The column is washed with 300 ml of the equilibration buffer, and developed with a linear gradient (4 liters from 0–1.5 M LiCl in 10 mM pyridine-HCl at pH 5.4). Fractions are monitored for uridine absorbance by adding 10 μl to 1.0 ml of 30 mM NaOH and measuring the absorbance at 280 nm. TFA-hexanolamine-UMP elutes from the column in a sharp peak at approximately 0.90 M LiCl but is contaminated with unreacted UMP and hexanolamine-phosphate. To remove these impurities, the TFA-blocked hexanolamine-UDP is deblocked by titrating to pH 11.5 with 4 N LiOH and allowing the solution to stand for 4 hr. The pH is adjusted to 5.4 with 1 N HCl, and, after concentrating to approximately 100 ml under reduced pressure, the sample is desalted with a column (4.0 × 100 cm) of Sephadex G-15. Fractions are monitored by UV absorbance (262 nm), and those absorbing at 262 nm are pooled and concentrated to 50 ml. Conductivity of the fractions is carefully monitored to assure removal of excess salt; preparations are often desalted twice on the Sephadex G-15 column. The desalted UDP-hexanolamine is chromatographed on a column (2.5 × 50 cm, Cl^- cycle) of Dowex 1-X2 equilibrated with 10 mM pyridine-HCl at pH 5.4. The column is washed with 200 ml of the equilibration buffer and developed with a 2-liter linear 0–1.0 M LiCl gradient in the equilibration buffer. UDP-hexanolamine is eluted at a LiCl concentration of approximately 0.5 M and is free of UMP and hexanolamine-phosphate. The fractions con-

taining UDP-hexanolamine are pooled, concentrated to 50 ml under reduced pressure and desalted with Sephadex G-15. The yield of UDP-
hexanolamine, based on the amount of UMP originally used, is usually
greater than 60%.

Preparations of Sepharose Absorbents. Cyanogen bromide-activated
Sepharose 4B (Pharmacia) is prepared as described by Porath *et al.*[11] and
Paulson *et al.*[12] with slight modifications. Washed Sepharose 4B and 5 M
phosphate buffer (3.33 mol of K_3PO_4 + 1.67 mol of K_2HPO_4 per liter of
solution) are added together in equal volumes (1 : 1, v/v) and chilled on
ice. Appropriate amounts of UDP-hexanolamine are adjusted to give 10 to
15 μmol per ml of settled agarose in a volume of 0.1 M $NaCO_3$ at pH 9.5,
equal to the volume of settled agarose. CnBr (300 mg per g of packed gel)
is dissolved in anhydrous DMF to a final concentration of 1 g per ml.
When the temperature of the buffered agarose reaches 4°, CnBr is added
intermittently over 3 min. The solution is stirred for 8 min and washed on a
glass filter with 20 volumes cold water and 20 volumes of cold 0.1 M
$NaHCO_3$. The CnBr-activated Sepharose is immediately mixed with the
ligand, and the mixture gently stirred at 4° for 24 hr.

Quantitation of the extent of linkage is determined in the following
manner. Intermittently the stirring is stopped, allowing the agarose to
settle. The ligand concentration in the supernatant is assessed by recording the absorbance at 262 nm (pH 12.0). Routinely, greater than 95%
coupling was achieved; UDP-hexanolamine-agarose preparations contained 6–12 μmol of UDP per ml of settled agarose.

Acknowledgment

This research has been supported by NIH grant GM26221.

[11] J. Porath, K. Aspberg, H. Drevin, and R. Axén, *J. Chromatogr.* **86**, 53 (1973); see also this
series, Vol. 34, Article [2].
[12] J. C. Paulson, W. E. Beronek, and R. L. Hill, *J. Biol. Chem.* **252**, 2356 (1977).

[22] Bilirubin UDPglucuronyltransferase

By Brian Burchell

Bilirubin + UDPglucuronate ⇌ bilirubin glucuronide + UDP
(R—COO⁻)   (R—CO—O—glucuronide)

Hepatic microsomal bilirubin UDPglucuronyltransferase appears to be
the rate limiting enzyme in the excretion of toxic bilirubin in neonatal

jaundice[1] and Crigler-Najjar syndrome.[2] Genetic deficiency of this activity is also responsible for the hyperbilirubinaemia observed in Gunn rats. The apparent instability of isolated preparations of this transferase has delayed the purification of this important enzyme. The realization that very insoluble but polar bilirubin will strongly interact with polar phospholipids[3] in the environment of UDPglucuronyltransferase suggested that this transferase activity would not be supported solely by the nonionic detergents used for the purification of nitrophenol UDPglucuronyltransferase. Lecithin liposomes need to be added to each isolated transferase fraction to facilitate assay and purification of the hitherto elusive bilirubin UDPglucuronyltransferase.

This method of purification of bilirubin UDPglucuronyltransferase is based on that previously described.[4]

Assay Method

Principle. The assay is based on the selective diazotization of bilirubin glucuronides in the presence of bilirubin at pH 2.7. Conjugated bilirubin is converted into colored azoderivatives, which are easily extracted into organic solvents and measured spectrophotometrically. The assay procedure is based on the method previously described[5] and more recently modified.[6]

Reagents

Bilirubin–albumin: 0.85 mM bilirubin and 20 mg per ml bovine serum albumin in 0.4 M Tris-maleate, containing 60 mM magnesium chloride, adjusted to pH 7.7, is freshly prepared in the dark just before use

UDPglucuronate, triammonium salt (Sigma Chemical Co.), 20 mM, adjusted to pH 7.4 with sodium hydroxide

Glycine-HCl buffer: 0.4 M Glycine adjusted to pH 2.7 with hydrochloric acid

Diazo reagent: Finely disperse 0.1 ethyl anthranilate (Eastman-Kodak) in 10 ml 0.15 M hydrochloric acid and add 0.3 ml 0.07 M

[1] L. M. Gartner, K. Lee, S. Vaisman, D. Lane, and I. Zarafu, *J. Pediatr.* **90**, 513 (1977).
[2] R. Schmid and A. F. McDonagh, *in* "The Metabolic Basis of Inherited Disease" (J. B. Stanbury, J. B. Wyngaarden, and D. S. Fredrickson, eds.), 4th ed., p. 1221. McGraw-Hill, New York, 1978.
[3] R. Brodersen, *J. Biol. Chem.* **254**, 2364 (1979).
[4] B. Burchell, *FEBS Lett.* **111**, 131 (1980).
[5] F. P. Van Roy and K. P. M. Heirwegh, *Biochem. J.* **107**, 507 (1968).
[6] K. P. M. Heirwegh, M. Van der Vijver and J. Fevery, *Biochem. J.* **129**, 605(1972); see also this volume, Article [50].

sodium nitrite; mix and incubate for 5 min at 20°; add 0.1 ml 88 mM
ammonium sulfamate; mix and incubate for 3 min at 20° before use
Ascorbic acid, 0.57 M

Procedure. The standard incubation mixture prepared in the dark contains 100 μl bilirubin solution together with up to 225 μl of enzyme or water warmed to 37°. The reaction is initiated by the addition of 75 μl of 20 mM UDPglucuronic acid.

After incubation at 37° for 15 min in the dark, 1 ml of ice-cold glycine buffer is added and the mixture is rapidly vortexed and placed in ice for 5 min. Diazo reagent, 0.5 ml, is added and the mixture is incubated for 30 min at 20°. The diazotization reaction is stopped by addition of 0.25 ml of 0.57 M ascorbic acid, followed by 1 ml of pentan-2-one : 1-butyl acetate (17 : 3). The mixtures are shaken vigorously and the phases separated by centrifugation of 2000 g for 10 min. The absorbance of the diazotized glucuronides extracted into the upper organic phase is measured at 530 nm. The extinction coefficient of the azo-bilirubin glucuronide is 44.4 $\times$ 10³ cm² mol⁻¹. One unit of activity represents the formation of 1 nmol bilirubin glucuronide per minute. Protein concentrations are determined by the method of Bradford.[7]

Reactivation of UDPglucuronyltransferase with Lecithin Liposomes

Egg yolk lecithin (40 mg) in chloroform/methanol is transferred to a thick walled Pyrex tube and the solvent is removed by evaporation under nitrogen at 37°. Tris-maleate, 0.1 M, at pH 7.7 (4 ml) containing 15 mM magnesium chloride, is added to the lipid and the mixture is sonicated, in 30-sec periods, for a total of 4 min. The mixture is cooled for 2 min in ice water after each sonic burst to maintain its temperature below 20°.

The chilled phospholipid dispersion is added to purification fractions (2 mg lecithin/mg protein) and the mixture is incubated for 10 min at 5° before assay. The effectiveness of this procedure is illustrated by activation of bilirubin UDPglucuronyltransferase up to 9-fold by addition of >1.6 mg lecithin per mg protein,[4] to ammonium sulfate fractions of liver microsomes.

Purification Procedure

A summary of recent results is shown in the table. Steps 1 and 2 of this procedure are the same as described for the purification of nitrophenol UDPglucuronyltransferase.[8] The ammonium sulfate precipitate from two

[7] M. M. Bradford, *Anal. Biochem.* **72**, 248 (1976).
[8] B. Burchell, this volume, Article [20].

PURIFICATION OF RAT LIVER BILIRUBIN UDPGLUCURONYLTRANSFERASE

Step	Volume (ml)	Protein (mg)	Activity[a] (nmol/min)	Specific activity (units/mg)
1 Lubrol extract	68	1400	240	0.3
2 Ammonium sulfate	31	322	258	0.8
3 DE-52 fraction	20	42	126	3.0
4 UDP-hexanolamine Sepharose	9	0.9	5	5.1

[a] Bilirubin UDPglucuronyltransferase activity is measured after reconstitution of protein with phosphatidylcholine.

rat livers is dissolved in Buffer C (0.01% (w/v) Lubrol, 10 mM potassium phosphate, 5 mM 2-mercaptoethanol, at pH 8.0). This clear red fraction ($\cong$ 30 ml) is dialyzed against 6 liters of Buffer C overnight.

Step 3. DEAE-Cellulose. The clear, red ammonium sulfate fraction (up to 400 mg protein) is applied slowly (15 ml/hr) to a column (3.5 × 12 cm) of DEAE-cellulose (DE-52), previously equilibrated with Buffer C. Less than 5% of the transferase activity toward bilirubin applied to the DE-52 column is eluted with 300 ml of Buffer C. The major peak of bilirubin UDPglucuronyltransferase activity is eluted between 0.08 and 0.15 M KCl of a linear gradient (400 ml) from 0 to 0.3 M KCl in 5 mM potassium phosphate containing 0.1% (w/v) Lubrol and 5 mM 2-mercaptoethanol at pH 8.0; this activity is completely separated from 4-nitrophenol UDPglucuronyltransferase. More than 90% of the transferase activity toward bilirubin is recovered and fractions exhibiting enzyme activity greater than 0.7 nmol per min per ml are pooled, concentrated approximately 4-fold by vacuum dialysis, and dialyzed against 200 volumes of Buffer C overnight.

Step 4. UDP-Hexanolamine Sepharose. Dialyzed fractions from Step 3, about 10 ml containing approximately 2 mg protein/ml, are applied slowly (5 ml/hr) to two columns (2 × 6 cm) of UDP-hexanolamine Sepharose 4B.[8] After washing with Buffer C, bilirubin UDPglucuronyltransferase is eluted with 3 ml of 5 mM UDPglucuronic acid in Buffer C for each column. As a consequence of the affinity step, the transferase is completely delipidated. Assessment of the purification of the enzyme is dependent on the method of reconstitution of the transferase. The simple addition of lecithin liposomes to transferase protein facilitates the measurement of some bilirubin transferase activity (see table) but may not be sufficient to restore full catalytic activity to all of the enzyme protein. Lecithin liposomes are added to the final enzyme preparation to improve the stability of the transferase during storage in ice at 0°.

Purity

In most preparations, a single polypeptide band of about 57,000 daltons, is observed (by staining) to be present after SDS gel electrophoresis, which suggests that the enzyme is highly purified. Occasionally, preparations are contaminated by an unknown, high-molecular-weight ($>130,000$) protein species. The presence of the contaminant protein is dependent on the slightly variable elution of the transferase from the DE-52 column, appearing when a higher concentration ($>0.12\,M$) KCl is required to elute the enzyme.

[23] Dismutation of Bilirubin Monoglucuronide

By J. Roy Chowdhury and Irwin M. Arias

2 Bilirubin monoglucuronide → bilirubin diglucuronide + bilirubin

Bilirubin monoglucuronide is the major pigment in human and rat bile.[1] Formation of bilirubin diglucuronide from bilirubin requires two steps. In rat and human liver, the microsomal enzyme UDPglucuronosyltransferase (UDPglucuronate β-D-glucuronosyltransferase, EC 2.4.1.17), catalyzes the transfer of glucuronic acid from uridine diphosphate glucuronic acid (UDPGA) to bilirubin, forming bilirubin monoglucuronide and uridine diphosphate.[2] Two mechanisms of conversion of bilirubin monoglucuronide to diglucuronide have been demonstrated *in vitro*. In cat[3] and rat[4] liver microsomal systems, bilirubin diglucuronide is formed by UDPglucuronosyltransferase catalyzed transfer of the glucuronosyl moiety from UDPGA to bilirubin monoglucuronide. A second enzyme, provisionally named bilirubin-glucuronoside glucuronosyltransferase (EC 2.4.1.95), which is concentrated in plasma membrane-enriched fractions of rat liver, catalyzes the dismutation of bilirubin monoglucuronide to bilirubin diglucuronide.[5]

The dismutation reaction may involve transfer of a glucuronyl moiety from one molecule of bilirubin monoglucuronide to another, or an enzymatically catalyzed rearrangement of the two dipyrrolic halves of bilirubin

[1] B. H. Billing, P. G. Cole, and G. H. Lathe, *Biochem. J.* **65**, 774 (1957).
[2] J. Fevery, P. Leroy, M. van de Vijver, and K. P. M. Heirwegh, *Biochem. J.* **129**, 635 (1972). See also this volume [22].
[3] P. L. M. Jansen, *Biochim. Biophys. Acta* **338**, 170 (1974).
[4] N. Blanckaert, J. Gollan, and R. Schmid, *Proc. Natl. Acad. Sci. U.S.A.* **76**, 2037 (1979).
[5] P. L. M. Jansen, J. Roy Chowdhury, E. B. Fischberg, and I. M. Arias, *J. Biol. Chem.* **252**, 2710 (1977).

monoglucuronide. Neither mechanism has been firmly established. The products of the reaction have the IX_α configuration, which is the isomeric form of bilirubin in bile. Dismutation of bilirubin monoglucuronide occurs at a normal rate *in vitro* in the liver of UDPglucuronosyltransferase-deficient man and rat.[6] *In vivo,* bilirubin diglucuronide formation appears to involve both dismutation and UDPglucuronosyltransferase mechanisms.

Preparations

Step 1. Preparation of Rat Liver Microsomes. Male Wistar rats, 200–250 g, are killed by decapitation. All of the following procedures are performed at 0° to 4°. Livers are removed and perfused through the portal vein and inferior vena cava with ice-cold 5 mM Tris-acetate at pH 8.0, containing 0.25 M sucrose and 1 μM EDTA. Each liver is minced into fine pieces with scissors and homogenized with 3 ml per g of the same buffer as used in perfusion. Homogenization is performed in a Potter-Elvehjem homogenizer equipped with a motor-driven Teflon pestle. The homogenate is centrifuged at 9000 g for 15 min. The pellet is discarded, and the supernatant liquid is centrifuged at 105,000 g for 60 min in preweighed centrifuge tubes. The pellets are suspended by homogenization in 3 ml per g in the perfusion buffer. Microsomes may be stored in 8-ml aliquots at −70° for up to 2 weeks.

Step 2. Biosynthesis of Bilirubin Monoglucuronide. UDPglucuronosyltransferase activity in rat liver microsomes is activated as follows: Triton X-100, 0.025 ml, is dissolved in 2 ml of the perfusion buffer and added to 8 ml of microsomal preparation. Bilirubin, 4.4 mg, is dissolved in 0.5 ml of 0.05 N sodium hydroxide, 0.5 ml at room temperature, and diluted with 0.1 M sodium phosphate at pH 7.8 to 15 ml. UDPglucuronic acid, 100 μmol, and $MgCl_2$, 200 μmol, are dissolved in 0.1 M Tris-HCl at pH 7.8 (pH adjusted at 37°), and the microsome–detergent mixture (10 ml) and bilirubin solution (6 ml) are added. The mixture is deoxygenated by bubbling with nitrogen gas and incubated at 37° in the dark for 15 min. The reaction is stopped by addition of an equal volume of ice-cold SDS solution (2 mg/ml), and stirred with a magnetic stirrer at 4° in the dark for 20 min. The pH is lowered to 3.2 by addition of 10 N HCl, and ethyl acetate (40 ml) is added. Nitrogen gas is bubbled through the solution for 2 min; pigments are extracted in ethyl acetate by shaking in a separating funnel; the organic solvent extract is separated by centrifugation (900 g for 10 min); and ethyl acetate is evaporated under reduced pressure using a vacuum pump equipped with a glass organic solvent trap immersed in

---

[6] J. Roy Chowdhury, P. L. M. Jansen, E. B. Fischberg, A. Daniller, and I. M. Arias, *J. Clin. Invest.* **21,** 191 (1978).

liquid nitrogen. The bilirubin monoglucuronide preparation can be stored dry in nitrogen at −170° for 24 hr.

Step 3. Preparation of Ethyl Anthranilate Diazo Reagent.[7] Freshly prepared sodium nitrite solution (5 mg/ml), 0.3 ml, is mixed with 0.15 N HCl, 10 ml. Ethyl anthranilate, 0.1 ml, is slowly added while the mixture is stirred vigorously on a magnetic stirrer. After 5 min of stirring, 0.1 ml of ammonium sulfamate solution (10 mg/ml) is added. The diazo reagent is stored on ice for not longer than 30 min.

Assay Method

The enzyme suspension, 50 μl of a 2.5% liver homogenate or of a subcellular fraction, is incubated with 0.1 ml of 0.1 M sodium phosphate at pH 6.6 containing 10 mM glucaro-1,4-lactone at 25° for 1 hr. Bilirubin monoglucuronide is dissolved in a minimum volume of 0.1 M Tris-HCl at pH 7.8 and 0.05 ml is added to the enzyme–buffer mixture. The blank is an identical incubation mixture in which the enzyme is inactivated by heating at 100° for 10 min. After incubation at 37° for 3 min, the reaction is stopped with 2 ml ice-cold ethyl anthranilate diazo reagent. After incubation at 25° for 30 min, 1 ml of 20% ascorbic acid is added, and azopigments are extracted in 0.5 ml of 2-pentanone : butyl acetate (17 : 3, v/v). The organic solvent extract is separated by centrifugation (10,000 g, 10 min) and azopigments are analyzed by thin-layer chromatography (tlc) or high-performance liquid chromatography (HPLC) as described in the following sections.

Azopigment Analysis[7]

Principle. In the diazo reaction used here, unconjugated bilirubin does not react. Bilirubin monoglucuronide splits into equimolar amounts of unconjugated (α) and glucuronidated (δ) azodipyrroles. Bilirubin monoglucuronide is the only source of pigments in this system, and represents twice the concentration of unconjugated azodipyrrole. Bilirubin diglucuronide splits into two glucuronidated azodipyrroles (δ), and can be quantitated from $\delta - \alpha$.

THIN-LAYER CHROMATOGRAPHY

A 0.2 ml aliquot is applied to a tlc plate (Silicagel G-60, E. Merck) with chloroform : methanol : water (65 : 25 : 3, v/v/v) as the developing solvent.[7]

---

[7] F. P. Van Roy and K. P. M. Heirwegh, *Biochem. J.* **107**, 507 (1968); see also this volume, Article [50].

The faster moving purple pigment (unconjugated azodipyrrole, α) and the slower moving purple pigment (conjugated azodipyrrole, δ) are removed by scraping and extracted by shaking with 1 ml of methanol, the absorbance at 530 nm is determined and the amount of bilirubin monoglucuronide and diglucuronide may be calculated, assuming $E_{530}^{1\,cm} = 44.4 \times 10^3 \, M^{-1}$.[7]

High Performance Liquid Chromatography

Equipment and Solvents

Column: Waters' μ-Bondapak C-18

Detection system: Absorbance spectrometer set at 530 nm or 546 nm

Buffer: Sodium acetate, 0.1 M, pH 4.0, containing 1-heptane sulfonic acid, 5 mM

Method. The organic solvent is evaporated from a 0.2-ml azopigment extract under reduced pressure using a centrifugal evaporator (Speedvac, Savant Corp. NY). The pigment is dissolved in 0.1 ml methanol, and 75 μl are injected into the column. The pigments are eluted with a buffer: methanol mixture (2 : 8, v/v) at 1 ml per min and peaks are quantitated with an electronic integrator. Azopigments are quantitated from peak area using a standard curve prepared as follows: Bilirubin (1 μmol) is dissolved in 100 ml dimethyl sulfoxide and 0.01- to 0.4-ml samples are further diluted to 1 ml with dimethyl sulfoxide. Ethyl anthranilate diazo reagent, 0.5 ml, is added. After 10 min at room temperature, 1 ml of each of 2-pentanone and 0.15 N HCl are added sequentially. The organic solvent is separated by centrifugation (900 g; 2 min) and 0.5 ml is evaporated under reduced pressure with a centrifugal evaporator. The azopigment is dissolved in 0.1 ml methanol, and 0.05 ml, containing 5×10^{-11} to 2×10^{-9} mol, is analyzed by HPLC as previously described. In the presence of dimethyl sulfoxide, bilirubin is quantitatively converted to unconjugated azodipyrrole. The peak area : nmol ratio is used for quantitating azodipyrroles in other experiments.

Purification[8]

The bilirubin monoglucuronide dismutating enzyme is a basic (pI 7.9) protein that is concentrated in plasma membrane-enriched fractions of rat

[8] J. Roy Chowdhury, N. Roy Chowdhury, M. Bhargava, and I. M. Arias, *J. Biol. Chem.* **254**, 8336 (1979).

liver. After solubilization from these plasma membrane fractions, the protein can be purified by isoelectric focusing.

Stock Solutions

Buffer A: Tris-HCl, 5 mM at pH 8.0, containing 0.25 M sucrose
Buffer B: Tris-HCl, 5 mM at pH 8.0, containing 57% sucrose (w/w)
Buffer C: Tris-HCl, 5 mM at pH 8.0, containing 37% sucrose (w/w)
Buffer D: Tris-HCl, 5 mM at pH 8.0

Method

Step 1. Preparation of Rat Liver Plasma Membranes.[9] Male Wistar rats are killed by decapitation. The livers are removed, weighed, perfused with Buffer A, cut into fine pieces with scissors, and homogenized in 3 ml Buffer A/g tissue in an ice-cooled Potter-Elvehjem homogenizer with one stroke of a motor-driven Teflon pestle. All procedures are performed at 0° to 4°. The homogenate is centrifuged at 900 g for 10 min. The supernatant fluid is discarded and the pellet is washed twice by suspension and homogenization in the original volume of Buffer A. The resuspended pellet is centrifuged at 54,000 g for 7.5 min and the pellet is homogenized with 2.5 ml Buffer B per gram of original liver weight. The suspension, 15 ml, is placed in cellulose nitrate tubes for Beckman SW 27.2 rotors and carefully overlayered with 15 ml of Buffer C and 8 ml of Buffer A. After centrifugation at 27,000 g for 16 hr, membranes from the 37%–0.25 M sucrose interface are aspirated with a peristaltic pump. This fraction is diluted to a sucrose concentration of 12% with Buffer D, and centrifuged at 105,000 g for 1 hr.

Step 2. Solubilization of Bilirubin Monoglucuronide Dismutating Activity. The pellet is suspended in 0.3 ml Buffer D per gram original liver weight and sonicated in ice for nine 3-sec periods with a probe type of sonicator. The sonicated membrane preparation is centrifuged for 90 min at 105,000 g; the supernatant liquid, designated as "solubilized plasma membrane preparation," is concentrated by negative pressure dialysis to 0.1 ml per gram of original liver weight.

Step 3. Column Chromatography. An agarose (A-5M, BioGel) column is equilibrated with Buffer D. The solubilized plasma membrane preparation is applied and the proteins are eluted with the same buffer. The active fractions are pooled and concentrated to 0.1 ml per gram of original liver weight.

Step 4. Isoelectric Focusing. Focusing is performed at 4° with LKB

[9] O. Touster, N. H. Aronson, J. T. Dulaney, and H. Hendrikson, *J. Cell. Biol.* **47**, 604 (1970).

PURIFICATION AND RECOVERY OF BILIRUBIN MONOGLUCURONIDE DISMUTASE ACTIVITY
FROM RAT LIVER

	Total protein (mg)	Volume (ml)	Bilirubin diglucuronide (nmol/min)	Specific activity (units/mg protein)
Plasma membrane fraction	42.3	28	1200	28.4
Sonication	19.0	5	356	18.7
Agarose column	0.96	5	317	330
Isoelectric focusing	0.081	8	178	2200

electrofocusing equipment using a 50 to 0% sucrose gradient containing 1% carrier ampholytes with a pH range of 7 to 9. Fractions of 2 ml are collected in ice-cooled tubes, and those with enzyme activity are pooled and concentrated by negative pressure dialysis. The purification procedure is summarized in the table.

Properties

The enzyme has an apparent molecular weight of 160,000 by agarose gel chromatography, and a subunit molecular weight of 28,000 by SDS gel electrophoresis. The isoelectric point is 7.9. The K_m for bilirubin monoglucuronide is 33 μM.

[24] Aryl Sulfotransferases

By RONALD D. SEKURA, MICHAEL W. DUFFEL, and WILLIAM B. JAKOBY

$$\text{Phenol + adenosine 3'-phosphate 5'-phosphosulfate} \rightleftarrows$$
$$\text{aryl sulfate + adenosine 3',5'-bisphosphate}$$

This group of enzymes (EC 2.8.2.1) is active with a wide variety of phenolic compounds as substrates.[1-5] At least four homogeneous enzymes with this activity have been isolated from the liver of Sprague-

[1] W. B. Jakoby, R. D. Sekura, E. S. Lyon, C. J. Marcus, and J.-L. Wang, in "Enzymatic Basis of Detoxication" (W. B. Jakoby, ed.), Vol. 2, p. 199. Academic Press, New York, 1980.
[2] A. B. Roy, Handb. Exp. Pharmacol. [N.S.] 28, Part 2, 536 (1971).
[3] R. D. Sekura and W. B. Jakoby, J. Biol. Chem. 254, 5658 (1979).
[4] R. D. Sekura and W. B. Jakoby, Arch. Biochem. Biophys., in press (1981).
[5] R. D. Sekura, K. Sato, H. J. Cahnmann, J. Robbins, and W. B. Jakoby, Endocrinology 108, 454 (1981).

Dawley rats.[3,4] The aryl sulfotransferases fall into two families and are designated by roman numerals in the order of decreasing isoelectric points of the proteins. The first family, consisting of aryl sulfotransferases I and II, comprises enzymes having entirely similar specificity for phenols; with 2-naphthol they have a pH optimum at about pH 6.5.[3] Aryl sulfotransferases III and IV also resemble each other in their specificity for phenols but, in addition, act with organic hydroxylamines and with peptides bearing N-terminal tyrosine residues, as sulfate acceptors.[4] With a simple phenol as acceptor, i.e., 2-naphthol, this latter group of enzymes has optimal activity at pH 5.5.

Assay Method[6]

Principle. The assay[9] takes advantage of the partition into an organic solvent of the ion pair formed between certain organic sulfate monoesters and methylene blue. 2-Naphthol and PAPS[10] serve as substrates.

Reagents

Buffer: 0.5 M sodium phosphate at pH 7.4 for the assay of sulfotransferases I and II, or 0.5 M sodium acetate at pH 5.5 for the assay of sulfotransferases III and IV

---

[6] The assay presented here, with 2-naphthol as sulfate acceptor, is effective in following the course of purification but not for evaluating the kinetics of the reaction or with other substrates. A chromatographic assay[7] is more versatile and is particularly useful, because of its low blank values, in determining kinetic constants with poor substrates. Sulfate transfer to phenols is assayed in a total volume of 50 μl containing 0.25 M buffer at the appropriate pH; 5 mM 2-mercaptoethanol; the phenol at a concentration of 0.5 mM; less than 5% acetone that is added along with the phenol; and 0.1 mM [^{35}S]PAPS (about 120 μCi per μmol). The reaction is initiated by addition of enzyme and, after 10 min at 37°, is terminated by addition of 13 μl of 2 M acetic acid. An aliquot of 5 μl of the reaction mixture is applied to a thin-layer strip (1 × 10 cm) of cellulose on a plastic backing (Eastman No. 6064). Once the samples are dry, the chromatographic strips are developed with n-propanol : ammonia : water (6 : 3 : 1) at 4°. After the solvent has moved about 8 cm, the strips are removed and the solvent front is marked. With most substrates, a 2-cm section is cut below and including the solvent front and placed directly into a scintillation vial containing 10 ml of Aquasol. With more polar substrates, such as those containing free amino groups, product must be localized by autoradiography.

With [^{35}S]PAPS that has been purified by chromatography on DEAE-cellulose,[8] there is essentially no radioactive contaminant that migrates with the sulfate esters.

[7] R. D. Sekura, C. J. Marcus, E. S. Lyon, and W. B. Jakoby, *Anal. Biochem.* **95**, 83 (1979).

[8] Commercially available ^{35}S-labeled PAPS contains material that migrates to the solvent front when used in the chromatographic assay.[6] The radioactive PAPS can be readily purified by applying 250 μCi of the nucleotide in 50% ethanol to a column (0.9 × 10 cm) of DE-52 (Whatman) that has been equilibrated with 50 mM triethylamine-carbonate at pH

2-Mercaptoethanol, 0.1 M, prepared by adding 0.07 ml of the thiol to 10 ml water

2-Naphthol, 5 mM, dissolved in acetone[11]

PAPS, 4 mM

Methylene blue reagent: 250 mg methylene blue, 50 g anhydrous sodium sulfate, and 10 ml H_2SO_4 are dissolved in water to a final volume of 1 liter

Procedure. The following are added sequentially, resulting in a total volume of 0.4 ml: water; 0.2 ml buffer solution; 30 μl 2-mercaptoethanol; 20 μl 2-naphthol; 20 μl PAPS; enzyme.[12] Following the addition of enzyme and incubation at 37° for 10 min, the reaction is terminated with 0.5 ml of methylene blue reagent followed by 2 ml of chloroform. After agitation with a vortex mixer for 20 sec, the suspension is centrifuged at 2000 g for 10 min. Chloroform, the lower layer, is transferred to a tube containing between 50 to 100 mg of anhydrous sodium sulfate. Absorbance is determined at 651 nm; 10 nmol of either 4-nitrophenyl sulfate or 1-naphthyl sulfate, carried through the extraction procedure, yield an A_{651} of 0.3.

Definition of Unit. A unit of enzyme is the amount necessary for the catalysis of the formation of 1 nmol of product per minute. Specific activity is defined in terms of units per milligram of protein; protein is determined by the method of Lowry *et al.*[13] with bovine serum albumin as standard.

7.6. The column is washed with the same buffer and eluted with a linear gradient (150 ml total volume) established between the starting buffer and 1 M triethylamine-carbonate at the same pH; fractions of 2 ml are collected. PAPS elutes sharply with a peak concentration in about fraction 38. The major fractions are pooled and brought to dryness at 37° with a rotary evaporator; the samples are dissolved in water and evaporated again. This procedure is repeated several times to remove traces of triethylamine (yield: 67%). PAPS is stored in aqueous solution at −20°.

[9] Adapted[3] with modification from that of Y. Nose and F. Lipmann, *J. Biol. Chem.* **233**, 1348 (1958).

[10] Although adenosine 3'-phosphate 5'-phosphosulfate or 3'-phosphoadenylylsulfate are the recommended names for the compound, it was originally designated as 3'-phosphoadenosine 5'-phosphosulfate and abbreviated as PAPS. PAPS is retained here as the abbreviation. A convenient organic synthesis of this nucleotide is in this volume [53].

[11] Because 5% acetone is not inhibitory to the sulfotransferases, this solvent is a useful medium for stock solutions of the relatively hydrophobic phenols.

[12] Sulfotransferase IV is inhibited at high salt concentrations. The amount of enzyme solution used for its assay, particularly if it is from a salt gradient, should be below 0.25 ml in order to avoid inhibition.

[13] O. H. Lowry, N. J. Rosebrough, A. C. Farr, and R. J. Randall, *J. Biol. Chem.* **193**, 265 (1951).

TABLE I

SUMMARY OF THE PURIFICATION OF ARYL SULFOTRANSFERASES I AND II[a]

Step	Volume (ml)	Activity (units)	Protein (mg)	Specific activity (units/mg)
1. Extract	1390	19.700	43,200	0.46
2. Affi-Gel blue	1140	9,880	1,280	7.7
3. Salt precipitation	57	7,990	650	12
Sulfotransferase I				
4A. DEAE-cellulose	59	1,260	41	31
5A. Hydroxylapatite	26	750	5.2	144
7A. Gel filtration	30	600	3.2	187
8A. ATP-agarose	7.3	219	0.88	250
Sulfotransferase II				
4B. DEAE-cellulose	57	1,390	106	15
5B. Hydroxylapatite	22	700	8.6	82
6B. CM-cellulose	7.4	400	2.8	142
7B. Gel filtration	26	347	1.7	199
8B. ATP-agarose	7.3	138	0.49	282

[a] From Sekura and Jakoby,[3] by permission of the *Journal of Biological Chemistry*.

Purification Procedure

General. Purification of all four aryl sulfotransferases is conducted together (Steps 1 through 4) until the stage of partial separation upon elution from DEAE/cellulose. The procedure has been applied to lots of about 50 rat livers (0.5 kg) obtained from well-fed, male Sprague-Dawley[14] rats weighing about 200 g. Livers are maintained at −80° prior to use. Unless otherwise specified, all operations are conducted at 4°. The results of purification are summarized in Tables I and II.

Step 1. Extraction. Frozen rat livers are allowed to thaw in 1.5 liters of Buffer A (10 mM Tris-HCl, 0.25 M sucrose, and 3 mM 2-mercaptoethanol, adjusted to pH 7.4 at 25°) and homogenized for two 30-sec periods in a Waring blender maintained in a cold room. The homogenate is centrifuged at 140,000 g for 90 min and, after removal by aspiration of the low density material at the top of the tube, the supernatant fluid is collected.

Step 2. Affi-Gel Blue. Without further treatment, the protein solution is applied to a column (4 × 22 cm) of Affi-Gel Blue (BioRad, 100–200 mesh) that has been equilibrated with Buffer A. The column is sequentially

[14] The livers were obtained in a frozen state (Dri-Ice) from ARS Sprague-Dawley, Madison, WI.

TABLE II
SUMMARY OF PURIFICATION FOR ARYL SULFOTRANSFERASE IV

Step	Volume (ml)	Activity (units)	Protein (mg)	Specific activity (units/mg)
1. Extract	1430	77,000	41,000	1.9
2. Affi-Gel blue	67	18,300	2,310	7.9
3. DEAE-cellulose	53	4,450	107	107
4. Hydroxylapatite	17	3,640	20.8	175
5. ATP-agarose	12.9	1,740	3.2	540

eluted at a flow rate of about 120 ml per hour with each of the following solutions: (1) 1.1 liters of Buffer A; (2) a linear gradient between 1.4 liters of Buffer A and 1.4 liters of Buffer A containing $0.4 M$ NaCl; (3) a linear gradient consisting of 220 ml of Buffer A containing $0.4 M$ NaCl and an equal volume of Buffer A containing $2 M$ NaCl.

Although a significant amount of activity appears in the initial column washings, the major portion elutes as a broad peak, beginning in the later part of the first salt gradient; the second, higher salt gradient is necessary to fully remove the bound enzymes. Activities for aryl sulfotransferases I and II (assayed at pH 7.4) and III and IV (assayed at pH 5.5) elute in an entirely parallel pattern. The active fractions are pooled to yield about 1 liter.

Step 3. Salt Precipitation. The sulfotransferases are concentrated by precipitation upon addition of 313 g per liter of solid ammonium sulfate during a half-hour period. After stirring for another 30 min, the suspension is centrifuged at $16,000 g$ for 30 min and the precipitate is dissolved in 2.5 mM sodium phosphate at pH 6.8 containing 3 mM 2-mercaptoethanol and $0.25 M$ sucrose. The enzyme is dialyzed overnight against two 4-liter changes of the same buffer; the precipitate accumulated upon dialysis is removed by centrifugation.

Step 4. DEAE-Cellulose. The protein solution is adjusted to pH 7.9 with $1 M$ Tris base and is applied to a column (2.5×17 cm) of DEAE-cellulose (DE-52, Whatman) previously equilibrated with Buffer A. The column is washed with 500 ml of Buffer A and eluted with a linear gradient established between 400 ml of Buffer A and 400 ml of Buffer A that is 0.3 M in sodium chloride.

This procedure allows the elution of aryl sulfotransferases I and II, in that order, followed by aryl sulfotransferases III and IV as a single peak of activity. Thus, there is significant separation of the transferases, except separation of transferases III and IV from each other. The solution containing enzymes III and IV is concentrated about 2-fold with an Amicon ultrafiltration cell and a PM-10 membrane. Each of the three pools of enzyme is separately dialyzed overnight against two 2-liter changes of Buffer B (10 mM sodium phosphate, 0.25 M sucrose, and 3 mM 2-mercaptoethanol; adjusted to pH 6.8).

Sulfotransferases I and II

Step 5. Hydroxylapatite. The dialyzed protein solutions of transferase I and transferase II are charged onto separate columns (1.5 × 9.5 cm) of hydroxylapatite that are equilibrated with Buffer B. The columns are washed with 30 ml of the same buffer and eluted with a linear gradient established between 150 ml of Buffer B and 150 ml of Buffer B containing 300 mM potassium phosphate; fractions of about 2.5 ml are collected.

Step 6. Carboxymethyl-Cellulose. Transferase II from Step 5 is dialyzed overnight against two 2-liter changes of Buffer C (10 mM sodium phosphate, 3 mM 2-mercaptoethanol, 0.2 mM Na$_2$EDTA, and 0.25 M sucrose; adjusted to pH 6.5). The dialyzed material is applied to a column (0.9 × 9 cm) of CM52 (Whatman) that had been equilibrated with Buffer C. The column is washed with 30 ml of Buffer C and eluted with a linear gradient established between 75 ml of Buffer C and 75 ml of the same buffer containing 0.2 M NaCl. Fractions of 2 ml are collected at a flow rate of about 40 ml per hour. Contaminating protein appears in the initial column washings, whereas the enzyme elutes with the gradient as a symmetrical peak corresponding to the major protein peak. Because of appreciable losses in activity, without significant gain in purification, it is not recommended that transferase I be subjected to this step.

Step 7. Gel Filtration. Transferase I from Step 5 and transferase II from Step 6 are separately concentrated to about 10 ml with an Amicon ultrafiltration cell using a PM-10 membrane. The concentrated materials are applied to columns (2.0 × 85 cm) of Sephadex G-100 equilibrated with Buffer D (50 mM sodium phosphate, 0.25 M sucrose, 0.2 mM EDTA, and 3 mM 2-mercaptoethanol; adjusted to pH 6.5). Elution is conducted with the same buffer at a flow rate of about 20 ml per hour. Each enzyme appears as a slightly asymmetric protein peak that corresponds closely with enzyme activity. At this stage, both enzymes represent more than 97% of the protein based on patterns obtained after SDS-gel electrophoresis.

Step 8. ATP-Agarose. Half of the pooled fractions of transferases I and II are applied, at 25°, to columns (0.5 × 1.5 cm) of agarose-hexane-adenosine 5'-triphosphate (P-L Biochemicals) that had been equilibrated with Buffer D. The columns are washed with 10 ml of the buffer and eluted with a linear gradient established between 20 ml of Buffer D and 20 ml of

Buffer D containing 45 μM PAPS. Fractions of 1 ml are collected at a flow rate of 40 ml per hour. The enzymes are each eluted as single peaks in which activity is coincident with protein. Despite the high purity of each of the transferases at earlier stages of preparation as determined by SDS-gel electrophoresis, this step increases specific activity by more than 30%, suggesting that the affinity resin may be removing not only traces of contaminating protein, but also inactive aryl sulfotransferases.

Aryl Sulfotransferases III and IV

Step 9. Hydroxylapatite. The dialyzed and concentrated solution obtained after DEAE-cellulose chromatography (Step 4) is applied to a column of hydroxylapatite[10] equilibrated with Buffer B. The column is washed with 50 ml of Buffer B and eluted with a linear gradient established between 250 ml of Buffer B and 250 ml of Buffer E (0.3 M potassium phosphate, 0.25 M sucrose, and 3 mM 2-mercaptoethanol; adjusted to pH 6.8). Sulfotransferases III and IV are partially resolved on this column; transferase III elutes as a trailing peak immediately after transferase IV. Higher ionic strength is necessary to elute transferases I and II, if they are present.

Step 10. ATP-Agarose. Pooled fractions of sulfotransferase IV from Step 9 are combined and dialyzed against 2 liters of Buffer F (0.1 M sodium acetate, 0.25 M sucrose, and 3 mM 2-mercaptoethanol; adjusted to pH 5.5). The dialyzed sample (5 ml) is applied to a column (0.9 × 2.5 cm) of agarose–hexane–adenosine 5'-triphosphate (P-L Biochemicals) previously equilibrated with Buffer F at 25°. The column at 25° is washed with 30 ml of Buffer F and then with 20 ml of Buffer C containing 20 μM PAPS. Sulfotransferase IV is eluted from the column at the beginning of the PAPS wash as a slightly trailing peak.

Step 11. Isoelectric Focusing. For studies in which PAPS must be excluded from the enzyme, isoelectric focusing may be used instead of Step 10. After chromatography on hydroxylapatite, the pooled fractions of transferase IV are concentrated in an Amicon cell with a PM-10 membrane and then dialyzed against two 2-liter changes of a pH 6.8 buffer containing 0.25 M sucrose, 2.5 mM sodium phosphate, and 3 mM 2-mercaptoethanol. The resulting protein solution is mixed with a linear sucrose gradient in a 110-ml isoelectric focusing column (LKB), using 1.5% (v/v) Ampholines (pH 3.5–10). Electrofocusing is carried out for 16 hr at constant power (15 W). The peak activity of sulfotransferase IV is collected in fractions with a pH between 5.6 and 5.8. This is well separated from the peak of sulfotransferase III activity, which is at pH 6.4. The enzyme prepared by isoelectric focusing, rather than by Step 10, is also homogeneous and at the same specific activity.

Storage

The transferases may be stored at 4° in sodium phosphate at pH 7.0 in the presence of 0.25 M sucrose, 5 mM mercaptoethanol, and 3 mM sodium azide; transferases I, II, and IV lose about 5% of their activity per week at a protein concentration of between 0.5 and 1 mg per milliliter. Transferase III may deteriorate at $-80°$ at any stage of purification and is definitely unstable to isoelectric focusing.

Properties

Physical Properties.[1,3,4] The three sulfotransferases that have been studied in detail, I, II, and IV, appear to be homogeneous by the criteria of disc gel electrophoresis and SDS-gel electrophoresis. Molecular weight, subunit size, and the isoelectric point have been determined for all four enzymes (Table III). Amino acid analyses are available for transferases I, II, and IV.[3,4]

pH Optima and Salt Effects. The effects of pH and salt on these enzymes are complex and are detailed elsewhere.[3,4] The major pH optimum for 2-naphthol is at 6.5 for transferases I and II and at 5.5 for transferases III and IV. The two groups of enzymes are also distinguishable on the basis of their activity at high salt concentrations: transferases I and II are activated appreciably in the presence of 0.5 M sodium chloride or potassium chloride, whereas transferase IV is strongly inhibited.

The difficulty in evaluating a possible phenolic substrate is that homogeneous preparations of these enzymes may display, with some substrates, two pH optima for a single enzyme and have a range of three pH units within which the pH optimum for a phenol can fall for any one transferase. For example, at pH 5.5 (the pH optimum for 2-naphthol),

TABLE III

PROPERTIES OF ARYL SULFOTRANSFERASES[a]

Property	Aryl sulfotransferase			
	I	II	III	IV
Molecular weight[b]	64,000	64,000	61,000	61,000
Subunit molecular weight[c]	35,000	35,000	33,500	33,500
Isoelectric point	8.1	6.9	6.4	5.8

[a] Data from Refs. 9–11.

[b] Estimated by gel filtration with Sephadex G-100.

[c] Estimated by sodium dodecyl sulfate disc gel electrophoresis.

TABLE IV
KINETIC CONSTANTS FOR THE ARYL SULFOTRANSFERASES[a]

| | Aryl sulfotransferase | | | | | |
| | I[b] | | II[b] | | IV[c] | |
Substrate	K_m (mM)	k_{cat} (min^{-1})	K_m (mM)	k_{cat} (min^{-1})	K_m (mM)	k_{cat} (min^{-1})
2-Naphthol	0.06	47	0.09	54	0.10	42
Phenol	1.8	6.8	2.6	7.3	0.92	1.8
4-Chlorophenol	1.5	48	1.2	40	0.34	18
4-Methoxyphenol	4.2	93	6.5	120	1.4	27
Epinephrine[d]	N	N	N	N	0.43	7.3
Tyrosine methyl ester[d]	N	N	N	N	0.91	14
Tyramine[e]	N	N	N	N	0.46	2.8
2-Cyanoethyl-N-hydroxythioacetimidate[d]	N	N	N	N	6.9	14

[a] Apparent K_m values were obtained with 0.1 mM PAPS. Data from Refs. 9–11. N indicates that no detectable sulfation occurred.
[b] Data obtained at pH 6.5 unless otherwise indicated.
[c] Data obtained at pH 5.5 unless otherwise indicated.
[d] Determined at pH 7.9.
[e] Determined at pH 9.0.

transferase IV does not catalyze the formation of sulfate esters (within the limits of the assay method,[7] from tyrosine methyl ester, epinephrine, octopamine, tyramine, serotonin, and N-acetylserotonin. Yet all are active at pH 9.0 and some at pH 7.9. Only the last two of these compounds serve as substrates for transferase I, which has a pH optimum for 2-naphthol at pH 6.5; however, with this enzyme, serotonin is active between pH 6.5 and 9.0, as is N-acetylserotonin, although the latter has its pH optimum at pH 6.5. It will be obvious that the pH activity relationship must be determined for each substrate in which there is an interest.

Substrates. The enzymes are specific for PAPS with apparent K_m in the standard assay system of 6.5 μM, 12 μM, and 24 μM for transferases I, II, and IV, respectively.

The activity of these enzymes with simple phenols has been detailed.[1,3–5] Aminophenols are better substrates for transferase IV than for I or II. Specifically, tyrosine methyl ester, epinephrine, and such peptides that contain N-terminal tyrosine residues (e.g., certain of the enkephalins[4] and cholecystokinin heptapeptide[15]) are substrates for transferase IV but not for I and II. Neither tyrosine itself nor peptides in which tyrosine is in

[15] M. Beinfeld and M. W. Duffel, unpublished data.

a position other than N-terminal are substrates for any of the purified sulfotransferases.[4] Transferase IV, but not transferases I or II, is active with organic hydroxylamines such as N-hydroxy-2-acetylaminofluorene and 2-cyanoethyl-N-hydroxythioacetimidate.[1,4] Kinetic constants for a few phenols and one hydroxylamine derivative are presented in Table IV.

Detailed studies of the substrate specificity of aryl sulfotransferase IV have indicated that electron-withdrawing substituents decrease the maximal velocity of phenol sulfation.[16] Phenols with many strongly electron-withdrawing substituents (e.g., pentachlorophenol) are dead-end inhibitors.[16]

Inhibitors. With phenols showing a low K_m, there is generally pronounced substrate inhibition at high substrate concentrations. The enzymes are also inhibited by thiol reagents,[17] reaction products (i.e., adenosine 3',5'-bisphosphate or aryl sulfates) and dead-end inhibitors such as 2,6-dichloro-4-nitrophenol and pentachlorophenol.[18]

[16] M. W. Duffel and W. B. Jakoby, *J. Biol. Chem.*, in press.
[17] D. J. Barford and J. G. Jones, *Biochem. J.* **123**, 427 (1971).
[18] G. J. Mulder and E. Scholtens, *Biochem. J.* **165**, 553 (1977).

[25] Hydroxysteroid Sulfotransferase

By Ellen Sue Lyon, Carol J. Marcus, Jun-Lan Wang, and William B. Jakoby

$$ROH + \text{adenosine 3'-phosphate 5'-phosphosulfate} \rightarrow$$
$$ROSO_3H + \text{adenosine 3',5'-bisphosphate}$$

The homogeneous preparations of hydroxysteroid sulfotransferase (EC 2.8.2.2) from rat liver catalyze the formation of sulfate monoesters from PAPS[1] with a very large number and variety of primary and secondary alcohols that include among them primary and secondary hydroxysteroids.

Assay Method

Principle. The assays for hydroxysteroid sulfotransferase are based on the separation of radiolabeled substrate from product on the basis of polarity. A filter method,[2] using radiolabeled steroid as substrate, is sim-

[1] Although adenosine 3'-phosphate 5'-phosphosulfate or 3'-phosphoadenylylsulfate are the recommended names for the compound, it was originally designated as 3'-phosphoadenosine 5'-phosphosulfate and abbreviated as PAPS. This abbreviation is retained here.
[2] R. D. Sekura, C. J. Marcus, E. S. Lyon, and W. B. Jakoby, *Anal. Biochem.* **95**, 82 (1979).

ple, rapid, and effective for following enzyme purification and is presented here in detail. A chromatographic assay[3] requires [35S]PAPS[4] but is more flexible because a larger number of substrates, including aliphatic alcohols, can be evaluated.

Reagents

Sodium acetate, 0.25 M, pH 5.5

Radioactive steroid, 10 mM : [14C]dehydroepiandrosterone (10 μCi/ μmol) is used here, although the ^{3}H-labeled steroid is equally effective. The steroids are prepared as stock solutions in acetone because a concentration of 5% or less of this solvent in the incubation mixture is not inhibitory. The solution also contains benzene, which is the solvent for the radioactive steroid as purchased

Buffer–steroid solution: 20 μl of the radioactive steroid solution are evaporated to dryness. To the vessel are added 200 μl of acetone and 1.8 ml of the sodium acetate buffer, in that order

PAPS,[5] 1.8 mM

Carrier steroid solution, 0.5 mM: nonradioactive steroid, dehydroepiandrosterone, in absolute ethanol

Procedure. In a total volume of 50 μl, the following are added in the indicated sequence: 20 μl buffer-steroid solution, 5 μl PAPS, and 0.1 to 1 unit of enzyme. After 30 min at 37°, the reaction is stopped by placing the containers in a boiling water bath for 2 min. Carrier dehydroepiandro-

---

[3] The chromatographic method is somewhat more time consuming but is a necessity for kinetic studies because blank values of close to zero can be attained[2] if commercial preparations of [35S]PAPS are purified[4] before use. With sterols as acceptor, the incubation mixture in a total volume of 50 μl includes 0.25 M sodium acetate at pH 5.5, 50 μM sterol, 60 μM [35S]PAPS (200 μCi/μmol), and 4% acetone. (The steroid is dissolved in acetone, which at this concentration is not inhibitory.) After addition of enzyme and incubation at 37° for 30 min, the reaction is terminated by placing the covered reaction vessels into a boiling water bath for 2 min.

With aliphatic alcohols, transferase activity is measured in a total volume of 100 μl containing 0.1 M Tris-chloride at pH 7.5; 80 μM [35S]PAPS (200 μCi/μmol), and the appropriate alcohol at a concentration of 5 mM. After addition of enzyme, the reaction is allowed to proceed for 20 min at room temperature and is terminated with 50 μl of acetone.

Aliquots, 10 to 15 μl, of the reaction mixtures are applied to a thin-layer strip (1 × 10 cm) of silica-coated gel (Eastman No. 13181) and developed with 2-propanol : chloroform : methanol : water (10 : 10 : 5 : 2) at room temperature. After the solvent has moved about 8 cm, the strips are removed and the solvent front is marked. A 1.2-cm section is cut below and including the solvent front and placed directly into a scintillation vial containing 10 ml of Aquasol.

[4] See footnote 7 of Article [24], this volume, for the purification of [35S]PAPS.

[5] A convenient organic synthesis of this nucleotide is presented by Sekura in this volume [53].

sterone (50 μl) is added and the vessel subjected to vortex mixing. An aliquot of 20 μl from this mixture is spotted onto a 3-cm square of a silica gel-impregnated glass fiber sheet that has been numbered with pencil for identification. After drying (hair dryer), as many as 50 squares may be soaked in 25% dioxane (v/v) in hexane contained in a 500-ml beaker. After 5 min, the wash fluid is replaced and, after another 5 min, the procedure is repeated. The steroid substrate is removed by the washing procedure, whereas the steroid sulfate remains bound to the square. Filters are dried and may be counted by adding them to 10 ml of Aquasol (New England Nuclear). Blank values are determined by using a parallel incubation mixture that is free of PAPS.

Definition of Units. A unit of activity is that amount of enzyme required for the formation of 1 nmol of product per minute in that standard assay. Specific activity is defined in terms of units of activity per milligram of protein; protein is determined colorimetrically.[6]

Purification Procedures[7-9]

Three hydroxysteroid sulfotransferases can be prepared in homogeneous form from the livers of female Sprague-Dawley rats (ARS Sprague-Dawley, Madison, WI). The enzymes are identified by arabic numerals in the order in which they are eluted from DEAE-cellulose. Only sulfotransferases 2 and 3 appear to be isolated in quantity from *male* rats of similar age.[10,11] All operations are carried out at 4°. A summary of the results of typical purifications is presented in Table I.

Step 1. Extraction. Livers from 25 female rats, about 250 g, are homogenized in a Waring blender for 45 sec in 500 ml of 10 mM Tris-HCl at pH 7.5 containing 0.25 M sucrose, 1 mM EDTA, and 3 mM 2-mercaptoethanol. The homogenate is centrifuged at 140,000 g for 90 min, the supernatant liquid removed by suction and the pellet discarded.

Step 2. DEAE-Cellulose. The supernatant liquid is applied directly to a column (5 × 30 cm) of DEAE-cellulose (DE-52; Whatman) that has been equilibrated with homogenization buffer. The column is washed with 2 liters of the buffer and is eluted thereafter with a 3-liter gradient of

[6] O. H. Lowry, N. J. Rosebrough, A. L. Farr, and R. J. Randall, *J. Biol. Chem.* **193**, 265 (1951).

[7] E. S. Lyon and W. B. Jakoby, *Arch. Biochem. Biophys.* **202**, 474 (1980).

[8] C. J. Marcus, R. D. Sekura, and W. B. Jakoby, *Anal. Biochem.* **107**, 296 (1980).

[9] E. S. Lyon, J.-L. Wang, and W. B. Jakoby, in preparation.

[10] S. S. Singer, D. Giera, J. Johnson, and S. Sylvester, *Endocrinology* **98**, 963 (1976).

[11] W. B. Jakoby, R. D. Sekura, E. S. Lyon, C. J. Marcus, and J.-L. Wang, *in* "Enzymatic Basis of Detoxication" (W. B. Jakoby, ed.), Vol. 2, p. 199. Academic Press, New York, 1980.

TABLE I
SUMMARY OF PURIFICATION OF HOMOGENEOUS RAT LIVER HYDROXYSTEROID
SULFOTRANSFERASES 1, 2, AND 3

Step	Volume (ml)	Total activity (units)	Total protein (mg)	Specific activity (units/mg)
Sulfotransferase 1[7]				
1. Extract	560	35,000	18,600	1.8
2. DEAE-cellulose	275	8,550	1,630	5.2
3. Salt precipitation	50	8,060	1,550	5.2
4A. Hydroxylapatite	110	1,030	1.8	570
5A. Gel filtration	45	910	1.6	570
Sulfotransferase 2[8]				
1. Extract	665	90,200	23,500	3.8
2. DEAE-cellulose	512	32,200	2,160	15.0
3. Salt precipitation	58	12,700	1,540	8.2
4B. Hydroxylapatite	30	782	4.4	178
Sulfotransferase 3[9]				
1. Extract	450	60,000	19,000	3.1
2. DEAE-cellulose	250	19,400	1,250	15.6
3. Salt precipitation	50	15,000	1,100	13.6
4C. Hydroxylapatite	110	9,450	82	115
5C. Isoelectric focusing	15	7,080	18	390
6. Dehydroepiandrosterone-Sepharose	6	1,200	1.6	740

homogenization buffer that increases linearly in KCl concentration from 0 to 0.3 M. Three peaks of sulfotransferase activity are evident that represent sulfotransferases 1, 2, and 3; each is pooled separately, although the peaks usually overlap.

Step 3. Salt Precipitation. To each of the pooled solutions of sulfotransferase is added 50 g of ammonium sulfate with stirring. After 1 hr, precipitated sulfotransferases 1 and 3 are collected separately by centrifugation and are dissolved in 50 ml of Buffer A (10 mM potassium phosphate, 0.25 M sucrose, 1 mM EDTA, and 3 mM 2-mercaptoethanol, adjusted to pH 6.8). The precipitate from sulfotransferase 2 is dissolved in 0.1 M potassium phosphate, pH 6.8, containing 0.25 M sucrose and 3 mM 2-mercaptoethanol.

Hydroxysteroid Sulfotransferase 1[7]

Step 4A. Hydroxylapatite. The enzyme preparation of sulfotransferase 1 is dialyzed overnight against two to three changes of 4 liters of Buffer A and clarified, if necessary, by centrifugation for 30 min at 40,000 g. The dialyzed preparation of sulfotransferase 1 is charged into a column (2 × 20 cm) of hydroxylapatite that has been equilibrated with Buffer A. The column is washed with 200 ml of Buffer A. A 300-ml salt gradient (in Buffer A), which increases linearly from 10 mM to 250 mM in potassium phosphate, is applied and is followed with 100 ml of Buffer A containing 250 mM potassium phosphate. The enzyme is eluted with a 600-ml linear salt gradient (in Buffer A), which increases in potassium phosphate from 250 to 600 mM (pH 6.8). Active fractions are pooled and concentrated by addition of 48 g of ammonium sulfate per 100 ml of solution. The precipitate is suspended in 2 ml of Buffer B (0.1 M potassium 3,6-endomethylene-1,2,3,6-tetrahydrophthalate, pH 6.5, containing 0.25 M sucrose, 1 mM EDTA, and 3 mM 2-mercaptoethanol).

Step 5A. Gel Filtration. The enzyme preparation is applied to a column (2.5 × 70 cm) of Sepharose 6B that is equilibrated and eluted with Buffer B. Enzyme is homogeneous and appears at a position to be expected for a globular protein of 180,000 daltons.

Sulfotransferase 2[8]

4B. Hydroxylapatite. The preparation of sulfotransferase 2, dialyzed overnight against three 2-liter volumes of Buffer C (10 mM potassium phosphate, 0.25 M sucrose, and 3 mM 2-mercaptoethanol; adjusted to pH 6.8), is applied to a column (2.5 × 19 cm) of hydroxylapatite that has been equilibrated with the same buffer. The column is washed with 350 ml of this buffer and is eluted with a 2-liter gradient established between Buffer C and Buffer C containing 0.7 M potassium phosphate. The enzyme elutes at about 13 mmho (4°) (about 0.5 M potassium phosphate) in a homogeneous form and can be concentrated 3-fold with an Amicon PM-10 membrane and stored at 4° in 0.05% sodium azide. Before use, the preparation may be dialyzed against 20 mM sodium 3,6-endomethylene-1,2,3,6-tetrahydrophthalate at pH 6.8, containing 0.25 M sucrose, 0.1 mM EDTA, and 3 mM 2-mercaptoethanol.

Sulfotransferase 3[9]

4C. Hydroxylapatite. The procedure is carried out as described for sulfotransferase 1 except that, after washing with 200 ml of Buffer A, enzyme is eluted directly with a linear gradient of 800 ml of Buffer A that increases from 10 mM to 400 mM in potassium phosphate. The pooled active fractions are concentrated and dissolved in a small volume of Buffer B.

Step 5C. Isoelectric Focusing. The preparation from Step 4C is inserted into a 0 to 40% sucrose gradient in the presence of 2.5% Ampholines (pH 5 to 7) contained in a 110-ml isoelectric focusing column (LKB). Focusing is carried out for 24 hr at constant power of 15 W. The active fractions are in the region of pH 6.1 and are pooled.

Step 6C. Dehydroepiandrosterone-Sepharose. To 5 ml of packed dehydroepiandrosterone-Sepharose[12] is added 10 ml of Buffer D (0.1 M Tris chloride, 0.2 M potassium chloride, 0.25 M sucrose, 0.1 mM EDTA, and 3 mM 2-mercaptoethanol; adjusted to pH 7.5). The enzyme solution from Step 5C is added and mixed with this suspension, and the mixture allowed to equilibrate for 1 hr with occasional stirring. After washing with 100 ml of Buffer D on a sintered glass funnel, the packed matrix is treated with Buffer D that is adjusted to pH 8.5 and that also contains 50 μM dehydroepiandrosterone[13] and 100 μM PAPS. The suspension is placed in a stoppered 25 ml graduate that is tumbled, end over end, for 3 hr at 4°. The eluting buffer is collected by filtration with a sintered glass filter and the gel washed with another 50 ml of the elution buffer. The filtrates are concentrated by Amicon ultrafiltration (PM-10 membrane) and dialyzed overnight against two changes of 500 ml of the homogenization buffer.

Properties[7-9,11]

Stability. Homogeneous sulfotransferases 1 and 3 are stable for several months at −80° but lose 10% of their activity per week upon storage at 4°. At no stage of purification is sulfotransferase 2 stable to freezing; at 4°, the enzyme loses from 10 to 20% of its activity per week.

---

[12] The affinity matrix is prepared by coupling the steroid amine to aminohexyl-agarose as previously described [A. H. J. Gijzen, *J. Mol. Med.* **2**, 99 (1977); J. B. Adams and D. McDonald, *Biochim. Biophys. Acta* **567**, 144 (1979)]. Dehydroepiandrosterone, 2 g, and α-aminoxyacetic acid (Calbiochem), 2.6 g, are dissolved in 500 ml ethanol and the pH of the mixture adjusted to 10 with 5 N NaOH. The mixture is refluxed for 8 hr, allowed to cool overnight, and concentrated to 50 ml by flash evaporation at 30°. After 50 ml of water are added, the mixture is extracted three times, each with an equal volume of ethyl ether. The pH of the aqueous layer (about pH 10) is adjusted to pH 6.5 with 1 N HCl to precipitate the oxime; it is collected by centrifugation. Additional oxime may be harvested by raising the pH of the supernatant solution to 10 and then lowering it again to pH 6.5. The combined yield is dissolved in 40 ml of tetrahydrofuran and allowed to evaporate overnight at room temperature. The residue is dissolved in water and extracted into ether from which dehydroepiandrosterone-17-(O-carboxymethyl)oxime crystallizes.

The oxime, 100 mg, is dissolved in 10 ml dimethylformamide and 2 ml of water are added. This solution is added to 1 g of AH Sepharose 4B in 200 ml of 0.5 M sodium chloride and the pH adjusted to 4.5 with 1 N HCl. 1-Ethyl-3-(3-dimethylaminopropyl)carbodiimide hydrochloride is added with stirring and the suspension mixed end-over-end at 20° overnight. The resultant gel is washed sequentially with one liter of 50% aqueous dimethylformamide and one liter of 10 mM Tris chloride at pH 7.5 containing 0.02% sodium azide. The gel may be stored at 4°.

[13] The steroid is added as a solution in 2 ml of acetone.

TABLE II
PROPERTIES OF THE HYDROXYSTEROID SULFOTRANSFERASES[a]

	Hydroxysteroid sulfotransferase		
	1	2	3
Molecular weight	180,000	290,000	120,000
Subunit weight	28,000	32,000	60,000
Isoelectric point	5.0	7.9	6.1
pH optimum	6.0	5.5	5.5
Specific activity (in nmol/min/mg)	570	180	750
Apparent K_m (in μM) for PAPS	12	47	20
Reaction with antibody to sulfotransferase 1	+	−	−

[a] All activity assays were conducted with PAPS and dehydroepiandrosterone as substrates. Data taken from Refs. 7–9 and 11.

Physical Characteristics. Each of the three hydroxysteroid sulfotransferases is obtained as a homogeneous protein. As noted in Table II, the enzymes differ in molecular weight, subunit structure, and isoelectric point. Sulfotransferase 2 has a strong tendency to aggregate[8] when present at concentrations above 0.3 mg/ml and this probably explains its low yield

TABLE III
HYDROXYSTEROID SULFOTRANSFERASES: APPARENT K_m AND OTHER
KINETIC PARAMETERS[a]

	Sulfotransferase 1		Sulfotransferase 2		Sulfotransferase 3	
	K_m' (μM)	k_{cat}/K_m' (mM^{-1} min^{-1})	K_m' (μM)	k_{cat}/K_m' (mM^{-1} min^{-1})	K_m' (μM)	k_{cat}/K_m' (mM^{-1} min^{-1})
Dehydroepiandrosterone	12	10,000	24	5600	22	4400
Testosterone	70	400	11	2000	29	280
β-Estradiol	35	1,000	15	1700	27	270
Cortisol	290	38	84	340	70	180
Ethanol	42,000	0.2	40,000	0.1	52,000	0.3
1-Butanol	3,000	5.6	3,300	1.6	3,000	2.4
Isoamyl alcohol	1,200	21	2,000	10	1,500	16
1-Amyl alcohol	1,700	16	—	—	1,200	18

[a] Data from Refs. 7–9 and 11.

upon purification. All three have a different amino acid composition[7-9] and the antibody to sulfotransferase 1 does not cross-react with sulfotransferases 2 or 3.

Substrates. With dehydroepiandrosterone and PAPS, the pH optimum is at 6.0, 5.5, and 5.5 for sulfotransferases 1, 2, and 3, respectively; with 1-butanol, the three enzymes have the same optimum at pH 7.5. PAPS serves as a sulfate donor with apparent K_m of 12, 45, and 20 μM for sulfotransferases 1, 2, and 3, respectively. Adenosine $3',5'$-bisphosphate is a competitive inhibitor of PAPS with each of the sulfotransferases.

Despite their disparate physical properties, the three enzymes are very similar in their substrate specificity. In addition to many hydroxysteroids, in which the functional hydroxyl group is at several different locations on the steroid nucleus (testosterone, corticosterone, hydrocortisone, aldosterone, 11-deoxycorticosterone, Δ^5-androstene-$3\beta,17\beta$-diol), the enzymes are active with primary and secondary aliphatic alcohols. Among these are ascorbic acid, chloramphenicol, ephedrine, retinol, ouabain, 2-propanol, and L- and D-propranolol. Apparent kinetic constants for a few substrates are presented in Table III.

Sulfotransferases 1, 2, and 3 do not utilize the following compounds as sulfate acceptors: cholesterol, 2-naphthylamine, 2-naphthol, taurolithocholic acid, estrone, progesterone and N-hydroxy-2-acetylaminofluorene.[7-9]

[26] Bile Salt Sulfotransferase

By LEE J. CHEN

Bile salt + adenosine $3'$-phosphate $5'$-phosphosulfate $\rightarrow$
bile salt sulfate + adenosine $3',5'$-bisphosphate

Assay Methods

Principle. Enzyme activity is determined by the incorporation of radioactive sulfate from adenosine $3'$-phosphate $5'$-phosphosulfate (abbreviated as PAPS) into a bile salt. Taurolithocholate is routinely used as substrate due to its higher solubility in water than glycolithocholate or lithocholate.[1] Two assay procedures are described here for the determina-

---

[1] When assayed with glycolithocholate or other bile salts, the steroid is prepared in methanol and the solution evaporated to dryness in the assay vessel prior to the addition of reaction mixture and enzyme.

tion of bile salt sulfotransferase. Method A,[2] which determines the incorporation of radioactive sulfate from PAPS into bile salt, can be used with any bile salt. Method B[3] employs glycolithocholate, coupled to Sepharose, as substrate; the assay is carried out directly in scintillation counting vials and does not require chromatographic separation of the reaction products. For routine assays, including purification steps, Method B provides a considerable saving in time over chromatographic methods. However, this procedure has limitations in studies of substrate specificity or of enzyme kinetics.

METHOD A

Reagents

Sodium phosphate, 0.1 M, pH 6.5, containing 1 mM MgCl$_2$

Taurolithocholate (1 mM): Dissolve 50.57 mg of taurolithocholate in 100 ml of distilled water; store at 4°

PAPS (0.5 mM)[4]

[^{35}S]PAPS (5 × 10^5 cpm/10 μl)

Procedure. Place 50 μl of the phosphate buffer, 10 μl of taurolithocholate, 10 μl cold PAPS, and 10 μl radioactive PAPS in a test tube (12 × 75 mm). Water is added to a final volume of 90 μl. The reaction is started by adding 10 μl of enzyme solution and incubated at 37° for 15 min until terminated with 0.4 ml methanol. After 5 min on ice, the precipitate is removed by centrifugation at 2600 rpm in a Sorvall GLC-1 centrifuge for 5 min at room temperature. The precipitate is washed with another 0.4 ml of methanol, and the combined supernatant solutions are evaporated to dryness under reduced pressure. The dried residue is dissolved in 50 μl of 60% ethanol, and 10 μl of this solution is applied to a thin-layer plate (Adsorbosil-5; Applied Science Lab. Inc.).

The plate is developed in a solvent system containing butanol : acetic acid : water (10 : 1 : 1) for 6 hr. Sulfated and nonsulfated bile salt standard are visualized by spraying the plate with 10% phosphomolybdic acid and heating the plate at 110° for 2 min. Radioactive areas are located by autoradiography. The radioactive taurolithocholate sulfate formed is scraped from the plate and counted in a Beckman SL-230 scintillation counter with 10 ml of Bray's scintillant.[5]

[2] L. J. Chen, R. J. Bolt, and W. H. Admirand, *Biochim. Biophys. Acta* **480**, 219 (1977).

[3] L. J. Chen, *Anal. Biochem.* **105**, 170 (1980).

[4] Preparation of PAPS, see J. Gregory and F. Lipman, *J. Biol. Chem.* **229**, 1081 (1957) and Ref. 1. Also available commercially from P-L Biochemicals, Milwaukee, WI. See also this volume, Article [53].

[5] G. A. Bray, *Anal. Biochem.* **1**, 279 (1960).

METHOD B

Reagents

Sodium phosphate (0.1 M), pH 6.5, containing 1 mM MgCl$_2$
Glycolithocholate-Sepharose 4B (GLC-Sepharose)[6]
PAPS (0.5 mM) containing 4 × 10^5 cpm [^{35}S]PAPS per 10 μl

Procedure. The reaction is carried out directly in a scintillation counting vial (Pico Vials, Packard Instruments). One-tenth milliliter of the
sodium phosphate solution is added into the vial first, followed by 50 μl of
GLC-Sepharose (wet gel: H$_2$O = 1:1) using an SMI micro/pettor. GLC-
Sepharose is agitated with a Vortex mixer prior to addition. The reaction
is initiated by adding enzyme. After 10 min of incubation at 37°, the
reaction is terminated by placing the reaction vessel in boiling water for 30
sec. Upon cooling, 3 ml of 0.3 M NaCl is added to the suspension. After
centrifugation at full speed in a clinical centrifuge (International Equipment Co., Needham, MA) for 5 min, the clear supernatant fluid is removed by aspiration and discarded. The washing procedure is repeated
twice. Radioactivity in the washed gel is measured in 5 ml of Bray's
scintillant.

Purification[2,7]

The procedure described in this section is based on 4 ml of rat liver
cytosol. However, so little protein remains after Step 4 PAP-Agarose
chromatography, six separate preparations are usually pooled at this point
for further purification. All operations are carried out at 4° unless otherwise stated (see table).

Step 1. Extract. Minced rat liver, 4 g, is homogenized in 20 ml of 0.25
M sucrose containing 5 mM Tris-HCl (pH 7.5), 1 mM EDTA, and 10 mM
2-mercaptoethanol, with a Potter-Elvehjem homogenizer and a Teflon pestle. The homogenate is centrifuged at 105,000 g for 60 min in a Beckman

---

[6] Preparation of glycolithocholate-Sepharose 4B is carried out at room temperature. AH-
Sepharose 4B (Pharmacia, Piscataway, NJ), 1.25 g, is allowed to swell in 100 ml 0.5 M
NaCl. It is washed on a Büchner funnel with 250 ml 0.5 M NaCl and then with 100 ml water
to remove excess NaCl. The Sepharose gel is added to a solution containing 45.56 mg
glycolithocholate dissolved in 10 ml 50% dioxane. The pH of the resulting mixture is
adjusted to 4.5 with 1 N HCl. 1-Ethyl-3-(3-dimethylaminopropyl)carbodiimide (BioRad
Lab. Richmond, CA), 300 mg, is added in portions to the solution. After stirring overnight,
the mixture is washed successively with 200 ml 1 M NaCl (pH 10) and 200 ml 1 M NaCl
(pH 3). Finally, the gel is washed with 1 liter of distilled water. About 1 μmol of
glycolithocholate per milliliter of wet gel is found to be coupled, as estimated by adding a
small amount of [24-^{14}C]glycolithocholate to the unlabeled glycolithocholate before attaching to Sepharose.

[7] L. J. Chen, T. Imperato, and R. J. Bolt, *Biochim. Biophys. Acta* **522**, 443 (1978).

SUMMARY OF PURIFICATION PROCEDURE[a]

Step	Volume (ml)	Total protein (mg)	Total activity (units)[b]	Specific activity (units/mg protein)
1. Supernatant fraction	4	33.0	760	23
2. DEAE-Sephadex	2	5.70	616	108
3. Sephadex G100	14	1.40	490	350
4. PAP-agarose	16	0.034	198	5,865
5. Isoelectric focusing	1	0.004	61	16,422

[a] Preparation based on 4 ml liver cytosol. Actually, data were obtained with 24 ml of supernatant solution.

[b] One unit of enzyme activity is defined as the amount of enzyme necessary to catalyze the sulfation of 1 pmol of bile salt per minute.

Spinco L2B ultracentrifuge and the resultant supernatant solution is passed through two layers of cheesecloth to remove lipid.

Step 2. DEAE-Sephadex A-50. Four milliliters of the resulting solution are applied to a column (1.5 × 20 cm) of DEAE-Sephadex A-50, which has been previously equilibrated with 5 mM Tris-HCl buffer (pH 7.5), containing 1 mM EDTA, and 10 mM 2-mercaptoethanol (TME buffer). The column is eluted with a linear gradient of NaCl in the same buffer. The reservoir contains 50 ml of 0.5 M NaCl in TME buffer, and the mixing flask contains 50 ml of the same buffer without NaCl. Fractions of 2 ml are collected at a rate of 1 ml/min; enzyme is eluted at a NaCl concentration of 0.12 M. The fractions containing the major enzyme activity are combined and 47.2 g/100 ml of powdered ammonium sulfate (70% saturation) are added with stirring to the pooled eluate. The precipitate is collected by centrifugation. At this stage the enzyme may be stored at −20° for several months without loss of enzyme activity.

Step 3. Sephadex G-100. The enzyme precipitate from Step 2 is dissolved in 1 ml TME buffer. The solution is charged onto a column (2 × 90 cm) of Sephadex G-100 that has been equilibrated with the same buffer. The enzyme is eluted after 44 ml of the buffer have passed through the column. Fractions containing the enzyme are combined and concentrated to one-fourth of the original volume by ultrafiltration with an Amicon apparatus and a PM-10 membrane.

Step 4. Agarose-Hexane-Adenosine 3′,5′-Bisphosphate (PAP-Agarose).[8] The concentrated enzyme preparation from Step 3 is placed on a column (1.5 × 20 cm) of PAP-agarose that has been equilibrated with TME buffer.

[8] Purchased from P-L Biochemicals, Milwaukee, WI.

The column is washed with 100 ml of the same buffer followed by a 200-ml linear gradient of NaCl from 0 to 1.0 M. The enzyme is eluted at approximately 0.25 M NaCl. The fractions containing the enzyme activity are pooled and dialyzed against 4 liters TME buffer for 16 hr.

Step 5. Isoelectric Focusing. Isoelectric focusing is performed according to the method originally described by Svensson[9] using a 110-ml column (LKB Instruments, Inc.) maintained at 4° by a circulating water bath. A 2% carrier Ampholine with a pH range of 3.5 to 10 is used. The pH gradient is established during electrophoresis following the sequential addition of ampholyte solution in a 0 to 47% (w/v) sucrose gradient. The dialyzed solution is applied in the center of the column. Electrophoresis is initiated at 200 V, increased to 1000 V during the next 12 hr, and continued at this voltage for at least another 12 hr. Upon completion of electrophoresis, fractions of 2 ml are collected and assayed for enzyme activity and pH. The active enzyme fractions are pooled and dialyzed against TME buffer and concentrated to one-tenth of the original volume by ultrafiltration with a PM-10 membrane.

Properties

Purity. The purified enzyme preparation shows one major and two minor proteins upon polyacrylamide disc gel electrophoresis.

Specificity. The enzyme catalyzes the sulfation of a number of different types of bile acids, conjugated as well as unconjugated. However, the rate of sulfation is higher with conjugated bile salts. Furthermore, the enzyme catalyzes sulfation of both primary (cholate and chenodeoxycholate) and secondary (deoxycholate and lithocholate) bile salts. The rates of sulfation, in decreasing order, are monohydroxylated > dihydroxylated > trihydroxylated. The purified enzyme does not react with estrone, estradiol, testosterone, dehydroepiandrosterone, cholesterol, or cortisol. The apparent K_m value for 3'-phosphoadenosine-5'-phosphosulfate is 8 μM and that for taurolithocholate is 50 μM.

Inhibitors and Activators. p-Chloromercuribenzoate and iodoacetate at 0.1 mM completely inhibit the enzyme. It is also inhibited by adenosine 3',5'-bisphosphate and to a lesser extent by ATP, EDTA, and NaN$_3$. Inhibition by Cu^{2+}, Fe^{2+}, and activation by Mg^{2+}, Mn^{2+}, and Co^{2+} have also been reported.[10]

Other Properties. A molecular weight of 130,000 has been estimated by gel filtration. The isoelectric point is at 5.3.

[9] H. Svensson, *Acta Chem. Scand.* **16**, 456 (1962).
[10] L. Lööf and S. Hjerten, *Biochim. Biophys. Acta* **617**, 192 (1980).

Distribution. The enzyme is present in the liver of rat,[2] human,[10] guinea pig,[11] rabbit,[12] and hamster,[13] but only in kidney of certain animals including the rat and, possibly, man. The enzyme also has been demonstrated in the proximal intestine and the adrenal gland of the hamster.[13]

[11] L. J. Chen, M. Thaler, R. J. Bolt, and M. Golbus, *Life Sci.* **22,** 1817 (1978).
[12] J. B. Watkins and E. Goldstein, *Gastroenterology* **76,** 1267 (1979).
[13] S. Barnes, P. Burhol, R. Zander, G. Haggstrom, R. Settin, and B. Hirschowitz, *J. Lipid. Res.* **20,** 952 (1972).

[27] Glutathione *S*-Transferases (Rat and Human)

By WILLIAM H. HABIG and WILLIAM B. JAKOBY

$$(1)$$

Although the glutathione *S*-transferases are the enzymes catalyzing conjugation reactions with glutathione as the first step in mercapturic acid synthesis [Reaction (1)], they are viewed in better perspective as the catalysts of any reaction in which glutathione (GSH) acts as a nucleophile, i.e., in which the glutathione thiolate ion (GS$^-$) is a participant.[1,2] This statement implies that any compound bearing a sufficiently electrophilic atom may be attacked: reactions with C, S, N, and O atoms have been demonstrated.[3-5] Some limitations apply since certain compounds fulfilling the aforementioned requirements do not serve as substrate; in these instances, other charged groups on the electrophilic substrate interfere.[4] In addition to their catalytic activity, these proteins, also known as ligandins, bind a large number of nonsubstrate ligands.[6-8]

[1] W. B. Jakoby, *Adv. Enzymol.* **46,** 383 (1978).
[2] W. B. Jakoby and W. H. Habig, *in* "Enzymatic Basis of Detoxication" (W. B. Jakoby, ed.), Vol. 2, p. 63. Academic Press, New York, 1980.
[3] J. H. Keen, W. H. Habig, and W. B. Jakoby, *J. Biol. Chem.* **251,** 6183 (1976).
[4] J. H. Keen and W. B. Jakoby, *J. Biol. Chem.* **253,** 5654 (1978).
[5] J. R. Prohaska and H. E. Ganther, *Biochem. Biophys. Res. Commun.* **71,** 952 (1977).
[6] G. Litwack, B. Ketterer, and I. M. Arias, *Nature (London)* **234,** 466 (1971).
[7] W. H. Habig, M. J. Pabst, G. Fleischner, Z. Gatmaitan, I. M. Arias, and W. B. Jakoby, *Proc. Natl. Acad. Sci. U.S.A.* **71,** 3879 (1974).
[8] A. W. Wolkoff, R. A. Weisiger, and W. B. Jakoby, *Prog. Liver Dis.* **6,** 213 (1979).

Assay

Elsewhere in this volume (Article [51]), the details of assays for the several activities of the glutathione transferases are described. Although these enzymes may be distinguished from each other by their characteristic substrate–activity spectrum, all of the transferases are active with 1-chloro-2,4-dinitrobenzene (CDNB) as substrate [Reaction (1)]. As described in Article [51], the reaction of this substrate and GSH occurs spontaneously and the assay, therefore, requires the use of a control from which enzyme is absent. It will be noted that not all of the data collected in the tables summarizing the purification procedure were obtained with 1-chloro-2,4-dinitrobenzene. Methyl iodide, 1,2-dichloro-4-nitrobenzene, and 1,2-epoxy-3-(p-nitrophenoxy)propane have also been used; in each case, the substrate is identified and reference to Article [51] will allow conversion of activity values from one substrate to another.

Definition of Units

All activities are expressed in terms of micromoles of product formed per minute and specific activity is defined as units of activity per milligram of protein; protein is determined colorimetrically[9] with bovine serum albumin as standard.

Enzyme Purification

Glutathione Transferases from Rat Liver

Rat liver is a most convenient source for purification because of the high concentration[10] of the transferases in this tissue; the transferases represent approximately 10% of the total soluble protein. Livers from male Sprague-Dawley rats of about 200 g in weight have been used after storage at $-70°$ for as long as several months. It is of importance that each of the transferases has been observed to be present in a single liver.[11]

Purification[12]

Step 1. Extraction. About 500 g of liver are partially thawed and homogenized with 1.5 liters of water in a Waring blender for 1 min. All steps are carried out at 4°. The homogenate is centrifuged for 1.5 hr at

[9] O. H. Lowry, N. J. Rosebrough, A. L. Farr, and R. J. Randall, Jr., *J. Biol. Chem.* **193**, 265 (1951).
[10] W. B. Jakoby, J. N. Keltey, and W. H. Habig, *in* "Glutathione: Metabolism and Function" (I. M. Arias and W. B. Jakoby, eds.), p. 213. Raven, New York (1976).
[11] W. H. Habig, M. J. Pabst, and W. B. Jakoby, *Arch. Biochem. Biophys.* **175**, 710 (1976).
[12] W. H. Habig, M. J. Pabst, and W. B. Jakoby, *J. Biol. Chem.* **249**, 7130 (1974).

10,000 g and the supernatant solution is filtered through a plug of glass wool to remove floating lipids.

Step 2. DEAE-Cellulose. The extract is passed through a column (15 × 22 cm) of DEAE-cellulose (Whatman DE-52) that has been equilibrated with 10 mM Tris-HCl at pH 8.0. The column is rinsed with the same buffer until no additional activity emerges. Typically, 85 to 90% of the activity with CDNB passes through this column. However, at least one glutathione transferase, transferase M,[13] is retained.[14]

Step 3. Salt Precipitation. The transferase activity passing through the DEAE-cellulose column is precipitated by addition of 66 g of ammonium sulfate per 100 ml of enzyme solution. After centrifugation at 10,000 g for 30 min, the precipitate is dissolved in 150 ml of 10 mM potassium phosphate at pH 6.7 and dialyzed against two changes of 2 liters each of this buffer for 10 hr.

Step 4. CM-Cellulose. The dialyzed preparation is applied to a column (5 × 40 cm) of CM-cellulose (Whatman 52), equilibrated with the same buffer. After rinsing with 2 liters of the buffer, the column is eluted with a 4-liter linear gradient of 0 to 200 mM KCl, in 10 mM potassium phosphate at pH 6.7. It is at this stage that the major transferases are separated from each other; a representative elution pattern is shown in Fig. 1. The transferases are named in the reverse order of their elution from this column.

The most readily eluted enzymes are glutathione transferases D and E; purification of transferase E to homogeneity[15] is presented in a subsequent section of this chapter; transferase D has not been purified. The preparation of homogeneous transferases AA, A, B, and C requires similar purification steps that are described in following sections.

GLUTATHIONE TRANSFERASES AA[11], A[16], B[12], AND C[12]

Step 5. Hydroxylapatite 1. The active fractions in each enzyme peak (AA, A, B, and C) are separately concentrated by ultrafiltration to about 35 ml using an Amicon PM-10 membrane. After dialysis against Buffer A (10 mM potassium phosphate, 30% glycerol (v/v), 1 mM GSH, and 0.1 mM EDTA, adjusted to pH 6.7), each pool is applied to a column (2 × 20 cm) of hydroxylapatite that has been equilibrated with Buffer A. The

[13] B. Gillham, *Biochem. J.* **121**, 667 (1971).

[14] We have partially purified transferase M by eluting the DEAE-cellulose column with a 2-liter linear gradient of 0 to 150 mM potassium chloride. Subsequent steps include salt precipitation, another DEAE-cellulose step, and elution from hydroxylapatite. While not homogeneous, transferase M had a specific activity of 0.11 μmol min^{-1} mg^{-1} of protein with α-menaphthyl sulfate, a substrate[13] having high activity with this enzyme.

[15] T. A. Fjellstedt, R. H. Allen, B. K. Duncan, and W. B. Jakoby, *J. Biol. Chem.* **248**, 3702 (1973).

[16] M. J. Pabst, W. H. Habig, and W. B. Jakoby, *J. Biol. Chem.* **249**, 7140 (1974).

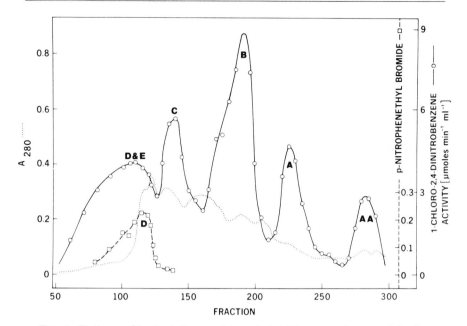

FIG. 1. Elution profile of rat liver protein and glutathione transferase activity from a column of carboxymethyl cellulose (Step 4). The salt gradient was begun at fraction 1. Peaks are labeled with letters used to designate individual transferases (transferase A, transferase AA, etc.). (· · · ·), Absorbance at 280 nm; (O——O), 1-chloro-2,4-dinitrobenzene activity.

column is rinsed with 200 ml of this buffer and eluted with a 600-ml linear gradient composed of Buffer A and Buffer A supplemented to contain 400 mM potassium phosphate at pH 6.7.

Step 6. Hydroxylapatite 2. The activity peaks from each column are pooled and concentrated by ultrafiltration. After dialysis, each pool is again applied to an identical hydroxylapatite column and eluted in a similar manner. The active fractions are pooled, concentrated, and dialyzed against Buffer A. The purification of each of these enzymes is summarized in Table I.

GLUTATHIONE TRANSFERASE E[15]

Transferase E has been purified by a somewhat different method, partly because of the low concentration and greater lability of this species of transferase. The assay substrate is 1,2-epoxy-3-(*p*-nitrophenoxy)propane (see Article [51]). The preparation described here is based on 2 kg of rat liver.

TABLE I
Summary of Purification of Major Rat Liver Glutathione Transferases

Step		AA			A			B			C	
	P	A[b]	SA	P	A[c]	SA	P	A[d]	SA	P	A[c]	SA
1. Extract[e]	60000	22750	0.4	61000	1100	0.018	6100	360	0.006	61000	1100	0.018
2. DEAE-cellulose	8200	19800	2.4	8700	940	0.11	8700	665	0.08	8700	940	0.11
3. Ammonium sulfate	Not recorded			7100	860	0.12	7100	665	0.09	7100	860	0.12
4. CM-cellulose	180	740	4.1	330	250	0.75	700	140	0.20	570	230	0.40
5. Hydroxylapatite 1	41	570	13.9	42	150	3.6	98	65	0.66	56	100	1.8
6. Hydroxylapatite 2	38	532	14.0	37	150	4.1	68	55	0.81	48	96	2.0

[a] P, Protein in milligrams; SA, Specific activity in micromoles per minute per milligram.

[b] Activity with 1-chloro-2,4-dinitrobenzene in micromoles per minute.

[c] Activity with 1,2 dichloro-4-nitrobenzene in micromoles per minute.

[d] Activity with methyl iodide in micromoles per minute.

[e] Extract prepared from 500 g of liver.

Step 1. Extraction. Rat liver is extracted in batches of 1 kg with water (2.5 ml/g) using a Waring blender. The combined homogenates are centrifuged for 30 min at 14,000 g and the supernatant solution is adjusted to pH 8.3 with 2 M Tris base. The resultant extract is diluted with distilled water to decrease the conductivity to below 1.3 mmho, yielding approximately 10 liters of solution.

Step 2. DEAE-Cellulose. One-half of the extract is applied to each of two columns (14.5 × 30 cm) of DEAE-cellulose (Whatman DE-52) previously equilibrated in 10 mM Tris-HCl, pH 8.25. After application of the extract, 4 liters of the equilibration buffer are used to wash each column. The total effluent from each column, about 10 liters, is adjusted to 50 mM phosphate by the addition of 1 M sodium phosphate at pH 6.5. Glutathione is added to 5 mM and the pH adjusted to 6.5 if necessary.

Step 3. Ammonium Sulfate. Ammonium sulfate (245 g/liter) is added to the effluent from the DEAE-cellulose columns. After stirring for 15 min, the precipitate is removed by centrifugation at 14,000 g for 20 min and is discarded. Additional ammonium sulfate (245 g/liter) is added to the supernatant solution. The resultant precipitate is suspended in 700 ml of 0.1 M sodium phosphate, pH 6.5, containing 5 mM GSH and is dialyzed against the same buffer overnight. After centrifugation for removal of insoluble material, the solution is concentrated to 250 ml by ultrafiltration using an Amicon U-2 membrane.

Step 4. Gel Filtration 1. The concentrated protein solution is divided into eight equal portions and applied to columns (5 × 97 cm) of Sephadex G-100 equilibrated with 0.1 M sodium phosphate, pH 6.5, containing 5 mM GSH. The most active fractions from each column are pooled and concentrated to approximately 15 ml each by ultrafiltration. The concentrates are themselves pooled.

Step 5. CM-Cellulose. The preparations from Step 5 are diluted to 25 mM sodium phosphate by the addition of three volumes of 5 mM GSH. The diluted protein is adjusted to pH 6.0 with acetic acid and concentrated to about 100 ml by ultrafiltration. One-half of this preparation is charged onto each of two columns (2.5 × 50 cm) of CM-cellulose (Whatman CM-52) previously equilibrated with 25 mM sodium phosphate, pH 6.0, containing 5 mM GSH. After applying the sample, each column is washed with 200 ml of starting buffer followed by a 1.1-liter linear gradient of 0 to 0.1 M sodium chloride in the same buffer. Three peaks of activity, as determined with the epoxide substrate, emerge: one in the wash, a major peak at the beginning of the gradient, and a third peak near the end of the gradient. The major peak, containing over 50% of the total activity is combined from both columns and is used in subsequent steps; the last peak of activity appears to be transferase B, which is also active with the epoxide substrate.

Step 6. Hydroxylapatite 1. The protein solution is diluted to 10 mM phosphate by addition of 5 mM GSH and adjusted to pH 6.8 with NaOH. After concentration to about 20 ml with an Amicon UM-2 membrane, the sample is applied to a column (2.5 × 30 cm) of hydroxylapatite equilibrated with 10 mM sodium phosphate, pH 6.8, containing 5 mM GSH. The column is washed with 50 ml of starting buffer, and developed with a 900-ml linear gradient of 0 to 300 mM potassium phosphate, pH 7.0, containing 5 mM GSH. The active fractions are pooled and concentrated to about 12 ml by ultrafiltration.

Step 7. Isoelectric Focusing. The preparation from Step 6 is further purified in two portions by isoelectric focusing using an LKB-110 column. Focusing is carried out for 40 hr at 500 V in a 1% solution of pH 3–10 Ampholine as ampholyte. The column is stabilized with a glycerol gradient (0–60% v/v) containing 1 mM mercaptoethanol. Fractions of 2 ml are collected; the active fractions between pH 7.0 and 7.5 are retained and adjusted to 50 mM sodium phosphate, pH 6.5, and 5 mM GSH.

Step 8. Gel Filtration 2. The solution from Step 7 is concentrated to about 3 ml and applied to a column (1.5 × 90 cm) of Sephadex G-100 equilibrated with 50 mM sodium phosphate, pH 6.5, 5 mM GSH, and 30% glycerol. The active fractions are pooled and concentrated by ultrafiltration to 2 ml.

Step 9. Hydroxylapatite 2. The last traces of contaminating protein are removed by applying the protein to a column (1.5 × 30 cm) of hydroxylapatite equilibrated with the buffer used in Step 8. After washing with 50 ml of the buffer, a 400-ml linear gradient of 0 to 300 mM potassium phosphate, pH 6.5, containing 5 mM GSH and 30% glycerol is used to elute the transferase activity. A similar degree of purity can be attained by repeating the gel filtration (Step 8). The enzyme solution is concentrated by ultrafiltration.

This purification procedure is summarized in Table II.

Human Glutathione Transferases

TRANSFERASES FROM LIVER[17]

In contrast to their counterparts from rat liver, the glutathione transferases in human liver seem to be the product of a single gene and form a more homogeneous group with respect to their spectrum of enzymatic activities.[17] The purification of several of these enzymes from human liver to the stage of homogeneity is described here. The enzymes have been designated as glutathione transferases α, β, γ, δ, and ϵ on the basis of

[17] K. Kamisaka, W. H. Habig, J. N. Ketley, I. M. Arias, and W. B. Jakoby, *Eur. J. Biochem.* **60**, 153 (1975).

TABLE II

SUMMARY OF PURIFICATION OF RAT LIVER GLUTATHIONE TRANSFERASE E

Step	Volume (ml)	Protein (mg)	Activity[a] (μmol/min)	Specific activity (μmol/ min/mg)
1. Extract	7,500	321,000	4,035	0.013
2. DEAE-cellulose	10,400	65,000	3,543	0.054
3. Ammonium sulfate	250	44,000	5,150	0.12
4. Gel filtration 1	90	9,400	4,140	0.44
5. CM-cellulose (main peak)	37	830	1,200	1.4
6. Hydroxylapatite 1	12.5	101	1,113	11.0
7. Isoelectric focusing	11.7	ND[b]	461	ND
8. Gel filtration 2	26.6	12	256	21.1
9. Hydroxylapatite 2	3.8	9	262	29.2

[a] Activity with 0.5 mM 1,2-epoxy-3-(p-nitrophenoxy)propane and 10 mM GSH in 0.2 M potassium phosphate, pH 6.5.

[b] ND, Not determined.

increasing isoelectric points. Another species of very low isoelectric point, transferase ρ, has been obtained in homogeneous form from human erythrocytes, but its preparation differs and is detailed separately.

The prototype purification, which is summarized in Table III, was carried out at 4° with the liver from a normal male who had died as the result of acute injury.[18] Preliminary studies have indicated that livers from patients with other pathological processes contain similar enzymes, although in different proportions.[19]

Step 1. Extract. Human liver, 375 g, that has been stored at −80°, is pulverized while frozen and then allowed to thaw in four volumes of distilled water. The suspension is homogenized in a Waring blender for 80 sec and centrifuged at 10,000 g for 2 hr. Floating lipid is removed from the extract by filtration through a plug of glass wool.

Step. 2. DEAE-Cellulose and Salt Precipitation. The extract is applied to a column (5 × 45 cm) of DEAE-cellulose (Whatman DE-52) that has been equilibrated with 10 mM Tris-HCl at pH 8.0. The column is washed with the same buffer until the elute is free of transferase activity. About 10% of

[18] The investigator contemplating purification from human tissue should be aware of the danger from hepatitis virus infection. The risk is particularly acute with tissue from individuals receiving multiple blood transfusions.

[19] W. H. Habig, K. Kamisaka, J. N. Ketley, M. J. Pabst, I. M. Arias, and W. B. Jakoby, *in* "Glutathione: Metabolism and Function" (I. M. Arias and W. B. Jakoby, eds.), p. 225. Raven, New York, 1976.

TABLE III

SUMMARY OF PURIFICATION OF HUMAN LIVER GLUTATHIONE TRANSFERASES[a]

Step	Preparation IA			Preparation IB			Preparation II			Preparation III			Preparation IV		
	P	A	SA	P	A	SA	P	A	SA	P	A	SA	P	A	SA
1. Extract	36400	16200	0.4				264	380	1.4	431	2660	6.2	1030	1070	1.0
2. Ammonium sulfate	6270	10400	1.7												
3. CM-cellulose	978	3210	3.3	112	614	5.5	59	341	5.8	177	3300	18.6	223	1680	7.6
4. Hydroxylapatite 1	138	800	5.8	70	517	7.4	34	296	8.8	148	3050	20.6	188	1520	8.1
5. Hydroxylapatite 2	104	744	7.2	43	325	7.6	25	254	10.1	76	2190	28.8	46	880	21.1
6. Gel filtration 1	39	410	10.5	30	262	8.6				62	1950	31.1	29	590	20.4
7. Gel filtration 2	27	406	15.1	3.3	56	19									
8. Isoelectric focusing[b]	5.6	91	16	4.9	77	16	2.1	35	17	20	730	37	5.1	170	34

[a] P, Protein in milligrams; A, activity with 1-chloro-2,4-dinitrobenzene in micromoles per minute; SA, specific activity in micromoles per minute per milligram.

[b] The preparations used here gave rise to specific transferases, each of which is denoted by a greek letter: preparation IA, β; preparation IB, α and β, in that order; preparation II, γ; preparation III, δ; preparation IV, ε.

the activity remains bound to the column; this presumably acidic group of transferases has not been investigated. Ammonium sulfate (660 g per liter) is added to the opalescent eluate (2.3 liters) from the column. The resultant suspension is centrifuged for 30 min at 10,000 g, and the precipitate dissolved in 220 ml of Buffer B (10 mM potassium phosphate at pH 6.7). Dialysis overnight is against the same buffer.

Step 3. CM-Cellulose. The dialyzed preparation is applied to a column (5 × 40 cm) of CM-cellulose (Whatman CM-52) that has been equilibrated with Buffer B. After washing the column with 1 liter of Buffer B, a 2-liter linear gradient of 0 to 150 mM KCl in Buffer B was used to elute the column. At this stage, four peaks of activity are evident: two peaks are in the initial wash and two are within the salt gradient. The active factions in each region are pooled and designated as preparations I, II, III, and IV in the order of elution.

Step 4. Hydroxylapatite 1. Subsequent purification steps are identical for each of the pools. Each preparation is concentrated to about 50 ml by ultrafiltration through an Amicon PM-10 membrane and dialyzed against Buffer C (10 mM potassium phosphate, 2 mM GSH, 0.1 mM EDTA, and 30% glycerol (v/v), adjusted to pH 6.7). These preparations are then applied to separate columns (2 × 20 cm) of hydroxylapatite, previously equilibrated with Buffer C. After washing each column with the same buffer, the transferases are eluted with a 600-ml linear gradient of 10 to 400 mM potassium phosphate, pH 6.7, containing 2 mM GSH, 0.1 mM EDTA, and 30% glycerol. Preparation I is partially resolved by this step into a peak of activity with a prominent shoulder; the shoulder and the peak tubes are pooled separately and designated as IA and IB; only a single pool of active fractions is obtained for preparations II, III, and IV.

Step 5. Hydroxylapatite 2. Each pool is concentrated to about 25 ml by ultrafiltration and dialyzed overnight against Buffer D (1 mM potassium phosphate, 30% glycerol, 1 mM GSH, 3 mM KCl, and 0.1 mM EDTA, adjusted to pH 6.7). Pools from IA, IB, III, and IV are each charged onto separate columns of hydroxylapatite (2 × 15 cm) equilibrated with Buffer D. The columns are washed with 200 ml of this buffer and then eluted, using a more shallow linear gradient (1 liter of 1 to 200 mM potassium phosphate, pH 6.7, in Buffer D).

Steps 6 and 7. Gel Filtrations 1 and 2. The more active fractions from each hydroxylapatite column are pooled and concentrated to about 15 to 20 ml by ultrafiltration. These pools (and the preparation II pool from the previous hydroxylapatite column) are separately applied to columns (5 × 90 cm) of Sephadex G-75 that have been equilibrated with 100 mM potassium phosphate at pH 7.4. After development with the same buffer, the active fractions are each concentrated to 10 ml by ultrafiltration, and the entire gel filtration step is repeated. The pooled solutions of the active fractions are adjusted to 1 mM GSH and 30% glycerol (v/v) prior to again concentrating them to 10 ml by ultrafiltration.

At this stage of purification, no preparation contains more than about 3% inactive protein as judged by gel isoelectric focusing. However, in some cases, more than one active species is found. Accordingly, a final preparative isoelectric focusing step is included to allow separation of each isoelectric species.

Step 8. Isoelectric Focusing. Approximately 10 mg of protein from each preparation is applied to an LKB-110 column containing 110 ml of a mixture of ampholytes prepared with 1% pH 7–10 Ampholine and 0.1% pH 3–10 Ampholine. The column is stabilized with a 0–40% sucrose gradient. Isoelectric focusing is carried out at 400 V for the first 8 to 12 hr and at 600 V for an additional 3 to 4 days. The final current is approximately 0.7 mA. Fractions of 1 ml are collected and analyzed for activity. The isoelectric points of the purified enzymes are as follows: α, 7.8; β, 8.25; γ, 8.55; δ, 8.75; and ϵ, 8.8. Those fractions of identical isoelectric point are pooled, concentrated, and applied to a column of Sephadex G-25 equilibrated in 0.1 M potassium phosphate at pH 6.5 containing 30% glycerol. The column is developed with the same solution. After concentration to about 5 ml, the preparations are supplemented to 1 mM GSH.

A summary of the results of purification of transferases α, β, γ, δ, and ϵ is presented in Table III.

TRANSFERASE ρ FROM ERYTHROCYTES[20]

A major impediment to the purification of glutathione transferase ρ from human erythrocytes, as with most other enzymes from this source, is the removal of the large amount of hemoglobin present in the lysates. This is reflected by the greater than 25,000-fold purification required to obtain homogenous transferase ρ. Identical purification is obtained from outdated units of packed erythrocytes and from freshly drawn blood.[18]

Step 1. Lysis. Two units (1 liter) of packed red blood cells are washed 5 times by centrifugation at 1500 g for 15 to 30 min with a 5-fold excess of cold 0.9% NaCl. The washed cells are lysed by adding distilled water (1.25 times the volume of packed cells) and shaking vigorously. After standing for 1 hr, the lysate is centrifuged for 1 hr at 13,000 g. If storage is desired, the supernatant solution may be frozen but should be clarified by centrifugation prior to continuing the procedure.

Step 2. SP-Sephadex. Because of the large amount of protein present, the following chromatographic step is carried out in portions. Hemolysate, 75 to 100 ml, is applied to a column (5 × 60 cm) of SP-Sephadex (Pharmacia C-50) that has been equilibrated with 60 mM sodium phosphate at pH 6.46. Each column is washed with the same buffer until eluate

[20] C. J. Marcus, W. H. Habig, and W. B. Jakoby, *Arch. Biochem. Biophys.* **188**, 287 (1978).

is free of transferase activity. The enzyme is eluted, whereas hemoglobin is retained.

Step 3. Salt Preparation. The active fractions from several SP-Sephadex columns are pooled and adjusted to contain 0.1 mM EDTA and 0.1 mM GSH prior to the addition of 16.4 g of ammonium sulfate per 100 ml. After stirring for 1 hr, the mixture is centrifuged and the precipitate discarded. Additional ammonium sulfate (40.2 g/100 ml) is added and, after stirring for 1 hr and centrifuging, the precipitate is dissolved in approximately 100 ml of Buffer E (10 mM potassium phosphate, 0.1 mM EDTA, and 0.1 mM GSH, adjusted to pH 7.8).

Step 4. DEAE-Cellulose 1. After thorough dialysis against Buffer E (three changes of 2 liters each of the buffer), half of the protein solution is applied to each of two columns (1.5 × 30 cm) of DEAE-cellulose (Whatman DE-52), previously equilibrated with the same buffer supplemented with 1 mM GSH. After washing the column with 1 liter of this buffer, the transferase is eluted with 2 liters of a linear gradient of 0 to 300 mM KCl in the same buffer.

Step 5. Hydroxylapatite. The active fractions from both DEAE-cellulose columns are pooled and dialyzed against Buffer F (10 mM potassium phosphate, 0.1 mM EDTA, 5 mM GSH, and 30% glycerol, adjusted to pH 6.7) and then concentrated to 60 ml by ultrafiltration using an Amicon PM-10 membrane. This entire sample is applied to a column (2 × 23 cm) of hydroxylapatite that has been equilibrated with Buffer F. The enzyme is eluted by simply washing with Buffer F.

Step 6. DEAE-Cellulose 2. The active fractions from the previous column are pooled, dialyzed against Buffer G (10 mM Tris-HCl, 5 mM GSH, 0.1 mM EDTA, and 30% glycerol, adjusted pH 7.75) and applied to a small column (0.9 × 26 cm) of DEAE-cellulose equilibrated with Buffer G. The column is washed with 100 ml of the buffer and then eluted with an 800-ml linear gradient of 0 to 250 nM KCl in Buffer G.

Step 7. Sephadex G-75. After concentration of the active fractions to about 2 ml by ultrafiltration, the transferase activity is applied to a column (1.5 × 44 cm) of Sephadex G-75 in 0.1 M potassium phosphate, pH 7.0, containing 5 mM GSH, 0.1 mM EDTA, and 30% glycerol. The column is washed with the same buffer and most active fractions are pooled.

The purification procedure is summarized in Table IV.

Properties[1,2]

Substrates and Ligands

The glutathione transferases form a family of proteins that have the capability of binding lipophilic compounds and, at a different site, glutathione. It is the close juxtaposition of the two substrates on the enzyme

TABLE IV
SUMMARY OF THE PURIFICATION OF GLUTATHIONE TRANSFERASE ρ

STEP	Volume (ml)	Protein (mg)	Activity[a] (μmol/min)	Specific activity (μmol/min/mg)
1. Hemolysate[b]	710	145,000	328	0.002
2. SP-Sephadex	1175	4,340	191	0.044
3. Ammonium sulfate	92	1,610	169	0.105
4. DEAE-cellulose 1	231	162	125	0.77
5. Hydroxylapatite	37	5.6	126	22.5
6. DEAE-cellulose 2	16.2	2.22	71	31.9
7. Gel filtration	12	0.56	37	66.0

[a] Activity with 1-chloro-2,4-dinitrobenzene as substrate.
[b] From erythrocytes in 1 liter of blood.

(a proximity effect mechanism) that is thought to account for the catalytic activity of the protein.[4] With this simple catalytic mechanism, a large number and variety of reactions are within the scope of these enzymes: thioether formation,[1,2] reduction of nitrate esters[3]; cyanide formation from organic thiocyanates[3]; isomerization[21]; thiolysis[4]; disulfide interchange[4]; and glutathione perioxidase activity without selenium.[5] Details of assays for most of these reactions are in Articles [51] and [44]; specific activities are presented in Article [51].

By reason of its binding ability, much like that of albumin in the circulation, this group of proteins seems to act as storage proteins intracellularly and, in this role, have been named as ligandins.[6-8] Bilirubin (as well as its glucuronides), indocyanin green, and hematin are all tightly bound with binding constants in the micromolar range[22]; none of these ligands, however, has a sufficiently electrophilic carbon to serve as a substrate.

Physical Properties

The homogeneous transferases prepared from human and rat liver represent a group that, aside from overlapping substrate specificity, resemble each other in size ($M_r = 45,000$ to $49,000$) and subunit number (2). Although transferases A and C are similar to a greater degree, by reason of cross-reactivity to antibody raised against either protein, the other transferases from rat liver seem to be quite separate species. This is in contrast to the enzyme from human liver, all of which may be derived

[21] A. M. Benson, P. Talalay, J. H. Keen, and W. B. Jakoby, *Proc. Natl. Acad. Sci. U.S.A.* **74**, 158 (1977).
[22] J. N. Ketley, W. H. Habig, and W. B. Jakoby, *J. Biol. Chem.* **250**, 8670 (1975).

from a single gene product by successive deamidation.[17] Evidence for the closer relationship among the human liver transferases includes cross-reactivity of all enzyme species with antibody raised against the most basic of the charge isomers (transferase ϵ) and the general lower degree of diversity in catalytic activity with a range of substrates.[17]

The characterization of each of the transferases is presented in the specific publication reporting purification, although this information has been summarized in reviews.

[28] Glutathione Transferase (Human Placenta)

By BENGT MANNERVIK and CLAES GUTHENBERG

Glutathione transferases are a group of ubiquitous enzymes believed to serve essential functions in the biotransformation of xenobiotics.[1] The major part of the enzyme activity is located in the cytosol of all tissues investigated and the concentration of the enzymes amounts to several percent of the cytosolic proteins in many tissues. Glutathione transferases with basic isoelectric points have been purified from human liver,[2] and an acidic form from human erythrocytes.[3] An acidic form, which may be identical with the erythrocyte enzyme, has been found in high concentrations in human placenta.[4,5]

Assay Method

Principle. Enzyme activity during purification is determined spectrophotometrically at 340 nm by measuring the formation of the conjugate of glutathione (GSH) and 1-chloro-2,4-dinitrobenzene (CDNB).[6,7]

[1] W. B. Jakoby and W. H. Habig, *in* "Enzymatic Basis of Detoxication" (W. B. Jakoby, ed.), Vol. 2, p. 63. Academic Press, New York, 1980.

[2] K. Kamisaka, W. H. Habig, J. N. Ketley, I. M. Arias, and W. B. Jakoby, *Eur. J. Biochem.* **60**, 153 (1975).

[3] C. J. Marcus, W. H. Habig, and W. B. Jakoby, *Arch. Biochem. Biophys.* **188**, 287 (1978).

[4] C. Guthenberg, K. Åkerfeldt, and B. Mannervik, *Acta Chem. Scand.* **B33**, 595 (1979).

[5] G. Polidoro, C. Di Ilio, G. Del Boccio, P. Zulli, and G. Federici, *Biochem. Pharmacol.* **29**, 1677 (1980).

[6] W. H. Habig, M. J. Pabst, and W. B. Jakoby, *J. Biol. Chem.* **249**, 7130 (1974).

[7] W. H. Habig and W. B. Jakoby, this volume, Article [27].

PURIFICATION OF GLUTATHIONE TRANSFERASE FROM HUMAN PLACENTA

Step	Volume (ml)	Total protein (mg)[a]	Total activity (units)	Specific activity (units/mg)
1. Supernatant fraction	1800	24000	5300	0.22
2. Sephadex G-25	2550	25000	4800	0.19
3. DEAE-cellulose	750	2800	2000	0.71
4. CM-cellulose	870	409	1740	4.25
5. Affinity chromatography	27	17.6	1370	77.8
6. Sephadex G-75	102	11.9	1250	105

[a] Protein was estimated from the absorbance at 260 and 280 nm,[8] except in the last step where the lower (and more accurate) microbiuret method[9] was used.

Reagents

Sodium phosphate, $0.2 M$, pH 6.5
GSH, 20 mM, in deionized water
CDNB, 20 mM in 95% ethanol

Procedure. To a 1-ml cuvette are added 500 μl of buffer, 50 μl of GSH, a suitable amount of enzyme and deionized water to final volume of 1 ml. The reaction, which is carried out at 30°, is started by addition of 50 μl of CDNB. The reaction is monitored spectrophotometrically by the increase in absorbance at 340 nm ($\epsilon_{340} = 10$ mM^{-1} cm^{-1}). A correction for the spontaneous reaction is made by measuring and subtracting the rate in the absence of enzyme.

Definition of Unit and Specific Activity. A unit of enzyme activity is defined as the amount of enzyme that catalyzes the formation of 1 μmol of S-2,4-dinitrophenylglutathione per minute at 30° using 1 mM concentrations of GSH and CDNB. Specific activity is expressed as units per milligram of protein. Protein concentrations given in the purification scheme are based on absorbance at 260 and 280 nm[8] after correction for hemoglobin in the supernatant fraction. For solutions of pure enzyme, protein concentration was determined by a microbiuret method.[9]

Purification Procedure

The enzyme should be maintained at about 5° during the entire purification procedure (see table). All buffers contain 1 mM EDTA.

Step 1. Extraction from Human Placenta. Full term placentas are kept at 5° and processed within 12 hr after parturition. A total of 900 g of placental tissue (corresponding to about 1.5 to 2 placentas), cleared of amniotic

[8] H. M. Kalckar, *J. Biol. Chem.* **167**, 461 (1947).
[9] J. Goa, *Scand. J. Clin. Lab. Invest.* **5**, 218 (1953).

membranes, is washed with cold 0.25 M sucrose. The tissue is cut into small pieces with scissors and homogenized in 0.25 M sucrose in a blender. The homogenate is diluted with sucrose solution to give a concentration of 35–40% (w/v) and centrifuged at 15,000 g for 60 min. The supernatant fraction is filtered through cheesecloth (or any other suitable tissue) to remove floating lipid material.

Step 2. Sephadex G-25. The supernatant fraction from Step 1 is chromatographed on a column (12.5 × 80 cm) of Sephadex G-25 Coarse equilibrated with 10 mM Tris-HCl at pH 7.8. The active fractions are pooled.

Step 3. DEAE-Cellulose. The pooled fractions from Step 2 are applied to a column (9 × 13 cm) of DEAE-cellulose equilibrated with 10 mM Tris-HCl at pH 7.8. The bed is rinsed with 1500 ml of the same buffer. The adsorbed enzyme is eluted with a linear concentration gradient formed by linear mixing of 6.2 liter starting buffer with 6.2 liter of 0.2 M NaCl in the same buffer. Fractions showing enzyme activity are pooled.

Step 4. CM-Cellulose. The pooled effluent from Step 3 is dialyzed overnight against 3 × 10 liter of 10 mM sodium phosphate buffer (pH 6.2). The dialyzed solution is applied to a column of CM-cellulose [9 (diameter) × 6 cm], equilibrated and eluted with the same buffer. The enzyme passes unadsorbed through the bed.

Step 5. Affinity Chromatography on S-Hexylglutathione Coupled to Epoxy-Activated Sepharose 6B. The pooled effluent from step 4 is applied to an affinity column (2 × 15 cm) prepared as described later and packed in 10 mM Tris-HCl at pH 7.8. Nonspecifically adsorbed material is eluted with 200 ml of 0.2 M NaCl in the starting buffer. Glutathione transferase is eluted with a linear concentration gradient formed by mixing 200 ml of 0.2 M NaCl in 10 mM Tris-HCl (pH 7.8) with 200 ml of the same buffer system also containing 5 mM S-hexylglutathione. The active fractions are pooled and concentrated to 25 ml by ultrafiltration with an Amicon PM-10 membrane.

Step 6. Sephadex G-75. The pooled fractions from Step 5 are charged onto a column (4 × 140 cm) of Sephadex G-75 that is equilibrated and eluted with 10 mM sodium phosphate at pH 6.7, containing 0.1 mM dithioerythritol. The active fractions of effluent are pooled.

Preparation of Affinity Gel[10]

Epoxy-activated Sepharose 6B from Pharmacia Fine Chemicals is coupled with ligand essentially as prescribed by the manufacturer. The gel (15 g) is swollen in deionized water for 15 min and washed with 1.5 liter of

[10] A.-C. Aronsson, E. Marmstål, and B. Mannervik, *Biochem. Biophys. Res. Commun.* **81**, 1235 (1978).

deionized water on a sintered glass filter. S-Hexylglutathione (1.5 g) is stirred in 70 ml of deionized water and adjusted with 2 M NaOH to about pH 12. The resulting solution is combined with the gel suspension and the coupling reaction is allowed to proceed for 16 hr at 30° with gentle shaking. Excess ligand is washed from the gel with deionized water on a sintered glass filter. The gel is then treated with 50 ml of 2 M ethanolamine (pH 9) for 4 hr of 30° with gentle shaking. After additional washing with deionized water, the gel is treated three times, alternatingly, with 0.1 M sodium acetate at pH 4.0, and 0.1 M Tris-HCl buffer at pH 8.0; both buffers are 0.5 M in NaCl. The gel is stored at 4° in the presence of 0.02% (w/v) sodium azide to prevent microbial growth.

Synthesis of S-Hexylglutathione

Method A of Vince *et al.*[11] is used. GSH (3.0 g) is dissolved in 10 ml of deionized water and mixed with 10 ml of 2 M NaOH. Ethanol (95%) is added to the cloud point; approximately 60 ml are required. Iodohexane (2.0 g) is added in portions with vigorous stirring. After standing overnight, or for a minimum of 3 hr at 22°, the reaction mixture is adjusted to pH 3.5 by dropwise addition of 47% HI and the solution is chilled in a refrigerator. The precipitated glutathione derivative is collected on a sintered glass filter and washed with cold water. The product is dried over silica gel under reduced pressure.

Properties

Purity of the Enzyme. The purified glutathione transferase is homogeneous in several analytical electrophoretic and chromatographic systems. In the presence of SDS, the enzyme dissociates into subunits of equal size. Various immunological tests based on electrophoresis and diffusion against antibodies raised against the purified enzyme also provide evidence for homogeneity.

Molecular Properties. The enzyme (M_r = 47,000) is composed of two subunits and has an isoelectric point of 4.8. The enzyme does not give precipitates with antibodies raised against any of the following: rat liver transferases A, B, or C; the basic human liver transferases[2] ($\alpha-\epsilon$); and the newly discovered liver transferase (μ)[12,13] that is distinct from $\alpha-\epsilon$. On the other hand, antibodies raised against the placental enzyme reacts with

[11] R. Vince, S. Daluge, and W. B. Wadd, *J. Med. Chem.* **14**, 402 (1971).
[12] M. Warholm, C. Guthenberg, B. Mannervik, C. von Bahr, and H. Glaumann, *Acta Chem. Scand.* **B34**, 607 (1980).
[13] M. Warholm, C. Guthenberg, B. Mannervik, and C. von Bahr, *Biochem. Biophys. Res. Commun.* **98**, 512 (1981).

the transferase in human erythrocytes.[3] This finding and the similarities in substrate specificity strongly indicate that the placental and erythrocyte enzymes are very similar or identical. The amount of blood contained in the placental tissue used for purification is such that less than 1% of transferase could originate from blood rather than placenta.

Kinetic Properties. Analysis of the steady-state kinetics, using a simple sequential model for data obtained with GSH and CDNB as substrates, gave K_m values of 0.5 and 2.1 mM, respectively, for these reactants. Specific activities, in micromoles per minute per milligram of protein, for alternative electrophilic substrates used at a single concentration[13] are as follows: 1-chloro-2,4-dinitrobenzene, 105; 3,4-dichloro-1-nitrobenzene, 0.11; *trans*-4-phenyl-3-buten-2-one, 0.01; styrene oxide, 0.07; 1,2-epoxy-3-(*p*-nitrophenoxy)propane, 0.37; and cumene hydroperoxide, 0.03; *p*-nitrophenyl acetate. Sulfobromophthalein (<0.002) and hydrogen peroxide (<0.01) were not substrates within the noted detection limit.

[14] C. Guthenberg and B. Mannervik, *Biochim. Biophys. Acta* (in press).

[29] Purification of Glutathione S-Transferases by Glutathione-Affinity Chromatography

By Peter C. Simons and David L. Vander Jagt

The glutathione S-transferases from a variety of sources have been purified by use of conventional chromatographic procedures as well as by affinity chromatography.[1–4] When affinity chromatography is used, it is generally necessary to apply additional separation procedures because most of the glutathione S-transferases contain several similar forms of the enzyme.

Here we describe the preparation of a glutathione-affinity chromatography column that has been successfully used for the purification of the glutathione S-transferases from human liver.[4] This procedure can serve as the initial step in the isolation of the family of glutathione S-transferases from a crude mixture, or it may be used as a final purification step after the several forms of the glutathione S-transferases are separated from one another.

[1] W. H. Habig, M. J. Pabst, and W. B. Jakoby, *J. Biol. Chem.* **249**, 7130 (1974).

[2] T. A. Fjellstedt, R. H. Allen, B. K. Duncan, and W. B. Jakoby, *J. Biol. Chem.* **248**, 3702 (1973).

[3] A. G. Clark, M. Letoa, and W. S. Ting, *Life Sci.* **20**, 141 (1977).

[4] P. C. Simons and D. L. Vander Jagt, *Anal. Biochem.* **82**, 334 (1977).

METHODS IN ENZYMOLOGY, VOL. 77

Preparation of the Glutathione-Affinity Matrix

Epoxy-activated Sepharose 6B (Sigma), 4 g, is washed on a Büchner funnel with 500 ml of water. The gel is washed with 40 ml of a 44 mM phosphate buffer, pH 7.0, prepared by mixing 8 ml of 0.2 M KH$_2$PO$_4$ and 28 ml of 0.1 M Na$_2$HPO$_4$ per 100 ml of solution. The gel is transferred to a 50-ml flask where the volume is adjusted to 20 ml with the same buffer. Nitrogen is passed through the flask and suspension for 5 min; 4 ml of a solution of glutathione (400 mg of glutathione in 4 ml of water, adjusted to pH 7 with KOH) is added to the suspension. After 24 hr at 37° with gentle shaking on a Dubnoff metabolic incubator (ca. 60 rpm), the coupled gel is washed with 100 ml of water, and the remaining active groups are blocked by allowing the gel to stand in 1 M ethanolamine for 4 hr. The gel is washed sequentially with 100 ml each of 0.5 M KCl in 0.1 M sodium acetate (pH 4.0), 0.5 M KCl in 0.1 M sodium borate (pH 8.0), and the starting buffer for the column.

General Procedure for the Purification of the Glutathione S-Transferases Using Glutathione-Affinity Chromatography

The glutathione-affinity matrix prepared as described has a capacity of about 0.2–0.4 mg of glutathione S-transferase per milliliter of bed volume. This capacity is observed both with crude preparations and with partially purified enzyme. The following is a representative procedure: A partially purified glutathione S-transferase, 25 ml of solution with specific activity 4.1 μmol min^{-1} mg^{-1}, total protein 48 mg, is added to a glutathione-affinity column (1 × 15 cm) that had been equilibrated with 22 mM potassium phosphate buffer, pH 7.0. The column is washed with this buffer until no protein is recorded in the effluent passing through an LKB Uvicord detector. At that stage, the column is developed with 0.05 M Tris-HCl (pH 9.6 at 4°) containing 5 mM glutathione. Activity appears as a single sharp peak of activity. For the example presented the purified enzyme (specific activity 83 μmol min^{-1} mg^{-1}) was obtained in 98% yield.

Comments

1. The glutathione-affinity column described here has been used to purify glutathione S-transferases from human, rat, and porcine liver. In all cases, the column profile is similar to that described by us for the human liver enzymes.[4] Recoveries of activity are generally greater than 90%.

2. Several glutathione-requiring enzymes, including yeast glyoxalase I (S-lactoyl-glutathione methylglyoxal-lyase), wheat germ glutathi-

one reductase, and porcine kidney γ-glutamyltranspeptidase, did not bind to the affinity column. The stability of the column is probably the result of its low affinity for the transpeptidase.

3. The affinity column can be regenerated by washing with 3 M NaCl, followed by equilibration with the starting buffer. After repeated use, the column material tends to clump. At this point, the column material is removed and forced through a fine mesh screen. The product can be used to pack a new column, which shows similar capacity and flow rate to the original.

4. This affinity procedure provides glutathione S-transferase that shows a specific activity comparable to the highest values reported in the literature for purified enzyme.

[30] γ-Glutamyl Transpeptidase

By Alton Meister, Suresh S. Tate, and Owen W. Griffith

Introduction

γ-Glutamyl transpeptidase catalyzes transfer of the γ-glutamyl (γ-glu) moiety of glutathione, S-substituted glutathione derivatives, and other γ-glutamyl compounds to a number of acceptors.[1] When the acceptor is an amino acid (or a dipeptide), a γ-glutamyl amino acid (or a γ-glutamyl dipeptide) is formed [Reaction (1)]. If the γ-glutamyl donor is also the acceptor, autotranspeptidation occurs [Reaction (2)]. When the nucleophile is water, the overall result is hydrolysis [Reaction (3)]. The most active amino acid acceptors of the γ-glutamyl moiety include L-cystine[2] and L-glutamine.[3] L-Methionylglycine, L-glutaminylglycine, L-alanylglycine, L-cystinylbisglycine, L-serylglycine, and glycylglycine are among the most active dipeptide acceptors.[4] Glutathione disulfide is also a γ-glutamyl donor substrate of the enzyme, although it is somewhat less active than glutathione.

[1] Several reviews on glutathione and γ-glutamyl transpeptidase have been published: A. Meister, (1965), "Biochemistry of the Amino Acids," 2nd ed., Vol. 1, p. 452. Academic Press, New York, 1965; A. Meister, *in* "Metabolic Pathways" (D. M. Greenberg, ed.), 3rd ed., Vol. 7, p. 101. Academic Press, New York, 1977; A. Meister and S. S. Tate, *Annu. Rev. Biochem.* **45**, 559 (1976); S. S. Tate, *in* "Enzymatic Basis of Detoxication" (W. B. Jakoby, ed.), Vol. 2, p. 95. Academic Press, New York, 1980.
[2] G. A. Thompson and A. Meister, *Proc. Natl. Acad. Sci. U.S.A.* **72**, 1985 (1975).
[3] S. S. Tate and A. Meister, *J. Biol. Chem.* **249**, 7593 (1974).
[4] G. A. Thompson and A. Meister, *J. Biol. Chem.* **252**, 6792 (1977).

$$\text{Glutathione} + \text{amino acid} \rightleftharpoons \gamma\text{-glu-amino acid} + \text{CYSH-GLY} \qquad (1)$$
$$2 \text{ Glutathione} \rightleftharpoons \gamma\text{-glu-glutathione} + \text{CYSH-GLY} \qquad (2)$$
$$\text{Glutathione} + \text{H}_2\text{O} \rightarrow \text{glutamate} + \text{CYSH-GLY} \qquad (3)$$

Kinetic studies indicate a ping-pong mechanism involving two half reactions, the first of which leads to formation of a covalent γ-glutamyl-enzyme intermediate. This may react with an acceptor to form γ-glutamyl-acceptor or may undergo hydrolysis leading to the formation of glutamate.[3,5]

γ-Glutamyl transpeptidase also exhibits apparent "glutathione oxidase" activity.[6–8] This phenomenon may be ascribed to the production of cysteinylglycine, which is formed in the transpeptidase-catalyzed reaction of glutathione. This dipeptide oxidizes rapidly and nonenzymatically to form cystinylbisglycine, and the oxidation of glutathione takes place by nonenzymatic transhydrogenation between glutathione and cystinylbisglycine and between glutathione and the mixed disulfide of cysteinylglycine and glutathione.[8] Other thiols, e.g., cysteine, may participate in similar nonezymatic reactions. It is notable that compared to such compounds as cysteinylglycine and cysteine, glutathione reacts rather sluggishly with oxygen; however, nonenzymatic oxidation of glutathione may be sufficiently rapid as to complicate determination of transpeptidase activity.

γ-Glutamyl transpeptidase occurs in many anatomical locations in the mammal and is highly concentrated in the renal brush border.[9–12] The enzyme is predominantly membrane bound and is localized on the external surface of proximal renal tubular and lymphoid cells.[13] Activity has also been found in the cytosol of certain cells, and significant but very low levels of activity are found in human blood plasm.[14] The enzyme has been found in the intestinal brush border, pancreatic acinar, and ductile epithelial cells, epididymal epithelium, thyroid follicular epithelium, bile duct epithelium, bile canalicular epithelium, choroid plexus epithelium, sper-

[5] G. A. Thompson and A. Meister, *J. Biol. Chem.* **254**, 2956 (1979).

[6] S. S. Tate, E. M. Grau, and A. Meister, *Proc. Natl. Acad. Sci. U.S.A.* **76**, 2715 (1979).

[7] S. S. Tate and J. Orlando, *J. Biol. Chem.* **254**, 5573 (1979).

[8] O. W. Griffith and S. S. Tate, *J. Biol. Chem.* **255**, 5011 (1980).

[9] A. Meister, S. S. Tate, and L. L. Ross, *in* "The Enzymes of Biological Membranes" (A. Martinosi, ed.), Vol. 3, p. 315. Plenum, New York, 1976.

[10] M. Orlowski and A. Szewczuk, *Clin. Chim. Acta* **7**, 755 (1962).

[11] G. G. Glenner, J. E. Folk, and P. J. McMillan, *J. Histochem. Cytochem.* **10**, 481 (1962).

[12] A. M. Seligman, H. L. Wasserkrug, R. E. Plapinger, T. Seito, and J. S. Hawker, *J. Histochem. Cytochem.* **18**, 542 (1970).

[13] A. Novogrodsky, S. S. Tate, and A. Meister, *Proc. Natl. Acad. Sci. U.S.A.* **73**, 2414 (1976).

[14] S. B. Rosalki, *Adv. Clin. Chem.* **17**, 53 (1965).

matocytes, oocytes, and nonpigmented iridial epithelial cells of the posterior pupilary margin, visual receptor cells, retinal epithelium, cerebral astrocytes or their capillaries, cytoplasm of cerebellar Purkinje cells, anterior horn cells, and in other mammalian cells. The enzyme has also been found in insects, plants, hydra, and in several types of microorganisms.[15] The enzyme is especially concentrated in mammalian kidney, which has often been used as a source for purification of the enzyme. The enzyme has also been purified from tumors,[16] rat seminal vesicles,[17] pancreas,[18] and liver.[19,20]

γ-Glutamyl transpeptidase plays a key role in the γ-glutamyl cycle, which is the pathway for the synthesis and degradation of glutathione.[21,22] Glutathione is synthesized from its constituent amino acids by the successive actions of γ-glutamylcysteine and glutathione synthetases.[23] These reactions facilitate the storage of cellular cysteine as glutathione, and γ-glutamyl transpeptidase catalyzes the first step in the pathway that leads to release of the cysteine moiety from the tripeptide. Glutathione is translocated (as GSH) across cell membranes, and thus serves as a substrate for membrane-bound γ-glutamyl transpeptidase. γ-Glutamyl amino acids, formed in close association with the cell membrane, are transported into the cell.[24] The role of the γ-glutamyl cycle and γ-glutamyl transpeptidase in this pathway of amino acid transport has been reviewed.[21,22]

The formation of mercapturic acids follows a pathway involving the reaction of a foreign compound with the sulfhydryl group of glutathione catalyzed by glutathione S-transferase (Article [27]) to form a glutathione conjugate whose γ-glutamyl moiety is removed by γ-glutamyl transpeptidase.[25] Cleavage of the glycine moiety followed by N-acetylation leads to the formation of a mercapturic acid. Reactions of this type occur in endogenous metabolism. For example, the formation of leukotriene D, a slow reacting substance of anaphylaxis, involves leuko-

[15] A. Meister, in "Microorganisms and Nitrogen Sources" (J. W. Payne, ed.), p. 493. Wiley, New York, 1980.

[16] N. Taniguchi, J. Biochem. Tokyo 75, 473 (1974).

[17] L. W. DeLap, S. S. Tate, and A. Meister, Life Sci. 16, 691 (1975).

[18] B. Nash, Fed. Proc., Fed. Am. Soc. Exp. Biol. 39, 1866 (1980).

[19] L. M. Shaw, J. W. London, and L. E. Petersen, Clin. Chem. 24, 905 (1978).

[20] N. Taniguchi, K. Saito, and E. Takakuwa, Biochim. Biophys. Acta 391, 265 (1975).

[21] A. Meister and S. S. Tate, Annu. Rev. Biochem. 45, 559 (1976).

[22] A. Meister, Curr. Top. Cell. Regul. 18, 21 (1981).

[23] A. Meister, in "The Enzymes" (P. D. Boyer, ed.), 3rd ed., Vol. 10, p. 671. Academic Press, New York, 1974.

[24] O. W. Griffith, R. J. Bridges, and A. Meister, Proc. Natl. Acad. Sci. U.S.A. 76, 6319 (1979).

[25] W. B. Jakoby, ed., "Enzymatic Basis of Detoxication," Vol. 2. Academic Press, New York, 1980.

triene A (an epoxide derived from arachidonic acid), which reacts with glutathione to form leukotriene C. Removal of the γ-glutamyl moiety of leukotriene C by transpeptidase yields leukotriene D.[26] Glutathione and γ-glutamyl transpeptidase also function in analogous reactions involved in the metabolism of prostaglandins,[27] steroids,[28] certain melanins,[29] and probably other compounds. The transpeptidase-catalyzed removal of the γ-glutamyl moiety of such glutathione conjugates takes place in the presence of amino acids and is generally accelerated by amino acids, suggesting that this reaction is facilitated by transpeptidation and coupled with the formation of γ-glutamyl amino acids. A variety of *in vitro* and *in vivo* studies indicate that transpeptidation is a significant physiological function of the enzyme; these findings do not exclude the possibility that the enzyme also acts as a hydrolase *in vivo*.

Methods of Assay

The most widely used substrate is L-γ-glutamyl-p-nitroanilide.[30,31] p-Nitroaniline, released during transpeptidation and hydrolysis, is readily determined from the increase in absorbance at 410 nm. Because the enzyme exhibits strict L-stereospecificity toward acceptor substrates, the use of D-γ-glutamyl-p-nitroanilide prevents autotranspeptidation and has been employed to separately study the hydrolytic reaction.[5,32] The apparent K_m values for the L and D-isomers of γ-glutamyl-p-nitroanilide are of the same order of magnitude as that of glutathione and the V_{max} values for hydrolysis of the two isomers are the same. Because the L-isomer is poorly bound as an acceptor, it can also be used to determine the hydrolysis reaction provided its concentration is kept below 10 μM.[32] Separate determinations of the hydrolytic and transpeptidase activities of the enzyme have proved of value in probing the mechanism of differential modulation of these two activities by maleate, hippurate, and related compounds.[5,33,34] The procedures described in the following sections have been used to monitor transpeptidase activity during purification.

[26] L. Örning, S. Hammarström, and B. Samuelson, *Proc. Natl. Acad. Sci. U.S.A.* **77**, 2014 (1980).

[27] L. M. Cagen, H. M. Fales, and L. J. Pisano, *J. Biol. Chem.* **251**, 6550 (1976).

[28] E. Kuss, *Hoppe-Seyler's Z. Physiol. Chem.* **350**, 95 (1969).

[29] G. Prota, *in* "Natural Sulfur Compounds" (D. Cavallini, G. E. Gaull, and V. Zappia, eds.), p. 391. Plenum, New York.

[30] M. Orlowski and A. Meister, *Biochim. Biophys. Acta* **73**, 679 (1963).

[31] M. Orlowski and A. Meister, *J. Biol. Chem.* **240**, 338 (1965).

[32] G. A. Thompson and A. Meister, *Biochem. Biophys. Res. Commun.* **71**, 32 (1976).

[33] S. S. Tate and A. Meister, *Proc. Natl. Acad. Sci. U.S.A.* **71**, 3329 (1974).

[34] G. A. Thompson and A. Meister, *J. Biol. Chem.* **255**, 2109 (1980).

Reagents

Buffer: 0.1 *M* Tris-HCl, pH 8.0 at 25°

Glycylglycine, 0.1 *M*: 1.32 g of glycylglycine is dissolved in 80 ml of water and the pH of the solution is adjusted to 8.0 by adding 2 *M* NaOH. The final volume is made to 100 ml with water. This solution is divided into 20-ml portions and stored at −15°; it can be kept up to 3 months

L-γ-Glutamyl-*p*-nitroanilide, 5 m*M*: 80 mg of L-γ-glutamyl-*p*-nitroanilide is added to 20 ml of 1 *M* HCl and stirred at 25° until the compound dissolves. Then, 30 ml of water and 0.73 g of Tris base are added (in this order), and the pH is adjusted to 8.0 by adding 2 *M* HCl. The final volume is made to 60 ml with water. The solution is divided into 10-ml portions and stored frozen at −15°; it can be kept up to 3 months. Prior to use, defrosting is carried out by warming in a water bath at about 55°.

D-γ-Glutamyl-*p*-nitroanilide, 5 m*M*: The D-isomer is synthesized[30] and solutions are prepared and stored as described for the L-isomer

Procedure

A. Transpeptidase Activity. Additions are made to a spectrophotometer cuvette (semimicro; 1 cm light path) as follows: 0.2 ml of L-γ-glutamyl-*p*-nitroanilide (final concentration, 1 m*M*), 0.2 ml of glycylglycine (final concentration, 20 m*M*), and 0.6 ml of Tris-HCl buffer. (The volume is adjusted to allow for that of the enzyme solution added; the final volume of the reaction mixture is 1.0 ml.) The solution is brought to 37° in a spectrophotometer equipped with a thermostatted cuvette holder. Reaction is initiated by adding a suitable amount of enzyme and the rate of release of *p*-nitroaniline is recorded at 410 nm ($\epsilon = 8800$ M^{-1} cm^{-1}). The activity is usually expressed as micromoles of *p*-nitroaniline released per minute (units) and the specific activity as the units per milligram of protein. The presence of glycylglycine (20 m*M*) effectively suppresses hydrolysis of γ-glutamyl-*p*-nitroanilide and the autotranspeptidation reaction, and thus, the major reaction that occurs initially under these conditions is

L-γ-Glutamyl-*p*-nitroanilide + Gly-Gly → L-γ-glutamyl-Gly-Gly + *p*-nitroaniline

When several assays are to be performed, a mixture containing L-γ-glutamyl-*p*-nitroanilide, glycylglycine, and Tris-HCl buffer, 4:4:12 (v/v) is prepared; 1 ml of this solution is used per assay as previously described. This mixture can be stored at −15° for 4 days and at 4° for 24 hr.

B. Hydrolytic Activity. The hydrolytic activity of transpeptidase can be conveniently assayed with D-γ-glutamyl-p-nitroanilide ($K_m = 31$ μM for the rat kidney enzyme).[32] The reaction mixture (prepared as in Method A) contains 0.1 ml of D-γ-glutamyl-p-nitroanilide (final concentration, 0.5 mM) and 0.9 ml of Tris-HCl buffer. After equilibration at 37°, the reaction is initiated by adding the enzyme, and p-nitroaniline release is recorded. The L-isomer can also be used to assay the hydrolytic activity of the enzyme provided the final concentration of this substrate is 10 μM or lower ($K_m = 5$ μM for rat kidney transpeptidase).[32]

Other Methods of Assay

A variety of other procedures have been used to determine the activity of γ-glutamyl transpeptidase. Other chromogenic substrates include L-γ-glutamylanilide[35] and L-γ-glutamylnaphthylamides.[10-12] Assay procedures in which glutathione is used are complicated by its oxidation to glutathione disulfide; oxidation can be retarded by including EDTA (1 mM) in the reaction mixtures (EDTA has no effect on transpeptidase).[6] The reaction may be followed by determining the disappearance of glutathione,[3] or the rate of formation of its hydrolytic or transpeptidation products. Thus, glutamate can be conveniently quantitated using glutamate dehydrogenase[4,36] and transpeptidation can be determined by use of a radioactive amino acid acceptor; the γ-glutamyl amino acid can be separated by various chromatographic and electrophoretic techniques.[3,33,36]

Convenient spectrophotometric assays involving use of S-substituted glutathione derivatives are also available. Such derivatives include S-pyruvoylglutathione and S-acetophenoneglutathione (GS-CH$_2$-COCOOH and GS-CH$_2$COC$_6$H$_5$, respectively). The corresponding Cys-Gly derivatives produced by the action of transpeptidase cyclize spontaneously to yield products that exhibit high absorbance at 300 to 305 nm.[3] The S-acyl derivatives of glutathione (e.g., S-acetyl- and S-benzoylglutathione) provide sensitive assays for the enzyme, because the S-acyl-Cys-Gly products rapidly undergo S $\rightarrow$ N transfer of the acyl moiety producing N-acyl-Cys-Gly derivatives.[37] The free sulfhydryl groups can be readily quantitated with 5,5'-dithiobis(2-nitrobenzoate) [mammalian transpeptidase is not affected by sulfhydryl reagents such as 5,5'-dithiobis(2-nitrobenzoate)].

[35] J. A. Goldbarg, O. M. Friedman, E. P. Pineda, E. E. Smith, R. Chatterji, E. H. Stein, and A. M. Rutenberg, *Arch. Biochem. Biophys.* **91**, 61 (1960).

[36] T. M. McIntyre and N. P. Curthoys, *J. Biol. Chem.* **254**, 6499 (1979).

[37] S. S. Tate, *FEBS Lett.* **54**, 319 (1975).

Purification Procedures

Because γ-glutamyl transpeptidase is bound to plasma membranes, the enzyme must be brought into a soluble form before attempting purification. The various methods of solubilization that have been used include treatment of the particulate enzyme with detergents, organic solvents, and proteinases. The most highly purified preparations of transpeptidase have been obtained from mammalian kidney,[1] which contains a very high activity. The rat kidney enzyme, which has been extensively studied, has been purified following its solubilization with either proteinases (e.g., papain and bromelain) or detergents (e.g., Triton X-100 and Lubrol WX). Both procedures are described later. Much of the work in this laboratory has been carried out with bromelain-solubilized rat kidney transpeptidase. The method of purification results in about a 30% yield of the enzyme and involves extraction of the 100,000 g pellet obtained from kidney homogenates with Lubrol WX, acetone fractionation, digestion with bromelain, chromatography on DEAE-cellulose, affinity chromatography on concanavalin A covalently attached to Sepharose, followed by gel filtration.[7,38] Also, a relatively rapid procedure in which papain is used for solubilization of the enzyme from renal brush borders has been devised.[39] This method gives greater than 50% yields, and such preparations exhibit properties similar to the bromelain-solubilized enzyme.

Purification of Papain-Solubilized Rat Kidney γ-Glutamyl Transpeptidase

In this procedure, kidney brush border membranes are isolated by the method of Malathi *et al.*[40]

Step 1. Homogenate. Fresh or frozen kidneys are homogenized in 15 volumes (v/w) of 2 mM Tris-HCl at pH 7.4 (4°) containing 50 mM D-mannitol. For homogenization, either a Potter-Elvehjem homogenizer equipped with a Teflon pestle (ten strokes) or a Waring blender (30 sec at top speed) may be used. All procedures are carried out at 4°.

Step 2. Brush Border Membranes. To the homogenate, 1 M CaCl₂ solution is added to achieve a final concentration of 10 mM, and the mixture is placed on ice with occasional stirring for 10 min. The homogenate is then centrifuged at 3,000 g for 15 min. The supernatant fluid is carefully decanted and centrifuged at 43,000 g for 20 min. The pellet thus obtained is suspended in the Tris-mannitol buffer (one-half the volume of the original

[38] S. S. Tate and A. Meister, *J. Biol. Chem.* **250**, 4619 (1975).
[39] E. M. Kozak and S. S. Tate, *FEBS Lett.* **122**, 175 (1980).
[40] P. Malathi, H. Preiser, P. Fairclough, P. Mallett, and R. K. Crane, *Biochim. Biophys. Acta* **554**, 259 (1979).

TABLE I
PURIFICATION OF RAT KIDNEY γ-GLUTAMYL TRANSPEPTIDASE[a]

Step	Volume (ml)	Total protein[b] (mg)	Total activity[c] (units)	Specific activity[c] (units/mg)
1. Homogenate	441	3704	9920	2.7
2. Brush border membranes	29.5	264	6140	23.3
3. Papain treatment and gel filtration	2.7	6.5	5150	792

[a] From 29.4 g of rat kidney.[39]

[b] Protein is determined by the method of Lowry et al.[41]

[c] Transpeptidase activity is determined with 1 mM L-γ-glutamyl-p-nitroanilide and 20 mM glycylglycine.

homogenate) using an all-glass Dounce homogenizer with a loose pestle. The suspension is again centrifuged at 43,000 g for 20 min and the pellet is suspended (using a Dounce homogenizer) in 0.01 M sodium phosphate at pH 7.4, containing 0.15 M NaCl; the volume used is about equal to the original weight of the kidneys (v/w).

Step 3. Papain Treatment and Gel Filtration. 2-Mercaptoethanol is added to the suspension to achieve a final concentration of 20 mM. Papain (18 units/mg; Sigma) is dissolved in phosphate-NaCl buffer containing 20 mM 2-mercaptoethanol to give a solution containing 10 mg papain/ml and then incubated at 25° for 30 min. The papain solution is added to the membrane suspension to achieve a final concentration of 1 mg papain/15 mg of total membrane proteins,[41] and the mixture is stirred gently at 25° for 2 hr. The suspension is centrifuged at 43,000 g for 30 min. The supernatant fluid is treated with 65 g of $(NH_4)_2SO_4$ per 100 ml (90% of saturation) and the precipitate obtained by centrifugation (18,000 g for 30 min) is dissolved in the minimal volume [about one-tenth the weight of kidneys (v/w)] of 0.05 M Tris-HCl at pH 8.0. The solution is chromatographed on a Sephadex G-150 column (2.5 × 100 cm column for a preparation from 10 to 50 g of kidneys), equilibrated and developed with 0.05 M Tris-HCl at pH 8.0. Fractions containing transpeptidase are pooled, dialyzed for 18 hr against 100 volumes of water, and lyophilized; the residue is dissolved in a small volume of water. About 300-fold purification with over 50% yield is obtained (Table I).

[41] O. H. Lowry, N. J. Rosebrough, A. L. Farr, and R. J. Randall, *J. Biol. Chem.* **193**, 265 (1951).

Purification of Triton-Solubilized Rat Kidney γ-Glutamyl Transpeptidase

The procedure described in this section is based on that used by Hughey and Curthoys.[42]

Steps 1 and 2. Homogenization and Triton Extraction. The brush border membranes (from 50 g of kidney), isolated as described above are suspended in 25 ml of 0.05 M imidazole-HCl at pH 7.2 containing 1% Triton X-100 (imidazole-Triton) and the mixture is homogenized with a Potter-Elvehjem homogenizer. After stirring at 25° for 1 hr, the suspension is centrifuged at 43,000 g for 1 hr.

Step 3. Acetone and Triton Extractions. The supernatant liquid is treated at 4° with 10 volumes of acetone (pre-cooled to −15°). The mixture is centrifuged at 8000 g for 15 min in a stainless steel centrifuge bottle. The pellet is suspended in about 100 ml of acetone (−15°) and the mixture is homogenized in a Waring blender (low speed). The suspension is again centrifuged (8000 g, 15 min). The pellet is homogenized in 25 ml of 50 mM imidazole-HCl at pH 7.2 that is free of Triton (to remove water-soluble proteins) and then centrifuged at 8000 g for 15 min. The pellet is homogenized in 10 ml of the imidazole-Triton buffer and centrifuged. The supernatant liquid, which contains less than 10% of the total activity, is discarded and the pellet is again homogenized in 40 ml of the imidazole-Triton buffer. After stirring at 25° for 30 min, the mixture is centrifuged at 28,000 g for 30 min and the supernatant fraction concentrated to 5 ml in an Amicon ultrafiltration apparatus using a XM-50 membrane.

Step 4. Sephadex G-150. The solution is then applied to a Sephadex G-150 column (2.5 × 100 cm) previously equilibrated with imidazole-Triton buffer. The major portion of the activity elutes at about 1.2 times the void volume.

Step 5. Con A-Sepharose. The active fractions from Step 4 are pooled and incubated with 10 ml (settled volume) of Con A-Sepharose (concanavalin A covalently bound to Sepharose 4B; Pharmacia) for 2 hr at 37° in a reciprocating shaker bath. The slurry is packed into a column and eluted with imidazole-Triton buffer until no more protein elutes. The column is then eluted with imidazole-Triton buffer containing 0.1 M α-methylmannoside. The transpeptidase activity elutes as a broad peak.

Step 6. Hydroxylapatite. The active fractions from Step 5 are pooled and applied to a hydroxylapatite column (1 × 5 cm; BioRad) that has been equilibrated with the imidazole-Triton buffer. Most of the transpeptidase activity washes through. The combined eluates are concentrated by ultrafiltration as in Step 3. About 260-fold purification is achieved in about 16% yield (Table II).

[42] R. P. Hughey and N. P. Curthoys, *J. Biol. Chem.* **251**, 7863 (1976).

TABLE II
PURIFICATION OF TRITON-SOLUBILIZED γ-GLUTAMYL TRANSPEPTIDASE FROM
RAT KIDNEY[a]

Step	Volume (ml)	Protein[b] (mg)	Total activity (units)	Specific activity (units/mg)
1. Homogenate	758	6601	16,500	2.5
2. Triton extract of brush border membranes	25	359	9,735	27.1
3. Acetone treatment followed by Triton extraction	5	61	7,216	118.3
4. Chromatography on Sephadex G-150	35	21	4,485	213.6
5. Chromatography on Con A-Sepharose	28	9.5	3,740	394
6. Hydroxylapatite column	3	4	2,610	652

[a] From 50 g of rat kidney. Brush border membranes are isolated by the method of Malathi *et al.*[40] The remainder of the procedure is based on that described by Hughey and Curthoys.[42]

[b] Protein is determined by the method of Lowry *et al.*[41]

Some Chemical, Physical, and Catalytic Properties of the Enzyme

The purified preparations of rat kidney γ-glutamyl transpeptidase described in the preceding sections are stable, often for several years, when stored at 4°. The papain- (or bromelain-) solubilized and the Triton-purified enzymes exhibit similar catalytic properties but differ in certain physical characteristics. Thus, the papain-solubilized enzyme is soluble in aqueous solutions, but the Triton-purified enzyme is soluble only in the presence of detergents[42] and can associate with unilamellar lecithin vesicles.[43] Both forms of the enzyme are composed of two unequal subunits, which are both glycopeptides.

The proteinase-solubilized enzyme ($M_r = 68,000$) consists of two unequal subunits: $M_r = 46,000$ and $22,000$.[44] The purified enzyme exhibits considerable heterogeneity on polyacrylamide gels and is separable by isoelectric focusing into 12 enzymatically active isozymes ranging in pI from 5 to 8. The isozymes are similar with respect to catalytic properties and amino acid composition. The hexose and aminohexose content of the isozymes are also similar (about 745 and 700 nmol/mg of protein, respectively); however, the sialic acid content varies from 14 to 61 nmol/mg of protein (average for the enzyme, 48 nmol/mg). Neuraminidase treatment

[43] R. P. Hughey, P. J. Coyle, and N. P. Curthoys, *J. Biol. Chem.* **254,** 1124 (1979).
[44] S. S. Tate and A. Meister, *Proc. Natl. Acad. Sci. U.S.A.* **73,** 2599 (1976).

followed by isoelectric focusing indicates that the multiple forms are primarily due to different degrees of sialylation.[44]

The light subunits of papain- and Triton-solubilized rat kidney transpeptidase exhibit identical molecular weights and amino acid compositions.[45] However, the heavy subunit of the latter is larger than that of the papain-solubilized enzyme by about 52 amino acid residues. The amino terminal residues of the heavy and light subunits of the Triton-solubilized enzyme are methionine and threonine, respectively, whereas those of papain-enzyme are glycine and threonine, respectively. Treatment of the Triton-solubilized enzyme with papain results in a form that is identical in all respects to the papain-purified enzyme. On the basis of these data, it has been suggested that the papain-sensitive amino terminal sequence of amino acids of the large subunit is responsible for association of transpeptidase with the brush border membranes.[45]

Treatment of rat kidney transpeptidase with dissociating agents such as urea and SDS at neutral pH values results in extensive proteolytic degradation of the large subunit.[46] The proteinase activity, which is not evident with the native dimeric enzyme, appears to be a catalytic function associated with the light subunit and involves the active center residue (a hydroxyl group) at which γ-glutamylation of the enzyme also occurs during γ-glutamyl transfer reactions (see later).[47,48] However, intact subunits can be isolated upon urea treatment of the enzyme in 1 M acetic acid followed by gel filtration in the same medium. The denatured subunits are inactive. Renaturation of individual subunits, by dialysis against pH 6.8 buffers containing no urea, does not restore activity. Mixing of the renatured subunits also fails to restore activity. Significant reconstitution of transpeptidase activity (up to 15% of the native enzyme) is achieved by prior mixing of denatured subunits followed by removal of urea by dialysis.[49,50] The reconstituted active species has been shown to be similar to the native enzyme.

Highly purified preparations of renal γ-glutamyl transpeptidase have also been obtained from other mammalian species: rabbit,[46] cattle,[46] human,[51] sheep,[52] and hog.[53] These enzymes contain carbohydrate and are dimeric proteins. The light subunits of the several enzymes have molecu-

[45] A. Tsuji, Y. Matsuda, and N. Katunuma, J. Biochem. (Tokyo) 87, 1567 (1980).
[46] S. J. Gardell and S. S. Tate, J. Biol. Chem. 254, 4942 (1979).
[47] S. S. Tate and A. Meister, Proc. Natl. Acad. Sci. U.S.A. 74, 931 (1977).
[48] S. S. Tate and A. Meister, Proc. Natl. Acad. Sci. U.S.A. 75, 4806 (1978).
[49] S. J. Gardell and S. S. Tate, Fed. Proc., Fed. Am. Soc. Exp. Biol. 39, 1692 (1980).
[50] S. J. Gardell and S. S. Tate, J. Biol. Chem. 256, 4799 (1981).
[51] S. S. Tate and M. E. Ross, J. Biol. Chem. 252, 6042 (1977).
[52] P. Zelazo and M. Orlowski, Eur. J. Biochem. 61, 147 (1976).
[53] S. S. Tate and M. Maack, unpublished.

lar weights of about 22,000, and the light subunits of the rat and human kidney transpeptidases exhibit similar amino acid compositions.[51] However, the several enzymes fall into two categories with respect to the size of the heavy subunit: those with heavy subunits exhibiting M_r between 46,000 and 50,000 (rat and rabbit), and those with heavy subunits of M_r in excess of 60,000 (cattle, sheep, hog, and human).[54] Apparently homogeneous preparations of γ-glutamyl transpeptidase have also been obtained from rat hepatoma,[16] rat pancreas,[18] and human liver.[19] The pancreas enzyme exhibits physicochemical properties similar to those of the rat kidney enzyme. Antibodies, prepared in rabbits against purified rat kidney transpeptidase, cross-react with the enzymes from other rat tissues[55,56] but are inactive against the enzyme from human tissues.[54] Similarly, antibodies prepared against purified human kidney transpeptidase cross-react with the enzyme from other human tissues (including lymphoid cells)[57] but not with the rat enzyme.

The hydrolysis reaction catalyzed by γ-glutamyl transpeptidase exhibits a broad optimum between pH 6 and 8, whereas optimum transpeptidation is seen between pH 8 and 9.[33,58] The relative rates of utilization of a γ-glutamyl substrate by hydrolysis and transpeptidation depend on the presence and concentration of an acceptor as well as on the pH. The V_{max} for hydrolysis of glutathione in the absence of added acceptor is about 8% of that for transpeptidation in the presence of L-methionine.[36] Similar findings have been obtained with L-alanine.[36,59] However, at pH 7.4, in the presence of 50 μM glutathione and an amino acid mixture that closely approximates the amino acid composition of blood plasma, about 50% of the glutathione utilized participates in transpeptidation.[59]

Earlier studies on the acceptor specificity of transpeptidase were carried out under conditions where hydrolysis and autotranspeptidation of the γ-glutamyl substrates were largely suppressed by using high concentrations of substrates.[3] Later kinetic studies,[4,32] in accord with the earlier work, show that the best amino acid acceptors are neutral amino acids; L-cystine and L-glutamine are especially active, but a number of other amino acids can also serve as acceptors. The acceptor site exhibits absolute L-stereospecificity. L-Proline and α-substituted amino acids are inac-

[54] S. S. Tate, in "Enzymatic Basis of Detoxication" (W. B. Jakoby, ed.), Vol. 2, p. 95. Academic Press, New York, 1980.

[55] L. W. DeLap, Doctoral Thesis, Cornell University Medical College, Ithaca, New York (1975).

[56] G. V. Marathe, B. Nash, R. H. Haschemeyer, and S. S. Tate, FEBS Lett. 107, 436 (1979).

[57] G. V. Marathe, N. S. Damle, R. H. Haschemeyer, and S. S. Tate, FEBS Lett. 115, 273 (1980).

[58] E. G. Ball, J. P. Revel, and O. Cooper, J. Biol. Chem. 221, 895 (1956).

[59] D. Allison and A. Meister, J. Biol. Chem. 256, 2988 (1981).

tive and branched-chain amino acids, as well as threonine, aspartate, and histidine, are relatively poor acceptors. The high acceptor activity of methionylglycine and certain other dipeptides (aminoacylglycines) has been noted earlier.[4] The dipeptide acceptors appear to bind to the portion of the glutathione binding site of the enzyme that attaches to the Cys-Gly moiety of glutathione, and the free amino acid acceptors bind to the cysteinyl portion of the Cys-Gly site.

The γ-glutamyl binding site exhibits a broad optical and steric specificity.[4,54] Thus, L- and D-γ-glutamyl and L-γ-(α-methyl)glutamyl compounds can serve as γ-glutamyl donor substrates. Several S-derivatives of glutathione,[3,36,37] several γ-glutamyl amino acids,[3] glutamine,[33,38] and other γ-glutamyl compounds [such as γ-glutamyl-p-nitroanilide (see earlier)] can serve as γ-glutamyl donors. Indeed, a number of S-derivatives of glutathione are significantly more active than glutathione. The K_m values for these S-derivatives of glutathione, like that for glutathione, are in the range of 5–10 μM.[36]

Inhibition of γ-Glutamyl Transpeptidase

The central role of γ-glutamyl transpeptidase in glutathione metabolism has led to attempts to develop selective inhibitors of the enzyme. Both chemically reactive irreversible inhibitors and noncovalently bound reversible inhibitors have been described; the advantages and disadvantages attending the use of each *in vivo* are discussed in the following sections, and a detailed procedure for the synthesis of the isomers of γ-glutamyl-(o-carboxy)-phenylhydrazide, which are very useful noncovalent inhibitors, is given.

Irreversible Inhibitors

Transpeptidase is slowly inactivated by incubation with iodoacetamide, but not by incubation with N-ethylmaleimide, DTNB, or pCMB. Inhibition by iodoacetamide exhibits neither the rapidity nor the selectivity required in *in vivo* studies. The enzyme is effectively inactivated by incubation with O-diazo-acetyl-L-serine, (L-azaserine),[47] 6-diazo-5-oxo-L-norleucine (DON),[47] or L-(αS,5S)-α-amino-3-chloro-4,5-dihydro-5-isoxazoleacetic acid (AT-125).[8,60,61] These compounds, which are also glutamine antagonists, react covalently at the glutamyl portion of the substrate binding site. With purified enzyme, the relative rates of inactivation

[60] L. Allen, R. Meck, and A. Tunis, *Res. Commun. Chem. Pathol. Pharmacol.* **27,** 175 (1980).

[61] O. W. Griffith, and A. Meister, *Proc. Natl. Acad. Sci. U.S.A.* **77,** 3384 (1980).

by L-azaserine, DON, and AT-125 are in the ratio 2 : 5 : 100. The rates of inactivation by these compounds are decreased in the presence of other compounds that bind to the glutamyl site (e.g., glutathione, glutamine, serine plus borate). The rate of enzyme inactivation by DON and AT-125 is accelerated by maleate, a compound that enhances the hydrolytic and glutaminase activity of the enzyme. Hippurate has an effect similar to that of maleate, but, in contrast to maleate, may be used *in vivo*.[34]

Inhibition of transpeptidase in intact animals prevents the normal extracellular breakdown of glutathione and, if the inhibition is sufficiently extensive, produces a glutathionuria.[62] Urine from mice administered DON (2 mmol/kg) contains 9 to 35 μM glutathione (2–5 μM normal); the extent of enzyme inhibition can be enhanced about 2-fold if hippurate (2 mmol/kg) is administered at the same time as DON. Urine from mice administered AT-125 (2.5 mmol/kg) contains 15 to 30 mM glutathione. The much greater effectiveness of AT-125 derives from its higher affinity for the enzyme and thus its ability to more effectively compete with endogenous glutathione for the enzyme active site. AT-125 has the additional advantage of being less toxic than DON. AT-125 and DON have the disadvantage of being nonspecific; both inhibit several glutamine amidotransferases and both have an intrinsic chemical reactivity, suggesting that they can react nonspecifically with cellular nucleophiles. Inhibition by DON and AT-125 is effectively irreversible *in vitro*. *In vivo*, as judged by the extent of glutathionuria, the inhibition is slowly reversible, and is thus markedly diminished after 8 to 24 hours. It has not yet been determined whether such reversal *in vivo* reflects synthesis of new enzyme or breakdown of the covalent enzyme–inhibitor complex, or both. Reactivation or resynthesis of only a small percentage of the enzyme would be expected to reverse the glutathionuria because kidney contains a large excess of γ-glutamyl transpeptidase activity.

Reversible Inhibitors

L-Serine in the presence of borate has long been known to inhibit the transpeptidase[63]; D-serine is also active in the presence of borate.[64] Serine appears to occupy the active site region that accepts the α-amino and α-carboxyl groups of glutathione, and the borate anion bridges between an active site hydroxyl group and the hydroxyl group of serine. The tetrahedral borate complex thus serves as a transition-state analog at the catalytic center of the enzyme.[64] *In vitro* inhibition by serine plus borate is reversible upon dilution or dialysis; increasing the concentration of γ-

[62] O. W. Griffith, and A. Meister, *Proc. Natl. Acad. Sci. U.S.A.* **76**, 268 (1979).
[63] J. P. Revel and E. G. Ball, *J. Biol. Chem.* **234**, 577 (1959).
[64] S. S. Tate and A. Meister, *Proc. Natl. Acad. Sci. U.S.A.* **75**, 4806 (1978).

glutamyl donor decreases the extent of inhibition. Serine plus borate is a weak inhibitor *in vivo*. Administration of L-serine (32 mmol/kg) and Na borate (32 mmol/kg) to mice does not produce glutathionuria, but does slow somewhat the metabolism of D-γ-glutamyl-L-α-aminobutyrate, a compound which can be metabolized only by transpeptidase.[62] As an inhibitor, serine plus borate has the unique advantage that its specificity can be tested by omitting either serine or borate; effects not dependent on the simultaneous presence of both compounds are not referable to inhibition of transpeptidase.

L-γ-Glutamyl-(*o*-carboxy)phenylhydrazide (L-OC) is a tightly bound reversible inhibitor of transpeptidase. Its apparent K_i value is 8.2 μM as compared to 1.45 mM for L-serine plus borate.[62] When administered to mice, L-OC (0.5 mmol/kg) induces marked glutathionuria; urinary glutathione concentrations range between 3.6 and 5.6 mM. D-γ-Glutamyl-(*o*-carboxy)phenylhydrazide (D-OC) is also a strong inhibitor (K_i = 22.5 μM) and also induces glutathionuria (urinary glutathione, 1.8 to 3.9 mM).[62] Both L-OC and D-OC are more specific than the irreversible inhibitors previously described and would not be expected to inhibit glutamine amidotransferases. The phenylhydrazide derivatives are, however, rather toxic; doses greater than about 0.5 mmol/kg and 1.25 mmol/kg for the L- and D-isomers, respectively, are not well tolerated by mice. The toxicity is probably due to slow release of *o*-carboxyphenylhydrazine, a potent convulsant. Following the administration of L- or D-OC to mice, the urine is found to contain, in addition to a large amount of unchanged L- or D-OC, a metabolite eluting on amino acid analysis near cystine. The metabolite has not been identified, but it is noted that the elution time is about that expected for an S-substituted GSH derivative of the inhibitors.

It is of particular interest that administration of L-OC, D-OC, or AT-125 also lead to extensive urinary excretion of γ-glutamylcysteine and cysteine moieties,[61] indicating that a significant function of γ-glutamyl transpeptidase is associated with the metabolism or transport (or both) of these sulfur-containing compounds as well as glutathione.[61,65]

Synthesis of L- or D-γ-Glutamyl(*o*-Carboxy)Phenylhydrazide

Synthesis of N-Phthaloyl-L-Glutamic Anhydride[66]

L-Glutamic acid (200 g, 1.36 moles) and finely powdered phthalic anhydride (200 g, 1.36 moles) are placed in a 1-liter, 3-neck, round bottom

[65] O. W. Griffith, R. J. Bridges, and A. Meister, *Proc. Natl. Acad. Sci. U.S.A.* **78**, 2777 (1981).
[66] F. E. King and D. A. A. Kidd, *J. Chem. Soc.* p. 3315 (1949).

flask fitted with a strong mechanical stirrer; the side necks are left open. The flask is placed in an oil bath at 150° and the contents are slowly stirred. As the mixture melts, it becomes extremely viscous; a spatula inserted through the side neck is used to knead the mixture as well as to push down subliming phthalic anhydride. After 20 to 30 min, the melt becomes clear and less viscous. The temperature of the melt should not exceed 135°–140°; it is held at that temperature for 10 min and then allowed to cool to about 115° in a 100° oil bath. To the cooled mixture is added 250 ml of acetic anhydride and the resulting solution is stirred at 85°–95° for 10 min. The clear solution is poured while still hot into 750 ml of xylene, which is prewarmed to 50°–60°. That mixture is briefly agitated and then quickly filtered under reduced pressure through a 500-ml medium porosity filter. The filtrate is allowed to crystallize overnight at 4°. The crystals are collected on a filter and washed with a mixture of 50 ml acetic anhydride and 150 ml xylene; the damp product is dried in a vacuum desiccator over P_2O_5. The product, which may be stored in tightly capped bottles at room temperature, weighs 275–300 g (78–85%) and melts at 199°–200°. The D-isomer is made in a similar manner.

γ-Glutamyl-(o-Carboxy)Phenylhydrazide[62,67]

In a 250-ml round bottom flask, fitted with a $CaCl_2$ drying tube and a magnetic stirring bar, is placed N-phthaloyl-L-glutamic anhydride (6.48 g, 0.025 mole), o-hydrazinobenzoic acid hydrochloride (4.72 g, 0.025 mole, Eastman Kodak Co.), and anhydrous sodium acetate (2.05 g, 0.025 mole). To the solids is added 50 ml glacial acetic acid, and the mixture is stirred in a 100° oil bath for 20 min. The solution, which contains insoluble NaCl, is allowed to cool to 25° and the solvent is removed by rotary evaporation at reduced pressure (bath temperature 35°–40°). The gummy residue is washed twice by adding 50 ml ethanol and repeating the rotary evaporation. The residue is dissolved in 50 ml of methanol, filtered, and the filtrate mixed with 2.5 ml each of hydrazine hydrate and triethylamine. The solution is left at 25° for 48 hr, during which time a voluminous precipitate forms. Without removing the precipitate, the solvent is evaporated under reduced pressure and the residue dissolved in 50 ml of water. The pH is adjusted to 6 by adding NaOH and the solution is filtered. The filtrate (which contains glutamate, some phthaloyl hydrazide, and the product) is applied to a column (3.5 × 40 cm) of Dowex-1 (acetate). The column is washed with 1 liter of water and eluted with a linear gradient formed from 4 liters each of water and 1.5 M acetic acid. The fractions are monitored

---

[67] This compound has also been isolated from the culture medium of *Penicillium oxalicum* [S. Minato, *Arch. Biochem. Biophys.* **192**, 235 (1979)].

by carrying out tests with ninhydrin or *o*-phthalaldehyde; glutamate elutes in about 0.3 *M* acetic acid, the product elutes at 1–1.2 *M* acetic acid. Fractions containing the product are concentrated by rotary evaporation under reduced pressure. The resulting oil usually crystallizes without difficulty and the product is recrystallized from ethanol:water. The yield of material, which melts at 185°–187°, is about 50%.

[31] Cysteine Conjugate β-Lyase

By MITSURU TATEISHI and HIROTOSHI SHIMIZU

$$R-S-CH_2-CH(-NH_2)-COOH \rightarrow R-SH + NH_3 + CH_3-C(=O)-COOH$$

Assay Method

Principle. Cysteine conjugates of aromatic compounds (R) serve as the substrate of this enzyme. Utilizing a substrate labeled with ^{35}S, the product of the reaction, R-[^{35}S]H, is measured after converting to R-[^{35}S]CH$_3$ with methyl iodide. The derivatized product is purified by thin-layer chromatography and its radioactivity is determined.[1]

Reagents

Potassium phosphate buffer, 20 m*M*, pH 7.4
Enzyme: Prepared in 20 m*M* potassium phosphate buffer, pH 7.4
2,4-Dinitrobenzenethiol (DNP-SH), 1% (w/v) in ethanol
^{35}S-labeled 2,4-dinitrophenylcysteine ([^{35}S]DNP-cysteine) in methanol. Due to its instability, this substrate must be prepared within a few days of use. The method of preparation is based on the report of Saunders[2] for the synthesis of nonradioactive DNP-cysteine. 1-Fluoro-2,4-dinitrobenzene (20 μmol) in 200 μl of ethanol is added to 100 μl of 0.2 *M* sodium acetate, pH 5.2, containing L-[^{35}S]cysteine (6 μmol, 1 μCi) and sodium acetate (60 μmol). After mixing at room temperature, the reaction medium is immediately applied to a plate of silica gel tlc, which is developed with chloroform:ethanol:acetic acid:water (50:30:10:5). The silica gel at the UV-absorbing zone at $R_f = 0.64$ is scraped from the plate and the product extracted with methanol

Procedure. A portion of the methanol solution containing 0.4 μmol of [^{35}S]DNP-cysteine is placed in a test tube and the methanol is evaporated

[1] M. Tateishi, S. Suzuki, and H. Shimizu, *J. Biol. Chem.* **253**, 8854 (1978).
[2] B. S. Saunders, *Biochem. J.* **28**, 1934 (1977).

to dryness under a gentle stream of nitrogen. Potassium phosphate buffer, 0.1 ml, and 0.2 ml of enzyme are added and the reaction mixture incubated for 8 min at 37°. The reaction is stopped by cooling of the mixture in ice and by immediate addition of 0.05 ml of a 1.0% ethanol solution of DNP-SH.

Analytical Method. The reaction mixture is diluted with 1.0 ml of water and the DNP-SH immediately extracted into 10 ml of chloroform. The organic solvent is evaporated under a stream of nitrogen at room temperature and the residue is dissolved in 0.5 ml of ethanol. To the ethanol solution are added 100 mg of methyl iodide and 30 μl of 0.1 N sodium hydroxide. After standing for 3 hr at room temperature, excess methyl iodide and ethanol are removed by evaporation to dryness. The residue is dissolved in methanol and streaked on a thin-layer chromatoplate (Kieselgel, Merck), which is developed with benzene. A band at $R_f = 0.63$, corresponding to DNP-SCH$_3$, is located under ultraviolet light; the derivative is extracted with 4 ml of chloroform after scraping from the gel. Following complete evaporation of the extract, the dried residue is dissolved in 3.0 ml of ethanol. A 2.5-ml portion of the solution is mixed with 10 ml of toluene containing 80 mg of 2-(4'-*tert*-butylphenyl)-5-(4''-biphenyl)-1,3,4-oxadiazole for determination of radioactivity in a liquid scintillation spectrometer. Another 0.5-ml portion is used for monitoring the recovery of DNP-SH by measuring the absorbance at 332 nm, the absorption maximum of DNP-SCH$_3$ ($\epsilon_{332\,nm}^{EtOH} = 9670$). The radioactivity is corrected for the individual recoveries.

Purification Procedure[1]

Rat liver is employed as a source of this enzyme, but the liver of most experimental animals contains the enzyme at a similar specific activity. All steps, unless otherwise noted, are performed at 0° to 4°.

Step 1. Extraction. About 40 g of fresh rat liver is collected in ice-cold 1.2% KCl solution, minced with scissors, and homogenized with a motor-driven glass homogenizer in 80 ml of an ice-cold buffer containing 10 mM potassium phosphate, 5 mM MgCl$_2$, 150 mM KCl, and 0.5 mM dithiothreitol, all adjusted to pH 7.4. The homogenate is centrifuged at 9,000 g for 20 min and the supernatant fluid is centrifuged at 105,000 g for 60 min. Approximately 1.5 kg of rat liver is processed in the same manner, and the resultant supernatant fluid is combined and used for further purification.

Step 2. Heat Treatment. About 350-ml portion of the extract (120 g liver equivalent) is transferred into a 1-liter flask equipped with a thermometer, and the flask agitated in a water bath maintained at 58 $\pm$ 0.5°

PURIFICATION OF CYSTEINE CONJUGATE β-LYASE FROM RAT LIVER

Step	Volume (ml)	Protein (mg)	Enzyme activity (units)	Specific activity (units/mg)
1. Extraction	2200	265,000	51,300	0.19
2. Heat treatment	1970	26,100	82,000	3.14
3. Salt fractionation	150	6,040	33,200	5.49
4. DEAE-cellulose 1	410	1,480	30,700	20.7
5. DEAE-cellulose 2	15.0	122	22,400	183
6. Sephadex G-200	114	58.0	11,700	202

until the temperature in the cytosol equilibrates with that in the water bath, about 10 min. After standing for an additional 2 min in the water bath, the suspension is cooled to 4° in ice-chilled water and centrifuged at 9000 g for 20 min to remove the inactive precipitate.

Step 3. Ammonium Sulfate. Powdered ammonium sulfate, 210 mg/ml, is added to the enzyme solution to result in 35% of saturation. The mixture is centrifuged (20,000 g, 15 min), and the precipitate discarded. Additional ammonium sulfate, 200 mg/ml, is added to the supernatant fluid to increase the concentration to 65% of saturation. The precipitate (equivalent to 120 g of liver) is dissolved in 20 ml of 10 mM potassium phosphate, pH 7.4, and then subjected to gel filtration on a column (3.2 × 30 cm) of Sephadex G-25 for desalting.

Steps 4 and 5. DEAE-Cellulose. The enzyme solution, in portions of about 100 ml, is concentrated to 20 ml with a Diaflo PM-10 membrane filter and is applied to a column (2.5 × 23 cm) of DEAE-cellulose equilibrated with the same buffer. The column is first eluted with 280 ml of the same buffer and then with a linear gradient of the buffer increasing from 10 to 200 mM in potassium phosphate at pH 7.4. Fractions of 20 ml are collected at a flow rate of 0.4 ml/min. β-Lyase activity emerges as a single peak during the gradient elution at about fraction 30. Repetition of this step is necessary to achieve effective purification.

Step 6. Sephadex G-200. The combined enzyme preparation is concentrated to approximately 15 ml with a Diaflo PM-10 membrane and is applied to a Sephadex G-200 column (3.1 × 97 cm) previously equilibrated with 10 mM potassium phosphate at pH 7.4. The initial 325 ml of the eluate is discarded, and the following 110 ml, i.e., containing the enzyme, is collected.

Overall, 1100-fold purification is achieved through the Sephadex step (see table).

Properties[1,3]

Activators and Inhibitors. Hydroxylamine inhibits the enzyme, suggesting the involvement of pyridoxal phosphate in the reaction. Probably because of tight binding of the pyridoxal cofactor to the enzyme, activation by pyridoxal phosphate is not observed unless the preparation is subjected to chromatography on hydroxylapatite.[4] Although this treatment results in a loss of more than 90% of the activity, the enzyme prepared in this manner is activated 280% by addition of 0.02 mM pyridoxal phosphate to the assay mixture.

p-Chloromercuribenzoate at 0.3 mM inhibits the enzyme by 83%, but potassium cyanide (1 mM) and N-methylmaleimide (1 mM) are not inhibitory.

Substrate Specificity. Thioethers of L-cysteine, in which the amino acid's amino and carboxyl groups remain free, are the obligatory configuration. Derivatives of D-cysteine and DL-homocysteine are essentially inactive as substrates. L-Cysteine conjugates of aromatic compounds are good substrates, whereas derivatives of aliphatic (e.g., methyl, ethyl) or alicyclic (e.g., cyclohexyl) S-substituents are inactive. Thus, L-cysteine, S-methyl-L-cysteine, and cystationine do not serve as substrate. *p*-Nitrobenzyl-L-cysteine, possessing both the aromatic and aliphatic substituent structures, is 10-fold less effective as substrate than DNP-cysteine. At the optimum pH, the K_m value for DNP-cysteine was estimated to be 0.5 mM.

pH Optimum of Reaction. The enzyme has a broad pH optimum in the range of pH 7.4 to 8.5 with either Tris-HCl or potassium phosphate.

Distribution of β-Lyase. The enzyme is almost exclusively present in the cytosol. In the rat, only liver and kidney contain appreciable amounts of the enzyme. The livers of rat, mouse, hamster, guinea pig, and rabbit are similar in their β-lyase activity, whereas human liver, obtained at necropsy, is somewhat lower.

An analogous enzymatic activity has been found in certain bacteria and is designated as L-methionine γ-lyase (EC 4.4.1.11). However, unlike cysteine conjugate β-lyase of mammalian origin, L-methionine γ-lyase attacks aliphatic as well as aromatic derivatives of homocysteine and cysteine.[5]

---

[3] M. Tateishi and H. Shimizu, *in* "Enzymatic Basis of Detoxication" (W. B. Jakoby,ed.), Vol. 2, p. 121. Academic Press, New York, 1980.

[4] The Sephadex eluate (Step 5) is directly applied to a column (2.2 × 23 cm) of hydroxylapatite gel previously equilibrated with 1 mM potassium phosphate at pH 8.0 and containing 0.5 mM dithiothreitol. Protein is eluted with a linear gradient of the same phosphate buffer increasing to a phosphate concentration of 200 mM. Following the first and the largest protein peak, a sharp peak of enzyme activity emerges in a range of ionic strength between 80 and 105 mM.

[5] H. Tanaka, N. Esaki, and K. Soda, *Biochemistry* **16**, 100 (1977).

[32] Thiol S-Methyltransferase

By RICHARD A. WEISIGER and WILLIAM B. JAKOBY

$$\text{RSH} + S\text{-adenosyl-L-methionine} \rightarrow \text{RSCH}_3 + S\text{-adenosyl-L-homocysteine}$$

The enzyme is present in liver microsomes[1] but has been solubilized and partially purified by use of detergents.[2] The procedure described here is for enzyme from rat liver and provides a homogeneous protein without the use of detergents in the solubilization process.[3]

Assay Method

Principle. Thiol S-methyltransferase activity is determined by measuring the hydrophobic, radioactive S-methyl thioethers formed by the action of the enzyme on thiol substrates in the presence of [*methyl*-³H]S-adenosyl-L-methionine. EDTA is used to block the activity of magnesium-dependent O-methyltransferases that may be present in crude preparations. The lipophilic, radioactive product can be separated from unreacted S-adenosylmethionine by extraction into toluene at pH 10 (for neutral or basic products) or at pH 1 (for acidic products), and quantitated by scintillation counting. The assay may be used for all of the known substrates[4] although specific steps are necessary for avoiding loss of product during incubation with more volatile compounds.[5]

Reagents

Thiol substrate (0.1 M in 95% ethanol): 2-Mercaptoacetanilide (*not* "mercaptoacetanilide") has been used due to its ease of purification and minimal odor[6]; this solution is prepared daily

[*methyl*-³H]S-adenosyl-L-methionine (5 μM); 5 to 20 Ci per mmol in 10 mM sulfuric acid; this solution is stable for several months when stored at $-20°$

---

[1] J. Bremer and D. M. Greenberg, *Biochim. Biophys. Acta* **46**, 217 (1961).

[2] R. T. Borchardt and C. F. Cheng, *Biochim. Biophys. Acta* **522**, 340 (1978).

[3] R. A. Weisiger and W. B. Jakoby, *Arch. Biochem. Biophys.* **196**, 631 (1979).

[4] The method is unsuitable with compounds resulting in hydrophilic products. Of the limited number of such compounds that have been tested, e.g., cysteine and GSH, none are substrates.

[5] R. A. Weisiger, L. B. Pinkus, and W. B. Jakoby, *Biochem. Pharmacol.* **29**, 2885 (1980).

[6] Although 2-mercaptoacetanilide is used as the standard substrate, similar results are obtained with 4-chlorothiophenol, phenyl sulfide, and 4-nitrothiophenol. Any one of them is satisfactory if the standard substrate is not available.

METHODS IN ENZYMOLOGY, VOL. 77

Stopping solution: Use either 2 M sodium borate at pH 10 or 0.1 N HCl

Assay buffer: 0.1 M Potassium phosphate, pH 7.9, 1 mM EDTA, 0.5% Triton X-100; the Triton helps stabilize the purified enzyme against surface or sheer denaturation and may be omitted when assaying crude tissue preparations

Enzyme: The amount of activity used should convert approximately 1 to 10% of the total SAM present to products

Toluene (scintillation grade)

Procedure. Assays are performed in 15-ml glass-stoppered conical centrifuge tubes incubated in a 37° water bath. To 350 μl of the assay buffer are added, in sequence, 20 μl of the thiol substrate, 10 μl of the enzyme solution, and 20 μl of S-adenosylmethionine. After brief vortex mixing, the mixture is incubated for 10 min before the reaction is stopped by the addition of 1 ml of basic stopping solution (for 2-mercaptoacetanilide as substrate). Toluene (6 ml) is added and the stoppered tube shaken vigorously for 10 sec. The phases are separated by brief centrifugation (or prolonged standing) and 5 ml of the upper, toluene phase are added to 10 ml of scintillation fluid or quantitation of radioactivity.

Definition of Units. One unit of activity is defined as that required for the production of 1 pmol of product per minute under the assay conditions described in the preceding section. Specific activity is expressed as units per milligram of protein; protein is determined colorimetrically.[7] This assay is extremely sensitive, allowing quantitation of as little as 1 ng of purified enzyme; because S-adenosylmethionine is not present at saturating concentrations in the standard assay, even greater sensitivity is attainable by increasing the amount of nucleotide in the incubation mixture.

Purification Procedure[3]

The following steps must be followed carefully in order to reproducibly yield homogeneous, soluble S-methyltransferase at an overall yield of 2 to 5%. After the initial chromatographic step, the activity becomes very labile and, under the best conditions of storage, is lost with a half-life of 4 days. Freezing or vigorous mixing results in complete loss of activity. Thus, extreme care[8] should be exercised during the purification so as to avoid foaming or excessive stirring of enzyme solutions, and all steps

[7] O. H. Lowry, N. J. Rosebrough, A. L. Farr, and R. J. Randall, Jr., *J. Biol. Chem.* **193,** 265 (1951).

[8] To minimize problems due to enzyme instability, fractions are routinely pooled by drawing the contents into a 10-ml pipet and carefully allowing the liquid to flow directly into the liquid pool without splashing. In all ultrafiltration steps, pressure within an Amicon apparatus are maintained below 10 psi so as to avoid later foaming when the pressure is released; the stirring bar is maintained below the surface of the concentrating liquid and

TABLE I

THIOL METHYLTRANSFERASE: SUMMARY OF PURIFICATION (660 g LIVER)[a]

Step	Volume (ml)	Total protein (mg)	Total activity (pmol min^{-1})	Specific activity (pmol min^{-1} mg^{-1})
1. Extract	1150	76,400	3500	0.046
2. DEAE-cellulose	740	4,573	2400	0.52
3. Salt precipitation	85	3,400	2100	0.63
4. Hydroxylapatite (pH 6.7)	73	168	500	3.0
5. Hydroxylapatite (pH 6.2)	7	49	160	3.4
6. Isoelectric focusing	4.1	1.9	100	55
7. Sephadex G-75	4.7	0.3	94	310

[a] Data from Ref. 3.

should be completed as rapidly as possible over the course of 9 to 10 days. A typical purification is summarized in Table I.

Step 1. Solubilization. About 600 g of rat livers, frozen on dry ice,[9] are thawed by washing with tepid water and homogenized in batches of 150 g for 30 sec with 275 ml of ice-cold water in a Waring blender. After centrifugation for 1 hr at 100,000 g; the clear, red supernatant is carefully aspirated and pooled.[10]

Step 2. DEAE-Cellulose. The extract is adjusted to pH 7.7 with 1 M Tris base and diluted with cold water until the conductivity is 1 mmho. A column [14(diameter) × 11 cm] of DEAE-cellulose is prepared by suspending 1 kg of DE-52 (Whatman) in 0.05 M potassium phosphate at pH 7.7, readjusting to this pH, and then washing the poured column with 10 liters of Buffer A (20 mM Tris, 10 μM EDTA, adjusted to pH 7.7 with HCl). After the enzyme solution is applied to the column, it is washed with 4 liters of Buffer A before applying a linear, 4-liter gradient consisting of Buffer A and Buffer A containing 0.2 M NaCl. Active fractions (1 to 1.4 liters) are pooled.

---

turned at the minimum rate (less than 2 Hz). These precautions are of increasing importance with increasing purity of the enzyme.

[9] Livers from male Sprague-Dawley rats (ARS Sprague-Dawley, Madison, WI) of about 250 g, were frozen in dry ice immediately after excision and stored at $-80°$. Livers that were homogenized immediately after freezing were of similar activity and had the same elution patterns as those homogenized after storage for 2 months at $-80°$.

[10] Approximately 30% of the activity present in microsomes is solubilized by this method of freezing and extraction. Sepharose 4B chromatography of the extract indicates that the enzyme exists in a very high molecular weight form that is excluded by the gel, perhaps as a membrane fragment. During the first chromatographic step, however, the enzyme is converted to a low molecular weight form that elutes with the same R_f as the purified enzyme on Sephadex G-75 ($M_r = 28,000$). It is this conversion to the monomer form that appears to result in the enzyme's instability and the unusual kinetic properties.

Step 3. Salt Precipitation. Ammonium sulfate (516 g per liter) is added and, after 1 hr at 4°, the precipitate is harvested by centrifugation. The paste is suspended in 100 ml of 0.1 M potassium phosphate at pH 7.9, containing 0.1 mM EDTA, and dialyzed sequentially against 20 volumes each of the following: 0.1 mM disodium EDTA (4 hr); 0.1 mM disodium EDTA (10 hr); and 5 mM potassium phosphate at pH 6.7, containing 0.1 mM EDTA (4 hr). The enzyme solution is clarified by centrifugation for 20 min at 12,000 g.

Step 4. Hydroxylapatite (pH 6.7). Potassium phosphate (1 M) at pH 6.7 is added to the protein solution to raise the phosphate concentration to 20 mM, and the solution is charged onto a column (5 × 12 cm) of hydroxylapatite[7] that has been equilibrated with Buffer B (10 mM potassium phosphate and 0.1 mM EDTA, adjusted to pH 6.7). A 3-liter linear gradient of 0.01 to 0.2 M potassium phosphate in Buffer B is applied and active fractions collected and concentrated (with minimal stirring and pressure[8]) to 40 ml with an Amicon PM-10 membrane.

Step 5. Hydroxylapatite (pH 6.2). After overnight dialysis of the concentrate against Buffer C (0.01 M potassium phosphate and 0.1 mM EDTA, adjusted to pH 6.2), chromatography is repeated on a column (1.5 × 7 cm) of hydroxylapatite previously equilibrated with the same buffer. A 600-ml linear gradient is applied that consists of Buffer C and Buffer C supplemented to contain 0.2 M potassium phosphate. The most active fractions are pooled (20 ml).

Step 6. Preparative Isoelectric Focusing. The enzyme preparation is subjected to isoelectric focusing in a 110-ml column containing 2% pH 5–7 Ampholines (LKB) and stabilized by a 0 to 40% sucrose gradient. Focusing is for a period of 72 hr at up to 1000 V. Fractions of 1 ml are collected and those few fractions of highest specific activity (6 ml), found at about pH 6.2, are retained; attempts to pool more than about 60% of the activity leads to increased risk of inhomogeneity after Step 7.

Step 7. Gel Exclusion Chromatography. The pooled fractions are charged directly onto a column (1.5 × 90 cm) of Sephadex G-75 (Superfine) and eluted with 10 mM Tris-HCl, 0.1 M sodium chloride, and 0.1 mM EDTA, adjusted to pH 7.5. The enzyme elutes as a single peak of constant specific activity that exhibits only a single band on SDS–gel electrophoresis.

Properties[11]

Physical Characteristics and pH Optima.[3] The enzyme is in a homogeneous form by the criterion of SDS–gel electrophoresis and appears to be

---

[11] R. A. Weisiger and W. B. Jakoby, *in* "Enzymatic Basis of Detoxication" (W. B. Jakoby, ed.), Vol. 2, p. 131. Academic Press, New York, 1980.

composed of a single subunit of 28,000 daltons. In the standard assay system with 2-mercaptoacetanilide, optimum activity is at about pH 7.5 with half of maximal activity at pH 6.6 and 8.6. The lower activity at higher pH values may reflect, in part, the lability to alkali of S-adenosylmethionine.

Specificity and Kinetics. The nucleotide binding site appears to be highly specific: the homocysteine moiety is most important to binding whereas the enzyme is less sensitive to modification of the sugar or base.[2] The apparent K_m for S-adenosyl-L-methionine in the standard assay is temperature dependent with the following respective values at 15°, 30°, and 37°: 42 nM, 0.6 μM, and 1 μM[3].

Specificity for the thiol substrate is broad although apparently limited to the more lipophilic compounds (Table II); such polar compounds as GSH, cysteine, homocysteine, and dithiothreitol, are not substrates,[1-3] although the methyl ester of cysteine does serve as a thiol substrate.[3]

Although microsomal suspensions[1] and detergent-solubilized[2] enzyme generate product linearly with time, the homogeneous enzyme displays an initial burst of product formation followed by a slower linear phase that has been interpreted as indicating the rate-limiting dissociation of the product, S-adenosyl-L-homocysteine, from the enzyme.[3] This behavior is

TABLE II
SUBSTRATES OF THIOL METHYLTRANSFERASE[a]

Substrate	Apparent K_m (mM)	V_{max} (nmol mg^{-1} min^{-1})
p-Chlorothiophenol	0.00054	6.2
Phenyl sulfide	0.0011	6.1
4-Nitrothiophenol	0.0028	7.5
Diethylthiocarbamyl sulfide	0.012	3.5
2-Thioacetanilide	0.043	7.6
Hydrogen sulfide	0.064	7.8
2-Benzimidazole thiol	0.11	2.4
Thioglycolic acid	0.19	3.7
L-Cysteine methyl ester	0.21	1.4
Methane thiol	0.24	0.9
N-Acetyl-L-cysteine	0.40	1.0
6-Propyl-2-thiouracil	1.0	6.4
1-Methylimidazole-2-thiol	1.4	2.0
2,3-Dimercaptopropanol	1.6	5.4
3-Mercaptopropionic acid methyl ester	4.7	3.2
2-Mercaptoethanol	8.1	5.4

[a] Data from Refs. 3 and 5.

probably due to the 7000-fold change in the K_I for S-adenosyl-L-homocysteine following nondetergent solubilization: K_I of 144 μM and 0.2 μM for the detergent and the homogeneous preparation, respectively. Consistent with the proposed mechanism, in which the limiting rate is the dissociation of adenosylhomocysteine from the protein, is the finding that the maximal velocity for a wide range of substrates falls well within one order of magnitude whereas K_m values vary over a range of 10,000 (Table II).

Inhibitors. Sulfhydryl reagents such as p-chloromercuribenzoate and N-ethyl maleimide result in loss of activity.[3] As a product of the reaction, S-adenosyl-L-homocysteine is an effective competitive inhibitor ($K_D = 0.2$ μM).

Distribution. The liver enzyme is associated with the microsomal fraction,[1,2] although a soluble form has been reported in extracts of liver, kidney, intestine, and spleen.[9] Liver, lung, and kidney, tissues usually associated with detoxication, have high levels in the rat.[2,5] The highest specific activity is found in the colon mucosa, supporting a role for the enzyme in the detoxication of hydrogen sulfide that is generated by the colonic flora.[5,11]

Comments

The broad specificity of the enzyme for hydrophobic thiols and the generally toxic nature of these compounds supports a detoxication function for the enzyme. Indeed, a number of drugs have been found to be S-methylated *in vivo,* including diethyldithiocarbamate,[12–14] 2-thiouracil,[15,16] 6-thiopurine,[15] and 6-propyl-2-thiouracil.[16] However, the largest class of substrates may prove to be products of the hepatic enzyme, cysteine conjugate β-lyase, which cleaves the thioether linkage of the glutathione conjugates of electrophilic compounds to generate thiolated derivatives.[17] The major endogenous substrates for thiol S-methyltransferase may be hydrogen sulfide and methyl mercaptan, both products of intestinal metabolism. Failure of the enzyme to detoxify these compounds may contribute to the coma associated with hepatic failure.[11]

[12] M. Jakubowski and T. Gessner, *Biochem. Pharmacol.* **21**, 3073 (1972).

[13] J. Cobby, M. Mayershsohn, and S. Selliah, *Life Sci.* **21**, 937 (1977).

[14] T. Gessner and M. Jakubowski, *Biochem. Pharmacol.* **21**, 219 (1972).

[15] C. M. Remy, *J. Biol. Chem.* **238**, 1078 (1963).

[16] R. H. Lindsay, B. S. Hulsey, and H. Y. Aboul-Enein, *Biochem. Pharmacol.* **34**, 463 (1975).

[17] M. Tateishi, S. Suzuki, and H. Shimizy, *J. Biol. Chem.* **253**, 8854–8859 (1978); see also this volume, Article [31].

[33] Arylamine *N*-Methyltransferase

By Ellen Sue Lyon and William B. Jakoby

Tryptamine + *S*-adenosyl-L-methionine →

                   N-methyltryptamine + *S*-adenosyl-L-homocysteine

Although a number of specific *N*-methyltransferases are known,[1-3] the preparation described by Axelrod[4] from rabbit lung has broad specificity and appears to be related to the detoxication process. Its presence in rabbit liver led to purification of the enzyme from this issue and resulted in the homogeneous enzyme of somewhat narrower specificity than observed in crude preparations.

Assay Method

Principle. The assay depends on the highly radioactive, methyl-labeled, *S*-adenosylmethionine that is readily available commercially.[5] Tryptamine serves as the methyl group acceptor in the standard assay. After reaction, the methylated product is separated from the polar radioactive substrate by extraction of the product into an organic solvent.

Reagents

Potassium phosphate, 0.2 *M*, pH 7.8
Tryptamine hydrochloride, 40 m*M*
[*methyl*-^{3}H]*S*-Adenosyl-L-methionine, 340 μM, at a specific activity of about 10 μCi per μmol
Potassium borate, 0.1 *M*, pH 9.5

Procedure. The reaction mixture, in a total volume of 200 μl, contains the following added in the listed sequence: 100 μl of the phosphate buffer, 10 μl tryptamine, 20 μl methyl-labeled *S*-adenosylmethionine, and 0.1 to 2 units of enzyme. The incubation is carried out for 20 min at 37° and is arrested by addition of 1 ml of the borate solution. The radioactive product of the reaction is extracted with 5 ml of toluene containing 3% (v/v) isoamyl alcohol. A 4-ml portion of the organic phase is added to 10 ml

[1] N. Kirshner and McC. Goodall, *Biochim. Biophys. Acta* **24**, 658 (1957).
[2] H.-S. Lee, A. R. Schulz, and R. W. Fuller, *Arch. Biochem. Biophys.* **185**, 222 (1978).
[3] D. D. Brown, J. Axelrod, and R. Tomchick, *Nature (London)* **183**, 680 (1959).
[4] J. Axelrod, *J. Pharmacol. Exp. Ther.* **138**, 28 (1962).
[5] This procedure is modified from that of J. M. Saavedra, J. T. Carfe, and J. Axelrod, *J. Neurochem.* **20**, 743 (1973).

TABLE I
SUMMARY OF PURIFICATION FOR N-METHYLTRANSFERASE

Step	Volume (ml)	Total protein (mg)	Total activity (units)	Specific activity (units/mg)
1. Extract	1500	24,000	ND[a]	
2. DEAE-cellulose 1	325	5,200	320	0.06
3. DEAE-cellulose 2	78	390	245	0.6
4. Hydroxylapatite	35	245	172	0.7
5. Aminohexyl-Sepharose	80	6.0	106	17
6. Isoelectric focusing	30	3.0	66	22
7. Gel filtration	12	1.2	53	44

[a] ND, Not determined because of the presence of inhibitors; see text.

Hydrofluor (National Diagnostics) and radioactivity is measured with a scintillation spectrometer. A background value is obtained by measuring radioactivity in a similar reaction mixture that differs only by the absence of tryptamine.

The assay is linear for 30 min within the range of 0.3 to 3 mg of dialyzed extract from rabbit liver. The assay is meaningless with crude extracts because of the presence of an inhibitor in such preparations; overnight dialysis against 200 volumes of 0.1 M potassium phosphate at pH 7.8 is effective in removing most of the inhibitor.

Definition of Unit. A unit of activity is defined as that amount of enzyme required to catalyze the formation of 1 nmol of product per minute. Specific activity is defined in terms of units of activity per milligram of protein; protein is measured colorimetrically[6] with crystalline bovine serum albumin as standard.

Purification Procedure[7]

General. The enzyme is obtained from livers of young adult New Zealand rabbits; the livers may be stored at $-80°$ for as long as 2 years without apparent effect. All procedures are conducted at ice-bath temperatures. The results of a typical fractionation are presented in Table I.

Step 1. Extraction. Rabbit liver, about 500 g, minced in 1500 ml of Buffer A (10 mM potassium phosphate, 0.1 mM EDTA, 0.1 mM dithiothreitol, and 0.02% sodium azide, adjusted to pH 7.8) is homogenized with

[6] A. Bensadoun and D. Weinstein, *Anal. Biochem.* **70**, 241 (1976).
[7] E. S. Lyon and W. B. Jakoby, in preparation.

a Waring blender acting for two 30-sec intervals. The homogenate is centrifuged at 100,000 *g* for 90 min and the residue is discarded.

Step 2. DEAE-Cellulose I. The supernatant liquid is charged onto a column (9 × 22 cm) of DEAE-cellulose (DE52; Whatman) that has been equilibrated with Buffer A. The column is washed with 2 liters of Buffer A and eluted with a linear gradient created from 1.5 liters of Buffer A and 1.5 liters of Buffer A that is 300 m*M* in potassium phosphate at the same pH. Fractions containing the enzyme activity, eluted at a conductivity of approximately 4 mmho, are pooled and concentrated with an Amicon ultrafiltration apparatus using a PM-10 membrane. The concentrated enzyme is dialyzed overnight against 6 liters of Buffer A.

Step 3. DEAE-Cellulose II. The dialysate is charged onto a second column (2.5 × 40 cm) of DEAE-cellulose (DE52) that has been equilibrated with Buffer A. The column is washed with 1 liter of Buffer A and the enzyme is eluted with a 2-liter linear gradient of Buffer A that is 10 to 300 m*M* in potassium phosphate at the same pH. The active fractions are pooled, concentrated by ultrafiltration, and dialyzed overnight against 2 liters of Buffer B. Buffer B has the same constituents as Buffer A but is adjusted to pH 7.2.

Step 4. Hydroxylapatite.[8] The dialysate is applied to a column (1.5 × 20 cm) of hydroxylapatite that has been equilibrated in Buffer B. The column is washed with 500 ml of the same buffer; enzyme is not retained on the column.

Step 5. Aminohexyl-Sepharose. The active fractions are pooled and applied to a column (1.5 × 10 cm) of aminohexyl-Sepharose that has been equilibrated in Buffer A. The column is sequentially washed with 100 ml of Buffer A; with 300 ml of Buffer A adjusted to a conductivity of 2 mmho with 1 *M* K_2HPO_4; and with 500 ml of Buffer A adjusted to a conductivity of 4 mmho with 1 *M* K_2HPO_4. The enzymatically active fractions eluted in the last wash are pooled, concentrated with a PM-10 membrane, and dialyzed overnight against Buffer A that is 5 m*M* in phosphate.

Step 6. Isoelectric Focusing. The enzyme is applied to an isoelectric focusing column containing 2.5% of pH 3 to 9.5 Ampholine (Pharmacia) and stabilized by a 0–40% sucrose gradient. Focusing is carried out for 16 hr at constant power of 15 W. The active fractions, centered about pH 5.0, are pooled and the pH is adjusted to 7.5 with 1 *M* K_2HPO_4.

Step 7. Sephadex G-75. The enzyme solution is concentrated over a PM-10 membrane and applied to a column (1.5 × 80 cm) of Sephadex equilibrated in Buffer A that is 0.1 *M* in potassium phosphate. The active fractions are pooled, concentrated, and may be stored at −20° in 10% glycerol for up to 6 weeks without loss of activity.

[8] Prepared as described by D. Levin, this series, Vol. 5, Article [2].

TABLE II
AMINE SUBSTRATES FOR ARYLAMINE N-METHYLTRANSFERASE

Substrate	Relative activity[a]	K_m^b (mM)	V_m^b (nmol min/mg)
Tryptamine	100	0.1	50
N-Methyltryptamine	34	0.09	6.9
L-Tryptophan methyl ester	95	0.2	43
Serotonin	6	0.5	3.2
Aniline	54	1.8	44
Imidazole	30	0.6	23
Histamine	2	1.6	3.0

[a] Measured with the methyl group acceptor at a concentration of 1 mM.

[b] Apparent K_m and V_m values were determined with 42 μM S-adenosyl-L-methionine.

The enzyme is homogeneous by electrophoretic criteria: discontinuous polyacrylamide gels, pH 8.9, and SDS–polyacrylamide gels reveal one protein band; protein is coincident with enzyme activity in the non-denaturing gel system.

Properties[7]

Physical Properties. The molecular weight of the enzyme, as determined by gel filtration, sucrose density gradient centrifugation, and SDS–polyacrylamide gel electrophoresis, is estimated at 27,000. Preparative isoelectric focusing in sucrose gradients gave a measure of the pI as 4.8.

pH Optima and Isoelectric Point. With either tryptamine or aniline as the methyl group acceptor, the pH optimum is 7.5. As determined by preparative isoelectric focusing in a sucrose gradient, the pI was found to be 4.8.

Specificity. The enzyme will utilize a variety of aryl amines. Monomethylated amines will serve as acceptors forming the dimethylated species, but formation of quaternary amines has not been observed. A list of several amines that serve as acceptors is presented in Table II along with apparent K_m and V_{max} values for each. The following amines, at a concentration of 1 mM in an otherwise standard assay system, do not serve as methyl group acceptors: ethanolamine, octylamine, phenethylamine, spermine, nicotinic acid, dopamine, and norepinephrine.

S-Adenosyl-L-methionine serves as the methyl donor with an apparent K_m of 5 μM in the otherwise standard assay system.

[34] Catechol *O*-Methyltransferase

By RONALD T. BORCHARDT

Catechol *O*-methyltransferase catalyzes the transfer of a methyl group from *S*-adenosylmethionine (AdoMet) to a catechol substrate resulting in the formation of the *meta*- and *para*-O-methylated products. It is widely distributed in mammalian tissue and plays a primary role in the extraneuronal inactivation of endogenous catecholamines (dopamine, norepinephrine, epinephrine) as well as in the further metabolism of oxidized catecholamine metabolites (3,4-dihydroxymandelic acid, 3,4-dihydroxyphenylglycol, 3,4-dihydroxyphenylacetic acid). In addition, it is involved in the detoxication of catechol drugs (isoproterenol, α-methyldopa, L-dopa), which are used in the treatment of hypertension, asthma, and Parkinson's disease.[1]

Assay Method

Principle. The most convenient, rapid, and simple assay for catechol *O*-methyltransferase activity is a radiochemical technique using either $[^{14}CH_3]AdoMet$ or $[C^3H_3]AdoMet$. The resulting $^{14}CH_3$- or C^3H_3-labeled O-methylated products are isolated by simple solvent extraction.[1,2] The assay procedure is based on the use of $[^{14}C]AdoMet$ and 3,4-dihydroxybenzoic acid as substrates, followed by solvent extraction of the $^{14}CH_3$-labeled O-methylated products (3-hydroxy-4-methoxybenzoic acid and 4-hydroxy-3-methoxybenzoic acid).[3]

Reagents

Sodium phosphate, 500 mM, pH 7.6
S-Adenosyl-L-methionine iodide (Sigma), 10 mM
$[^{14}CH_3]S$-Adenosylmethionine (New England Nuclear, 55 mCi/mmol), 10 μCi/ml

---

[1] R. T. Borchardt, *in* "Enzymatic Basis of Detoxication" (W. G. Jakoby, ed.), Vol. 2, p. 43. Academic Press, New York, 1980.

[2] J. Axelrod, this series, Vol. 5, p. 748.

[3] B. Nikodejevic, S. Senoh, J. W. Daly, and C. R. Creveling, *J. Pharmacol. Exp. Ther.* **174**, 83 (1970).

Dihydroxybenzoic acid (Aldrich), 20 mM
Dithiothreitol (Sigma), 40 mM (DTT)
MgCl$_2$, 60 mM
HCl, 1 N
Toluene : isoamyl alcohol (7 : 3)

Procedure. The reaction mixture has a total volume of 250 μl containing the following components (final concentrations): 25 μl of AdoMet (1.0 mM), 5 μl of [^{14}CH$_3$]AdoMet (0.05 μCi), 25 μl of dihydroxybenzoic acid (2.0 mM), 25 μl of DTT (4 mM), 5 μl of MgCl$_2$ (1.2 mM), 50 μl of phosphate buffer, pH 7.6 (100 mM), and the enzyme sample. The reaction is started by addition of the enzyme. After incubating the reaction mixture at 37° for 5–30 min, the reaction is stopped by addition of 100 μl of 1.0 N HCl. The assay mixture is extracted by vortexing with 2 ml of toluene : isoamyl alcohol (7 : 3). After centrifugation to separate the phases, a 1-ml aliquot of the organic phase is used to measure radioactivity. The results are corrected using a blank free of dihydroxybenzoic acid. The activity can be expressed as amount of O-methylated products formed per milligram of protein per minute. Protein concentrations are determined by the method of Lowry *et al.*[4]

In biological samples in which the activity of catechol O-methyltransferase is low, e.g., brain or heart, the final concentration of AdoMet in the assay may be reduced to 0.1 mM, thereby increasing the specific activity of [^{14}CH$_3$]AdoMet 10-fold.[5] Alternatively, [C^{3}H$_3$]-AdoMet, which is commercially available at higher specific activities, can be used. When assaying the membrane-bound form of catechol O-methyltransferase, dithiothreitol should be excluded from the assay because of the presence of a thiol S-methyltransferase in this subcellular fraction.[5] When assaying the enzyme in erythrocytes, steps should be taken to remove endogenous calcium, an inhibitor of the catechol O-methyltransferase reaction.[6,7]

Purification Procedure

Because the soluble form of catechol O-methyltransferase from rat liver is most abundant and has been most extensively studied, the procedure for its purification is described in this chapter (see table). This procedure is a modification of that described earlier by Nikodejevic *et al.*[3] and

[4] O. H. Lowry, N. J. Rosebrough, A. L. Farr, and R. J. Randall, *J. Biol. Chem.* **193**, 265 (1951).
[5] R. T. Borchardt and Chao Fu Cheng, *Biochim. Biophys. Acta* **522**, 49 (1978); see Article [32], this volume.
[6] R. M. Weinshilboum and F. A. Raymond, *Biochem. Pharmacol.* **25**, 573 (1976).
[7] F. A. Raymond and R. M. Weinshilboum, *Clin. Chim. Acta* **58**, 185 (1975).

SUMMARY OF PURIFICATION OF THE RAT LIVER O-METHYLTRANSFERASE

Step	Volume (ml)	Total protein (mg)	Total activity (nmol/min)	Specific activity (units/mg)
1. Extract	3050	86,000	186,000	2.16
2. Acid precipitation	3150	46,000	166,000	3.6
3. Salt fractionation	330	10,500	110,000	10.5
4. Sephadex G-25	290	5,830	101,000	17.3
5. CaHPO$_4$ absorption	135	1,350	67,000	49.6
6. Affinity chromatography[a]	5	5.1	2,640	518
7. Sephadex G-100[b]	1	0.47	447	950

[a] A part of the CaHPO$_4$-purified enzyme (97.8 mg) was chromatographed on the 3,4-dimethoxy-5-hydroxyphenylethylamine-agarose conjugate.

[b] A part of the affinity purified enzyme (1.17 mg) was chromatographed on the Sephadex G-100.

Borchardt *et al.*[8] It has also been modified for the purification of the enzymes from rat brain and heart.[5]

Step 1. Subcellular Fractionation. Rat livers (Sprague-Dawley, 180–200 g) are rinsed in ice-cold sodium phosphate buffer (10 mM), pH 7.0 containing 0.25 M sucrose, to remove blood, trimmed free of fat and adjoining tissues, cut into small pieces, weighed, and homogenized with four volumes (v/w) of the same buffer in a glass homogenizer equipped with a loose fitting, motor-driven Teflon pestle. The homogenate is filtered through two layers of gauze and the filtrate centrifuged for 10 min at 3000 g to remove nuclei and unbroken cells. The supernatant liquid is centrifuged at 12,000 g for 10 min, 50,000 g for 30 min, and 100,000 g for 60 min to obtain sequentially the mitochondrial fraction, the intermediate fraction, and the light microsomal fraction, as well as the resulting cytosol. Greater than 95% of the catechol O-methyltransferase activity is associated with the cytosolic fraction. All subsequent steps in the purification are carried out at 4°.

Step 2. Acid Precipitation. The pH of the cytosolic fraction is slowly adjusted to 5.0 using 1 N acetic acid. The solution is stirred for 15 min, followed by centrifugation at 14,000 g for 20 min. The pellet is discarded and the pH of the supernatant fluid is adjusted to 6.8 by slowly adding 0.5 M sodium phosphate at pH 7.0.

Step 3. Ammonium Sulfate Fractionation. To the supernatant solution at pH 6.8 is added 196 g of solid ammonium sulfate per liter (33% saturation) in small quantities with mechanical stirring over a period of 60 min.

[8] R. T. Borchardt, C. F. Cheng, and D. R. Thakker, *Biochem. Biophys. Res. Commun.* **63**, 69 (1975).

The suspension is centrifuged at 14,000 g for 20 min and the precipitate discarded. An additional 142 g of ammonium sulfate per liter of supernatant liquid (55% saturation) is added in the same manner. The suspension is centrifuged and the residue is retained. The protein sedimented at 55% salt is suspended in cold 50% ammonium sulfate, 10 mM in sodium phosphate at pH 7.0, and gently stirred for 20 min. The suspension is centrifuged at 14,000 g for 20 min and the supernatant liquid discarded. This procedure is repeated with 48% ammonium sulfate solution. The sediment remaining after washing with 48% ammonium sulfate is washed twice with 100 ml of 33% ammonium sulfate. The supernatant liquids from the 33% salt washes are pooled and an additional 142 g of ammonium sulfate per liter are added. The suspension is centrifuged at 14,000 g for 20 min and the sediment dissolved in a minimum amount of 10 mM sodium phosphate at pH 7.0.

Step 4. Sephadex G-25. The protein solution is applied to a column (3 × 65 cm) of Sephadex G-25 equilibrated with 10 mM sodium phosphate at pH 7.0. The same buffer is used to elute the enzyme. The pooled fractions are concentrated, if necessary, by ultrafiltration through an Amicon PM-10 membrane to a final protein concentration of 15–20 mg/ml.

Step 5. Calcium Phosphate Gel. Calcium phosphate (Sigma) is suspended in 10 mM phosphate buffer, pH 7.0, so that the final suspension contains 50 mg of solid per milliliter. Pilot experiments with gel-to-protein ratios of 1.4, 1.6, and 1.8 are performed to establish optimal conditions. The optimal amount of gel is added to the protein solution from Step 4 with gentle stirring for 15 min at 4°. The mixture is centrifuged at 14,000 g for 20 min and the pale yellow supernatant fluid decanted. This solution is concentrated by ultrafiltration through an Amicon PM-10 membrane to a final protein concentration of 10 mg/ml. This O-methyltransferase preparation, with a specific activity of about 50, is very stable and quite useful for routine inhibitor studies.

Step 6. Affinity Chromatography. The affinity chromatography system developed for purifying this O-methyltransferase consists of a 3,4-dimethoxy-5-hydroxyphenylethylamine-Sepharose 4B conjugate, which has the ligand separated from the insoluble matrix by a spacer of approximately 30 Å.[8] The ligand (3,4-dimethoxy-5-hydroxyphenylethylamine) can be prepared from 3,4-dimethoxy-5-hydroxybenzaldehyde[9] by conversion of the benzaldehyde to 3,4-dimethoxy-5-benzyloxybenzaldehyde, by reaction with nitromethane to yield the nitrostyrene derivative, by reduction of the nitrostyrene with lithium aluminum hydride, and by removal of the benzyl ether protecting group by catalytic hydrogenation.[10]

Cyanogen bromide activation of Sepharose 4B (Sigma) and subsequent

[9] F. Mauthner, *Justus Liebigs Ann. Chem.* **449**, 102 (1926).
[10] R. T. Borchardt and D. R. Thakker, *Biochemistry* **14**, 453 (1975).

attachment of the hydrocarbon side chain, which consisted of units of 3,3'-iminobispropylamine and succinate, can be performed essentially as described by Lefkowitz et al.[11] The ligand 3,4-dimethoxy-5-hydroxy-phenylethylamine is attached to the succinylated gel using 1-(3-dimethyl-aminopropyl)-3-ethylcarbodiimide (Aldrich). In a typical preparation, 100 ml of the succinylated gel is suspended in 100 ml of distilled water. To the suspension is added 3,4-dimethoxy-5-hydroxyphenylethylamine · HCl (1 g, 4.26 mmol) and 1-(3-dimethylaminopropyl)-3-ethylcarbodiimide (0.82 g, 4.26 mmol); the pH is adjusted to 4.8. During the first 3 to 4 hr of the reaction a pH of 4.8 is maintained using 1 N HCl. When the pH is stabilized, the reaction mixture is stirred at ambient temperature for 2 days, after which the gel is filtered and washed with 4 liters of distilled water, followed by 500 ml of 5 mM phosphate buffer at pH 7.2. To determine the extent of coupling, samples of the substituted agarose are hydrolyzed with 2 N NaOH at 100°. An equivalent amount of the succinylated gel is used as a reference. The ultraviolet absorption of these samples at 285 nm can be measured (λ_{max} for 3,4-dimethoxy-5-hydroxyphenylethylamine = 285 nm, ϵ = 2007), from which the amount of ligand incorporated may be calculated. A typical preparation shows an incorporation of 10 μmol of the ligand per milliliter of packed volume of the gel.

The enzyme from Step 5 (97.6 mg) may be additionally purified by chromatography on a column (2.5 × 18 cm) of 3,4-dimethoxy-5-hydroxyphenylethylamine-Sepharose 4B. The affinity column is equilibrated with 5 mM sodium phosphate at pH 7.2 containing 0.2 mM EDTA and 0.2 mM magnesium chloride. After elution of the major protein peak, the catechol O-methyltransferase activity is eluted with 60 mM sodium phosphate at pH 7.2 containing 0.2 mM EDTA and 0.2 mM magnesium chloride. Fractions with high enzyme activity are pooled and concentrated by ultrafiltration (Amicon PM-10 membrane).

Step 7. Sephadex G-100. The enzyme (1.17 mg) is subjected to chromatography on a column (1 × 55 cm) of Sephadex G-100, eluting with 10 mM sodium phosphate, pH 7.2. Fractions of 1.5 ml are collected with the catechol O-methyltransferase activity eluting in fractions 27 through 36. The fractions are pooled and concentrated to yield an enzyme having a specific activity of 950. This method of fractionation (see table) results in 900-fold purification.

Purity and Stability

Electrophoretic studies in SDS–polyacrylamide gels show that the purified enzyme preparation contains one major band, representing 60 to 70% of the protein, and 2 or 3 minor bands. The major band has a molecu-

[11] R. J. Lefkowitz, E. Haber, and D. O'Hara, *Proc. Natl. Acad. Sci. U.S.A.* **69**, 2828 (1972).

lar weight of 23,000. The preparation after Step 7 is relatively stable and may be stored at $-4°$ for several weeks without substantial loss of activity.

Properties

The enzyme requires Mg^{2+} for activity, with Mg^{2+} concentrations greater than 2 mM causing inhibition; Ca^{2+} is inhibitory. The enzyme is very sensitive to inhibition by the product S-adenosylhomocysteine. In addition, it is sensitive to inhibition by various phenolic and polyphenolic compounds, as well as structurally related compounds, e.g., tropolone, 8-hydroxyquinoline, and 3-hydroxy-4-pyrone. The enzyme exhibits a strict requirement for AdoMet as methyl donor and a catechol as methyl acceptor substrate. Monophenolic compounds are not O-methylated by the enzyme. The ratio of meta- and para-O-methylated products generated *in vitro* is dependent on the nature of the catechol substrate and the pH of the reaction mixture. The pH optimum for the enzyme varies slightly with the nature of the buffer, but in general, is between pH 7.3 and 8.2. The enzyme contains sulfhydryl groups essential for catalytic activity as is evident by the inhibitory effects of p-chloromercuribenzoate, N-ethylmaleimide, and iodoacetic acid.

[35] N-Acetyltransferase and Arylhydroxamic
Acid Acyltransferase

By WENDELL W. WEBER and CHARLES M. KING

Introduction

Acetyl-CoA (CoASAc)-dependent N-acetylation (Reaction 1) is a major route of biotransformation of arylamine and hydrazine drugs and other foreign compounds in most animal species. Acetyl transfer as seen in the biotransformation of arylhydroxamic acids to reactive N-acetoxy-arylamines by N,O-acyl transfer (Reaction 2) also commonly occurs with arylamine carcinogens. The CoASAc-dependent N-acetyltransferase and arylhydroxamic acid N,O-acyltransferase activities are both implicated in the conversion of arylamines to reactive intermediates with acute, muta-genic and carcinogenic toxic potential. In the rabbit, both of these activi-ties in liver are properties of the same enzyme, and they are subject to wide genetic variation.

$$\text{CoASAc} + \text{R—NH}_2 \rightarrow \text{CoASH} + \text{R—NH—Ac} \qquad (1)$$
$$\text{Amine} \qquad\qquad\qquad \text{Amide}$$

$$\text{Aryl—N—Ac} \rightarrow \text{Aryl—N—H} \qquad\qquad (2)$$
$$\qquad\quad |\qquad\qquad\qquad\quad |$$
$$\qquad\text{OH}\qquad\qquad\qquad\text{O—Ac}$$
Aryl-   N-Acetoxy-
hydroxamic   arylamine
acid

Assay Methods

Procedures for determining CoASAc-dependent N-acetyltransferase activity and N,O-acyltransferase activity are described.

Radioassay Procedure for N-Acetyltransferase Using CoASAc as the Acetyl Donor[1]

Principle. This method determines the extent to which the labeled acetyl group of CoASAc is transferred to the primary arylamine acceptor substrate to form an arylacetamide.

Reagents

CoASAc, 2 mM, labeled with tritium at a specific activity of 4.5 × 10^{-3} μCi/nmol in water

2-Aminofluorene, 1 mM in dimethyl sulfoxide (CAUTION: Carcinogen!)

Potassium phosphate, 0.1 M, pH 7.4; 1 mM dithiothreitol

N-Ethylmaleimide, 2 mM in 1,2-dichloroethane

Toluene-based liquid scintillation counting fluid

Procedure. Incubation mixtures are prepared by placing 100 μl of suitably diluted cytosolic enzyme preparation in 1.5-ml capped polypropylene tubes, followed by 50 μl of the aminofluorene solution and a volume of buffer so that the final volume of the mixture is 0.20 ml. The amine substrate is omitted from the control tubes. Reaction mixtures are preincubated at 37° for 3 min, and the reaction is initiated by addition of 50 μl of the labeled CoASAc solution and then incubated further for specified times (15 sec, 30 sec, 1, 2, 3, and 5 min). The reaction is stopped by vortexing for at least 30 sec after the addition of 1 ml of the N-ethylmaleimide/1,2-dichloroethane solution. Acetylated product is extracted into the organic phase by shaking for 20 min on a platform rotator and centrifuged for 1 min to separate the phases. The upper (aqueous) layer is aspirated and 500 μl of the lower (organic) layer is pipetted into

[1] I. B. Glowinski, H. E. Radtke, and W. W. Weber, *Mol. Pharmacol.* **14,** 940 (1978).

5-ml glass scintillation vials and evaporated to dryness under vacuum. Three milliliters of toluene-based scintillation fluid is added, and the radioactivity is determined by liquid scintillation spectroscopy.

Initial reaction velocities are determined by extrapolating the time–activity curves to the time of addition of the CoASAc and determining the slope of the tangent to the curve at that point. Under the conditions of the assay, 5500 dpm corresponds to the acetylation of 1 nmol of aminofluorene after correction for extraction efficiency (98%) of acetylated product.

Colorimetric Procedure for N-Acetyltransferase Using CoASAc as the Acetyl Donor[2]

Principle. Acetylation is determined in this procedure from the decrease in the concentration of a primary aromatic amine in the incubation mixture; this decrease results from the transfer of the acetyl group of CoASAc to the acceptor amine.

Reagents

CoASAc, 10 mM, prepared in water
p-Aminobenzoic acid, 0.2 mM, prepared in water
Potassium phosphate, 0.1 M, pH 7.4, for dilution of enzyme
Trichloroacetic acid, 5%
Sodium nitrite, 0.1%, freshly prepared in water
Ammonium sulfamate, 0.5% in water
N-1-Naphthylenediamine dihydrochloride, 0.05% in water

Procedure. Incubation mixtures, 90 μl, are prepared by placing 20 μl of the CoASAc solution and 20 μl of the p-aminobenzoic acid solution in 0.4-ml capped polyethylene microtest tubes and then adding 50 μl of suitably diluted cytosolic enzyme preparation to initiate the reaction. CoASAc is omitted from the control tubes. Tubes are incubated at 37° for specified periods (5, 10, 15, 20, and 30 min). The reaction is stopped by addition of trichloroacetic acid solution and centrifuged (10,000 g, 1 min) to precipitate the protein. Diazotization of the residual nonacetylated amine in the supernatant fraction is carried out by addition of 0.020 ml of the sodium nitrite solution followed by mixing and allowing 3 min to elapse. This is followed by addition of 0.020 ml of the ammonium sulfamate solution (to remove excess nitrite) and mixing. After allowing 3 min to elapse, 0.1 ml of the N-1-naphthylenediamine dihydrochloride solution is added for development of color. Absorbance is measured at 540 nm in 400-μl cuvettes for 1-cm light path against a water blank.

[2] W. W. Weber, J. N. Miceli, D. J. Hearse, and G. S. Drummond, *Drug. Metab. Dispos.* **4,** 94 (1976).

The extent of acetylation is obtained by subtracting the experimental from the control absorbance. Under the conditions of this assay, a decrease of 1 absorbance unit corresponds to the acetylation of 6.08 nmol of p-aminobenzoic acid. Protein content of the enzyme preparation is determined spectrophotometrically.

N-Acetyltransfer Using Arylhydroxamic Acids as Acetyl Donors [3]

Principle. This procedure is based on that of Booth,[3] in which the N-acetyl moiety of an arylhydroxamic acid is transferred irreversibly to a primary arylamine to form a stable acetamide.

Reagents

N-Hydroxy-2-acetylaminofluorene, 25 mM in 95% ethanol (CAUTION: Carcinogen!)
4-Aminoazobenzene, 2.5 mM in 95% ethanol
Tetrasodium pyrophosphate, 54 mM, adjusted to pH 7.0 with HCl; 1 mM dithiothreitol
Trichloroacetic acid, 20% in 50% ethanol

Procedure. Incubation mixtures are prepared by incorporating cytosolic enzyme preparations in 2.3 ml of the pyrophosphate buffer and adding 0.1-ml aliquots of the acetyl acceptor (4-aminoazobenzene) and donor (N-hydroxy-2-acetylaminofluorene). After incubating these solutions for 15 min at 37°, an equal volume of the trichloroacetic acid solution is added to stop the reaction and precipitate the protein. The trichloroacetic acid protonates the residual free 4-aminoazobenzene and enhances the absorption at 497 nm, which, after clarification by centrifugation or filtration, is used to estimate the loss of acetyl acceptor through enzymatic acetyl transfer. The molecular extinction coefficient of 4-aminoazobenzene at 497 nm under these conditions is 13,200. A decrease in extinction at 497 nm of 0.264 is equivalent to the acetylation of 0.1 μmol of 4-aminoazobenzene.

Arylhydroxamic Acid N,O-Acyltransferase [4]

Principle. This assay determines the extent to which ring-labeled arylhydroxamic acids are incorporated into nucleic acid adducts as a consequence of the production of reactive N-acyloxyarylamines by N,O-acyl transfer.

[3] J. Booth, *Biochem. J.,* **100,** 749 (1966).
[4] C. M. King, *Cancer Res.* **34,** 1503 (1974).

Reagents

N-Hydroxy-2-acetylaminofluorene, labeled with tritium or carbon-14 in the ring at a specific activity equal to or greater than 1 mCi/mmol. 4.2 mM, in dimethyl sulfoxide (CAUTION: Carcinogen!)

tRNA from yeast, 150 $A_{260\,nm}$ units per milliliter in the pyrophosphate buffer

Tetrasodium pyrophosphate, 50 mM, adjusted to pH 7.0 with HCl; 1 mM dithiothreitol

Phenol that has been saturated with water

Potassium acetate, 2% in 95% ethanol

Procedure. Dithiothreitol is added to each of the aqueous solutions immediately before use. An aliquot (0.1 ml) of the tRNA solution is added to a glass centrifuge tube, followed by 10 μl of the labeled substrate and a volume of the buffer such that, with addition of the enzyme preparation to initiate the reaction, the final volume will be 0.8 ml. After mixing, the solution is incubated for 20 min at 37°. The reaction is stopped and the protein removed by the addition of 1 ml of water-saturated phenol. After centrifugation, an aliquot of the aqueous layer is added to 5 ml of the potassium acetate in ethanol that is contained in a glass filtration apparatus over a glass fiber filter. The tRNA precipitate is collected on the filters and then subjected to successive washes with 70 and 95% ethanol, acetone, and ether. The dry filter is placed in a scintillation vial, moistened with 15 μl of water, and then shaken with 1 ml of tissue solubilizer (NCS, Amersham-Searle). The isotopic content of the nucleic acid is determined by liquid scintillation spectroscopy after dilution with a toluene-based scintillation fluid.

Controls should include both enzyme-free incubations, as well as experiments in which the tRNA is added after the enzyme reaction has been stopped by the addition of phenol. The latter procedure allows estimation of the possible contamination of the tRNA with metabolites that are not covalently bound to the polynucleotide. The activity of the enzyme is usually expressed in terms of the moles of arylamine moiety bound to nucleic acid per minute per milligram protein.[4,5] It should be recognized, however, that this activity is dependent upon the concentration of tRNA, because by increasing the concentration of tRNA in the incubation mixture, there is an increase in the apparent activity of the enzyme as a result of trapping of a greater percentage of the reactive N-acyloxyarylamine by the nucleic acid.

[5] See also this volume, Article [7] for a general discussion of covalent binding to macromolecules.

PURIFICATION OF *N*-ACETYLTRANSFERASE FROM RABBIT LIVER (6)

Purification step	Protein (mg/ml)	Volume (ml)	*N*-Acetyltransferase activity[a] (units/mg)	*N,O*-Acyltransferase activity[b] (units/mg)	NAT/ AHAT
Supernatant fraction	23.4	125	4.6	0.80	5.8
Salt precipitation	38.0	12	6.1	0.94	6.4
DEAE-cellulose	0.31	66	55	6.5	8.5
Gel filtration	0.13	10	198	34	5.9

[a] *N*-Acetyltransferase (NAT) activity, determined by estimating the CoASAc-dependent acetylation of sulfamethazine, expressed as the nanomoles of product per minute per milligram protein.

[b] *N,O*-Acyltransferase (AHAT) activity, expressed as in footnote *a*, reflects the incorporation of the tritiated moiety of *N*-hydroxy-2-acetylaminofluorene into nucleic acid.

Enzyme Purification

Instability can be a major problem encountered in the purification of acyltransferases from tissues of both rabbit and rat, presumably caused by oxidation of essential sulfhydryl groups. It has been found useful to employ a chelating agent (pyrophosphate) as a buffer, and a sulfhydryl protective agent (dithiothreitol) during purification. A further precaution is the use of argon atmospheres to exclude oxygen. Use of these techniques has permitted the isolation of acyltransferases in approximately 15% yield from both rabbit[6] and rat[7] by the application of the procedures described in the following sections. The results of a typical purification of the rabbit liver enzyme are shown in the table.

Step 1. Solubilization. Centrifugation (105,000 *g*, 1 hr) of tissue homogenates prepared by use of a Waring blender, rotating homogenizer, or Polytron device have yielded supernatant fractions with high levels of *N*-acetyltransferase activity. While the homogenization of each gram of tissues with 5 to 9 volumes of 50 m*M* sodium pyrophosphate, which had been adjusted to pH 7.0 and contained 1 m*M* dithiothreitol, serves to preserve the activity, other buffers such as phosphate or Tris have also been used successfully to solubilize the enzyme.

Step 2. Fraction Precipitation with Neutralized Ammonium Sulfate. Saturated neutralized ammonium sulfate in pyrophosphate buffer[8] is

[6] I. B. Glowinski, W. W. Weber, J. M. Fysh, J. B. Vaught, and C. M. King, *J. Biol. Chem.* **255**, 7883 (1980).

[7] W. T. Allaben and C. M. King, *Fed. Proc., Fed. Am. Soc. Exp. Biol.* **36**, 349 (1977).

[8] Sodium pyrophosphate, 50 m*M*, containing 1 m*M* dithiothreitol, at pH 7.0, is saturated with solid ammonium sulfate at room temperature.

added with stirring at room temperature to the 105,000-g supernatant fraction to 45% of saturation. The precipitate is removed by centrifugation at ambient temperature and discarded. The precipitate, which forms subsequent to raising the ammonium sulfate concentration to 65% of saturation, is collected by centrifugation. This precipitate can be frozen and stored for long periods without loss of activity, or it can be dissolved in 2 mM sodium pyrophosphate (adjusted to pH 7.0) and 1 mM dithiothreitol and dialyzed overnight against the same buffer in preparation for further purification. Dialysis and all subsequent procedures are carried out at 4° under argon.

Step 3. Ion-Exchange Chromatography on DEAE-Cellulose. The dialysate (250–300 mg protein) is applied to a column (2.5 × 10 cm) of DEAE-cellulose that has been equilibrated with the 2 mM sodium pyrophosphate buffer. The enzyme is eluted with a linear NaCl gradient (0.0–0.3 M) in the same buffer; activity emerges at a NaCl concentration of approximately 0.15–0.17 M.

Cytosolic preparations in similar buffers of low concentration can be subjected directly to ion-exchange chromatography on DEAE-cellulose without prior purification by fractional precipitation with ammonium sulfate.

Step 4. Gel Filtration on Sephacryl S-200. The fractions from the DEAE-cellulose chromatography that contain the enzyme can be pooled and concentrated by ultrafiltration by use of a Millipore apparatus fitted with a type PT membrane. The concentrate is applied to a Sephacryl S-200 column (2.5 × 60 cm) and eluted with 20 mM sodium pyrophosphate, pH 7.0, containing 1 mM dithiothreitol.

Step 5. Electrophoresis on Polyacrylamide Gel. The fractions from gel filtration can be concentrated, if necessary, by ultrafiltration and subjected to electrophoresis on polyacrylamide gel. Gels (5 × 80 mm) that contain a final concentration of 7% acrylamide, 0.2% bisacrylamide, 0.14% ammonium persulfate, 0.083% N,N,N',N'-tetramethylethylenediamine and 0.089 M Tris-borate at pH 8.3. The gels are cast in quartz tubes and subjected to a current of 2 mA per gel for 2 hr after applying a sample containing 20 to 50 μg of protein, the same current is used for an additional 75 min. The gels are scanned at 280 nm and sliced without freezing. Enzyme activity is detected by homogenizing the slices in an appropriate enzyme assay system and carrying out an incubation in the presence of the gel particles.

In preparations obtained from the liver of a rapid acetylator rabbit, two major bands of protein are observed, only one of which is associated with N-acetyltransferase and arylhydroxamic acid N,O-acetyltransferase activities. This band, when subjected to electrophoresis on sodium dodecyl sulfate-containing gels, disclosed the presence of a single peptide of 33,000 molecular weight.

Properties[9,10]

Species and Hereditary Variation. Species differences in the capacity to N-acetylate primary arylamine and hydrazine drugs and other foreign compounds are large. The hamster and rabbit possess comparatively high N-acetylating capacities, whereas the dog has little or no capacity for acetylating these compounds. Other species such as the primates (including man), rat, and mouse have intermediate capacities.

Hereditary variation in individual N-acetylating capacity has been recognized in several species including man, rabbit, mouse, hamster, and rat. These differences are attributable to genetically determined differences in liver *N*-acetyltransferase activity enabling individuals of these species to be classified as rapid or slow acetylators. It appears that such variation extends to acetyl transfer involved in the generation of reactive *N*-acetoxyarylamines from arylhydroxamic acids by *N,O*-acyltransferase as seen in liver obtained from rapid and slow acetylator rabbits.

Distribution. The enzyme is cytosolic and is widely distributed among tissues, with highest activities occurring in liver and intestinal tract. Distribution along the gastrointestinal tract is heterogeneous, with maximal activity in proximal portions of the small intestine and very low activities in the stomach and colon. The activity level in the small intestine reflects the liver phenotype in man, rabbit, and mouse. No appreciable activity has been detected in plasma, brain, or muscle.

Substrate Specificity. The enzyme catalyzes CoASAc-dependent N-acetylation of drugs and foreign compounds that contain NH_2 or $NH-NH_2$ groups attached directly, or via a carbonyl group, to unsaturated rings; it can also catalyze N-acetylation of NH_2 groups attached to ring systems via a short aliphatic carbon side chain. A large array of these compounds are substrates including 2-aminofluorene, aniline, benzidine, methyl bis-2-chloroaniline, *p*-phenetidine, *p*-aminobenzoic acid, *p*-aminosalicyclic acid, *p*-aminosulfonic acid, α-naphthylamine, β-naphthylamine, sulfadiazine, sulfamethazine, sulfanilamide, sulfapyridine, diaminodiphenylsulfone, isonicotinyl hydrazide (isoniazid), hydralazine, phenelzine, procainamide, and thiazolsulfone. Certain aliphatic amines are also N-acetylated including histamine, tyramine, and 5-hydroxytryptamine. Evidence for lack of specificity for *p*-nitroaniline,

[9] W. W. Weber and I. B. Glowinski, *in* "Enzymatic Basis of Detoxication" (W. B. Jakoby, ed.), Vol. 2, p. 169. Academic Press, New York, 1980.

[10] C. M. King and W. T. Allaben, *in* "Enzymatic Basis of Detoxication" (W. B. Jakoby, ed.), Vol. 2, p. 187. Academic Press, New York, 1980.

phenylalanine, cyclohexylamine, and glucosamine has been obtained also. Although CoASAc is the most effective acetyl donor, acetylthiocholine or N-diacetylcysteamine can also serve as acetyl donors for the enzyme; the rate of N-acetylation of isoniazid with either of these two acetyl donors is less than 10% of that observed with CoASAc at concentrations 10 times greater than CoASAc.

The specificity of the N,O-acyltransferase reaction involves two successive steps, i.e., N-deacylation and O-acylation. Demonstration of the latter has depended so far on the production of N-acyloxyarylamines that are capable of reaction with nucleic acid. Studies undertaken primarily with preparations from rat liver or mammary gland have shown that a wide variety of aryl moieties provide suitable substrates for both of these reactions (e.g., fluorenyl, biphenyl, benzidine, stilbene, phenanthrene). The suitability of acyl groups as substrates, as reflected in their acylation of 4-aminoazobenzene, the production of mutagens (i.e., the production of arylhydroxamines), and the formation of arylamine-substituted nucleic acids, appears to be more complex. Mutagenicity experiments suggest that acyltransferase preparations may be capable of N-deacylation, but are unable to transfer the acyl group to 4-aminoazobenzene or to induce nucleic acid adduct formation. Thus, while the monochloroacetyl or formyl group may be hydrolyzed by purified preparations of rat liver acyltransferase, they are much less able to yield adducts that are acetylated substrates. It should be noted that enzymes capable of specifically activating formylated N-arylhydroxamines have been demonstrated, but their mechanism of action is not known. The specificity involved in the metabolism of propionyl derivatives are qualitatively similar to those of the acetylated substrates, but the propionyl compounds are usually less rapidly metabolized.

Inhibitors and Inducers. N-Ethylmaleimide and p-chloromercuribenzoate are potent inhibitors of the N-acetyltransferase. Neither cadmium chloride, arsenite, diisopropylfluorophosphate, nor paraoxon have any effect on the activity. Various metal ions including Cu^{2+}, Zn^{2+}, Mn^{2+}, and Ni^{2+} are also inhibitory. No systemic stimulatory effects on the acetyltransferase have been shown to occur with either barbiturates (pentobarbital) or isoproterenol.

Effect of pH. Irreversible inactivation of the enzyme occurs at pH values below pH 6. The pH for optimum activity and the shape of the pH–activity curve both vary with the acetyl acceptor substrate used: with isoniazid there is a broad optimal range between pH 5 and 9; with sulfamethazine the pH profile is bimodal with optima at pH 7.2 and 5.6; with sulfadiazine the optimum pH is 5.7 and activity decreases rapidly above and below that pH. These observations suggest that the enzyme interacts preferentially with the uncharged form of the acetyl acceptor substrate.

[36] Thioltransferase

By BENGT MANNERVIK, KENT AXELSSON, and KERSTIN LARSON

$$RSH + R'SSR' \rightleftharpoons RSSR' + R'SH \qquad (1)$$

Thioltransferases catalyze the reversible thiol–disulfide interchange reactions (Eq. 1) where RSH is a thiol and R'SSR' a disulfide. Reactions involving the endogenous thiol, glutathione (GSH), are of great importance because they are considered to have a major role in the intracellular reduction of endogenous as well as exogenous disulfides and thiosulfate esters.[1-4] Two consecutive reactions [Eqs. (2) and (3)], each catalyzed by

$$GSH + RSSR \rightarrow GSSR + RSH \qquad (2)$$
$$GSH + GSSR \rightarrow GSSG + RSH \qquad (3)$$

thioltransferase, reduce a disulfide to the thiol state. By coupling with the reaction catalyzed by glutathione reductase [Eq. (4)], reduced glutathione is regenerated and the overall reaction [Eq. (5) = Eqs. (2) + (3) + (4)] is linked to the pool of reduced pyridine nucleotides.

$$GSSG + NADPH + H^+ \rightarrow 2GSH + NADP^+ \qquad (4)$$
$$RSSR + NADPH + H^+ \rightarrow 2RSH + NADP^+ \qquad (5)$$

Assay Method

Principle. Thioltransferase activity is monitored spectrophotometrically at 340 nm by following the overall reaction [Eq. (5)] obtained by coupling with glutathione reductase. Because thiol–disulfide interchange takes place spontaneously, the velocity of a blank reaction, run in the absence of thioltransferase, must be subtracted from the velocity determined in the presence of thioltransferase. The thiosulfate ester, S-sulfocysteine ($CySSO_3^-$), is used as substrate in the standard assay because of the favorably low spontaneous reaction rate.[5]

$$GSH + CySSO_3^- \rightleftharpoons GSSCy + HSO_3^- \qquad (6)$$
$$GSH + GSSCy \rightleftharpoons GSSG + CySH \qquad (7)$$

[1] B. Mannervik, *in* "Enzymatic Basis of Detoxication" (W. B. Jakoby, ed.), Vol. 2, p. 229. Academic Press, New York, 1980.

[2] B. Mannervik and S. A. Eriksson, *in* "Glutathione" (L. Flohé, H. C. Benöhr, H. Sies, H. D. Waller, and A. Wendel, eds.), p. 120. Georg Thieme, Stuttgart, 1974.

[3] B. Mannervik and K. Axelsson, *in* "Functions of Glutathione in Liver and Kidney" (H. Sies and A. Wendel, eds.), p. 148. Springer-Verlag, Berlin and New York, 1978.

[4] K. Axelsson and B. Mannervik, *Biochim. Biophys. Acta* **613**, 324 (1980).

[5] K. Axelsson, S. Eriksson, and B. Mannervik, *Biochemistry* **17**, 2978 (1978).

Reagents

Potassium phosphate buffer, 0.2 M, pH 7.6, containing 2 mM EDTA

NADPH, 2 mM, prepared in the phosphate buffer

GSH, 10 mM, prepared in deionized water

Glutathione reductase, 200 units per milliliter (commercially available)

$CySSO_3^-$, 60 mM, prepared in deionized water

Procedure. To a 1-ml cuvette are added, at 30°, 450 μl of buffer, 50 μl of NADPH, 50 μl of GSH, 10 μl of glutathione reductase, and a volume of deionized water giving a total volume of 1 ml in the cuvette after addition of the enzyme solution and 50 μl of $CySSO_3^-$. The reaction is initiated by addition of $CySSO_3^-$. The spontaneous reaction is measured separately in the absence of thioltransferase. The steady-state velocity of the system, which should be reached within 2 min, is used as a measure of activity.

Definition of Unit and Specific Activity. A unit of thioltransferase activity is defined as the amount of enzyme that catalyzes the formation of 1 μmol of GSSG per minute at 30° under steady-state conditions. It is essential that the coupling enzyme, glutathione reductase, is present at a concentration high enough to prevent Eq. (4) from being rate limiting. Specific activity is expressed as units per milligram of protein.

Purification Procedure

The enzyme prepared from rat liver should be kept at about 5° during the entire purification procedure. All buffers contain 1 mM EDTA. Use of plastic instead of glass in columns and test tubes was found to increase the yield in purification steps succeeding Step 4.

Step 1. Preparation of Cytosol Fraction. Fifty male Sprague-Dawley rats (200–250 g) are decapitated and exsanguinated. The livers are removed, placed in ice-cold 0.25 M sucrose, and homogenized twice for 1 min each in a blender. The liver homogenate is made to a concentration of 20% (w/v). The homogenate is centrifuged for 45 min at 11,000 g. The pellet is suspended in a small amount of sucrose and again centrifuged. The first and the second supernatant fractions are combined and adjusted to pH 5.5 by slow addition of ice-cold 5% acetic acid (<10 ml) under continuous stirring; the low pH leads to aggregation of microsomes. The supernatant fraction (approximately 1 liter) is centrifuged for 60 min at 11,000 g and the new supernatant fraction is filtered through cheese cloth (or any other suitable tissue) to remove a layer of lipid at the surface of the liquid.

Step 2. Sephadex G-25. The enzyme from Step 1 is chromatographed on a column (12.5 × 80 cm) of Sephadex G-25 Coarse, equilibrated with 10 mM sodium phosphate at pH 6.1. The protein emerges in a total of about 2 liters after 3 liters of effluent.

Step 3. CM-Cellulose. The pooled effluent from Step 2 is loaded on a column (9 × 13.5 cm) of CM-cellulose (Whatman) equilibrated with 10 mM sodium phosphate at pH 6.1. The bed is rinsed overnight with 4.5 liters of the same buffer. The adsorbed enzyme is eluted by use of a linear concentration gradient formed by linear mixing of 5 liters of the starting buffer with 5 liters of 50 mM sodium phosphate at the same pH containing 0.2 M NaCl. The peak of thioltransferase activity emerges about 1.9 liters after the start of the gradient elution. Fractions containing the enzyme are pooled and thereafter concentrated by ultrafiltration on a Hollow Fiber Membrane Cartridge (Amicon, type H1DP10).

Step 4. Sephadex G-75. The third day the concentrate from Step 3 is chromatographed on a column (9 × 95 cm) of Sephadex G-75 in 10 mM sodium phosphate buffer (pH 6.7). Thioltransferase emerges in the effluent after 4 liters, far behind most proteins. The enzyme is pooled from about 30 fractions (15 ml each).

Step 5. CM-Sepharose. The pooled fractions from Step 4 are applied to a column (1.5 × 13 cm) of CM-Sepharose equilibrated with 10 mM sodium phosphate at pH 6.7. After eluting unadsorbed material with 3–5 bed volumes of the starting buffer, the enzyme is eluted with a linear concentration gradient formed by linear mixing of 100 ml of starting buffer with 100 ml of 50 mM sodium phosphate, pH 6.7, containing 0.2 M NaCl. The fractions with the highest specific activity of thioltransferase are pooled.

Further Purification

Steps 1 to 5 reproducibly yield enzyme with specific activity of about 200 units per milligram. A representative preparation is presented in the table. Occasionally, higher specific activities have been obtained after two additional steps; the highest value obtained has been 705 units per milligram.[5] The small amounts of protein involved in the last stages make protein concentrations difficult to measure and somewhat uncertain. In the particular purification cited in the table, the specific activity rose from 167 to 237 units/mg by addition of the following steps. However, for many applications Steps 1 through 5 would suffice.

Step 6. Hydroxylapatite. The pooled fractions from Step 5 are diluted with 2 volumes of deionized water and passed through a column (1 × 2 cm) of hydroxylapatite (BioRad Laboratories) equilibrated with 10 mM sodium phosphate buffer (pH 6.7). The enzyme passes through the bed unretarded.

PURIFICATION OF THIOLTRANSFERASE FROM RAT LIVER CYTOSOL

Step	Volume (ml)	Total protein (mg)[a]	Total activity (units)	Specific activity (units/mg)
1. Cytosol fraction	1070	58400	2080	0.0356
2. Sephadex G-25	2090	28900	1280	0.0443
3. CM-cellulose	1490	4830	565	0.117
4. Sephadex G-75	510	13.7	213	15.5
5. CM-Sepharose	16.5	0.52	87	167

[a] Protein was estimated from the absorbance at 260 and 280 nm[6] and the apparent decrease in protein content from Step 1 to Step 2 may, in part, be due to removal of interfering low-molecular-weight substances.

Step 7. CM-Sepharose. The pooled fractions from Step 6 are adsorbed on a column (1 × 4 cm) of CM-Sepharose equilibrated with 10 mM sodium phosphate at pH 6.7. The bed is rinsed with 2–3 bed volumes of the same buffer and the enzyme eluted with 50 mM sodium phosphate (pH 6.7) containing 0.2 M NaCl. This last step serves mainly as a means of concentrating the enzyme and enables good recovery.

Properties

Stability. At a state of purification corresponding to a specific activity of about 200 units/mg, thioltransferase is stable for several weeks at $-20°$. The purified enzyme is best stored in plastic tubes; the basic protein is probably adsorbed onto glass, which has negative surface charge.

Molecular Properties.[5] The enzyme from rat cytosol is composed of one subunit of 11,000 daltons and contains 8.6% (w/w) carbohydrates and with an isoelectric point at 22° of 9.6.

Kinetic Properties. The enzyme has optimal activity at pH 7.5.[5] Quantitative kinetic parameters are difficult to define because of coupled reactions in the assay system. The available apparent parameters[5] indicate that the k_{cat} value varies little with the nature of disulfide substrate. The apparent K_m value is about 50 μM for mixed disulfides containing a glutathione moiety and higher for other disulfides. The former substrates involve two [Eqs. (3) and (4)] and the latter three [Eqs. (2), (3), and (4)] coupled reactions, conditions that may explain the differences. All disulfides and thiosulfate esters that have sterically accessible S—S bonds appear to serve as substrates.[1,4,5] Even proteins containing disulfide bonds

[6] H. M. Kalckar, *J. Biol. Chem.* **167**, 461 (1947).

are substrates.[4,5,7,8] It has, therefore, been suggested that the enzyme may have a function in cellular regulation by thiol–disulfide interchange.[3,4,7,9] The purified thioltransferase has also been shown to be active with reductants other than glutathione in the reduction of S—S bonds.[1,4]

[7] B. Mannervik and K. Axelsson, *Biochem. J.* **190**, 125 (1980).
[8] K. Axelsson and B. Mannervik, *Acta Chem. Scand.* **B34**, 139 (1980).
[9] B. Mannervik and K. Axelsson, *Biochem. J.* **149**, 785 (1975).

[37] Thiosulfate : Cyanide Sulfurtransferase (Rhodanese)

By JOHN WESTLEY

$$SSO_3^{2-} + CN^- \rightarrow SO_3^{2-} + SCN^-$$

Rhodanese (EC 2.8.1.1) was discovered as the result of a search for the active principle in the tissues of animals known to be able to convert cyanide to the less toxic thiocyanate. Various aspects of rhodanese research have been reviewed.[1-4]

Assay Method

Principle. The thiocyanate product is determined colorimetrically as its red complex with Fe(III) in acid solution. The method given is adapted from that reported by Sörbo.[5]

Reagents

KCN, 0.25 M
KH$_2$PO$_4$, 0.20 M
Na$_2$S$_2$O$_3$, 0.25 M
Formaldehyde, 15%, prepared by adding 1.5 volumes of water to one of formalin (38% commercial formaldehyde reagent)
Ferric nitrate reagent: 100 g of Fe(NO$_3$)$_3$ · 9H$_2$O and 200 ml of 65% HNO$_3$ per 1500 ml

[1] J. Westley, *Adv. Enzymol.* **39**, 327 (1973).
[2] B. Sörbo, *in* "Metabolic Pathways" (D. M. Greenberg, ed.), 3rd ed., Vol. 7, p. 433. Academic Press, New York, 1975.
[3] J. Westley, *in* "Bioorganic Chemistry" (E. E. Van Tamelen, ed.), Vol. 1, p. 371. Academic Press, New York (1977).
[4] J. Westley, *in* "Enzymatic Basis of Detoxication" (W. B. Jakoby, ed.), Vol. 2, p. 245. Academic Press, New York, 1980.
[5] B. H. Sörbo, this series, Vol. II, p. 334.

Procedure. Phosphate and cyanide solutions, 0.2 ml of each, are added to a small test tube, followed by 0.4 ml of water that may contain any inhibitors or other desired test substances. Thiosulfate, 0.2 ml, is then added. The assay tube and an identical "blank" are equilibrated at 25°. The final volume is 1.0 ml; the pH is 8.6. The reaction is initiated by rapid addition of 1 to 10 μl of enzyme solution directly into the assay solution in the sample tube. After exactly 1 min, the reaction is stopped by rapid addition of 0.5 ml of formaldehyde. The "blank" tube also receives formaldehyde, and 1.5 ml of ferric nitrate reagent is then added to each tube. The absorbance is measured at 460 nm; color develops immediately and is stable for hours. As a rule of thumb, 1 μg of recrystallized rhodanese produces an absorbance of 1.0 in this system when the absorbances are determined in 13 × 100 mm tubes. When samples from crude extracts containing high concentrations of protein are assayed, the final solutions may be turbid. They may be clarified by centrifugation. In such cases, and whenever there is any question that the sample may contain preformed thiocyanate, it is wise to add a like aliquot of enzyme to the "blank" tube *after* the formaldehyde. The assay is specific and there are no known interfering side reactions, but low level nonenzymic catalysis of this reaction by thiols and by some metal ion complexes is a potential problem when dealing with sources that contain very low levels of the enzyme.

Definition of Unit and Specific Activity. One unit of enzyme is defined as that amount that catalyzes the production of one micromole of thiocyanate (A_{460} = 1.40) per minute in this system. Specific activity is expressed in units per milligram of protein.

Purification Procedure

All three of the procedures in common use for isolating bovine liver rhodanese derive from that reported by Horowitz and DeToma.[6] All published procedures for the bovine liver enzyme are based ultimately on Sörbo's method.[5] The procedure given in this chapter contains only minor modifications of the steps in the Horowitz and DeToma procedure. It has been used successfully in this form in at least three different laboratories. The quite different procedure that is required for isolating rhodanese from human liver tissue has been published elsewhere.[7]

All procedures are carried out at 0° to 5°.

Step 1. Preparation of the Extract. Approximately 2.7 kg of bovine liver is sliced and cubed (1-cm cubes), after which it is spread out flat in plastic bags and frozen. The frozen tissue is easily separated by hand and

---

[6] P. Horowitz and F. DeToma, *J. Biol. Chem.* **245**, 984 (1970).
[7] R. Jarabak and J. Westley, *Biochemistry* **13**, 3233 (1974).

placed, still frozen, into a large Waring blendor (3 batches) with a total of 3 liters of cold 10 mM $Na_2S_2O_3$. After being blended at top speed for 1 min, following a brief low-speed blending to break up the larger clusters of frozen cubes, the homogenate is poured into centrifuge bottles and subjected to 13,700 g at 4° for 30 min. The separation is rather poor, but the liquid supernatant layer is decanted and retained. The sediment in the bottles is suspended in 10 mM $Na_2S_2O_3$, and the centrifugation is repeated. The combined supernatant fluids total about 5 liters, containing most of the rhodanese present in the liver.

Step 2. Salt Fractionation. To the combined supernatant fluids is added 54.5 g of solid ammonium sulfate per liter, gradually (500 g/hr) and with good bottom stirring. The pH is adjusted to 3.8 with cold 2 M glycine-sulfate at pH 2.5. An additional 153 g of solid ammonium sulfate is added per liter of preparation, as before, and stirring is continued for 2 to 3 hr after addition is completed. The preparation is centrifuged and the supernatant liquid is retained. It should contain at least two-thirds of the rhodanese in the original extract. This supernatant fluid receives 85 g of solid ammonium sulfate per liter, after which it is allowed to stand for 2 hr (or overnight) to permit complete precipitation of the enzyme. After centrifugation, approximately 15% of the rhodanese should remain in the supernatant solution.

Step 3. pH-Salt Fractionation. The bulk of the activity is in the precipitate from the previous step. This precipitate is extracted by stirring with 80 ml of 10 mM Tris-sulfate at pH 7.6 for 1 hr and centrifuging at 48,000 g for 15 min. The supernatant fluid is adjusted to pH 7.6 by addition of cold 2 M glycine-hydroxide at pH 9.8, and a volume of cold 3.6 M ammonium sulfate equal to the volume of the preparation is added dropwise and with good cooling and stirring. The preparation is clarified by centrifugation if it is not entirely clear, and the pH is adjusted to 6.0 with a solution prepared by mixing equal volumes of 3.6 M ammonium sulfate and 2 M glycine-sulfate at pH 2.5. After 1 hr, the suspension is centrifuged at 48,000 g for 15 min. Rhodanese is precipitated by adjusting to pH 4.5 with the 1 M glycine-sulfate/1.8 M ammonium sulfate solution prepared as previously outlined. The suspension is centrifuged; less than 5 mg of rhodanese should remain in the supernatant fluid.

Step 4. Second pH-Salt Fractionation. The precipitate from the previous step is suspended in 25 ml of cold 1.8 M ammonium sulfate and the pH is adjusted carefully to 4.9 with a solution prepared by mixing equal volumes of 3.6 M ammonium sulfate and 2 M glycine-hydroxide, pH 9.8. The preparation is centrifuged; less than 10 mg of rhodanese should be in solution. The precipitate is again suspended in 15 to 20 ml of cold 1.8 M ammonium sulfate, and the pH is adjusted to 7.9 with cold 1 M glycine-

SUMMARY OF RHODANESE PURIFICATION[a]

Preparation	Volume (ml)	Activity (units $\times 10^{-3}$)	Protein (g)	Specific activity (units/mg)
1. Liver extract	5000	650	150	4.3
2. Precipitate from salt fractionation	80	450	5.2	86.5
3. Precipitate from pH-salt fractionation I	25	400	1.2	333
4. Crystals from pH-salt fractionation II	3	330	0.57	580
5. a. Washed crystals	3	255	0.385	660
b. Recrystallized enzyme	2	230	0.320	720

[a] The values given summarize the accumulated experience of many preparations over several years.

hydroxide/1.8 M ammonium sulfate. The rhodanese goes into solution at this pH and ionic strength; much of the contaminant protein does not. After centrifugation, the supernatant fluid receives an equal volume of cold 3.25 M ammonium sulfate, dropwise and with good stirring.

Step 5. Crystallization and Recrystallization. After standing overnight and being recovered by centrifugation, the amorphous precipitate of rhodanese is crystallized by suspending it in 2 ml of 1.8 M ammonium sulfate. The enzyme normally appears obviously crystalline at this stage; its specific activity is 80 to 85% that of the best preparations. Simply washing the crystals several times by repeated suspension in 1.8 M ammonium sulfate, followed by centrifugation, is usually sufficient to increase the specific activity to the 95 to 100% level. Recrystallization is done by dissolving the crystal mass in 4 to 5 ml of 10 mM Tris-sulfate at pH 7.6 and then adding an equal volume of 3.6 M ammonium sulfate very slowly with good stirring (see table).

Alternative Procedures

Although much of the modern research has been with rhodanese prepared by the Horowitz and DeToma procedure, there are two important variants to this procedure. Heinrikson's research that resulted in the elucidation of the complete primary structure[8,9] was with enzyme prepared by such a variant. Following Step 3 of the procedure given here, the precipitated protein was dissolved in 5 mM Tris-sulfate, pH 7.5, contain-

[8] L. Weng. J. Russell, and R. L. Heinrikson, *J. Biol. Chem.* **253**, 8093 (1978).
[9] J. Russell, L. Weng, P. S. Keim, and R. L. Heinrikson, *J. Biol. Chem.* **253**, 8102 (1978).

ing 1 mM sodium thiosulfate, dialyzed to remove residual ammonium sulfate, and chromatographed on DEAE-Sephadex. Highly purified rhodanese was eluted in two bands in which the enzyme appeared to have otherwise identical properties, including molecular weight and specific activity.[10] Crystallization was accomplished after concentrating pooled fractions from the column by ultrafiltration. Most of the sequence work was conducted with that form which elutes later from the ion exchanger.

The second important variant was reported by Horowitz.[11] In this procedure, following the salt fractionation step No. 2, the preparation is dialyzed and fractionated on a column of agarose-immobilized Cibacron Blue. After elution with sodium thiosulfate, the enzyme is recovered by precipitation with ammonium sulfate, subjected to one pH-salt fractionation procedure, and crystallized. The yield is typically less than half that obtained by the first of the procedures given above, but the ease and quickness of the blue agarose method are such that much future work is likely to be done with it.

The special significance of the preparative procedures used for rhodanese has been highlighted by Volini's demonstration that different molecular forms of the enzyme may be isolated even in different preparations by the same procedure.[12] Some of the controversy surrounding various features of the molecular structure may have its roots in these differences.

Properties[1-4]

Rhodanese has a broad biological distribution. The mammalian liver enzyme occurs exclusively in the mitochondrial matrix. Complete references to the papers reporting these findings, to the source papers for the constants and other properties of rhodanese reported before 1977, and to studies of the detailed chemical mechanism of the enzyme have been cited.[1-4]

Physical Constants. Sedimentation velocity studies have given a sedimentation coefficient of 3.0 ±0.1 for untreated rhodanese in various systems. This value, together with a diffusion coefficient of 7.5×10^{-7} and a partial specific volume of 0.74, yielded a molecular weight of 37,000. Sedimentation equilibrium studies[12] have shown that crystalline rhodanese preparations of full specific activity may contain both a nondissociable species of 33,000 daltons and a 37,000-dalton species that undergoes slow dissociation to species of near 19,000 daltons. The absorbance

[10] K. M. Blumenthal and R. L. Heinrikson, *J. Biol. Chem.* **246**, 2430 (1971).
[11] P. M. Horowitz, *Anal. Biochem.* **86**, 751 (1978).
[12] M. Volini, D. Craven, and K. Ogata, *J. Biol. Chem.* **253**, 7591 (1978).

at 280 nm of a 1 mg/ml solution of the enzyme in a cuvette with a 1-cm light path is 1.75, a value repeatedly confirmed.

Kinetic Constants. Rhodanese catalyzes the cyanolysis of $S_2O_3^{2-}$ by a double displacement mechanism involving a covalent sulfur–enzyme intermediate. In phosphate buffer, pH 8.5, ionic strength 0.20, at 25° the rate constant limiting the maximum velocity has a value of 300 sec^{-1}, corresponding to a molar activity of 1.8×10^4 min^{-1}. The K_m value (shown to be equal to the K_s value) for thiosulfate in this system is 4 mM. The value of the second order rate constant for the reaction of cyanide anion with the sulfur-enzyme is 2.5×10^7 M^{-1} sec^{-1}. This value approaches 10^9 M^{-1} sec^{-1} as the reaction appears to near diffusion control at pH values below 6. The equilibrium constant for the overall enzyme-catalyzed reaction is greater than 10^{10}.

Specificity. Sulfur-donor substrates for rhodanese are anions containing sulfane-level sulfur. These include inorganic thiosulfate, organic thiosulfonates of the form $RS(O_2)S^-$, and persulfides (RSS^-). Polysulfides (RSS_xSR) may be active only after conversion to persulfides.[13] Sulfur acceptor substrates include a broad variety of thiophilic anions, of which CN^- is the most active, followed by SO_3^{2-} and thiosulfinates, dihydrolipoate, monothiols, and even BH_4^-.

Inhibition and Inactivation. Rhodanese is inhibited reversibly and competitively with respect to thiosulfate by most anions at rather high concentrations. The most active inhibitors are aromatic anions; the K_i value for 2-naphthalene sulfonate is 6 mM.

Sulfhydryl reagents inactivate rhodanese by either alkylation or oxidation of the active site sulfhydryl group. Unusual reagents that also oxidize this group are the mild aromatic oxidants dinitrobenzene and phenylglyoxal.

Molecular Dynamics. There is a body of evidence which shows that lyzed structurally is a single polypeptide chain of 293 residues (33,000 daltons) distributed into two domains.[9,14] According to the X-ray crystallographic structure at the 2.5 Å level of resolution, the domains are nearly identical in tertiary structure although their primary structures are dissimilar.[15] The overall structure of the molecule in the crystals is an approximate ellipsoid of dimensions $40 \times 50 \times 60$ Å.

Molecular Dynamics. There is a body of evidence which shows that rhodanese undergoes substantial conformational change during catalysis. The most direct evidence comes from observations of optical rotatory

[13] R. Abdolrasulina and J. L. Wood, *Bioorg. Chem.* **9**, 253 (1980).

[14] J. H. Ploegman, G. Drent, K. H. Kalk, W. G. J. Hol, R. L. Heinrikson, P. Keim, L. Weng, and J. Russell, *Nature (London)* **273**, 124 (1978).

[15] J. H. Ploegman, G. Drent, K. H. Kalk, and W. G. J. Hol, *J. Mol. Biol.* **123**, 577 (1978).

dispersion and circular dichroism spectra in experiments closely corre-
lated with kinetic studies.[16,17] To date comparable changes have not been
detectable in crystal studies.[15]

[16] S.-F. Wang and M. Volini, *J. Biol. Chem.* **248**, 7376 (1973).
[17] M. Volini and S.-F. Wang, *J. Biol. Chem.* **248**, 7386 (1973).

[38] 3-Mercaptopyruvate Sulfurtransferase

By REBECCA JARABAK

$$HS-CH_2-\overset{\overset{\displaystyle O}{\|}}{C}-COO^- + RSH \rightleftharpoons CH_3-\overset{\overset{\displaystyle O}{\|}}{C}-COO^- + RSSH$$

$$\text{or} \qquad\qquad\qquad\qquad \text{or}$$
$$CN^- \qquad\qquad\qquad\qquad SCN^-$$

3-Mercaptopyruvate sulfurtransferase (EC 2.8.1.2) catalyzes the
cleavage of a carbon–sulfur bond and transfer of the sulfur atom from
3-mercaptopyruvate to any of a variety of thiophiles, including thiols,
cyanide, sulfite, and sulfinates. The role of this enzyme in detoxication
has been discussed.[1]

Volume V of this series contains articles by Hylin[2] on the purification
and properties of "mercaptopyruvate : cyanide transsulfurase" from rat
liver acetone powders and by Sörbo,[3] who assayed the mercaptopyruvate : sulfite transsulfurase activity of enzyme prepared (also from rat liver
acetone powders) by the method of Kun and Fanshier.[4] Because these
preparations were not homogeneous, Sörbo pointed out the possibility
that more than one enzyme is responsible for the various transsulfuration
reactions of mercaptopyruvate. Subsequently, Kun and Fanshier have
modified their procedure for the rat liver enzyme,[5] Van Den Hamer has
partially purified the rat liver and rat erythrocyte enzymes,[6] and Vachek
and Wood[7] have purified the enzyme from *Escherichia coli* to homogeneity.

[1] J. Westley, *in* "Enzymatic Basis of Detoxication" (W. B. Jakoby, ed.), Vol. 2, p. 245.
Academic Press, New York, 1980.
[2] J. W. Hylin, this series, Vol. V, Article [132].
[3] B. H. Sörbo, this series, Vol. V, Article [133a].
[4] E. Kun and D. W. Fanshier, *Biochim. Biophys. Acta* **32**, 338 (1954).
[5] E. Kun and D. W. Fanshier, *Biochim. Biophys. Acta* **58**, 266 (1962).
[6] C. J. A. Van Den Hamer, A. G. Morell, and I. H. Scheinberg, *J. Biol. Chem.* **242**, 2514
(1967).
[7] H. Vachek and J. L. Wood, *Biochim. Biophys. Acta* **258**, 133 (1972).

This article deals with the purification of the 3-mercaptopyruvate sulfurtransferase from bovine kidney.[8] Homogeneous samples of this enzyme use both 2-mercaptoethanol and cyanide as sulfur acceptors; work with other acceptors has not been done.

Assay Methods

The two assays described here are modifications of those used by Vachek and Wood in their work with the enzyme from *E. coli*.[7]

Method A

Assay with 3-mercaptopyruvate as the sulfur donor and 2-mercaptoethanol as the sulfur acceptor substrate.

Principle. The production of pyruvate is measured after reaction with 2,4-dinitrophenylhydrazine.

Reagents

2-Methyl-2-aminopropanediol buffer, 1.125 M, pH 9.55
Sodium 3-mercaptopyruvate,[9] 0.15 M, prepared fresh daily
2-Mercaptoethanol, 0.71 M, prepared fresh daily
Cadmium chloride, 0.5 M
2,4-Dinitrophenylhydrazine reagent: 0.1% solution in 2 M HCl
Sodium hydroxide, 1.5 M

Procedure. Into a 13 × 100 mm test tube are placed 0.68 ml of water, 0.20 ml of buffer, 0.02 ml of 2-mercaptoethanol, and 0.10 ml of 3-mercaptopyruvate. The reaction is initiated by adding 2 to 20 μl of enzyme solution to this mixture at 30°, and is terminated after 5 min by the addition of 0.5 ml of cadmium chloride solution. After 10 min, the suspension is centrifuged for 1 min in a clinical centrifuge, and 0.5 ml of the supernatant liquid is mixed at room temperature with 0.5 ml of the 2,4-dinitrophenylhydrazine reagent. After 5 min, 2.5 ml of sodium hydroxide solution is added, and the precipitate is removed by centrifugation. The clear supernatant fluid is decanted into a calibrated colorimeter tube or cuvette. The absorbance at 435 nm is measured 35 min after addition of the 2,4-dinitrophenylhydrazine reagent. This assay is linear to at least 0.5 mM pyruvate, which corresponds to approximately 80 ng of enzyme. The assay is linear with time for at least 5 min.

---

[8] R. Jarabak and J. Westley, *Arch. Biochem. Biophys.* **185**, 458 (1978).

[9] Sodium 3-mercaptopyruvate may be purchased from ICN Pharmaceuticals, Inc. When examined with 5,5'-dithiobis(2-nitrobenzoic acid), its absorbance at 412 nm corresponded to the expected sulfhydryl content. It is used without further purification.

Definition of a Unit of Enzyme Activity. One unit of activity is defined as the amount of enzyme that catalyzes the production of 1 μmol of pyruvate per minute in this assay system.

Method B

Assay with 3-mercaptopyruvate as the sulfur donor and cyanide as the sulfur acceptor.

Principle. The assay is based on a colorimetric method for the determination of thiocyanate, one of the reaction products. Although less sensitive, the method is simpler to perform and is the one used in this laboratory for routine assay during purification.

Reagents

2-Methyl-2-aminopropanediol buffer, 1.125 M, pH 9.55
Sodium 3-mercaptopyruvate, 0.15 M, prepared fresh daily
Potassium cyanide, 0.25 M
Formaldehyde solution: 40 ml of the commercial 37% solution diluted to 100 ml with water
Ferric nitrate reagent: 50 g of $Fe(NO_3)_3 \cdot 9H_2O$ and 100 ml of concentrated nitric acid per 750 ml

Procedure. In a 13 × 100 mm test tube are placed 0.65 ml of water, 0.20 ml of buffer, 0.05 ml of potassium cyanide, and 0.10 ml of sodium 3-mercaptopyruvate. Enzyme solution, 2 to 20 μl, is added to this mixture at 30° to initiate the reaction, which is subsequently terminated with 0.5 ml of formaldehyde. Finally, 1.5 ml of ferric nitrate reagent is added, and the absorbance at 460 nm is measured promptly; color fades measurably within a few minutes.

Purification Procedure

Small amounts of enzyme can be prepared conveniently from 300 g of bovine kidney. All buffers used in the purification contain 20% glycerol (v/v) and 0.02% 2-mercaptoethanol (v/v). All purification steps are carried out at 4°.

Step 1. Homogenization and Ammonium Sulfate Fractionation. Frozen, cubed bovine kidney in 100-g portions is homogenized for 1.5 min at top speed in a Waring blendor with 200 ml of potassium phosphate buffer (0.1 M, pH 7.4) for each portion. Homogenate from 300 g of tissue is centrifuged for 90 min at 48,000 g and solid ammonium sulfate (200 g per liter) is added to the supernatant liquid slowly with efficient stirring.

After standing overnight at 4°, the suspension is centrifuged for 90 min at 48,000 g and more ammonium sulfate (177 g per liter) is added to the supernatant fluid. After standing overnight at 4°, the preparation is centrifuged as before for 45 min and the precipitate, which contains most of the enzyme activity, is resuspended in a small volume of the supernatant fluid. The pooled suspension can be frozen for storage at −40°.

Step 2. DEAE-Sephadex A-50 Chromatography. The protein precipitate from Step 1 is redissolved in 5 mM potassium phosphate at pH 7.4 and the solution is dialyzed against 3 changes of 2-liter volumes of the same 5 mM buffer. The dialyzed solution is centrifuged for 2 hr at 48,000 g, and then 100 ml of 20% glycerol, 0.02% in 2-mercaptoethanol, is added to the supernatant, dropwise with stirring. The resulting solution is applied to a column (5 × 25 cm) of DEAE-Sephadex A-50 equilibrated with 5 mM potassium phosphate at pH 7.4. The column is washed overnight with the same 5 mM buffer. The enzyme is eluted with 50 mM potassium phosphate, pH 7.4. Fractions containing enzyme activity are combined and concentrated to 15–20 ml in an Amicon ultrafiltration cell using a UM-10 membrane. The concentrated solution is frozen for storage at −40°.

Step 3. Sephadex G-100. The frozen solution from Step 2 is thawed and applied to a column (4 × 90 cm) of Sephadex G-100 equilibrated with 0.1 M potassium phosphate, pH 7.4. The column is developed with the same buffer, and fractions containing 3-mercaptopyruvate sulfurtransferase activity are frozen at −40°.

Step 4. CM-Sephadex C-50 Chromatography. Fractions from Sephadex G-100 chromatography are thawed, pooled, and dialyzed against 5 mM potassium phosphate at pH 6.0. The dialyzed solution is applied to a column (2.5 × 12 cm) of CM-Sephadex C-50 equilibrated with the same pH 6.0 buffer. After the column has been washed with starting buffer, the enzyme is eluted with 5 mM potassium phosphate at pH 7.4. Fractions containing the enzyme are stored at 4° overnight.

Step 5. Calcium Phosphate Chromatography. Peak fractions from the previous step are pooled and concentrated in a 10-ml Amicon ultrafiltration cell with a UM-10 membrane. The pH of the solution is changed by adding 5 ml of 5 mM potassium phosphate (pH 6.0), concentrating again and repeating this process 3 to 4 times. Finally, the sample of about 2.5 ml is applied to a column (1.2 × 2.7 cm) of calcium phosphate prepared by the method of Anacker and Stoy[10] and equilibrated with 5 mM potassium phosphate at pH 6.0. The column is washed with starting buffer, and the enzyme is eluted with 15 mM potassium phosphate at pH 6.0.

At this point, the enzyme is usually, but not always, homogeneous on disc polyacrylamide gel electrophoresis at pH 8.3.[11] Table I gives a summary of the data from a purification that did yield homogeneous material.

---

[10] W. F. Anacker and V. Stoy, *Biochem. Z.* **330**, 141 (1958).

[11] Steps 1–4 of the purification are quite reproducible. Step 5, calcium phosphate chromatography, gives more variable results.

TABLE I

PREPARATION OF 3-MERCAPTOPYRUVATE SULFURTRANSFERASE FROM 300 GRAMS OF BOVINE KIDNEY

Step	Volume (ml)	Units (μmol pyruvate/min)	Protein (mg)	Specific activity (μmol pyruvate min^{-1} mg protein^{-1})
1. Ammonium sulfate	120	31,600	20,000	1.58[a]
2. DEAE-Sephadex A-50	222	14,350	849	16.9[b]
3. Sephadex G-100	157	7,940	137	57.9[b]
4. CM-Sephadex C-50	24	3,660	15.2	241[b]
5. Calcium phosphate	6	1,430	1.15	1240[b]

[a] Protein determined by a biuret method [A. G. Gornall, C. J. Bardawill, and M. M. David, *J. Biol. Chem.* **177**, 751 (1949)].

[b] Calculated from A_{280} measurement, assuming that A_{280} of a 1 mg/ml solution in a cuvette with a 1-cm light path is equal to 1.0.

TABLE II
KINETIC CONSTANTS FOR THE 3-MERCAPTOPYRUVATE
SULFURTRANSFERASE FROM BOVINE KIDNEY

Substrates	Constants[a]
3-Mercaptopyruvate, 2-mercaptoethanol[b]	$K_M^{3-MP} = 2.77 \times 10^{-3}\ M$
	$K_M^{2-ME} = 3.13 \times 10^{-2}\ M$
	Molar activity[c] $= 2.88 \times 10^3\ \text{sec}^{-1}$
3-Mercaptopyruvate, cyanide	$K_M^{3-MP} = 0$
	$K_M^{CN^-} = 1.65 \times 10^{-3}\ M$
	Molar activity $= 2.47 \times 10^2\ \text{sec}^{-1}$

[a] Reaction conditions: pH 9.55, 30°. 3-MP, 3-Mercaptopyruvate; 2-ME, 2-mercaptoethanol.
[b] The values given are based on a fit of the data to a simple sequential mechanism; this is a reasonable approximation in the concentration range examined.
[c] The value given is based on a linear extrapolation that ignores the participation of 3-mercaptopyruvate as a sulfur acceptor at high concentrations.

Properties

The elution volume of the bovine kidney 3-mercaptopyruvate sulfurtransferase on Sephadex G-100 columns is close to that of bovine liver rhodanese; therefore, its molecular weight is estimated to be approximately 33,000.[12]

There has been interest in the copper content of 3-mercaptopyruvate sulfurtransferase prepared from various sources. Vachek and Wood found 0.5 mole of copper (and 1.1 mole of zinc) per mole of protein for the *E. coli* enzyme. Kun and Fanshier have found copper in the rat liver enzyme,[4,5] but Van Den Hamer *et al.* have reported that treatment of rat liver and rat erythrocyte enzymes with sodium diethyldithiocarbamate for removal of copper allowed retention of full activity.[6] Atomic absorption methods have shown that homogeneous samples of the bovine kidney 3-mercaptopyruvate sulfurtransferase, isolated by the procedure described here, contain less than 0.1 copper per 33,000 daltons.[13] No other structural or physical studies have been done with the bovine kidney enzyme.

Steady-state kinetic studies have shown that the 3-mercaptopyruvate sulfurtransferase from bovine kidney functions by sequential formal mechanisms. When 2-mercaptoethanol is the sulfur acceptor, addition of

[12] R. Jarabak and J. Westley, unpublished data.
[13] M. Volini and R. Jarabak, unpublished data.

the substrates is random. 3-Mercaptopyruvate can serve as a second sulfur acceptor as well as the sulfur donor in this system, and nonlinear double reciprocal plots result from the branched mechanism. Pyruvate is a product inhibitor competitive with respect to both substrates.[8]

When cyanide is the sulfur acceptor substrate, there is an equilibrium mixture of free 3-mercaptopyruvate, free cyanide, and 3-mercaptopyruvate cyanohydrin in the kinetic assay solutions. The K_D for the cyanohydrin equilibrium, determined polarographically, is 5.3 × 10^{-4} M. Analysis of steady-state kinetic data, using the concentrations of free cyanide and free 3-mercaptopyruvate as the actual substrate concentrations, shows that the formal mechanism is rapid equilibrium-ordered, with 3-mercaptopyruvate as the first substrate to bind to the enzyme.[14] Kinetic constants are given in Table II.

[14] R. Jarabak and J. Westley, *Biochemistry* **19**, 900 (1980).

[39] Glyoxalase I (Rat Liver)

By Bengt Mannervik, Anne-Charlotte Aronsson, Ewa Marmstål, and Gudrun Tibbelin

Glyoxalase I (EC 4.4.1.5 lactoyl-glutathione lyase) catalyzes the formation of S-2-hydroxyacylglutathione from a corresponding 2-oxoaldehyde and glutathione (GSH). A second enzyme, glyoxalase II (EC 3.1.2.6 hydroxyacylglutathione hydrolase) regenerates GSH by hydrolyzing the resulting thiolester.[1] Thus, the glyoxalase system may detoxify the electrophilic 2-oxoaldehydes, which can be formed in the cell from endogenous or exogenous precursors.[2]

Assay Method

Principle. The formation of S-lactoylglutathione from methylglyoxal and GSH is followed spectrophotometrically at 240 nm.[1]

[1] E. Racker, *J. Biol. Chem.* **190**, 685 (1951).
[2] B. Mannervik, *in* "Enzymatic Basis of Detoxication" (W. B. Jakoby, ed.), Vol. 2, p. 263. Academic Press, New York, 1980.

METHODS IN ENZYMOLOGY, VOL. 77

TABLE I
PURIFICATION OF GLYOXALASE I FROM RAT LIVER CYTOSOL

Step	Volume (ml)	Total activity (units)	Total protein (mg)	Specific activity (units/mg)
1. Cytosol fraction	830	3569	32,000	0.11
2. Sephadex G-25	980	3528	11,700	0.30
3. CM-cellulose	1020	3060	7,340	0.42
4. DEAE-cellulose	212	2926	1,460	2.00
5. Affinity chromatography	35	1910	ND[a]	ND
6. Sephadex G-100	91	1840	13.2	139
7. Affinity chromatography	30	1420	11.6[b]	122[b]
8. DEAE-Sepharose	66	858	0.5	1720

[a] ND, Not determined.

[b] Values uncertain as a result of the presence of the glutathione derivative in the pooled effluent.

Reagents

Sodium phosphate buffer, 0.1 M, pH 7.0
Methylglyoxal, 40 mM
Glutathione, 50 mM

Procedure. In a 1-ml quartz cuvette (1-cm light path) are placed 500 μl buffer, 50 μl methylglyoxal, 20 μl glutathione, and deionized water to a final volume of 1 ml. The increase in absorbance at 240 nm is followed spectrophotometrically at 30°. The reaction is initiated by addition of enzyme. The value of 3.37 mM^{-1} cm^{-1} is used for the absorption coefficient of the product, S-lactoylglutathione.

Definition of Unit and Specific Activity. A unit of glyoxalase I activity is defined as the amount of enzyme that catalyzes the formation of 1 μmol of S-lactoylglutathione per minute at 30°. Specific activity is expressed in units per milligram of protein. Protein concentrations given in Table I are calculated from the absorbance at 260 and 280 nm using the equation of Kalckar[3] or determined by a microbiuret method.[4]

Purification Procedure

All operations are performed at about 5° unless otherwise noted.

Buffer Solutions

Buffer A: Sodium phosphate, 10 mM, pH 6.1
Buffer B: Tris-HCl, 10 mM, pH 7.8

[3] H. M. Kalckar, *J. Biol. Chem.* **167**, 461 (1947).
[4] J. Goa, *Scand. J. Clin. Lab. Invest.* **5**, 218 (1953).

Step 1. Preparation of Cytosol Fraction. Livers are obtained from 25 male Sprague-Dawley rats killed by decapitation. The livers are cut into small pieces and homogenized (20%, w/v) in ice-cold 0.25 M sucrose with a Turmix blender. The homogenate is centrifuged at 11,000 g for 45 min. The sediment is washed with 150 ml of cold 0.25 M sucrose and centrifuged. The supernatant fractions are combined with the previous centrifugate (total volume 880 ml) and adjusted to pH 5.5 with ice-cold 5% acetic acid. The mixture is centrifuged for 60 min at 11,000 g.

Step 2. Sephadex G-25. The supernatant fraction from Step 1 (830 ml) is chromatographed on a Sephadex G-25 column (9 × 60 cm) equilibrated and eluted with Buffer A. Flow rate: 2.5–3 liters per hour.

Step 3. CM-Cellulose. The pooled effluent from Step 2 (980 ml) is applied to a CM-cellulose column (9 × 7 cm) previously equilibrated with Buffer A. The column is washed with the same buffer at a flow rate of about 700–750 ml/hr. Glyoxalase I is not absorbed on the column.

Step 4. DEAE-Cellulose. Pooled effluent containing glyoxalase I activity (1020 ml) is adjusted to pH 7.8 with ice-cold 2 M NH$_3$ and applied to a column (4 × 20 cm) of DEAE-cellulose equilibrated with Buffer B. The column is washed with the same buffer and eluted with a linear concentration gradient formed by linear mixing of 500 ml of starting buffer with 500 ml of 0.3 M NaCl in Buffer B. Flow rate: 400 ml/hr during sample application and washing and 40 ml/hr during elution of the enzyme.

Step 5. Affinity Chromatography with S-p-Bromobenzylglutathione-Sepharose 6B. The pooled fractions from Step 4 are applied to a column (2 × 15 cm) of *S-p*-bromobenzylglutathione-Sepharose 6B previously equilibrated with Buffer B containing 0.2 M NaCl. The column is washed with 100 ml of the buffer used for equilibration and eluted with a gradient formed by linear mixing of 100 ml of starting buffer with 100 ml of 10 mM glutathione and 6 mM *S-p*-bromobenzylglutathione in Buffer B that is 0.2 M in NaCl. Flow rate during the whole procedure is 40 ml/hr. The active fractions are pooled and concentrated to 9 ml with Millipore immersible-CX ultrafilters.

Step 6. Sephadex G-100. The concentrated enzyme from Step 5 is chromatographed on a Sephadex G-100 column (4 × 42 cm) equilibrated and eluted with Buffer B at a flow rate of 150 ml/hr.

Step 7. Affinity Chromatography with S-p-Bromobenzylglutathione-Sepharose 6B. To pooled effluent from Step 6 is added solid NaCl to a final concentration of 0.2 M and the solution is charged onto a column (1 × 12 cm) of *S-p*-bromobenzylglutathione-Sepharose 6B. The column is equilibrated and washed with Buffer B containing 0.2 M NaCl. Elution of glyoxalase I is carried out with the gradient containing glutathione and *S-p*-bromobenzylglutathione used in Step 5. Total volume 200 ml; flow rate of 30 ml/hr.

TABLE II
MOLECULAR AND KINETIC PROPERTIES OF GLYOXALASE I FROM
RAT LIVER[a]

Property	Value
Molecular weight	46,000
Number of subunits	2
Stokes radius	2.8 nm
$s_{20,w}$	4.0 S
Isoelectric point at 4°	4.7
Apparent steady-state kinetic parameters[b]	
K_m for methylglyoxal	0.14 mM
k_{cat} for methylglyoxal	71,000 min^{-1}
K_m for phenylglyoxal	0.06 mM
k_{cat} for phenylglyoxal	66,000 min^{-1}

[a] Data from Marmstål et al.[12]
[b] At 2 mM free GSH.

Step 8. Chromatography on DEAE-Sepharose. The enzyme preparation from Step 7 is diluted to 100 ml with Buffer B and applied to a column
(1 × 7 cm) of DEAE-Sepharose equilibrated with Buffer B. The enzyme is
eluted with a linear concentration gradient of 0 to 0.15 M NaCl in Buffer
B with a total volume of 600 ml at a flow rate of 30 ml/hr.

Preparation of Affinity Gel. The ligand, *S-p*-bromobenzylglutathione
may be synthesized according to Method A of Vince *et al.*[5] The procedure described in this volume[6] may be followed by substituting
p-bromobenzylbromide dissolved in 95% ethanol for iodohexane. The
ligand is immobilized on epoxy-activated Sepharose 6B according to the
instructions of the manufacturer (Pharmacia Fine Chemicals). Details are
given by Mannervik and Guthenberg.[6] The epoxy-activated matrix[7] was
found superior to the CNBr-activated gel that had been used previously.[8]

Properties

Purity. The purified glyoxalase I is homogeneous in several analytical
electrophoretic and chromatographic systems. In the presence of SDS, it
dissociates into subunits of equal weight. In the original purification procedure,[9] some preparations of enzyme were found by immunoelec-

[5] R. Vince, S. Daluge, and W. B. Wadd, *J. Med. Chem.* **14,** 402 (1971).
[6] B. Mannervik and C. Guthenberg, this volume, Article [28].
[7] A.-C. Aronsson, E. Marmstål, and B. Mannervik, *Biochem. Biophys. Res. Commun.* **81,** 1235 (1978).
[8] A.-C. Aronsson and B. Mannervik, *Biochem. J.* **165,** 503 (1977).
[9] E. Marmstål and B. Mannervik, *Biochim. Biophys. Acta* **566,** 362 (1979).

trophoresis to be slightly contaminated with albumin.[10] Enzyme prepared by the revised procedure presented here is free of albumin when analyzed by quantitative ("rocket") immunoelectrophoresis as described by Weeke.[11] The sensitivity was such that albumin, at a level of 0.5% of the amount of protein in the glyoxalase preparation, would clearly have been detected.

Molecular and Kinetic Properties. Some of the known properties of glyoxalase I from rat liver are presented in Table II.[12] Antibodies raised against the enzyme do not give precipitin lines with purified glyoxalase I from porcine erythrocytes, human erythrocytes, or *Saccharomyces cerevisiae* in Ouchterlony double-diffusion experiments. Neither do antibodies against the other three enzymes give positive reactions with the purified enzyme from rat liver.[10] The rat liver enzyme, like glyoxalase I from other sources, has a broad specificity for 2-oxoaldehydes.[2,13]

The enzyme is composed of two subunits and has a molecular weight of 46,000, which includes two zinc atoms. The amino acid composition has been determined.[12] Maximal enzymatic activity is obtained between pH 6.5 and 7.5.

[10] K. Larsen, E. Marmstål, A.-C. Aronsson, and B. Mannervik, unpublished experiments.
[11] B. Weeke, *Scand. J. Immunol.* 2, Suppl. 1, 37 (1973).
[12] E. Marmstål, A.-C. Aronsson, and B. Mannervik, *Biochem. J.* **183**, 23 (1979).
[13] L.-P.B. Han, L. M. Davison, and D. L. Vander Jagt, *Biochim. Biophys. Acta* **445**, 486 (1976).

[40] Benzoyl-CoA : Amino Acid and Phenylacetyl-CoA : Amino Acid *N*-Acyltransferases

By LESLIE T. WEBSTER, JR.

Benzoyl-CoA + amino acid → benzoyl-N-amino acid + CoASH
Phenylacetyl-CoA + amino acid → phenylacetyl-N-amino acid + CoASH

Introduction

Two closely related acyl-CoA : amino acid *N*-acyltransferases have been identified in liver mitochondria from cattle, rhesus monkeys, and humans.[1,2] One of these two enzymes uses benzoyl-CoA as the prototype acyl donor and is active also with certain medium chain fatty acyl-CoA

[1] D. L. Nandi, S. V. Lucas, and L. T. Webster, Jr., *J. Biol. Chem.* **254**, 7230 (1979).
[2] L. T. Webster, Jr., U. A. Siddiqui, S. V. Lucas, J. M. Strong, and J. J. Mieyal, *J. Biol. Chem.* **251**, 3352 (1976).

esters. The other is an arylacetyltransferase that employs phenylacetyl-CoA as the preferred thioester substrate and is active also with indoleacetyl-CoA. Whereas the benzoyltransferases from all of the aforementioned species and the phenylacetyltransferase from cattle prefer glycine as the acyl acceptor, the phenylacetyltransferase from rhesus monkeys and humans is specific for L-glutamine.[2] Assays, purification procedures, and properties of the essentially homogeneous bovine transferases are described in this chapter. Information concerning other transferase preparations of this type is presented elsewhere.[3]

Assay Methods

Principle. The most convenient transferase assays are those based on either the amino acid-dependent rate of appearance of CoA thiol trapped by reaction with 5,5'-dithiobis(2-nitrobenzoic acid) (abbreviated DTNB)[4] or the amino acid-dependent rate of disappearance of the thioester absorbance of the acyl-CoA substrate.[1,5] Both assays are continuous, rapid, and quite sensitive but may be unsuitable for crude enzyme preparations where high rates of non-amino acid-dependent esterolysis occur; the first method may also underestimate the activity of transferases inactivated by thiol binding reagents like DTNB. Both methods can be validated by discontinuous direct assays of peptide product formation, which are preferred for crude preparations with high nonspecific esterolytic activity.[1,6]

Reagents

Benzoyl-CoA or phenylacetyl-CoA, 1 mM: Concentrations of acyl-CoA substrates are determined with DTNB after enzymatic release of CoA[1,4]

Glycine, 2 M: Adjust the stock solution with KOH so that a 1:10 dilution gives a pH reading of 8 at 30°

5,5'-Dithiobis(2-nitrobenzoic acid) (DTNB), 1 mM: Use for the CoA thiol appearance assays only

Tris-HCl buffer, 250 mM: Adjust the stock solution of Tris base with HCl so that a 1:10 dilution gives a pH reading of 8 at 30°

Enzyme: Enzyme is diluted with the appropriate enzyme dilution buffer depending on the transferase activity to be assayed; enzyme protein is determined by optical absorption at 280 nm (1 $A \equiv 1$ mg/ml)

[3] P. G. Killenberg and L. T. Webster, Jr., *in* "Enzymatic Basis of Detoxification" (W. B. Jakoby, ed.), Vol. 2, p. 141. Academic Press, New York, 1980.
[4] G. L. Ellman, *Arch. Biochem. Biophys.* **82**, 70 (1959).
[5] D. Schachter and J. V. Taggart, *J. Biol. Chem.* **203**, 925 (1953).
[6] M. O. James and J. R. Bend, *Biochem. J.* **172**, 285 (1978).

Enzyme dilution buffers: For benzoyltransferase, the dilution buffer contains 25 mM Tris-HCl, pH 8 at 30°, and 0.2 mg bovine serum albumin per ml; for phenylacetyltransferase, the dilution buffer also contains 0.1 M KCl

Procedure. Two procedures are described. At 30°, the initial rate of glycine-dependent appearance of CoA thiol from either benzoyl-CoA or phenylacetyl-CoA is monitored in the presence of 0.1 mM DTNB in a multichannel recording spectrophotometer set at 412 nm as follows: To a 1.2-ml quartz cuvette with a 1-cm light path is added a solution containing 0.1 ml of Tris-HCl buffer/0.1 M KCl for the phenylacetyltransferase assay only, 0.1 ml of the appropriate acyl-CoA substrate and distilled water to bring the volume to 0.75 ml. The reagents are mixed and warmed to 30° for 2 min. Diluted enzyme (0.05 ml in this case) is introduced, the reaction mixture is remixed, and 0.1 ml of DTNB, warmed to 30°, is added with prompt mixing. The change in absorbance of the mixture is monitored continuously for 2 min with the full scale deflection of the recorder set at 0.1 A. Then, 0.1 ml of glycine at 30° is added, mixed, and the change in absorbance is recorded. The specific activity of the transferase (μmol min^{-1} mg^{-1}) is calculated as the difference in the initial linear rate of increase in absorbance (ΔA/min) in the presence and absence of glycine according to the following expression:

$$\frac{\Delta A}{\text{min}} \times \frac{1}{13.6 \ A/\text{ml}} \times \frac{1}{\text{mg protein/ml}} \equiv \Delta A \times 0.0735 \ \text{min}^{-1} \ \text{mg}^{-1}$$

where $E_{412 \ nm}$ DTNB = 13.6 mM^{-1} cm^{-1}.

For assays depending on the disappearance of absorption of the thioester bond, the procedure is nearly identical to that described earlier except that the DTNB reagent is omitted and the recorder is set at 280 nm for the benzoyltransferase assay or at 236 nm for the phenylacetyltransferase assay. Differences in the rate of decrease in absorbance of the thioester bond in the presence and absence of glycine are calculated as $\Delta A \times 0.143$ min^{-1} mg^{-1} protein based on $E_{280} = 7.0$ mM^{-1} cm^{-1} for benzoyl-CoA and $\Delta A \times 0.250$ min^{-1} mg^{-1} protein based on $E_{236} = 4.0$ mM^{-1} cm^{-1} for phenylacetyl-CoA.

For both assays, enzyme should be sufficiently diluted so that linear initial rates are obtained; in certain instances this may correspond to a ΔA of no more than 0.01 per minute.

Comment. Because the benzoyltransferase enzyme is quite stable, both the thiol release and the thioester bond cleavage assays yield similar results. In contrast, the bovine phenylacetyltransferase is sensitive to thiol binding reagents so that the thioester cleavage assay is preferred despite its lower sensitivity. Nevertheless, the thiol release assay gives better than 90% of the thioester cleavage activity when purified preparations of phenylacetyltransferase are used and the enzyme is added last.[1] Because purified preparations of bovine benzoyl- and phenylacetyltransferase cause little esterolysis of acyl-CoA substrates in the absence of glycine, glycine may be added to the initial incubation mixture and the reaction initiated by addition of enzyme.

Purification Procedure

Procedures for purifying the benzoyl- and phenylacetyltransferases from bovine liver mitochondria are quite similar except for the final separation on hydroxylapatite.

PURIFICATION OF BENZOYLTRANSFERASE

Step 1. Mitochondrial Supernatant Fraction. Fresh chilled bovine liver (360 g) is homogenized in 4 volumes of 0.13 M KCl kept at pH 8 and 4°. Mitochondria, isolated by sequential differential centrifugation (600 g for 10 min; 9000 g for 10 min), are suspended in 10% of the original volume of 0.13 M KCl at pH 8 and stored at $-70°$. At least 3 days later the suspension (250 ml $\cong$ 300 g liver) is thawed and centrifuged at 35,000 g for 2 hr at 4°; the supernatant fluid is pipetted away from the surface fat. Subsequent purification steps are carried out while maintaining the pH between 7.2 and 8.0 at 4°.

Step 2. Ammonium Sulfate. After the protein concentration is adjusted to 20 mg/ml, the mitochondrial supernatant is fractionated with solid ammonium sulfate. The precipitate obtained after addition of 288 mg of salt per milliliter is discarded after centrifugation at 10,000 g for 15 min; the precipitate obtained after addition of an additional 144 mg of salt is suspended in a small volume of 0.1 M KCl/20 mM Tris-HCl, pH 8, and clarified by centrifugation at 40,000 g for 10 min; this last manipulation avoids the necessity for dialysis.

Step 3. Aluminum Hydroxide Gel. The enzyme solution from Step 2 is adjusted to 10 mg protein/ml with 20 mM Tris-HCl at pH 8. Gel solution (Rehsorptar, Reheis Chemical Co.) diluted to 10 mg/ml with the same buffer is added with rapid stirring to achieve a protein/gel ratio of 2. After 5 min, the gel is removed by centrifugation (600 g for 2 min). Protein in the supernatant liquid is collected by centrifugation after precipitation with 540 mg of ammonium sulfate per milliliter of solution; the pellet is washed in aqueous neutralized[7] ammonium sulfate (313 mg/ml), centrifuged, and dissolved in a minimal volume of 50 mM KCl/20 mM Tris-HCl at pH 8.

---

[7] Ammonium sulfate solution adjusted to pH 7.0 with concentrated ammonium hydroxide.

TABLE I
SUMMARY OF PURIFICATION OF BENZOYLTRANSFERASE[a]

Step	Volume (ml)	Total activity (μmol/min)	Total protein[b] (mg)	Specific activity (μmol min^{-1} mg^{-1})
1. Mitochondrial supernatant fraction	124	2820	2820	1.0
2. (NH$_4$)$_2$SO$_4$ (40–60%)	18	2340	1020	2.3
3. Al(OH)$_3$ gel	3.6	1900	432	4.4
4. BioGel P-100	2.2	924	88	10.5
5. Sephadex G-100	2.7	504	28	18.0
6. Hydroxylapatite	17.5	168	7	24.0

[a] Data from Nandi et al.[1]
[b] Total weight of liver is 360 g.

Step 4. BioGel P-100. After adjustment of the protein concentration to 40 mg/ml, the gel fraction is placed on a column (2.5 × 41 cm) of BioGel P-100 equilibrated with 50 mM Tris-HCl at pH 8. The protein is eluted with the same buffer at a flow rate of 10 ml/hr. Fractions of about 1.2 ml each having specific activities of 8 or greater are combined. The protein is concentrated with ammonium sulfate as in the previous step.

Step 5. Sephadex G-100. The BioGel fraction is chromatographed on a column (2.5 × 90 cm) of Sephadex G-100 eluted with 50 mM KCl/20 mM Tris-HCl, pH 8, at a flow rate of 15 ml/hr. The 5 to 6 fractions (3 ml each) having the highest specific activities are pooled, concentrated twofold in a Diaflo cell equipped with a UM 2 membrane, and the protein is precipitated by addition of 540 mg ammonium sulfate per milliliter (75% saturation). The precipitate resulting from centrifugation is washed in neutralized[7] ammonium sulfate solution (390 mg/ml), centrifuged, and dissolved in 2 ml of 20 mM Tris-HCl at pH 7.2. The solution is dialyzed against 1 liter of this buffer on a rocking dialyzer for 1 hr.

Step 6. Hydroxylapatite. The Sephadex fraction is placed on a column (2.5 × 8 cm) of hydroxylapatite and washed with 60 ml of 20 mM Tris-HCl, pH 7.2. The enzyme is eluted by a linear gradient solution made with 175 ml of 20 mM Tris-HCl, pH 7.2, in the mixing vessel and 175 ml of the same buffer containing 0.15 M potassium phosphate at pH 8.0 in the reservoir. Fractions of 3 ml are collected at a flow rate of 15 ml/hr. The enzyme elutes after about 180 ml; fractions having the highest specific activities are stored at either 4° or −70° (see Properties; Table I).

Comment. Because affinity chromatography on blue dextran-Sepharose B has also been used to achieve a one-step 15-fold purification of the bovine benzoyltransferase,[8] it might conceivably be combined with one or more of the preceding steps to achieve a more efficient purification procedure.

PURIFICATION OF PHENYLACETYLTRANSFERASE

Step 1. Mitochondrial Supernatant Fraction. Follow Step 1 for benzoyltransferase; 650 g of fresh liver are used here.

Step 2. Ammonium Sulfate. Solid ammonium sulfate (288 mg/ml; 40% saturation) is added to the mitochondrial supernatant fraction after diluting it to 20 mg protein/ml. After centrifugation, the pellet is dissolved in 34 ml of 20 mM Tris-HCl, pH 8, containing 0.1 M KCl and centrifuged at 40,000 g for 30 min to remove an inactive precipitate.

Step 3. Aluminum Hydroxide Gel. Follow Step 3 for benzoyltransferase.

Step 4. BioGel P-100. The concentrated gel supernatant is layered onto a column (2.5 × 60 cm) of BioGel P-100 equilibrated and eluted with 0.1 M KCl/20 mM Tris-HCl buffer, pH 8. Fractions of about 1.4 ml are collected at a flow rate of approximately 0.14 ml per minute. Fractions containing the most phenylacetyltransferase activity are combined and concentrated about 7-fold by diafiltration over a UM 2 membrane. The enzyme is precipitated by addition of solid ammonium sulfate to 65% saturation (468 mg/ml) and, after centrifugation, the pellet is dissolved in 0.9 ml of 0.1 M KCl/20 mM Tris-HCl at pH 8.

Step 5. BioGel P-60. Enzyme solution from the previous step is layered onto a column (1.5 × 90 cm) of BioGel P-60 equilibrated and eluted with 0.1 M KCl/20 mM Tris-HCl at pH 8. Fractions of about 1.1 ml are collected at a flow rate of approximately 0.13 ml per minute. The fraction containing the most phenylacetyltransferase activity is combined with subsequent active fractions and concentrated about 5-fold by diafiltration over a UM 2 membrane. Protein is precipitated by addition of ammonium sulfate, 390 mg/ml, and collected by centrifugation before it is dissolved and dialyzed against 1 liter of 20 mM Tris-HCl at pH 7.2 on a rocking dialyzer for 1.5 hr.

Step 6. Hydroxylapatite. The enzyme solution from Step 5 is layered onto a column (2 × 6 cm) of hydroxylapatite (BioRad) and washed with 40 ml of 20 mM Tris-HCl, pH 7.2, the equilibrating buffer. A linear gradient is established between 80 ml of equilibrating buffer in the mixing chamber and 80 ml of the same buffer containing 150 mM potassium phosphate at pH 8.0 in the reservoir. Fractions of about 1.4 ml are collected at a flow rate of 15 ml/hr. Virtually complete separation of phenylacetyltransferase from benzoyltransferase activity occurs with the phenylacetyltransferase

[8] E. P. Lau, B. E. Haley, and R. E. Barden, *Biochemistry* **16**, 2581 (1977).

TABLE II
SUMMARY OF PURIFICATION OF PHENYLACETYLTRANSFERASE[a]

Step	Volume (ml)	Total activity (μmol/min)	Total protein[b] (mg)	Specific activity (μmol min^{-1} mg^{-1})
1. Mitochondrial supernatant fraction	224	319	5910	0.054
2. $(NH_4)_2SO_4$ (0–40%)	37.5	112	1650	0.068
3. $Al(OH)_3$ gel	10.5	85	508	0.17
4. BioGel P-100	1.1	56	16.1	3.5
5. BioGel P-60	1.5	27	4.5	5.9
6. Hydroxylapatite	9.1	15	0.3	50.0

[a] Data from Nandi et al.[1]

[b] Total weight of liver is 650 g.

eluting after about 170 ml. Fractions with the highest specific activities are pooled and kept at 4° (see Table II).

Properties

This summary pertains only to the transferases from bovine liver mitochondria.[1] Somewhat different properties are ascribed to analogous enzymes from liver mitochondria of other mammalian species.[2,3]

Homogeneity and Physical Properties. Hydroxylapatite fractions with the highest activity for either benzoyltransferase or phenylacetyl-transferase (Step 6) contain enzyme that is >95% homogeneous as judged by disc gel electrophoresis. Each enzyme consists of a single polypeptide chain with a molecular weight of about 33,000 and a single active site.[1,7]

Substrate Specificity. The benzoyltransferase catalyzes not only benzoyl but salicyl, acetyl, propionyl, butyryl, isobutyryl, isovaleryl, 3-methylcrotonyl, and tiglyl transfer from the corresponding acyl-CoA substrate to glycine. Phenylacetyltransferase prefers phenylacetyl-CoA but can also employ indoleacetyl-CoA as an acyl donor to glycine. Glycine is the favored amino acid acceptor for both enzymes but weak activity for both is noted also with high concentrations of L-asparagine or L-glutamine.

Substrate Concentration Dependence. Apparent values of K_m found for reactants at high fixed concentrations of the alternate substrate are about 10–20 μM and 3 mM for benzoyl-CoA and glycine, respectively, in the benzoyltransferase reaction. Comparable estimates are 20 μM and 20 mM for phenylacetyl-CoA and glycine, respectively, in the phenylacetyl-transferase reaction.

Inhibitors and Stimulators. Acyl-CoA substrates of one transferase act kinetically as competitive inhibitors with respect to the prototype acyl-CoA substrate of the other transferase. Both transferases are inhibited by high nonphysiological concentrations of divalent cations such as Mg^{2+}, Ni^{2+} and Zn^{2+}. In contrast to the benzoyltransferase, the phenylacetyltransferase is sensitive to thiol binding reagents such as p-chloromercuribenzoate (10^{-5} M) or DTNB (10^{-4} M); preincubation of the phenylacetyltransferase with phenylacetyl-CoA affords partial protection from DTNB inactivation. Monovalent cations such as K^+, Rb^+, Li^+, Na^+, Cs^+, or $(NH_4)^+$ appear to be required for phenylacetyltransferase activity, but such requirement has not been demonstrated for the benzoyltransferase.

Stability and pH Optimum. Both enzymes retain about 80% of their activity after 3 weeks storage at 4°. Phenylacetyltransferase fractions from the hydroxylapatite column are inactivated by dialysis or concentration by ultrafiltration. Both enzymes have a pH optimum of 8.4–8.6.

Reaction Mechanisms. Preliminary two-substrate and product inhibition kinetic studies of both enzymes in addition to photoaffinity labeling studies of the benzoyltransferase are consistent with a sequential reaction mechanism wherein the acyl-CoA associates first with the enzyme, glycine adds to the enzyme prior to dissociation of the first product (ternary complex), and the peptide product dissociates last.[1,8]

[41] Bile Acid–CoA: Amino Acid N-Acyltransferase

By Paul G. Killenberg

$$\text{Bile acid-CoA + amino acid} \rightarrow \text{bile acid-}N\text{-amino acid + CoASH}$$

Assay Methods

Principle. Enzymatic activity is determined either by measuring the rate of appearance of CoASH or by monitoring the rate of loss of the thiolester bond of the bile acid–CoA substrate. The assays presented here are most conveniently performed on a multichannel recording spectrophotometer; simultaneous recording of activity in the presence and absence of amino acid is necessary in order to control for amino acid-independent hydrolysis of the bile acid–CoA substrate by one or more bile acid–CoA hydrolases. A third method, which measures the appearance of CoASH by following the reduction of NAD in the presence of

METHODS IN ENZYMOLOGY, VOL. 77

α-ketoglutarate and α-ketoglutarate dehydrogenase, has also been described.[1]

Reagents

Bile acid–CoA,[2] 1 mM

Amino acid (usually glycine or taurine), 20–100 mM, pH 7.2

Bovine serum albumin, 50 mg/ml, treated to remove fatty acids[3,4]

2,6-Dichlorophenolindophenol, 0.9 mM: Dissolve an excess of the reagent in water, filter (0.45 μm) and adjust by dilution such that 0.05 ml, diluted to 1.0 ml results in an absorbance of 1.0 at 600 nm (1-cm light path). (This reagent is stable for approximately 10 days if refrigerated and protected from light.)

Phenazineethylsulfate, 0.4 mM: Prepare daily and protect from light

4-(2-Hydroxyethyl)-1-piperazineethanesulfonic acid (HEPES), 0.5 M: Adjust to pH 7.2 with potassium hydroxide

Sodium phosphate, 1.0 M, pH 7.2

Enzyme: Dilute enzyme with either HEPES or phosphate buffer depending on the procedure used. Protein concentration is determined by the biuret method[5]

NOTE: All reagents except enzyme should be filtered (0.45 μ) prior to addition to the reaction mixture. Buffer, water and amino acid solution should be warmed to 30° prior to addition to the reaction mixture.

Procedure A. To a solution containing 0.05 ml HEPES buffer, 0.05 ml dichlorophenolindophenol, 0.05 ml phenazineethylsulfate, 0.02 ml defatted bovine serum albumin, and 0.05 ml bile acid–CoA substrate, water is added in a cuvette to a volume of 0.85 ml. The reagents are mixed and warmed to 30° for 2 min. Diluted enzyme, 0.05 ml, is added and, after 1 min, 0.1 ml of either water or amino acid solution is added with rapid mixing. The change in absorbance at 600 nm is measured at 30° for 4 min on a recording spectrophotometer (0.2 A, full scale, is optimal). The specific activity of the enzyme (μmol min^{-1} mg^{-1}) is calculated as the difference in the rate of loss of absorbance in the presence and absence of amino acid; $\Delta A \times 0.0455$ min^{-1} mg^{-1} enzyme protein, assuming $E_{600\,nm} = 22 \times 10^3$ M^{-1} cm^{-1}.[6]

Comment. Because of its relatively high sensitivity and the fact that the

[1] D. A. Vessey, *J. Biol. Chem.* **254**, 2059 (1978).
[2] P. G. Killenberg and D. F. Dukes, *J. Lipid Res.* **17**, 451 (1976); see also this volume, Article [57].
[3] R. F. Chen, *J. Biol. Chem.* **242**, 173 (1967).
[4] R. W. Hanson and F. J. Ballard, *J. Lipid Res.* **9**, 667 (1978).
[5] A. G. Gornall, C. J. Bardawill, and M. M. David, *J. Biol. Chem.* **177**, 751 (1949).
[6] D. D. Hoskins, this series, Vol. 14, p. 110.

rate of reaction remains linear for at least 4 min under the preceding condition, this procedure is especially suited to measurement of initial rates of reaction with purified enzyme. In crude preparations, high rates of amino acid-independent reduction of the indophenol may be encountered. This is especially true with crude liver homogenate from rabbit or guinea pig but not a problem with rat liver protein. Under all circumstances, care must be taken to dilute the enzyme such that the final rate of reduction is no more than 1.2 nmol/min.[7]

Procedure B. To a solution containing 0.05 ml of the phosphate buffer and 0.05 ml bile acid–CoA, water is added to 0.85 ml. Diluted enzyme and amino acid are added sequentially as in Procedure A and the decrease in the concentration of the thiolester bond is estimated from the difference in the rate of loss of absorbance at 232 nm with and without added amino acid. Specific activity (μmol min^{-1} mg^{-1}) is calculated as $\Delta A \times 0.244$ min^{-1} mg^{-1} assuming $E_{232\,nm}$ of the thiolester bond as $4.1 \times 10^3\ M^{-1}\ cm^{-1}$.[2]

Comment. This procedure is less sensitive than Procedure A but is better suited for the assay with crude protein preparations. The rate of reaction is linear for up to 2 min after final addition of amino acid and at rates as high as 10 nmol of product formation per minute.[7]

Purification Procedure

The procedure given here has been previously applied to rat liver.[8] Another procedure, using DEAE-cellulose and glycocholate-Sepharose affinity chromatography, has been reported for bovine liver.[1]

Step 1. Extract. This step and all following procedures are conducted at 0°. Rat livers are homogenized with a Teflon-glass homogenizer in three volumes of 0.2 M potassium phosphate at pH 7.5. The resulting slurry is diluted to 6 times the original wet weight of the liver and centrifuged at 20,000 g for 12 min. The supernatant liquid is decanted and retained. The pellet is resuspended in 3 volumes of the same buffer, homogenized, and centrifuged. The two supernatant liquids are combined and centrifuged at 245,000 g for 65 min. Protamine sulfate, 50 mg/ml in 0.1 N HCl, is added dropwise to the ultracentrifugation supernatant fraction to a final concentration of 0.2 mg/ml. The pH is adjusted with dilute formic acid to 6.5, and after 10 min, the resulting suspension is centrifuged at 42,000 g for 12 min. The supernatant fluid is retained and adjusted to 15 mg protein/ml by addition of buffer and to pH 7.5 with 1 N NH$_4$OH.

Step 2. Salt Precipitation. To each 100 ml of protein solution is added 36 g of solid ammonium sulfate, pH being maintained at 7.5 with 1 N

---

[7] P. G. Killenberg, *J. Lipid Res.* **19**, 24 (1978).

[8] P. G. Killenberg and J. T. Jordan, *J. Biol. Chem.* **253**, 1005 (1978).

NH_4OH.[9] After 30 min, the resulting suspension is centrifuged at 42,000 g for 12 min. The pellet is centrifuged again to remove traces of the supernatant fluid and is dissolved in 20 mM potassium HEPES at pH 7.2, to a protein concentration of 15 mg/ml. To each 100 ml of this solution, 16 g of solid ammonium sulfate is added as before and, after 30 min equilibration, the suspension is centrifuged. The pellet is discarded, and an additional 12 g of salt per 100 ml original volume is added to the supernatant liquid.

The precipitate is retained, centrifuged to remove traces of supernatant fluid, and dissolved to 20 mg protein/ml in 20 mM potassium HEPES at pH 7.2.

Step 3. (Carboxymethyl)Sephadex 1. The protein solution, 32 to 40 ml, is applied to a column (2 × 50 cm) of CM-Sephadex, previously equilibrated with 20 mM potassium HEPES at pH 7.2. The column is washed with 50 ml of the same buffer and is eluted with a linear gradient formed by 250 ml of the starting buffer and 250 ml of 0.2 M KCl in the same buffer. Eluant fractions (2.5 ml) containing enzyme activity are pooled and precipitated by addition of 50 g solid ammonium sulfate per 100 ml. After 45 min, the precipitate is recovered by centrifugation and the pellet dissolved to 7 mg protein/ml in the HEPES buffer.

Step 4. Liquid Ammonium Sulfate. A saturated solution of ammonium sulfate at 4°, adjusted to pH 7.2 with NH_4OH, is added to the enzyme from Step 3 and the material precipitating between 33 and 41% of saturation is collected and dissolved in 5 ml of the potassium HEPES buffer containing 1.4 M ammonium sulfate, pH 7.2. After 1 hr, the resulting suspension is centrifuged and solid ammonium sulfate is added to the supernatant fluid to a final concentration of 50 g/100 ml. Following equilibration for 45 min, the pellet is recovered by centrifugation and dissolved in 2 ml of the potassium HEPES buffer. The solution is dialyzed for 2 hr against 500 ml of the same buffer.

Step 5. (Carboxymethyl)Sephadex 2. The dialyzed protein fraction is applied to a column (1 × 15 cm) packed with CM-Sephadex and equilibrated with the 20 mM potassium HEPES buffer. After application of an additional 5 ml of the buffer, the protein is eluted with a linear salt gradient formed from 50 ml of starting buffer and 50 ml of the starting buffer that is 0.2 M in KCl. Fractions of 1 ml with highest activity are pooled and, after precipitation of the protein by addition of 50 g solid ammonium sulfate per 100 ml, the precipitate is dissolved in a minimal volume of the potassium HEPES buffer and stored at −70°.

A summary of the purification of rat liver bile acid–CoA : amino acid N-acyltransferase is given in the table.

---

[9] pH determined by insertion of electrodes directly into the protein solution.

SUMMARY OF PURIFICATION PROCEDURE[a]

Step	Volume (ml)	Total protein (mg)	Taurine-dependent[b]		Glycine-dependent[b]	
			Total activity (μmol/min)	Specific activity (μmol min^{-1} mg^{-1})	Total activity (μmol/min)	Specific activity (μmol min^{-1} mg^{-1})
1. Extract	336	11,100	271	0.024	448	0.040
2. Salt precipitation	20	1,720	150	0.083	216	0.125
3. CM-Sephadex 1	152	48.2	41.9	0.868	74.5	1.55
4. Liquid ammonium sulfate	1.0	21.6	24.3	1.13	40.7	1.89
5. CM-Sephadex 2	10.3	8.0	17.4	1.99	34.0	3.88

[a] Prepared from 93 g of rat liver.[8]

[b] Assays were performed using Procedure A in the presence of either 20 mM taurine or 100 mM glycine.

Properties

The following pertain to rat liver enzyme prepared as described in the preceding sections.[8] Some properties of bovine liver enzyme have been reported elsewhere.[1,10]

Purity. Electrophoresis of the preparation from Step 5 on polyacrylamide gel resolves the protein into one major band (87% of the protein) and several minor bands. The major band possesses activity with both glycine and taurine as acyl acceptors.

Specificity. The purified enzyme is active with CoA thiolesters of cholate, lithocholate, chenodeoxycholate, deoxycholate, and 3β-hydroxy-5-cholenoate. No activity is observed with palmitoyl-CoA, succinoyl-CoA, acetyl-CoA, benzoyl-CoA, or phenylacetyl-CoA. Activity with amino acids is limited to glycine, β-alanine, aminomethanesulfonic acid, and taurine.

Kinetics. In the presence of 50 μM bile acid–CoA, the apparent K_m for taurine and aminomethanesulfonate is 1 mM, whereas the apparent K_m for glycine and β-alanine are 30 mM and 200 mM, respectively. Calculated V_{max} for aminocarboxylate acceptors is always higher than for aminosulfonates at all stages of purification. With the final fraction, these approximate values are glycine, 6.4 μmol min^{-1} mg^{-1}; β-alanine, 4.3 μmol min^{-1} mg^{-1}; taurine, 2.4 μmol min^{-1} mg^{-1}, and aminomethanesulfonate, 1.1 μmol min^{-1} mg^{-1}.

pH Optima. Enzyme activity for each of the substrates tested is optimal between pH 7.8 and 8.2. The pH activity curve rises more than 2-fold from pH 6.0 to 8.0 with glycine as the acceptor but rises only 50% over the same interval with taurine. The enzyme at 37° loses activity at pH 6.0 or 9.0; approximately 50% of original activity remains after 2 hr under these conditions.

Inactivation. The enzyme is stable at −70° for at least 6 months and at 22° for up to 4 hr. Activity is rapidly lost at temperatures above 50°. Significant inactivation occurs upon exposure of the enzyme to 5,5′-dithiobis(2-nitrobenzoic acid); preincubation of enzyme with cholyl-CoA protects from inactivation by this reagent.

[10] B. Czuba and D. A. Vessey, *J. Biol. Chem.* **255**, 5296 (1980).

[42] Formaldehyde Dehydrogenase[1]

By LASSE UOTILA and MARTTI KOIVUSALO

Formaldehyde + GSH + NAD$^+$ $\rightleftharpoons$ S-formylglutathione + NADH + H$^+$

Formaldehyde dehydrogenase (EC 1.2.1.1) is a specific, NAD-linked dehydrogenase, the only one with a specific requirement for glutathione.[2] The enzyme is apparently involved in the oxidation of formaldehyde produced, for example, in the microsomal demethylation of drugs. S-Formylglutathione, the product of the enzyme, is rapidly hydrolyzed by another specific enzyme, S-formylglutathione hydrolase (EC 3.1.2.12), which has been purified and characterized (Article [43]) from human liver.[3]

Assay Method

Principle. Formaldehyde dehydrogenase is assayed by recording the initial velocity spectrophotometrically at 340 nm and 25°. Either the formation of NADH in the forward reaction or the disappearance of NADH in the reverse reaction can be measured.

Reagents (Forward Reaction)

Sodium Phosphate (or pyrophosphate), 0.12 M, pH 8.0
Formaldehyde, 0.1 M
Glutathione, 0.1 M
NAD, 0.04 M

Formaldehyde is prepared by heating paraformaldehyde or by the hydrolysis of hexamethylenetetramine with H$_2$SO$_4$; concentration is standardized by the chromotropic acid method after steam distillation.[2]

Reagents (Reverse Reaction)

Sodium acetate, 0.12 M, pH 5.7
S-Formylglutathione,[4] 0.07 M
NADH, 5 mM

Procedure. The assay mixture for the *forward reaction* contains 0.75 ml of the phosphate or pyrophosphate buffer, 10 μl formaldehyde, 10 μl

---

[1] This research was supported in part by the Sigrid Jusélius Foundation, Finland.
[2] L. Uotila and M. Koivusalo, *J. Biol. Chem.* **249**, 7653 (1974).
[3] L. Uotila and M. Koivusalo, *J. Biol. Chem.* **249**, 7664 (1974).
[4] For synthesis of S-formylglutathione, see L. Uotila in this volume, Article [56].

glutathione, 20 μl NAD, enzyme, and water to 1.0 ml. The reaction can be started either by addition of the enzyme or by NAD; to ensure equilibration in the reaction between formaldehyde and GSH (see later), the other components should be incubated for 5 min after mixing at 25° before starting the reaction.

The assay mixture for the *reverse reaction* contains 0.75 ml of the acetate buffer, 20 μl S-formylglutathione, 20 μl NADH, enzyme, and water to 1.0 ml. The reaction is started by addition of the enzyme.

Definition of Unit. One unit of enzyme catalyzes the formation of 1 μmol of NADH per minute under the forward assay conditions described ($\epsilon = 6220\ M^{-1}\ cm^{-1}$ at 340 nm). Specific activity is defined as the number of units of activity per milligram of protein; protein is determined by the method of Lowry et al.[5]

Comment. For crude tissue preparations, only the forward reaction assay may be used because of the presence of a large excess of activity catalyzing the hydrolysis of S-formylglutathione.[3,6] The crude extracts should be freed from endogenous glutathione by either dialysis or gel chromatography. A blank without glutathione must be included in the assay; this blank measures the activity for formaldehyde of nonspecific aldehyde dehydrogenase(s) in the preparation, whereas the additional activity with glutathione represents formaldehyde dehydrogenase. At least for liver preparations, the assay mixture should also contain 3 mM pyrazole in order to block alcohol dehydrogenase. The latter would otherwise interfere by catalyzing the transformation of formaldehyde and NADH to methanol and NAD.

Purification Procedure

The enzyme has been purified to homogeneity from human liver,[2] *Candida boidinii,*[7] and pea seeds.[8] The purification procedure for human liver, which appears useful for other animal tissues with minor modifications, is described in the following sections. All manipulations are carried out at 0° to 4°.

Step 1. Extract. Liver tissue (about 600 g) is homogenized for 2 min in a Waring blender with 3 volumes of 50 mM potassium phosphate at pH 7.4 containing 5 mM 2-mercaptoethanol. The homogenate is centrifuged at 23,000 g for 60 min. The precipitate is discarded.

[5] O. H. Lowry, N. J. Rosebrough, A. L. Farr, and R. J. Randall, *J. Biol. Chem.* **193**, 265 (1951).
[6] L. Uotila, *Biochemistry* **12**, 3944 (1973).
[7] H. Schütte, J. Flossdorf, H. Sahm, and M.-R. Kula, *Eur. J. Biochem.* **62**, 151 (1976).
[8] L. Uotila and M. Koivusalo, *Arch. Biochem. Biophys.* **196**, 33 (1979).

Step 2. Salt Fractionation. The supernatant fluid is fractionated with ammonium sulfate while the pH is maintained at 7.4 by addition of 0.1 M ammonium hydroxide. The protein fraction precipitated by the addition of 258 g of solid ammonium sulfate per liter of the solution is removed by centrifugation. More ammonium sulfate, 139 g per liter, is added to the supernatant. The protein precipitates are collected by centrifugation, dissolved in 10 mM Tris-HCl, pH 7.6, containing 5 mM 2-mercaptoethanol, and dialyzed against the same buffer (about 18 hr with two buffer changes of 5 liters of buffer each time).

Step 3. DEAE-Cellulose 1. The dialyzed enzyme is applied to a DEAE-cellulose (DE22) column (5 × 85 cm) equilibrated with 10 mM Tris-HCl, pH 7.6, that is 5 mM in 2-mercaptoethanol. The column is eluted first with a linear gradient (total volume of 4 liters) established between the equilibration buffer and 40 mM Tris-HCl, pH 7.0, containing 5 mM 2-mercaptoethanol. A second linear gradient (2 liters) is prepared from the latter buffer to the same buffer that also contains 20 mM KCl; this gradient elutes formaldehyde dehydrogenase.

Step 4. DEAE-Cellulose 2. The active fractions are pooled and dialyzed against 10 mM Tris-HCl, pH 7.6, containing 5 mM 2-mercaptoethanol. The dialyzed solution is applied to a DEAE-cellulose (DE32) column (3 × 35 cm) equilibrated with the dialysis buffer. The column is eluted with a linear gradient (3 liters) established between the equilibration buffer and 40 mM Tris-HCl, pH 7.0, that is 5 mM in 2-mercaptoethanol. Formaldehyde dehydrogenase is sharply eluted at an ionic strength of about 0.03.

Step 5. Isoelectric Focusing. The enzyme from Step 4 is concentrated by ultrafiltration (200-ml Amicon cell; PM-10 membrane) to 40 ml and dialyzed against 10 mM Tris-HCl, pH 7.6, containing 5 mM 2-mercaptoethanol. The protein solution is applied to an isoelectric focusing column (LKB, model 8101, 110-ml volume) in the light gradient solution. The column contains a 0 to 40% sucrose gradient that is 1% pH 5–8 Ampholine and 1 mM dithiothreitol. The voltage is gradually increased stepwise from 400 to 700 V and the focusing continued for 50 hr at 4°. The contents of the column are collected by pumping and the five most active 2-ml fractions, near pH 6.35, are pooled.

Step 6. Sephadex G-100. The preparation from Step 5 is concentrated by ultrafiltration to 4 ml and passed through a Sephadex G-100 column (2.5 × 45 cm), equilibrated and eluted with 10 mM potassium phosphate, pH 6.8, containing 2 mM 2-mercaptoethanol. The enzyme is eluted in the midpart of a broad protein peak. Fractions (2 ml) containing at least one-half of the activity of the best fraction are pooled.

Step 7. Hydroxylapatite. The preparation from Step 6 is applied to a column (1.5 × 30 cm) of Hypatite C (Clarkson Chem. Co., Williamsport), equilibrated with 10 mM potassium phosphate at pH 6.8 containing 2 mM 2-mercaptoethanol. The column is eluted with a linear 700-ml gradient established between the equilibration buffer and 140 mM potassium phosphate at the same pH. Formaldehyde dehydrogenase is eluted at a phosphate concentration of about 90 mM.

Step 8. QAE-Sephadex. The enzyme, pooled from Step 7, is concentrated 10-fold by ultrafiltration, and dialyzed against 10 mM Tris-HCl at pH 7.6 containing 2 mM 2-mercaptoethanol (enzyme dialyzed for 9 hr; new dialysis buffer, 500 ml, changed every 3 hr). The dialyzed enzyme is applied to a column (1 × 26 cm) of QAE-Sephadex equilibrated with the dialysis buffer. The column is eluted with a linear 400-ml gradient consisting of the equilibration buffer and the same buffer supplemented with 60 mM KCl. Fractions containing formaldehyde dehydrogenase are pooled.

The final preparation (see table) is homogeneous in polyacrylamide electrophoresis under denaturing and nondenaturing conditions and represents about 1400-fold purification over the crude soluble fraction of human liver. The forward assay for purified preparations can be used after Step 3, because alcohol and aldehyde dehydrogenase activities have been removed. Hydrolase activities interfering in the assay of the reverse reaction are totally removed after Step 5. To obtain a hydrolase-free preparation, one may use Step 6 directly without the time-consuming Step 5 if homogeneous formaldehyde dehydrogenase is not needed. Affinity chromatography on 5'-AMP-Sepharose can also be used; formaldehyde dehydrogenase binds and can be eluted with 1 M NaCl[8] or 0.4 mM NAD.

SUMMARY OF PURIFICATION OF FORMALDEHYDE DEHYDROGENASE FROM HUMAN LIVER

Step	Volume (ml)	Total activity (units)	Total protein (mg)	Specific activity (units/mg)
1. Supernatant of homogenate	1570	167	74,100	0.0023
2. Ammonium sulfate (45–67%)	232	123	26,200	0.0047
3. DEAE-cellulose 1	360	43.5	1230	0.035
4. DEAE-cellulose 2	160	27.6	176	0.157
5. Isoelectric focusing	10.3	20.6	—	—
6. Sephadex G-100	21.4	8.82	5.50	1.60
7. Hydroxylapatite	143	4.95	1.80	2.75
8. QAE-Sephadex A 50	70	1.90	0.60	3.17

Properties

Molecular Properties. The human liver enzyme has an apparent molecular weight of 81,400 and represents a dimer of similar subunits (M_r 39,500).[2] Similar results have been reported for the enzymes purified from *Candida boidinii*[7] and pea seeds.[8] The human liver enzyme has a pI of 6.35.

Substrate Specificity. Besides formaldehyde, the enzyme is active with a number of α-ketoaldehydes as substrate. Methylglyoxal gives a maximum velocity that is 85% that of formaldehyde. Hydroxypyruvaldehyde, 3-ethoxy-2-hydroxybutyraldehyde (kethoxal), and glyoxal are used less effectively. Phenylglyoxal, acetaldehyde, and other aldehydes tested are not substrates. The enzyme is specific for glutathione; 2-mercaptoethanol, dithiothreitol, coenzyme A, cysteine, and homocysteine are inactive. The beef liver enzyme, tested for the specific determination of glutathione,[9] was also inactive when thioglycolate, thiomalate, thiolactate, cysteamine, and N-acetylcysteine replaced glutathione. In addition to glutathione, the human liver enzyme can use homoglutathione[10] in which a glycine residue is replaced by β-alanine.[11]

The forward reaction of human liver formaldehyde dehydrogenase has a pH optimum at 8.0 with NAD as the electron acceptor. Under the same conditions, NADP has a K_m of about 1000 times greater than NAD. NADP is used better at pH 6 but this reaction appears to be without physiological significance. The enzyme can use some of the analogs of NAD. The most effective of them, 3-acetylpyridine-adenine dinucleotide, has a maximal velocity 23 times that of NAD.[12]

At pH 8, the rate of the reverse reaction with S-formylglutathione and NADH is about the same as that of the forward reaction with formaldehyde, GSH, and NAD. At the optimum pH (5.7), the rate of the reverse reaction with S-formylglutathione and NADH is 3.9 times, and with S-formylglutathione and NADPH 2.0 times, that of the forward reaction rate with NAD at pH 8.0. S-Pyruvylglutathione is reduced by the enzyme to methylglyoxal and glutathione. S-Lactoylglutathione, S-acetylglutathione, and six other glutathione thiol esters tested do not react, nor are they inhibitory.

Kinetic Constants and Mechanism. The apparent K_m values of the enzyme at pH 8 are 8 μM for glutathione, 9 μM for formaldehyde, and 9 μM for NAD.[2] At pH 5.7, NADH has an apparent K_m of 4.2 μM; NADPH, 45

[9] M. Koivusalo and L. Uotila, *Anal. Biochem.* **59**, 34 (1974).
[10] L. Uotila and M. Koivusalo, unpublished results. We are grateful to Dr. P. R. Carnegie for a gift of purified homoglutathione.
[11] P. R. Carnegie, *Biochem. J.*, **89**, 459 (1963).
[12] L. Uotila and B. Mannervik, *Biochem. J.*, **177**, 869 (1979).

μM; and S-formylglutathione, 300 μM.[2] Extensive kinetic studies[12] have shown that the true substrate of the enzyme is, in agreement with earlier postulations,[13,14] S-hydroxymethylglutathione, i.e., the adduct of formaldehyde and GSH that is formed nonenzymically with a dissociation constant of 1.5 mM at pH 8.[2] The apparent K_m for the adduct is about 2 μM.[12] When the adduct concentration remains constant, another component of the equilibrium mixture of formaldehyde and GSH influences the enzymic rate: free glutathione appears to be an essential, probably allosteric activator of the enzyme with an apparent K_m of 1 to 2 μM.[12] Both initial-velocity[12] and product-inhibition[15] experiments suggest that NAD and the hemimercaptal adduct are bound to the enzyme in a random order.

Product. In contrast to the former belief,[13,14] formate is not a product of formaldehyde dehydrogenase. Rather, a thiol ester is formed with the purified enzyme and may be observed at 240 nm with a spectrophotometer. The other reaction product, NADH, absorbs less at 240 nm than does NAD, and their absorbance difference can quench a large part of the absorbance of the thiol ester product. In this experiment, an NADH oxidizing system[2] [2 mM sodium pyruvate and lactate dehydrogenase (EC 1.1.1.27) in a 100-fold or higher excess over formaldehyde dehydrogenase] is needed as a further supplement to the forward reaction assay mixture. This addition prevents the net oxidation/reduction of NAD(H). In addition to using impure preparations, apparently containing S-formylglutathione hydrolase activity, former investigators[13,14] who found formate as the reaction product used a method for formate assay that is inadequate for the separation of the sensitive S-formylglutathione from formate. When formate was determined by the formyltetrahydrofolate synthetase method,[3,16] it appeared that formate is not produced by formaldehyde dehydrogenase.[2]

Inhibitors and Stability. Formaldehyde dehydrogenase is not inhibited by EDTA or other chelating agents but is inhibited by reagents that react with amino or thiol groups.[2] HgCl$_2$ and p-hydroxymercuribenzoate are the most effective inhibitors (50% inhibition at 0.2 to 0.5 μM inhibitor). NAD and NADH protect the enzyme from SH reagents. NAD, and especially NADH, also protect from denaturation by high temperature (50–60°), acid pH, and storage at 0°.[2] The half-life of the homogeneous enzyme at 0° is only a few hours. NAD (1 mM) or glycerol (40% v/v) increases the half-life to about 1 month and NADH (0.5 mM) is even more effective.

[13] P. Strittmatter and E. G. Ball, *J. Biol. Chem.* **213**, 445 (1955).
[14] Z. B. Rose and E. Racker, *J. Biol. Chem.* **237**, 3279 (1962); see also this series, Vol. 9, Article [66].
[15] L. Uotila and B. Mannervik, *Biochim. Biophys. Acta* **616**, 153 (1980).
[16] J. C. Rabinowitz *in* "Methoden der enzymatischen Analyse" (H.-U. Bergmeyer, ed.), p. 1503. Verlag Chemie, Weinheim, 1970.

Among reversible inhibitors, α-NAD and ADP-ribose are the most effective.[12] S-Methylglutathione and its homologs with a longer alkyl chain are not inhibitory.[12]

Comments

Formaldehyde is produced from the methyl group of a variety of xenobiotics in a cytochrome P-450-linked demethylase reaction.[17] It has been shown that the nonspecific aldehyde dehydrogenase from mitochondria, which can use formaldehyde well, is involved in the oxidation of this formaldehyde.[18] A study with isolated hepatocytes showed that the glutathione-dependent formaldehyde dehydrogenase also plays a significant role in the oxidation of the formaldehyde produced in microsomes[19]: the rate of formaldehyde formation was estimated to be high (6 to 10 nmol per 10^6 cells per min).[19] The relative significance of the role of the two types of enzymes, cytosolic formaldehyde dehydrogenase and nonspecific mitochondrial aldehyde dehydrogenase, are not known. Because most tissues contain a high concentration of GSH that will inevitably bind small amounts of formaldehyde to form S-hydroxymethylglutathione, i.e., the substrate of formaldehyde dehydrogenase,[12] this pathway is a physiologically reasonable one. Formaldehyde dehydrogenase, which is present in all animal tissues thus far tested,[2] may have, apart from its role in detoxication, additional metabolic significance because of its capability in catalyzing the synthesis of glutathione thiol esters.

[17] J. R. Gillette, *Adv. Pharmacol.* **4**, 219 (1966).
[18] D. L. Cinti, S. R. Keyes, M. A. Lemelin, H. Denk, and J. B. Schenkman, *J. Biol. Chem.* **251**, 1571 (1976); see also H. Denk, P. W. Moldeus, R. A. Schulz, J. B. Schenkman, S. R. Keyes, and D. L. Cinti, *J. Cell Biol.* **69**, 589 (1976).
[19] D. P. Jones, H. Thor, P. Andersson, and S. Orrenius, *J. Biol. Chem.* **253**, 6031 (1978).

[43] S-Formylglutathione Hydrolase[1,2]

By LASSE UOTILA and MARTTI KOIVUSALO

S-Formylglutathione + H_2O → glutathione + formate

S-Formylglutathione is the product of the oxidation of formaldehyde by the glutathione- and NAD-dependent formaldehyde dehydrogenase.[3] S-Formylglutathione does not accumulate in crude tissue preparations but

[1] This research was supported in part by the Sigrid Jusélius Foundation, Finland.
[2] L. Uotila and M. Koivusalo, *J. Biol. Chem.* **249**, 7664 (1974).
[3] L. Uotila and M. Koivusalo, this volume, Article [42].

is rapidly hydrolyzed. A specific enzyme, S-formylglutathione hydrolase (EC 3.1.2.12), is responsible for most of the hydrolysis of this thiol ester.[2] The amount of the hydrolase activity is more than 700-fold compared to formaldehyde dehydrogenase activity in crude human liver preparations, i.e., the hydrolase is normally present in great excess.

Assay Method

Principle. S-Formylglutathione hydrolase is assayed by recording the absorbance decrease resulting from the hydrolysis of the thiol ester with a spectrophotometer at 240 nm and 25°.

Reagents

Potassium phosphate, 0.12 M, pH 7.1
S-Formylglutathione,[4] 10 mM

Procedure. The assay mixture contains 0.75 ml of the phosphate buffer, 0.05 ml S-formylglutathione, enzyme, and water to a final volume of 1.0 ml. A blank without enzyme is always included and the rate of hydrolysis without enzyme is subtracted from the enzymic rate. If the enzyme preparation contains a low-molecular-weight thiol as stabilizer, it is included in the blank at the same concentration. The rate of absorbance decrease at 240 nm is converted to moles of substrate hydrolyzed with $\Delta\epsilon = 3300$ M^{-1} cm^{-1} for S-formylglutathione.[4]

Definition of Unit. One unit of the enzyme catalyzes the hydrolysis of 1 μmol per minute of S-formylglutathione under the conditions described. Specific activity is defined as the number of units of activity per milligram of protein; protein is determined by the method of Lowry *et al.*[5]

Comment. The assay can also be used for crude preparations. From each of the sources that have been studied, the high activity of the enzyme has allowed the use of highly diluted preparations, which diminishes interference from the high absorbance of crude preparations. Glyoxalase II (EC 3.1.2.6, hydroxyacylglutathione hydrolase) from human liver can also catalyze the hydrolysis of S-formylglutathione (at a rate 40% of that with the most active substrate of glyoxalase II, S-lactoylglutathione[6]). However, 97% of the total enzymic hydrolysis of S-formylglutathione in liver is from the specific hydrolase described here, and only 3% is from glyoxalase II. With an enzyme source that has not been previously studied, removal of the other glutathione thiol esterases may be necessary.

[4] For the synthesis of S-formylglutathione, see L. Uotila, this volume, Article [56].
[5] O. H. Lowry, N. J. Rosebrough, A. L. Farr, and R. J. Randall, *J. Biol. Chem.* **193**, 265 (1951).
[6] L. Uotila, *Biochemistry* **12**, 3944 (1973).

However, human liver glyoxalase II can be inhibited specifically by the hemimercaptal of methylglyoxal and glutathione[6]; this compound does not inhibit S-formylglutathione hydrolase.[2]

Purification Procedure[2]

The procedure for extraction of the tissue, salt precipitation, and DEAE-cellulose chromatography (Steps 1, 2, and 3) are entirely similar to the first three purification steps used for formaldehyde dehydrogenase described in this volume.[3] S-Formylglutathione hydrolase is eluted from the DEAE-cellulose column, soon after formaldehyde dehydrogenase, with the same gradient.

Step 4. Hydroxylapatite. The pooled S-formylglutathione hydrolase preparation is concentrated by ultrafiltration (400-ml Amicon cell; PM-10 membrane) from about 400 to 80 ml. The solution is dialyzed for 10 to 12 hr, with one change of buffer, against 2 liters of 10 mM potassium phosphate at pH 6.8 containing 5 mM 2-mercaptoethanol. Half of the dialyzed enzyme solution is applied to a column (1.5 × 31 cm) of Hypatite C (Clarkson Chemical Co., Williamsport), previously equilibrated with the dialysis buffer. The column is eluted with the equilibration buffer at a flow rate of 40 to 50 ml per hour. The enzyme is adsorbed loosely but will be eluted slowly from the column with the equilibration buffer; about 15 column volumes of the elution buffer are needed. Before using the same column for the other half of the dialyzed fraction, the bulk of the impurities, bound tightly to the column, are eluted by sequentially washing with 250 ml of 0.5 M potassium phosphate (pH 6.8) and 250 ml of the starting buffer.

Step 5. Isoelectric Focusing. The enzyme pooled from two similar hydroxylapatite columns is concentrated by ultrafiltration to 40 ml and applied to a 110-ml isoelectric focusing column (LKB, model 8101). The gradient solutions for the column (for details, see Vesterberg[7]) contain a 0 to 40% sucrose gradient in 1 mM dithiothreitol and 1% pH 4–6 Ampholine. The enzyme is applied in the light gradient solution where it replaces a comparable amount of liquid. The initial voltage of 300 V is gradually increased stepwise to the final value of 800 V. After 48 hr, the contents of the column are collected by pumping and the five most active 2-ml fractions, near pH 5.4, are pooled.

Step 6. Sephadex G-100. The pooled enzyme from Step 5 is concentrated in a 12-ml ultrafiltration cell (Amicon; PM-10 membrane) to 4 ml and applied to a Sephadex G-100 column (2.5 × 45 cm) equilibrated and

[7] O. Vesterberg, this series, Vol. 22, Article [33].

SUMMARY OF PURIFICATION OF S-FORMYLGLUTATHIONE
HYDROLASE FROM HUMAN LIVER

Step	Volume (ml)	Total activity (units)	Total protein (mg)	Specific activity (units/mg)
1. Extract	1570	128,000	74,100	1.73
2. Salt precipitation	232	93,200	26,200	3.56
3. DEAE-cellulose	430	58,000	973	59.7
4. Hydroxylapatite	805	36,400	23.1	1570
5. Isoelectric focusing	12.7	17,500	—	—
6. Sephadex G-100	30.0	13,700	3.38	4050

eluted with 10 mM potassium phosphate, pH 6.8, containing 5 mM 2-mercaptoethanol. The first UV absorbing peak from the column contains the enzyme; the specific activity is constant across the peak.

The purification of the enzyme from 560 g of liver is summarized in the table. The final preparation is homogeneous by electrophoretic and ultracentrifugal criteria and represents about 2300-fold purification over the crude soluble fraction of human liver.

Properties

Molecular Properties. S-Formylglutathione hydrolase from human liver has a molecular weight of 52,500 as estimated by gel chromatography and 55,500 by sedimentation equilibrium analysis using an analytical ultracentrifuge. Polyacrylamide gel electrophoresis in the presence of sodium dodecyl sulfate results in a single protein band of M_r = 30,000. Thus the enzyme appears to be a dimer. The enzyme has a pI of 5.41 (at 4°). These properties clearly differ from those of other glutathione thiol esterases, glyoxalase II M_r; 22,900; pI, 8.35[6]) and S-succinylglutathione hydrolase M_r, 18,000; pI, 8.70[8]).

Substrate Specificity. The enzyme is highly specific for S-formylglutathione. Of the other glutathione thiol esters examined, S-acetylglutathione gives the highest rate,[9] which is only 0.6% of the rate with S-formylglutathione. S-Propionylglutathione (0.2%), S-pyruvylglutathione (0.1%), and S-glycolylglutathione (0.07%) react very slowly. Five other glutathione

[8] L. Uotila, *J. Biol. Chem.* **254,** 7024 (1979).
[9] All activity data are quoted for 0.5 mM thioester.

thiol esters, including S-lactoylglutathione, are not substrates (detection limit: 0.01% of the activity for S-formylglutathione). Of other formyl thioesters, S-formyl-N-acetylcysteine and formylmercaptoethanol give the same velocity as S-acetylglutathione (0.6%). Formyl-CoA is hydrolyzed very slowly (0.08%). Carboxyl esters or N-acetylimidazole are not hydrolyzed.

The reaction catalyzed by the enzyme is irreversible. A possible transferase function has been studied but the enzyme fails to catalyze a formyl transfer with the acceptors tested (other thiols, imidazole, tetrahydrofolate). Like formaldehyde dehydrogenase, S-formylglutathione hydrolase is a cytosolic enzyme in human liver.

K_m *Values.* The Michaelis constant for S-formylglutathione is 0.29 mM and for S-acetylglutathione, 0.12 mM. Both substrates give linear (Michaelis–Menten) kinetics. S-Acetylglutathione can easily be used as the substrate in the characterization of the purified enzyme, despite its much lower maximum velocity and has the advantages of easy preparation in purified form[4] and great stability to nonenzymic hydrolysis under the standard assay conditions.

pH Effects. The enzyme has optimal activity at pH 6.9 to 7.1; activity is highest in phosphate and acetate buffers. The enzyme is stable between pH 5 and 8 when stored for 24 hr at 4°.

Effect of Thiols. Purification of the enzyme is possible only in the continuous presence of low-molecular-weight thiols. 2-Mercaptoethanol (5 mM) and dithiothreitol (1 mM) are equally effective. Removal of the thiol by dialysis for 6 hr against thiol-free 10 mM Tris-HCl, pH 7.6, results in 85% loss in activity (in contrast to a quantitative yield in the presence of a thiol). Subsequent addition of 2 mM dithiothreitol stimulates the activity to 60% of the original; other thiols are less effective.

Inhibitors. The enzyme is highly sensitive to $HgCl_2$, p-hydroxymercuribenzoate, and 5,5′-dithiobis(2-nitrobenzoate); all three yield 50% inhibition at about 1 μM inhibitor when S-acetylglutathione (contaminated with less than 0.5% glutathione) is the substrate. Inhibition by these agents is rapidly reversed by dithiothreitol. As much as 50 to 100 μM N-ethylmaleimide is needed for a comparable inhibition. Arsenite is not inhibitory.

The enzyme is inhibited by the amino group reagents, 2,4,6-trinitrobenzene sulfonate and dansyl chloride, but not by chelating agents. Glutathione and its disulfide are inhibitory, but over 5 mM GSH is needed for significant inhibition at a substrate concentration of 0.5 mM. The tested derivatives of glutathione, e.g., S-methylglutathione and S-hydroxymethylglutathione, are not inhibitory. The enzyme is resistant to the organophosphates that inhibit "serine" esterases.

Other Sources

The enzyme has been partially purified and separated from other glutathione thiol esterases from human red cells,[10] livers and kidneys of several animals,[11] *Escherichia coli,*[11] and pea seeds.[12] *S*-Formylglutathione hydrolase also was purified to homogeneity from a methanol-utilizing yeast, *Kloeckera* sp. No. 2201.[13] Thus the enzyme appears to have a wide distribution similar to that observed for formaldehyde dehydrogenase.[3]

[10] L. Uotila, *Biochim. Biophys. Acta* **580**, 277 (1979).
[11] L. Uotila, unpublished results.
[12] L. Uotila and M. Koivusalo, *Arch. Biochem. Biophys.* **196**, 33 (1979).
[13] N. Kato, C. Sakazawa, T. Nishizawa, Y. Tani, and H. Yamada, *Biochim. Biophys. Acta* **611**, 323 (1980).

[44] Glutathione Peroxidase

By ALBRECHT WENDEL

$$\text{ROOH} + 2\,\text{GSH} \rightarrow \text{ROH} + \text{H}_2\text{O} + \text{GSSG}$$

Glutathione peroxidase (EC 1.11.1.9) was discovered to be an enzyme that, like catalase, metabolizes H_2O_2. Today, the enzyme, which is found in many animal tissues is regarded as a major protective system against endogenously and exogenously induced lipid peroxidation. The enzyme contains stoichiometric amounts of selenium and is reactive with a variety of organic hydroperoxides as well as with hydrogen peroxide.

Assay Methods

Principle. Two types of methods are used for determining the activity of glutathione peroxidase. One involves direct measurement of unconsumed GSH at fixed time periods by polarographic GSH analysis[1] (Method 1) or by the dithionitrobenzoic acid method (Method 2). The second approach takes advantage of the capability of glutathione reductase, with NADPH, to regenerate GSH from GSSG.[2] The decrease in NADPH is continuously measured spectrophotometrically while the GSH concentration in the enzymatic cycle remains essentially constant (Method 3).

[1] F. Schneider and L. Flohé, *Hoppe-Seyler's Z. Physiol. Chem.* **348**, 540 (1967).
[2] D. E. Paglia and W. N. Valentine, *J. Lab. Clin. Med.* **70**, 158 (1967).

Method 1

Reagents

GSH, 4 mM, in distilled water

H_2O_2, 5 mM, in distilled water (check manganometrically)

Potassium phosphate, 0.1 M, pH 7.0, containing 1 mM sodium azide to inhibit catalase and 2 mM EDTA

$HClO_4$, 0.7 M

Procedure. Buffered enzyme solution (0.1 M phosphate, pH 7.0; 2 mM EDTA), 1 ml, is incubated with 0.5 ml GSH solution for 10 min at 37°. The reaction is started by addition of 0.5 ml H_2O_2 solution and stopped after various times by addition of 2 ml perchloric acid. After addition of 10 μl octanol, the mixture is bubbled with nitrogen (99.999% purity) for 3 min to remove oxygen. For GSH determination a polarograph with mercury dropping electrode is used at 60 cm mercury and 30 drops per min. The reference electrode is a $Hg/HgSO_4$-electrode in 0.1 M K_2SO_4. A polarogram is recorded between 0 and -0.5 V with 5 μA full scale, recording at 0.2 V per minute. The calibration curve is prepared by adding perchloric acid before a series of different GSH concentrations. The spontaneous reaction rate between GSH and H_2O_2 is measured without enzyme and is subtracted for each determination.

Definition of Units. The glutathione peroxidase reaction involves a ping-pong mechanism and cannot be carried out under apparent saturation with respect to both substrates. Thus, first-order kinetics obtain and linearity of substrate consumption with time cannot be achieved. Therefore, a logarithmic definition of the enzyme activity is used[1]:

$$U = \Delta\log[GSH] \cdot min^{-1}$$

Practically, a plot of $\log[GSH]_0 - \log[GSH]_t$, representing the change in GSH concentrations versus time, yields a straight line with a slope that is proportional to the amount of enzyme and independent of the absolute GSH concentration. One enzyme unit is defined as the consumption of 90% GSH within 1 min.

Method 2

Determination of GSH with Ellman's reagent and its application to the glutathione peroxidase reaction has been described[3] and may be useful for serial measurements with purified enzyme.

---

[3] J. J. Zakowski and A. L. Tappel, *Anal. Biochem.* **89**, 430 (1978).

Method 3

Reagents

Potassium phosphate, $0.25\ M$, pH 7.0, containing 2.5 mM Na$_2$EDTA and 2.5 mM sodium azide

Glutathione reductase (from yeast) in the phosphate buffer; activity per ml: 6 μmol/min at 25°

GSH, 10 mM, in distilled water

NADPH, 2.5 mM, in 0.1% NaHCO$_3$ solution

ROOH, 12 mM, in distilled water (ROOH = H$_2$O$_2$, *tert*-butyl hydroperoxide, cumene hydroperoxide)

For determination in blood:

Drabkin's solution: 0.77 mM KCN and 0.6 mM K$_3$ [Fe(CN)$_6$] in 1 mM potassium dihydrogen phosphate

Transformation solution: 4.5 mM KCN and 0.45 mM K$_3$ [Fe(CN)$_6$] adjusted with $0.25\ M$ potassium dihydrogen phosphate to pH 7.0, prepared immediately before use

Procedure. One hundred microliters of buffer, GSSG reductase, GSH, and NADPH, respectively, are transferred into a 1-ml cuvette. Enzyme solution, 500 μl, containing up to 0.3 U/ml is added and incubated for 10 min at 37°. The reaction is started by addition of 100 μl hydroperoxide. The linear decrease in NADPH absorption is recorded at 366 nm; the spontaneous reaction is assessed without enzyme and is subtracted. With blood or in crude cell extracts containing hemoglobin, the following variation of the method is necessary in order to minimize and standardize the pseudoperoxidative activity of many hemoproteins[4]: Blood, 100 μl, is centrifuged and washed twice with 0.9% NaCl. The red cells are hemolyzed by addition of 1 ml H$_2$O. Hemoglobin concentration is determined by mixing 1 ml Drabkin's reagent with 0.1 ml of hemolysate. The absorption at 546 nm is measured ($A_{546} \times 16$ = mg Hb/ml) against a blank containing water instead of hemolysate. The hemolysate is diluted to exactly 3 mg hemoglobin per milliliter. From this solution, 1 ml is mixed with 0.5 ml of transformation solution. After 5 min, transformation to cyanomethemoglobin is complete at room temperature. Then, 500 μl of the hemolysate are transferred into the assay mixture as described. *tert*-Butyl hydroperoxide is a suitable substrate for red cell glutathione peroxidase.

Units. With GSH being regenerated during the coupled test procedure, pseudo-zero-order kinetics are obtained. Nevertheless, this apparently

---

[4] W. A. Günzler, H. Kremers, and L. Flohé, *J. Clin. Chem. Clin. Biochem.* **12**, 444 (1974).

linear indicator reaction, which measures a decrease in NADPH with time, still depends on the GSH concentration in the enzymatic cycle. The unit definition already given for Method 1 can be transferred to the assay here if:

$$\Delta\log[\text{GSH}] \cdot \min^{-1} = 1 \text{ unit} = 0.868 \cdot \Delta\text{NADPH} \cdot \min^{-1} \cdot [\text{GSH}]_0^{-1}$$

In the assay described here, the initial GSH concentration is unity.

Applicability. Method 1 allows substrate specificity and kinetic studies with purified enzyme. Inorganic halides interfere with the polarographic determination of GSH. Any other sulfhydryl compound will also interfere here, as they will with the DTNB determination.

The coupled test (Method 3), does not allow alternative donor substrates but is suitable for determination of the enzyme in crude homogenates. Although the spontaneous reaction rate with H_2O_2 is high, this substrate is necessary in order to differentiate Se-dependent glutathione peroxidase from a similar activity catalyzed by glutathione transferase. With organic peroxide substrates, the sum of glutathione peroxidase and non-Se-dependent activity is measured in organs that contain both activities, e.g., liver and kidney, but not in blood.

Purification

A convenient source for the preparation of glutathione peroxidase is bovine blood. The following procedure allows the isolation of the enzyme

TABLE I
PURIFICATION OF GLUTATHIONE PEROXIDASE FROM
BOVINE RED BLOOD CELLS

Step	Volume (ml)	Total protein (mg)	Specific activity (units/mg)	Total units
1. Hemolysate	8000	960,000	0.08	76,800
2. Organic solvent precipitation	1900	14,000	5.2	73,000
3. Phosphate precipitation	210	5,900	10.5	62,000
4. Hydrophobic chromatography	340	595	94.5	56,200
5. Ion-exchange chromatography	520	123	370	45,500
6. Gel exclusion chromatography	30	65	600	39,200

from as much as several hundred liters of starting material. All steps are conducted at 0° to 4° unless otherwise indicated (see Table I).

Step 1. Hemolysate. Bovine blood, containing 0.6% sodium citrate, is collected from the slaughterhouse and assayed according to Method 3. In our experience, the range of specific activity may vary from 0.5–180 units/g hemoglobin. Eight liters of active blood are centrifuged. The erythrocytes are washed twice with 0.9% NaCl and are hemolyzed by adding water to form 8 liters of hemolysate. Six liters of 0.6 M potassium phosphate at pH 6.6 are added and the mixture is stored in the cold for 2–12 hr.

Step 2. Organic Solvent Precipitation. A mixture of chloroform/ethanol (3 : 5, v/v), 4 liters, cooled to −20°, are added dropwise to the vigorously stirred hemolysate. It is not necessary to remove the stroma. The dark-brown precipitate is removed by centrifugation at 2000 g for 30 min at 0°. The yellowish supernatant liquid is filtered and immediately concentrated to about 1 liter with a Diaflo cell (Amicon) and a membrane with exclusion limits of 50,000 daltons. In our hands, a hollow fiber apparatus supplied by Berghof, FRG (internal fiber diameter, 0.6 mm; internal volume, 200 ml; area, 0.6 m², BM 500 fiber) driven with a flow of 2 liters/min and a filtration rate of 30 ml/min has proved to be optimal. Careful maintenance of the temperature near 0° is essential because the enzyme is in contact with organic solvents.

Step 3. Phosphate Precipitation. One liter of concentrated enzyme solution is mixed with 2 liters of 3.65 M potassium phosphate at pH 7.0. The floating precipitate is collected by filtration through a 1-cm Celite layer covered with a nylon gauze (aspirator). The precipitate is dissolved in 0.7 M potassium phosphate at pH 8.0. This step removes essentially all of the organic solvent.

Step 4. Phenyl-Sepharose. The enzyme is absorbed in a batch step to phenyl-Sepharose equilibrated with 0.7 M potassium phosphate at pH 8.0. In general, a capacity of 1000 to 2000 units per milliliter of packed gel is observed. The absorbed gel is added to a column (2 × 30 cm) previously prepared with 5 ml of fresh, equilibrated gel. The column is washed with 300 ml of 0.7 M phosphate buffer and then with 300 ml of 0.5 M phosphate buffer of the same pH. A linear, 300-ml gradient, from 100 to 10 mM potassium phosphate, pH 8.0, is used to elute the enzyme at a flow rate of 6 ml/hr. Enzyme is eluted rapidly at first under these conditions but then tails off slowly.

Step 5. DEAE-Sephadex. The pooled fractions are diluted to a final buffer concentration of 5 mM phosphate, pH 8.0 (roughly a 10-fold dilution; control is by measurement of conductivity), and applied to a column (4 × 20 cm) of DEAE-Sephadex equilibrated with the same buffer. After washing with 1 liter of starting buffer, activity is eluted with a linear gradient of 5 to 80 mM potassium phosphate, pH 8.0, that is saturated with nitrogen and contains 1 mM GSH.

Step 6. S-300 Sephacryl. The active, pooled fractions are concentrated by a Diaflo cell to 15 ml as in Step 2 and charged onto a S-300 Sephacryl (superfine) column (2.0 × 180 cm), previously equilibrated with 20 mM potassium phosphate, pH 8.0, containing 0.5 M NaCl; a flow rate of 12 ml/hr is maintained. The enzyme is eluted immediately after a single protein that has an apparent molecular weight of about 100,000.

Optional Step 7. Hydroxylapatite. In some preparation the product from Step 6 contains minor protein impurities as viewed by disc electrophoresis. Final purification, without further increase in specific activity, can be achieved. The enzyme is dialyzed against 5 mM potassium phosphate at pH 7.4, absorbed onto a hydroxylapatite column (1 × 10 cm) equilibrated with the same buffer, and eluted with a gradient of up to 80 mM phosphate.

Properties

Stability. Glutathione peroxidase is relatively stable in the range of pH 7 to 10 at protein concentrations above 1 mg/ml; as a precipitate in 2.5 mM potassium phosphate at pH 7.0; or in the crystalline state in 1.2 M potassium phosphate at pH 7.0.[5] Lyophilization is possible in the presence of 1% mannitol–Ficoll. Up of 30% organic solvents, e.g., methanol, ethanol, acetone, or dimethyl sulfoxide, are tolerated at −20°. The enzyme undergoes autoxidation to forms of different stability[6] but can be reactivated by incubation with GSH. This ability to reactivate is the rationale for the preincubation immediately prior to assay.

Purity. Although the purified enzyme exhibits a broad zone in discontinuous acrylamide gel electrophoresis, sometimes with three barely distinguishable bands, only one band is detected in the presence of SDS.[7] Enzyme with a specific activity higher than 400 units/mg readily crystallizes from 1.25 mM potassium phosphate at pH 7.0 in flat, rhombic crystals. The selenium content of the protein is a reliable index of purity (Table II).[8]

Substrates. The enzyme is highly specific for the donor substrate GSH, although a few sulfhydryl compounds, e.g., mercaptoacetic acid methyl ester and γ-glutamylcysteine methyl ester, are slowly metabolized.[9]

[5] R. Ladenstein and A. Wendel, *J. Mol. Biol.* **104,** 877 (1976).
[6] R. J. Kraus, J. R. Prohaska, and H. E. Ganther, *Biochim. Biophys. Acta* **615,** 19 (1980).
[7] L. Flohé, B. Eisele, and A. Wendel, *Hoppe-Seyler's Z. Physiol. Chem.* **352,** 151 (1971).
[8] R. Ladenstein, O. Epp, K. Bartels, A. Jones, R. Huber, and A. Wendel, *J. Mol. Biol.* **134,** 199 (1979).
[9] L. Flohé, W. Günzler, E. Schaich, and F. Schneider, *Hoppe-Seyler's Z. Physiol. Chem.* **352,** 159 (1971).

TABLE II
PHYSICAL PROPERTIES OF GLUTATHIONE PEROXIDASE
FROM BOVINE RED BLOOD CELLS[a,b]

Molecular weight	84,500
Subunits	$4 \times 21,000$; 222 symmetry
Isoelectric point	5.6–6.0
pH optimum	8.8
Temperature optimum	42°
Amino acids	4×178
Prosthetic group	4 Se atoms (selenocysteine)
Crystal space group	C 2 monoclinic
Monomer radius	19 Å
Protein crystal density	1.32 g cm^{-3}

[a] Flohé et al.[7]
[b] Ladenstein et al.[8]

Common peroxidase substrates such as guaiacol or benzidine are not substrates. In contrast, a large variety of peroxides are accepted: H_2O_2, ethyl, cumene, *tert*-butyl, linoleic, linolenic acid hydroperoxides and their methyl esters, allopregnanolone 17α-hydroperoxide, pregnenolone 17α-hydroperoxide, cholesterol 7β-hydroperoxide, and thymine hydroperoxide, and even peroxidized DNA.

Inhibitors. Specific inhibitors for the enzyme have not been found. In our experience, Ag^+ inhibits the enzyme instantly and completely at 5 μM. Iodo- and chloroacetate irreversibly alkylate the reduced enzyme's selenium moiety, whereas iodoacetamide shows no effect. Potassium phosphate in a complex manner is reversibly inhibitory.

Kinetic Parameters. Because the kinetics of the reaction follow a ping-pong mechanism, true Michaelis constants for either substrate cannot be given as they depend on each other. However, apparent saturation can be reached experimentally for the peroxide substrate but not for GSH. The rate law follows the expression[10]:

$$\frac{[E_0]}{v} = \frac{\phi_1}{[ROOH]} + \frac{\phi_2}{[GSH]}$$

with $K_m^{ROOH} = (\phi_1/\phi_2)[GSH]$ and $K_m^{GSH} = (\phi_2/\phi_1)[ROOH]$. From the data given in Table III, the apparent Michaelis constant for any of the substrates can be calculated. The extrapolated maximal velocity is identical for all peroxide substrates that have been investigated.

[10] L. Flohé, G. Loschen, W. A. Günzler, and E. Eichele, *Hoppe-Seyler's Z. Physiol. Chem.* **353**, 987 (1972).

TABLE III

KINETIC PARAMETERS FOR GLUTATHIONE PEROXIDASE FROM
BOVINE ERYTHROCYTES[a]

Substrate	$\phi_1\ 10^{-8}\ M$ sec	$\phi_2\ 10^{-6}\ M$ sec
Hydrogen peroxide[b]	0.56	1.27
Hydrogen peroxide[c]	0.94	0.83
Hydrogen peroxide[d]	1.70	2.19
Ethyl hydroperoxide[d]	3.3	2.24
Cumene hydroperoxide[d]	7.8	2.24
tert-Butyl hydroperoxide[d]	13.5	2.24

[a] Data from Flohé et al.[10]

[b] Potassium phosphate, 0.05 M, pH 7.0.

[c] Morpholinopropanesulfonic acid buffer, 0.25 M, pH 7.7.

[d] Morpholinopropanesulfonic acid buffer, 0.25 M, pH 6.7.

Protein Structure[8]

The enzyme consists of four nearly spherical monomers with a radius of about 19 Å. The subunit with 182 amino acids is built up from a central core of two parallel and two antiparallel strands of pleated sheet surrounded by four α-helices. One of the helices is running antiparallel to the neighboring β-strands. A selenocysteine residue is located within a flat depression on the subunit surface surrounded by aromatic amino acid residues and represents the active site.

Distribution

Glutathione peroxidase activity has been demonstrated and purified in many species and organs of animals but not in plants. In rats, high levels are present in liver[11] and lower concentrations in red cells, heart, lung, and kidney. Within the hepatocyte about 70% of the total activity is accounted for by the cytosol, with 30% localized within the mitochondrial matrix.

Nonselenium-Dependent Glutathione Peroxidase

This recently discovered activity,[12] seems to be due to a reaction catalyzed by glutathione S-transferases.[13] Unlike the peroxidase, the transferases do not metabolize H_2O_2 but are reactive with several organic hydroperoxides.

[11] A. L. Tappel, this series, Vol. 52, Article [53].

[12] R. A. Lawrence and R. F. Burk, *Biochem. Biophys. Res. Commun.* **71**, 952 (1976).

[13] This volume, Articles [27] and [51].

Acknowledgment

Substantial parts of the author's work were supported by the Deutsche Forschungsgemeinschaft (Grant We 686/5). A. Grossmann and W. Lödige contributed many details to the enzyme preparation.

[45] Carboxylesterases-Amidases

By Eberhard Heymann and Rolf Mentlein

Rat liver contains a number of carboxylesterases (EC 3.1.1.1) of the serine hydrolase type.[1-3] Their physiological role remains unknown, although most are capable of cleaving monoglycerides of long chain fatty acids[1] and, therefore, can also be classified as monoacylglycerol lipases (EC 3.1.1.23). Because most rat liver carboxylesterases are also active on aromatic amides,[2] we use here the term "carboxylesterase/amidase" and list the isoelectric point (e.g., pI 6.0) to discriminate between isoenzymes or multiple forms. This preliminary nomenclature seems to be more convenient than others that have been used before. Table I both summarizes the literature on purification procedures and compares the nomenclature used by the authors.

Here we report the simultaneous purification of five of the most prominent rat liver carboxylesterases-amidases. All of these enzymes are found in the microsomal fraction. As a side product of the procedure described, a dipeptidyl aminopeptidase (EC 3.4.14.—; dipeptidyl aminopeptidase IV, postproline dipeptidyl peptidase) is obtained that also belongs to the group of serine hydrolases.[4]

Assay Methods

For an optimal discrimination between the various carboxylesterases, the three substrates (methyl butyrate, 4-nitrophenyl acetate, and acetanilide) should be used throughout the isolation procedure although it is not necessary to test every fraction with each substrate. The acetanilide assay is described in Article [52] of this volume.

[1] R. Mentlein, S. Heiland, and E. Heymann, *Arch. Biochem. Biophys.* **200**, 547 (1980).

[2] E. Heymann, *in* "Enzymatic Basis of Detoxication" (W. B. Jakoby, ed.), Vol. 2, p. 291. Academic Press, New York, 1980.

[3] W. Junge and K. Krisch, *CRC Crit. Rev. Toxicol.* **3**, 371 (1975).

[4] E. Heymann and R. Mentlein, *FEBS Lett.* **91**, 360 (1978).

TABLE I

CORRELATION OF THE NOMENCLATURE FOR RAT LIVER CARBOXYLESTERASES-AMIDASES
AND REPORTS ON PURIFICATION PROCEDURES

Other reports	Enzymes in the nomenclature of this report				
	Carboxyl-esterase, pI $5.2^{m,n}$	Carboxyl-esterase, pI $5.6^{m,n}$	Carboxyl-esterase, pI 6.0^o	Carboxyl-esterase, pI 6.2^m	Carboxyl-esterase, pI 6.4^m
Earlier reports from our laboratory^{a,b,c}		Esterase/ amidase $EA_1/EA_2{}^n$	Esterase $E_1{}^o$	Esterase $E_2{}^p$	
Ljungquist and Augustinsson^d			Esterase $a_1(=e_1)^m$	Esterase $b_1{}^m$ e_3	e_2
Akao and Omura^e		Amidase^n			
Haugen and Suttie^f			Esterase A^q	Esterase $B_1{}^q$	
Ikeda et al.^{g,h}		Esterase I^o	Esterase II°	Monoacylglycerol lipase^r	
Raftell et al.^{i}		Esterase $e_1{}^s$			
Kaneko et al.^{j}	Esterase LI^t				
Ishitani et al.^{k}			Esterase $E-1^u$		
Oerlemans et al.^{l}				Glycerol monoester hydrolase^r	

^{a} R. Arndt and K. Krisch, *Hoppe-Seyler's Z. Physiol. Chem.* **353**, 589 (1972).

^{b} R. Arndt, E. Heymann, W. Junge, and K. Krisch, *Eur. J. Biochem.* **36**, 120 (1973).

^{c} R. Arndt, H.-E. Schlaak, D. Uschtrin, D. Südi, K. Michelssen, and W. Junge, *Hoppe-Seyler's Z. Physiol. Chem.* **359**, 641 (1978).

^{d} A. Ljungquist and K.-B. Augustinsson, *Eur. J. Biochem.* **23**, 303 (1971).

^{e} T. Akao and T. Omura, *J. Biochem. (Tokyo)* **72**, 1245 (1972).

^{f} D. Haugen and J. Suttie, *J. Biol. Chem.* **249**, 2717 (1974).

^{g} Y. Ikeda, K. Okamura, T. Arima, and S. Fujii, *Biochim. Biophys. Acta* **487**, 189 (1977).

^{h} Y. Ikeda, K. Okamura, and S. Fujii, *Biochim. Biophys. Acta* **488**, 128 (1977).

^{i} M. Raftell, K. Berzins, and F. Blomberg, *Arch. Biochem. Biophys.* **181**, 534 (1977).

^{j} A. Kaneko, Y. Yoshida, K. Enomoto, T. Kaku, K. Hirate, and T. Onoé, *Biochim. Biophys. Acta* **582**, 185 (1979).

^{k} R. Ishitani, H. Kin, T. Kuwae, K. Moroi, and T. Satoh, *Jpn. J. Pharmacol.* **29**, 413 (1979).

^{l} M. C. Oerlemans, M. M. Geelhoed-Mieras, and W. C. Hülsmann, *Biochem. Biophys. Res. Commun.* **78**, 1130 (1977).

$^{m-u}$ Substrates used for assay during purification: ^{m} 4-nitrophenyl acetate; ^{n} acetanilide; ^{o} methyl butyrate; ^{p} 2-nitrophenyl acetate; ^{q} phenyl butyrate; ^{r} 1-monooleylglycerol; ^{s} 1-naphthyl propionate; ^{t} 2-naphthyl acetate; ^{u} isocarboxazide.

Hydrolysis of Methyl Butyrate

In weakly buffered solutions, the amount of butyric acid released by enzymatic hydrolysis can be recorded continuously by the pH-stat technique.[5]

Five milliliters of 5 mM methyl butyrate[6] are pipetted into a stoppered vessel thermostatted to 30°. The pH is adjusted to 8.0 and maintained by continuous addition of 0.04 M NaOH from a 0.25-ml burette operated by an automatic titrator (e.g., TTT 1 from Radiometer, Copenhagen, Denmark, with autoburette ABU 13 and recorder). A blank, caused by spontaneous hydrolysis and CO_2 adsorption, is recorded for 3 min. Then the enzymatic reaction is started by addition of 10–500 μl of sample. One unit of carboxylesterase activity corresponds to 1 μmol of NaOH consumed per minute. The NaOH consumption is linear with time up to at least 10 μmol. With methyl butyrate concentrations above 2 mM, carboxylesterase pI 6.0 shows substrate inhibition.[5] However, the activities obtained with the 5 mM standard substrate are less than 10% below those obtained with the optimal 2 mM solution.

Hydrolysis of 4-Nitrophenyl Acetate

The yellow phenol liberated from 4-nitrophenyl acetate at alkaline pH is observed spectrophotometrically at 405 nm[7] in a filter photometer (e.g., Eppendorf, Hamburg, FRG) or spectrophotometer (e.g., Hitachi 100-40, Tokyo, Japan). This assay is well suited for on-line registration and calculation with a computer (e.g., Commodore 3032, Palo Alto, CA) equipped with a digital voltmeter (e.g., Micrologic 415, München, FRG). Another advantage of the assay is that all rat liver carboxylesterases cleave this ester.[1]

A 0.5 mM solution of 4-nitrophenyl acetate is made daily by dissolving 18.1 mg of the ester in 1 ml acetonitrile and adding water to 100 ml. Portions of 1.8 ml of the substrate solution at 30° are pipetted into a thermostatted cuvette with an optical path length of 1 cm. After addition of 0.2 ml of 0.5 M Tris-HCl, pH 8.0, the blank caused by spontaneous hydrolysis is registered for 60 sec. The enzymatic reaction is started by addition of 10–100 μl of sample; the release of 4-nitrophenol is registered for an additional 30 to 60 sec. If a recorder is used instead of the computer, the period of recording should be twice the time period mentioned. The increase in absorbance is linear with time up to a difference of 1.0. For the calculation, a molar absorbance of 16,400 liters mol^{-1} cm^{-1} is used.

---

[5] R. Arndt, E. Heymann, W. Junge, and K. Krisch, *Eur. J. Biochem.* **36**, 120 (1973).

[6] The ester is dissolved in water because traces of most organic solvents greatly influence the catalytic activity.

[7] Adapted from K. Krisch, *Biochim. Biophys. Acta* **122**, 265 (1966).

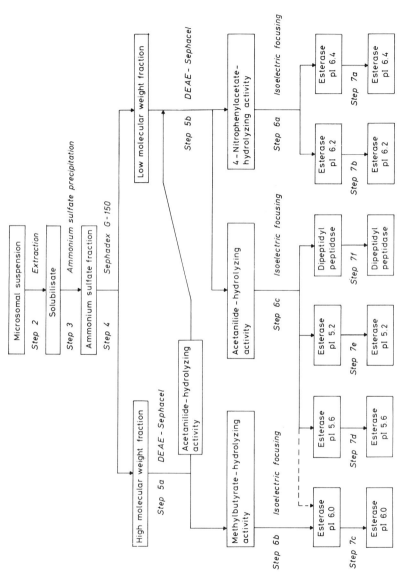

FIG. 1. Flow chart of procedure for simultaneous purification of five rat liver carboxylesterases.

Protein Estimation

In general, the absorbance at 280 nm provides a sufficient basis for the estimation of protein concentration during the purification procedure. We use a factor of 0.72 that has been obtained with a highly purified pig liver carboxylesterase.[8] A_{280} measured with a 1-cm light path and multiplied by the factor, gives the protein concentration in mg/ml. For the estimation of protein in microsomes, we use the biuret procedure as described by Alt et al.[9]

Procedure for the Simultaneous Purification of Five Rat Liver Carboxylesterases[1]

General. Figure 1 presents a general survey of the procedure. All steps are performed at temperatures of 0° to 4°. The columns for chromatography or isoelectric focusing are thermostatted at 4°. Although the procedure was developed with livers of male Wistar rats, similar results are obtained with females of this strain, and from male or female Sprague-Dawley rats. All gels used for preparative chromatography are prepared, equilibrated, and regenerated as described by the manufacturer (Pharmacia).

Step 1. Isolation of Microsomes. The procedure for the isolation of microsomes is essentially that of Krisch.[10] Male Wistar rats of 150–250 g, fasted for 24 hr, are killed by decapitation. The livers of 50–100 rats are collected in ice-cold 0.25 *M* sucrose. Fat and connective tissues are removed. To 150 g of liver, 500 ml of 0.25 *M* sucrose are added and the mixture is homogenized with a blender (e.g., Ultra-Turrax T 45/6, Janke und Kunkel, Staufen, FRG) at a rotatory speed of 10,000 rpm for 1 min, (the distance of the prongs that rotate in a shell of knives is 37 mm). The homogenate is diluted to 1 liter with 0.25 *M* sucrose, homogenized for another 20 sec, and centrifuged for 30 min at 16,000 *g*. The sediments are discarded and the supernatant liquid is centrifuged for 60 min at 105,000 *g*. Using a glass homogenizer with Teflon pestle, the sediment (microsomes) is suspended in 0.1 *M* Tris-HCl at pH 8.5 so that 1 g of suspension corresponds to 2 g of fresh liver. This microsomal suspension is kept at −20° until the enzyme isolation starts with Step 2.

Step 2. Extraction. The microsomal suspension, 125 ml, is thawed and treated in an ice bath with the blender described in Step 1 for 5 × 30 sec in order to reduce the size of the microsomal particles.[11] Pauses of 90 sec are

[8] M. Kunert and E. Heymann, *FEBS Lett.* **49**, 292 (1975).
[9] J. Alt, K. Krisch, and P. Hirsch, *J. Gen. Microbiol.* **87**, 260 (1975).
[10] K. Krisch, *Biochem. Z.* **337**, 531 (1963).
[11] E. Heymann, W. Junge, K. Krisch, and G. Marcussen-Wulff, *Hoppe-Seyler's Z. Physiol. Chem.* **355**, 155 (1974).

TABLE II
GRADIENTS FOR STEPS 5A AND 5B

Chamber no.	1 M NaCl in 10 mM Tris-HCl, pH 8.0 (ml)	10 mM Tris-HCl, pH 8.0 (ml)	Resulting NaCl concentration (mM)
1	12	388	30
2	28	372	70
3	60	340	150
4	160	240	400

necessary between each 30-sec pulse to avoid increasing the temperature above 10°. To the mixture are added 250 ml of 0.1 M Tris-HCl at pH 8.5, containing 3.75 g of saponin (Merck, Darmstadt, FRG), and the suspension is stirred for 60 min (final concentration: 1 g saponin/100 ml). After centrifugation of 2 hr at 105,000 g, the supernatant fluid contains the bulk of the carboxylesterase-amidase activities.

The yield of solubilized carboxylesterases-amidases can be improved if digitonin is used instead of saponin.

Step 3. Ammonium Sulfate. Within 1 hr of solubilization, 200 ml of ammonium sulfate solution, saturated at 0°, are added dropwise with stirring into 300 ml of the preparation from Step 2. The precipitate is collected by centrifugation (30 min at 10,000 g) and discarded. Solid ammonium sulfate, 103 g, is added in small portions to the supernatant fluid. The salt precipitation step should be completed within 60 to 90 min. The final ammonium sulfate concentration (70% saturation) is controlled by titration with BaCl$_2$.[12] After centrifugation (30 min at 15,000 g), the carboxylesterase-containing sediment is dissolved in 15 ml of 10 mM Tris-HCl at pH 8.

Step 4. Sephadex G-150. The solution from Step 3 is applied to a gel filtration column (5.0 × 96 cm) filled with Sephadex G-150. The column is equilibrated and eluted with 10 mM Tris-HCl at pH 8. The eluate is collected in fractions of 6 ml that are assayed with both methyl butyrate and 4-nitrophenyl acetate. 4-Nitrophenyl acetate activity is found in two peaks in the range of the fractions 60 through 120. The first peak with proteins of higher molecular weight (about fractions 60 to 90) contains the bulk of methyl butyrate-hydrolyzing activity and is further purified in Step 5a. The low-molecular-weight portion (fractions 91 to 115) is retained for Step 5b.

Step 5a. DEAE-Sephacel (High-Molecular-Weight Fraction). A column (2.6 × 33 cm) of DEAE-Sephacel is equilibrated with 10 mM Tris-HCl at

[12] H. V. Bergmeyer, G. Holz, E. M. Kander, H. Möllering, and O. Wieland, *Biochem. Z.* **333,** 471 (1961).

pH 8.0 and charged with the high-molecular-weight fraction obtained in Step 4. Four chambers of the gradient former as described by Peterson and Sober[13] are filled according to Table II, and the column is eluted with the resulting concave gradient of NaCl. If the volume of the applied enzyme solution is about 100 ml, and if fractions of 9.5 ml are collected, the 4-nitrophenyl acetate-hydrolyzing activity is found at about fractions 58 to 115, methyl butyrate-hydrolyzing activity in fractions 70 to 115, and acetanilide-cleaving enzymes in fractions 90 to 115. Based on the elution–activity profile, three ranges with enzyme activity are pooled.

1. The first 4-nitrophenyl acetate-hydrolyzing peak (about fraction 63) contains low amounts of carboxylesterases pI 6.2 and 6.4. Because the bulk of these esterases is obtained in Step 5b, this preparation may be discarded.

2. The steep methyl butyrate-hydrolyzing peak (about fraction 80) mainly consists of carboxylesterase-amidase pI 6.0. It is divided from the overlapping acetanilide-cleaving peak at the intersection of the steep decline of the first and the steep incline of the second activity. The pooled fractions are dialyzed against 5 mM Tris-HCl at pH 8.0 (2 × 5 liters) for 18 hr and are kept frozen (−20°) until the further separation in Step 6b.

3. The remaining acetanilide-hydrolyzing peak at about fraction 105 is combined with the corresponding peak of Step 5b. It contains carboxylesterase-amidase pI 5.6. The descending part of this peak also contains the main portion of the dipeptidyl aminopeptidase IV.

Step 5b. DEAE-Sephacel (Low-Molecular-Weight Fraction). The procedure in this step parallels exactly that of Step 5a, except that the low-molecular-weight portion obtained in Step 4 is applied to the ion-exchange column. When the eluate fractions are assayed with 4-nitrophenyl acetate, activity is found in fractions 55–130 and, with acetanilide, in fractions 80–130. The steep esterase peak at about fraction 63 is pooled, dialyzed as described in Step 5a, and further purified by isoelectric focusing (Step 6a). The acetanilide-hydrolyzing peak at about fraction 105 is combined with the corresponding peak of Step 5a and also dialyzed as previously described. After dialysis, the latter enzyme solution is frozen at −20° until further purified in Step 6c.

Step 6. Isoelectric Focusing

GENERAL PROCEDURE. An apparatus for carrier-free isoelectric focusing in sucrose gradients of a volume of 440 ml (LKB 8122) is used in Steps 6a–c. The apparatus is filled and emptied with a peristaltic pump at a speed of 50 ml/hr. Only specially purified sucrose "for density gradients" (Merck) is used. The lower electrode (anode) buffer is a mixture of 60 g sucrose, 60 ml water, and 4 ml of 1 M H$_3$PO$_4$. Over this, the enzyme and ampholyte-containing sucrose gradient is layered, and the cathode fluid

[13] E. A. Peterson and H. A. Sober, *Anal. Chem.* **31**, 857 (1959).

(50 mM NaOH) is pumped on top. The gradient is formed with a two-chamber gradient mixer (LKB): The first chamber contains a solution made of 108 g sucrose, 5 ml of ampholyte (Ampholine, LKB), and 136 ml water or dialyzed enzyme solution. The second chamber contains a solution of 10.8 g sucrose, 5 ml Ampholine, and 204 ml of the dialyzed enzyme fraction of Step 5. If the volume of the enzyme fraction is less than 204 ml, it is filled up with water; if greater, the excess volume is filled up to 136 ml with water and is applied to the first chamber. The enzymes are focused for 1–2 days at increasing voltage (max. 800 V), so that the power applied never exceeds 4 W. Most of the time, the power remains below 2 W. Focusing is terminated if precipitation of one of the enzyme bands becomes visible. The eluate is collected in fractions of 40 drops (about 2.2 ml), and the pH of each is determined at 4°.

Step 6a. Isoelectric Focusing of the Carboxylesterases with pI Values > 6. After dialysis, the first carboxylesterase peak of the DEAE-Sephacal column described in Step 5b is applied to the isoelectric focusing gradient. Each chamber of the gradient former contains 4.5 ml of Ampholine pH 5–7 and 0.5 ml of Ampholine pH 3.5–10. The eluate fractions are assayed with 4-nitrophenyl acetate. The largest activity peak is that of the pI 6.2 enzyme. The fractions in this peak are pooled as are those of the pI 6.4 enzyme; both are retained for the final gel filtration (Steps 7a and b). The small peak at pI 6.0 is discarded; it contains carboxylesterase-amidase pI 6.0 of lesser purity than that obtained in Step 6b.

Step 6b. Isoelectric Focusing of the Methyl Butyrate Cleaving Enzyme. The dialyzed methyl butyrate-hydrolyzing peak from Step 5a is applied in the same manner and with the same amounts and types of Ampholine as in Step 6a. A single methyl butyrate-hydrolyzing peak is found in the eluate at pH 6.0. This represents carboxylesterase pI 6.0 and is subjected to gel filtration in Step 7c.

Step 6c. Isoelectric Focusing of the Acetanilide-Cleaving Enzymes. The combined acetanilide-hydrolyzing fractions of Steps 5a and b are applied to the isoelectric focusing column as described earlier. Sometimes the total volume of these fractions exceeds the limited sample volume (340 ml) of the apparatus. In that event, the enzyme solutions must be concentrated by lyophilization or ultrafiltration. The gradient contains 10 ml of ampholyte pH 4–6.5, (e.g. Pharmalyte, Pharmacia; or a mixture of 5 ml Ampholine pH 4–6 and 5 ml Ampholine pH 5–7). The eluate fractions of this step are assayed with acetanilide and, if desired, with glycylprolyl-2-naphthylamide[14]; with the latter substrate a dipeptidyl aminopeptidase peak with a pI at about 4.8 is found. The acetanilide assay discloses a

---

[14] J. K. McDonald, P. X. Callahan, S. Ellis, and R. E. Smith, *in* "Tissue Proteinases" (A. J. Barett and T. J. Dingle, eds.), p. 69. Elsevier/ North-Holland Biomedical Press, Amsterdam, 1971.

TABLE III
PURIFICATION OF THE CARBOXYLESTERASE-AMIDASE ACTIVITIES

Fraction	Isoenzyme	Protein (mg)	Total activity (μmol min^{-1})			Specific activity (μmol min^{-1} mg^{-1})		
			Methyl butyrate	Acetanilide	4-Nitrophenyl acetate	Methyl butyrate	Acetanilide	4-Nitrophenyl acetate
Microsomes		8750	33,250	29.5	21,150	3.8	0.0034	2.42
Solubilized fraction		595	23,750	52.2	11,150	39.9	0.088	18.8
Salt fractionation		500	21,950	47.1	8,200	43.9	0.108	19.4
Sephadex G-150	Low-molecular-weight fraction	165	800	35.3	2,700	4.8	0.255	16.4
	High-molecular-weight fraction	295	20,950	5.9	3,700	71.0	0.020	12.5
DEAE-Sephacel	Esterases pI 5.2 and 5.6	8	140	20.6	92	17.0	2.58	11.5
	Esterase pI 6.0	58	11,310	—	1,850	195.0	—	31.9
	Esterases pI 6.2 and 6.4	30	—	—	1,700	3.0	—	55.7
Isolectric focusing followed by Sephacryl S-200	Esterase pI 5.2	1.3	25	0.71	—	18.8	0.55	5.5
	Esterase pI 5.6	3.2	82	19.6	72	25.6	6.13	22.4
	Esterase pI 6.0	16	6,000	—	710	375.0	—	44.4
	Esterase pI 6.2	5.9	—	—	550	5.1	—	94.0
	Esterase pI 6.4	2.6	—	—	335	2.0	—	131.0

major peak with p*I* 5.6 and a minor one with p*I* 5.2. For unknown reasons, the latter peak is sometimes absent. This focusing column always contains some carboxylesterase p*I* 6.0 that may be found if the eluate is assayed with methyl butyrate.

Step 7a–f. Sephacryl S-200. A column (2.6 × 95 cm) is packed with Sephacryl S-200 (Pharmacia) and equilibrated and operated with 10 m*M* Tris-HCl at pH 8.0. Each of the six enzymes obtained in Steps 6a–c are subjected to passage through this column. The column is designed to remove ampholyte and sucrose, but also separates remaining traces of the high-molecular-weight enzymes (carboxylesterase p*I* 6.0 and dipeptidyl peptidase) from those with lower molecular weight, and vice versa (compare Table V).

The yields and specific activities estimated after the individual purification steps are summarized in Table III.

Properties of the Carboxylesterases-Amidases

Purity of the Enzymes. The five isolated carboxylesterases-amidases have been completely separated from each other as can be demonstrated by analytical polyacrylamide gel electrophoresis and isoelectric focusing on flat gels that are stained for esterase activity (1). It is also evident from these gels that carboxylesterase-amidase p*I* 5.6 remains heterogenous; this is supported by an analysis of the terminal amino acids.[1] Only the enzymes with p*I* 5.6 and p*I* 6.0 appear as homogeneous proteins on

TABLE IV
SUBSTRATE SPECIFICITY OF HIGHLY PURIFIED RAT LIVER CARBOXYLESTERASES[1]

	Specific activities (μmol min^{-1} mg^{-1})				
	Esterase p*I* 5.2	Esterase p*I* 5.6	Esterase p*I* 6.0	Esterase p*I* 6.2	Esterase p*I* 6.4
Acetanilide	0.55	6.1	0.0	0.0	0.0
Butanilicaine[a]	0.0	2.0	8.3	0.0	0.0
Methyl butyrate	18.2	25.6	410[b]	5.1	2.0
4-Nitrophenyl acetate	5.5	22.4	44.4	94	131
1-Monobutyryl glycerol	0.0	8.2	11.9	6.0	0.0
1-Monolauryl glycerol	0.0	2.0	8.0	20.5	15.4
1-Monooleyl glycerol	0.29	1.59	0.67	3.21	3.55

[a] *N*-(Butylaminoacetyl)-2-chloro-6-methylanilide.

[b] At 2 m*M* substrate concentration.

TABLE V
PHYSICAL PROPERTIES OF THE FIVE CARBOXYLESTERASES-AMIDASES[1]

	Relative mobilities in polyacrylamide gel electrophoresis (gel concentration = 7.5%)	Subunit weight[a]	Molecular weight[b]	N-terminal amino acid	C-terminal amino acid
Esterase pI 5.2	0.62	58,000	60,000	Glycine	
Esterase pI 5.6	0.45 0.43 0.41	61,000	60,000[c]	Glycine, tyrosine, phenylalanine, and others	Leucine (and others?)
Esterase pI 6.0	0.21	58,000	180,000	Tyrosine	-Ala-Val-Leu
Esterase pI 6.2	0.35	61,000	60,000		
Esterase pI 6.4	0.32	61,000	60,000		

[a] Estimated by sodium dodecyl sulfate–polyacrylamide gel electrophoresis.

[b] Estimated by gel chromatography.

[c] This enzyme associates to trimers at higher concentrations; R. Arndt, H. E. Schlaak, D. Uschtrin, D. Südi, K. Michelssen, and W. Junge, *Hoppe-Seyler's Z. Physiol. Chem.* **359**, 641 (1978).

sodium dodecyl sulfate gels. Each of the other carboxylesterase-amidase preparations show minor impurities with this technique. Active site titration with diethyl(4-nitrophenyl) phosphate[5] reveals a purity[2] of >90% for carboxylesterase pI 6.0, and of about 80% for the pI 5.6 enzyme on the basis of a subunit weight of 60,000 and biuret protein estimation.[9]

Stability. A solution of purified carboxylesterase pI 6.0 shows no significant loss of activity in 6 hr at 30° (pH 8.0). Solutions of all carboxylesterases-amidases may be kept at −18° without loss of activity for months.

Inhibitors. Active site-directed inhibitors of serine hydrolases,[2,3] e.g., diethyl(4-nitrophenyl) phosphate, inactivate the five carboxylesterases-amidases rapidly and irreversibly.[5,15] The organophosphorus diesters, bis(4-nitrophenyl) phosphate and bis(4-cyanophenyl) phosphate,[15] act similarly, but in contrast to the well known toxic inhibitors of serine

[15] E. Brandt, E. Heymann, and R. Mentlein, *Biochem. Pharmacol.* **29**, 1927 (1980).

hydrolases, the diesters are rather specific for liver carboxylesterases-amidases and exhibit low toxicity.[16] Bis(4-cyanophenyl) phosphate shows a preference for the pI 5.6 esterase.[15]

Specificity and Physical Properties. Some data on the specificity of the five carboxylesterases/amidases are compiled in Table IV. A broader review also has been published.[2] See Ref. 4 for the specificity of the dipeptidyl peptidase.

Table V summarizes some physical properties of the six purified hydrolases. The pI 5.6, pI 6.0, and pI 6.2 carboxylesterases show differing peptide maps after cleavage with trypsin or CNBr. Carboxylesterase pI 6.4 is a glycosylated variant of carboxylesterase pI 6.2. All esterases except those with pI 5.2 and pI 6.4 are essentially free of bound carbohydrates (R. Mentlein and E. Heymann, unpublished).

[16] E. Heymann, K. Krisch, H. Büch, and W. Buzello, *Biochem. Pharmacol.* **18**, 801 (1969).

[46] Microsomal Epoxide Hydrolase

By Thomas M. Guenthner, Philip Bentley, and Franz Oesch

Microsomal epoxide hydrolase (EC 3.3.2.3) catalyzes the conversion of epoxides to glycols. Because many epoxides formed from exogenous compounds are potent electrophiles, capable of covalent interaction with cellular molecules, the enzyme represents a key step in the detoxication of reactive intermediates.[1-5] Although found primarily in the microsomal fraction, the enzyme is also present in lesser amounts in other cell membranes.[6] The microsomal enzyme should not be confused with another epoxide hydrolase activity, found primarily in the cytosolic fraction, which differs greatly from membrane-bound enzyme in substrate specificity[7-9] and immunological properties.[10] As with other membrane-

[1] F. Oesch, *Xenobiotica* **3**, 305 (1973).

[2] F. Oesch, *Prog. Drug Metab.* **3**, 253 (1979).

[3] A. Y. H. Lu and G. Miwa, *Annu. Rev. Pharmacol. Toxicol.* **20**, 513 (1980).

[4] T. M. Guenthner and F. Oesch, *in* "Polycyclic Hydrocarbons and Cancer" (H. V. Gelboin and P. O. P. Ts'o, eds.), Vol. 3. Academic Press, New York 1981.

[5] D. M. Jerina, P. M. Dansette, A. Y. H. Lu, and W. Levin, *Mol. Pharmacol.* **13**, 342 (1977).

[6] P. Stasiecki, F. Oesch, G. Bruder, E. Jarasch, and W. W. Franke, *Eur. J. Cell Biol.* **21**, 79 (1980).

[7] K. Ota and B. D. Hammock, *Science* **207**, 479 (1980).

[8] S. H. Mumby and B. D. Hammock, *Pestic., Biochem. Physiol.* **11**, 274 (1979).

[9] F. Oesch and M. Golan, *Cancer Lett.* **9**, 169 (1980).

[10] T. M. Guenthner, B. D. Hammock, U. Vogel, and F. Oesch, *J. Biol. Chem.* **256**, 3163 (1981).

bound enzymes, it is highly lipophilic and easily forms aggregates in solution.[11-13] Its physical properties, therefore, present special problems for purification.

The general scheme of purification presented here consists of solubilization from microsomes by a nonionic detergent; chromatography with DEAE-cellulose and phosphocellulose, which effect purification on the basis of charge; chromatography with butylsepharose, which effects purification on the basis of hydrophobicity; and removal of detergent by a second phosphocellulose step.[11] Several other purification procedures for this enzyme have been published;[12-16] they differ from the technique presented here mainly in the manner in which, or whether, the detergent is removed.

Assay Method

Principle. Enzyme activity is assayed by the conversion of radiolabeled epoxide substrates to dihydrodiols and the subsequent separation of products from substrate by simple solvent extraction.[17-19] We routinely use tritiated benzo[a]pyrene 4,5-oxide (BPO) and styrene 7,8-oxide (STO) to monitor enzyme activity during purification. One enzyme unit is defined as the amount catalyzing the hydration of 1 μmol of styrene oxide in 1 min under the defined conditions.[19]

Reagents

Tris-HCl buffer, 50 mM, pH 9.0

[³H]Styrene 7,8-oxide (specific activity, 0.34 μCi/μmol), 0.4 μmol in 10 μl acetonitrile

[³H]Benzo[a]pyrene 4,5-oxide (specific activity, 3.1 μCi/μmol), 75 nmol in 10 μl acetonitrile

Procedure. The normal assay system consists of 25 or 50 μl of enzyme (equivalent to 100 to 300 μg of microsomal protein) and 50 mM Tris-HCl at pH 9.0 in a final volume of 200 μl (STO) or 500 μl (BPO). In contrast to

[11] P. Bentley and F. Oesch, *FEBS Lett.* **59**, 291 (1975).

[12] P. Bentley, F. Oesch, and A. T. Sugita, *FEBS Lett.* **59**, 296 (1975).

[13] A. Y. H. Lu, D. Ryan, D. M. Jerina, J. W. Daly, and W. Levin, *J. Biol. Chem.* **250**, 8283 (1975).

[14] A. Y. H. Lu and W. Levin, this series, Vol. 52, p. 193.

[15] R. G. Knowles and B. Burchell, *Biochem. J.,* **163**, 381 (1977).

[16] F. P. Guengerich, P. Wang, M. B. Mitchell, and P. S. Mason, *J. Biol. Chem.* **254**, 12248 (1979).

[17] F. Oesch. D. M. Jerina, and J. Daly, *Biochim. Biophys. Acta* **227**, 685 (1971).

[18] H. U. Schmassmann, H. R. Glatt, and F. Oesch, *Anal. Biochem.* **74**, 94 (1976).

[19] F. Oesch, *Biochem. J.,* **139**, 77 (1974).

our initial protocol,[16] Tween 80 is no longer included.[18] Included are 0.4 μmol styrene oxide or 75 nmol benzo[a]pyrene 4,5-oxide, each in 10 μl acetonitrile. Samples are incubated for 5 min at 37°C in stoppered tubes, and the incubation is terminated and extracted with 3.5 ml light petroleum and 0.5 ml dimethylsulfoxide (for BPO), or 2.5 ml light petroleum (for STO). Samples are extracted once more with 2.5 ml light petroleum (STO), or twice more with 3.5 ml light petroleum (BPO). The diol product remaining in the aqueous layer is then extracted into 1 ml ethyl acetate. The ethyl acetate layer, 500 μl, is assayed for tritiated diol by liquid scintillation counting.

[³H]STO is available from the Radiochemical Centre, Amersham, Bucks., UK. [³H]BPO is not commercially available; both substrates are fairly easily synthesized.[17,20] Alternative nonradiometric assays have been published which measure a fluorescent or absorbant product,[21-23] or which oxidize the product with alcohol dehydrogenase and monitor the appearance of NADH spectrophotometrically.[24] References to additional assays can be found in publications quoted throughout this report.

Material. DEAE-cellulose (Whatman DE23) and phosphocellulose (Whatman P11) are obtained from Whatman Biochemicals, Maidstone, Kent, UK. Cutscum (isooctylphenoxypolyethanol) is obtained from Fisher Scientific, Pittsburgh, PA. Butyl-Sepharose is a product of Miles Laboratories, Elkhart, IN.

Preliminary Steps. Adult male Sprague-Dawley rats (200–300 g) are used. We have found that pretreatment of the animals with *trans*-stilbene oxide doubles or triples the yield of enzyme without affecting any of the properties we have investigated.[25] Therefore, we routinely inject rats intraperitoneally with 400 mg/kg *trans*-stilbene oxide (0.5 ml of 200 mg/ml in corn oil) 96, 72 and 48 hr before sacrifice. Fifty rats are used for a normal preparation. Columns are also prepared beforehand, because washing and pH equilibration of the phosphocellulose column can take 5 to 7 days, whereas preparation of the DEAE-cellulose column requires only 2 to 3 days. DEAE-cellulose (Whatman DE23) is suspended in water and packed in a glass column (5 × 100 cm) to a bed volume of 800 ml. The packing is washed with approximately 2.5 liters of 0.5 M NaCl/0.5 M NaOH, until the yellow discoloration disappears. The material should not remain overnight in this solution. The column is unpacked and the cel-

[20] P. Dansette and D. M. Jerina, *J. Am. Chem. Soc.* **96**, 1224 (1974).

[21] R. N. Armstrong, W. Levin, and D. M. Jerina, *J. Biol. Chem.* **255**, 4698 (1980).

[22] P. M. Dansette, G. C. DuBois, and D. M. Jerina, *Anal. Biochem.* **97**, 340 (1979).

[23] R. B. Westkaemper and R. P. Hanzlik, *Anal. Biochem.* **102**, 63 (1980).

[24] F. P. Guengerich and P. S. Mason, *Anal. Biochem.* **104**, 445 (1980).

[25] H. U. Schmassmann and F. Oesch, *Mol. Pharmacol.* **14**, 834 (1978).

lulose is washed in distilled water until the wash liquid is pH 9.0 or below. At this time, the fine grains are also decanted. The column is packed again and washed with 6 liters of 50 mM sodium phosphate, pH 7.0, followed by 4 to 6 liters of 5 mM phosphate buffer pH 7.0 or until the effluent reaches pH 7.0 and the ionic strength of the washing buffer is equal to that of the effluent. Phosphocellulose is prepared as a thick slurry in 200 ml of water that is stirred for 10 min with 0.5 M NaOH. It is washed with water on a Buchner funnel until approximately pH 8.0. After additional stirring for 10 min in 0.5 M HCl, the suspension is washed with water until the pH is greater than 5.0. It is packed in a glass column (5 × 20 cm) with a glass wool plug and washed with 2 liters of 50 mM sodium phosphate, pH 7.4, followed by 2 liters of 5 mM phosphate buffer, pH 7.4, or until a constant pH of 7.4 and a constant ionic strength is achieved in the effluent. This washing normally proceeds very slowly.

Purification Procedure

Steps 1–3. Preparation and Solubilization of Microsomes. Fifty pre-treated rats are killed by cervical dislocation and their livers are removed and placed in ice-cold homogenizing buffer (10 mM sodium phosphate, pH 7.0, containing 0.25 M sucrose and 0.3 mM EDTA). All further procedures are carried out either on ice or in a 4°C coldroom. After the buffer is decanted, the livers are weighed into beakers in 100-g portions. Each portion is homogenized in 300 ml of homogenizing buffer using a Braun MX 3 tissue homogenizer; bursts of 20 sec, interspersed with pauses of 20 sec, are used. The homogenate is centrifuged at 10,000 g for 15 min and the resultant supernatant fraction is centrifuged for 1 hr at 100,000 g; the microsomal fractions are combined. With two ultracentrifuges equipped with type 35 rotors (total volume 500 ml per rotor) we normally require about 6 hr for killing the animals, extracting and homogenizing the livers, and harvesting of microsomes. The microsomes are suspended in a Dounce homogenizer in 10 mM sodium phosphate, pH 7.0, containing 0.3 mM EDTA and 1% Cutscum (isooctylphenoxypolyethanol) in a volume equivalent to that of the original 10,000 g supernatant fraction. This suspension is stirred for 20 min. Finely ground ammonium sulfate is slowly added to a final concentration of 140 g per liter and the suspension is stirred 20 minutes longer. The suspension is centrifuged for 20 min at 10,000 g, yielding three phases: a floating thick brown layer, a red intermediate phase, and a pink pellet. The intermediate phase is siphoned off and discarded. The other two phases are combined and dissolved in a minimal volume of 5 mM sodium phosphate, pH 7.0, containing 0.3 mM EDTA, and dialyzed against 5 changes of 4 liters each of the same buffer.

SUMMARY OF PURIFICATION OF RAT MICROSOMAL EPOXIDE HYDROLASE[a]

Purification step	Total protein (mg)	Total enzyme (units)	Specific activity (units/mg)
1. Extract (10,000 g)	44,500	264	0.0059
2. Microsomes	15,800	286	0.0018
3. Salt precipitate	5,030	196	0.039
4. DEAE-cellulose	1,050	171	0.16
5. Phosphocellulose	260	72	0.28
6. Butyl-Sepharose	85	53	0.63
7. Phosphocellulose 2	51	38.7	0.76

[a] Enzyme purification from 50 rats treated with trans-stilbene oxide. Protein was measured colorimetrically.[26] One unit of activity is defined as that amount of enzyme that catalyzes the formation of 1 μmol of styrene glycol from styrene 7,8-oxide in 1 min under the incubation conditions described.[19] The purification factor noted here, about 130-fold, is about one-third that previously reported[11] when untreated rats were used as the source of enzyme. Therefore, the specific activity of the enzyme in the present study is approximately three times higher to begin with, resulting in the lower purification factor.

Step 4. DEAE-Cellulose Chromatography. The dialysate is centrifuged for 20 min at 10,000 g to remove sediment and applied to the DEAE-cellulose column over a period of about 2–3 hr. The enzyme is eluted with 5 mM sodium phosphate, pH 7.0-0.3 mM EDTA; it is not retained and elutes as a large initial peak of about 600 to 800 ml; fractions of between 50 and 100 ml are collected. Fractions containing the highest amounts of enzyme are combined.

Step 5. Phosphocellulose Chromatography. The combined fractions are dialyzed overnight against 5 mM sodium phosphate-0.3 mM EDTA, pH 7.4, applied to the phosphocellulose column, and washed with the same buffer. We have found that a buffer at pH 7.4 provides a better yield than the previously used pH of 7.0.[11] Normally, less then 10 percent of the enzyme washes through. The remainder, that which has been retained on the column, is eluted with 50 mM sodium phosphate, pH 7.4, containing 0.3 mM EDTA and 0.5 M NaCl. Those fractions, 50 ml each, with the highest specific activity are combined and dialyzed overnight against 2 liters of 50 mM sodium phosphate, pH 7.0, containing 1 mM EDTA and sufficient ammonium sulfate to bring the total concentration (inside and outside of the bag) to 430 g per liter. This procedure will concentrate and precipitate the enzyme inside the dialysis bag. The dialysate is centrifuged at 20,000 g for 15 min, and the resulting pellet is dissolved in a minimal

[26] O. H. Lowry, N. J. Rosebrough, A. L. Fair, and R. J. Randall, *J. Biol. Chem.* **193,** 265 (1951).

volume (~50 ml) of 50 mM sodium phosphate-0.3 mM EDTA at pH 7.4. The solution is dialyzed for 48 hr against several changes of the same buffer.

Step 6. Butyl-Sepharose Chromatography. The dialyzed enzyme is applied to a previously packed column (1.5 × 25 cm) of butyl-Sepharose, equilibrated with 5 mM sodium phosphate, pH 7.4, and containing 0.3 mM EDTA. The column is washed with the equilibration buffer until protein is no longer eluted. The retained enzyme is eluted with the same buffer containing 0.05% Cutscum with fractions of 5 ml collected; the fractions with the highest specific activities are combined.

Step 7. Phosphocellulose 2. The detergent is removed by chromatography over a small column (1.5 × 15 cm) of phosphocellulose which has been equilibrated with 5 mM sodium phosphate, pH 7.4, containing 0.3 mM EDTA. Active fractions are applied to the column, which is washed with 5 mM phosphate-0.3 mM EDTA, pH 7.4 and eluted with the same buffer containing 0.5 M NaCl; fractions of 1 ml are collected, and those containing protein (absorbant at 280 nm or slightly yellow) or, alternatively, enzyme activity, are combined and dialyzed overnight against 430 g per liter of ammonium sulfate. The following day, the dialysate is centrifuged at 20,000 g for 15 min and the resultant pellet is taken up in 50 mM sodium phosphate at pH 7.4. The solution is dialyzed overnight against the same buffer.

At this stage, the enzyme appears to be homogeneous and free of detergent. The preparation is stable at 4°C in an ice bath for up to 8 weeks or frozen at −70°C for months (see table).

Properties of the Homogeneous Enzyme

This preparative method yields a protein of M_r equal to 49,000 which is homogeneous by the criteria of SDS-polyacrylamide gel electrophoresis, ultracentrifugation in the presence and absence of detergent, immunological criteria, and amino acid analysis.[11,12] A normal yield from 50 rats that had been treated with *trans*-stilbene oxide, is approximately 50 mg of enzyme with a specific activity of approximately 0.5 units (0.5 μmol styrene glycol formed per minute) per milligram of protein.

Acknowledgment

This work was supported by the Deutsche Forschungsgemeinschaft.

Section III

Assay Systems

[47] Design of Solvent Extraction Methods

By MILTON T. BUSH

The quantitation of a drug (or metabolite) in biological material almost invariably requires that it be separated and purified to some degree before an analytical measurement can be made. Extraction with an aqueous-immiscible organic solvent is a common procedure for the initial recovery, partial purification, and concentration of such sought substances. Further (systematic) partitioning in solvent/aqueous system(s) has also been used widely for further purification and often in conjunction with a suitable assay method, identification, and quantitation. These methods can be called "liquid–liquid extraction analysis" or "analytical solvent extraction." Many variations continue to be useful both in themselves and as preliminaries to chromatographic procedures.

In this chapter I will not review the considerable literature in this field, because this has already been done extensively.[1-4] From a different point of view rather, I will consider the theoretical briefly and the practical in some detail, emphasizing some important points that are often not mentioned or are described inadequately. The handling of some of the practical problems will be illustrated by consideration of a few specific drugs and some of their metabolites.

General Considerations

For a stable, nonvolatile solute (pure substance) distributed between an organic solvent and an immiscible aqueous phase, the partition (distribution) coefficient (C) at equilibrium is defined as

$$C = \frac{\text{concentration in the solvent phase}}{\text{concentration in the aqueous phase}} \tag{1}$$

The volume of the solvent phase is V_x, of the aqueous phase V_y. The fraction of the total amount of solute in V_x is p, in V_y is q.

For the single solute

$$p + q = 1 \tag{2}$$

[1] L. C. Craig and D. Craig, in "Technique of Organic Chemistry" (A. Weissberger, ed.), 2nd ed., Vol. 3, p. 149. Wiley (Interscience), New York, 1956.
[2] C. Golumbic, Anal. Chem. 23, 1210 (1951).
[3] E. Titus, in "Toxicology" (C. P. Stewart and A. Stolman, eds.), Vol. 1, p. 392. Academic Press, New York, 1960.
[4] E. Titus, in "Concepts in Biochemical Pharmacology" (B. B. Brodie and J. R. Gillette, eds.), p. 123. Springer-Verlag, Berlin and New York, 1971.

METHODS IN ENZYMOLOGY, VOL. 77

Combining Eqs. (1) and (2),

$$p = CV_x/(CV_x + V_y) \tag{3}$$

and

$$C = pV_y/qV_x \tag{3'}$$

This is the formulation of Bush and Densen.[5] If the solute is a weak acid, it is helpful to utilize the well-known Henderson–Hasselbach equation (using "p" in the classical sense):

$$\text{pH} - \text{p}K_a' = \text{p[HA]} - \text{p[A}^-] \tag{4}$$

Combining Eqs. (1) and (4),

$$\text{pH} - \text{p}K_a' = \log\left[\frac{C_a}{C_0} - 1\right] \tag{5}$$

This is Butler's formulation[6] where pH and $\text{p}K_a'$ have their usual meaning, C_a is the partition coefficient of the practically unionized acid and C_0 is the partition coefficient of a particular mixture of unionized acid + its ionized salt at some pH $>$ $\text{p}K_a' - 2$. The assumption is implicit that the ionized form is insoluble in the organic phase. In practice, C_a is the partition coefficient when the pH $\leq$ $\text{p}K_a' - 2$ in the dilute aqueous solution.

For a weak base, analogous formulations can be made. For a neutral solute, variation of the pH (in dilute solution) has no effect on C.

The analytical or preparative application of these relationships depends on there being a difference between the values of C or C_0 for two or more substances. The larger the difference, the better. Differences can usually be optimized by suitable choice of solvent and/or pH, etc. It is rare, however, that a drug can be satisfactorily separated from its metabolites by a single partitioning. Even though their C's are likely to differ several fold, even 10- or 20-fold, a significant amount of metabolite is likely to be extracted along with the unmetabolized drug. Thus in the early days of drug-metabolism studies, the simple technique of buffer-washing of the extract gave more or less satisfactory results.[7] A much more powerful and often rewarding procedure, however, is the incorporation of the qualitative buffer-washing into a simple (or extended, if necessary) series of carefully controlled extractions and re-extractions, optimized insofar as possible in relation to two or more known partition coefficients. This is especially important if a large portion of the drug has been metabolized.

The first systematic use of this principle appears to have been de-

[5] M. T. Bush and P. M. Densen, *Anal. Chem.* **20**, 121 (1948).

[6] T. C. Butler, *J. Pharmacol. Exp. Ther.* **108**, 11 (1953).

[7] B. B. Brodie, *in* "Concepts in Biochemical Pharmacology" (B. B. Brodie and J. R. Gillette, eds.), p. 1. Springer-Verlag, Berlin and New York, 1971.

scribed by Jantzen[8] and was called "systematic multiple fractional extraction" (SME). Others applied and modified the method of Jantzen,[5,9] but the first really sophisticated development was that of Craig,[10] which he called countercurrent distribution (CCD). His development of apparatus[10-12] for carrying out this variation of SME and his application of the algebra of the binomial theorem to describe the movement of a solute through the series of many extractions[13] were great advances and stimuli to the use and further development of the method. I will not undertake additional discussion of CCD with large numbers of tubes (extractions) because this has been reviewed.[1-4]

The use of SME with a *small* number of extraction steps can often be optimized so that a drug and its major metabolite can be separated satisfactorily and both compounds then identified and quantitated. If a number of metabolites are produced, it is probably best also to apply some form of chromatography for separating them from each other.

The first step in optimizing SME was to generalize the mathematical aspects of the procedure,[5] following Craig's lead. Figure 1 illustrates the mechanics of the process, and Fig. 2 illustrates the "operational" expansion of Eq. (2) for a 4 × 4 SME. The binomial values for $(p + q)^n$ are seen on the horizontal lines; for $n = 2$: $p^2 + 2pq + q^2 = 1$, etc. These represent the compositions of the four phases before the combination of the pq fractions. The final algebraic terms represent the fractional amounts of the single original solute present in each final batch of the two phases; their sum is equal to 1, of course. This expansion can be carried on *ad infinitum* just as can the binomial expansion. For $(p + q)^n$ all the terms for $n = 6$ have been developed and all practical numerical values of the final terms for $n = 6$ and $n = 10$ have been given, as has the general expanded equation defining the algebraic terms.[5]

Long ago I guessed from arithmetical data that the most efficient separation of two solutes, A and B, would be achieved if the volumes of the two liquid phases (of a given solvent system) were adjusted so that $p_A = q_B$. At my request Dr. Thomas C. Butler derived from this premise the following equation (April 20, 1940)[9]:

$$\frac{V_x}{V_y} = \left(\frac{1}{C_A C_B}\right)^{1/2} \tag{6}$$

[8] E. Jantzen, DECHEMA-*Monogr.* 5, No. 48, 100 (1932).

[9] T. C. Butler and M. T. Bush, *J. Pharmacol. Exp. Ther.* 69, 236 (1940).

[10] L. C. Craig, *J. Biol. Chem.* 155, 519 (1944).

[11] L. C. Craig and O. Post, *Anal. Chem.* 21, 500 (1949).

[12] L. C. Craig, W. Hausmann, E. H. Ehrens, Jr., and E. J. Harfenist, *Anal. Chem.* 23, 1 (1951).

[13] B. Williamson and L. C. Craig, *J. Biol. Chem.* 168, 687 (1947).

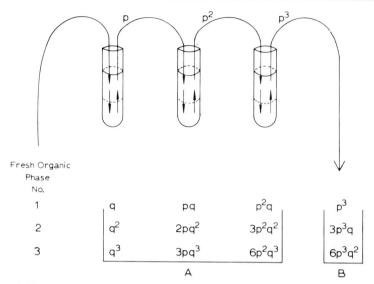

FIG. 1. The mechanics of systematic multiple fractional extraction. The fraction of a solute in the upper layer (organic solvent) of the first tube after equilibration, p, is transferred to a fresh aqueous phase in the second tube. These solutions are equilibrated and the solute in the upper phase, p^2, is transferred to the third tube, etc. After transfer of the first batch of upper phase from the first tube, a fresh batch (same volume) of solvent is introduced, equilibrated, and carried through the train, etc. The algebraic terms in "A" represent the fractional amounts of the solute remaining in the lower phases after the corresponding number of upper phases have been passed through. In "B" are the terms for the three upper phases after they have been passed through. The six final terms are those for the expansion of $(p + q)^n$ for $n = 3$. The mechanisms are further illustrated in Fig. 2.

That this adjustment, whether by volume ratio or choice of solvent composition or adjustment of pH, does indeed optimize the separation has been corroborated.[1,3,5,14]

But this is not necessarily the best choice of conditions for a particular problem. For an unknown mixture, Titus[4] defends a volume ratio of 1, but especially for small numbers of stages (2, 3, 4, etc.) we usually prefer ratios other than 1, and often other than defined by Eq. (6).

The Determination of Partition Coefficients

The measurement of a C requires a sample of the substance to be studied, a suitable solvent system, apparatus for the equilibration and separation of phases, and a method for assaying the concentration of the substance in the two phases. The sample is usually weighed in an amount

[14] E. Grushka, *Sep. Sci.* **7**, 293 (1972).

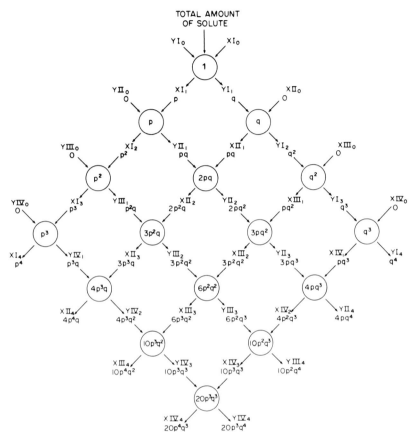

FIG. 2. The "4 × 4" systematic multiple fractional extraction. The circles represent shaking tubes (or separatory funnels, or units in a decantation apparatus). The amount of solute "1" is distributed between immiscible solvents XI_0 and YI_0. Solution YI_1 containing the fraction q of the solute is transferred to a second tube and equilibrated with a fresh batch of solvent, XII_0 (same volume as XI_0). In the meantime solution XI_1 containing the fraction p of the solute, in the original tube, is equilibrated with fresh solvent YII_0 (same volume as YI_0). Solution YI_2 is then removed from the second tube and YII_1 is transferred to it from the first tube. XII_1 and YII_1 are then equilibrated and separated. The process is continued until the "operational" expansion of $(p + q)^4$ has been completed. The eight final terms represent the compositions of the eight fractions for which the sum is 1. In practice it is convenient to have at hand tables of numerical values for these eight terms as well as intermediate terms for values of p at 0.05 intervals.[5] Such data are useful in guiding the design of separation procedures.

large enough to accommodate the assay procedure, as well as small enough to enable complete dissolution; the volumes of the two liquids measured into the apparatus are chosen in relation to these factors as well as to whether the amount of substance available is limited or ample. For present purposes we will assume that the substance is stable and not significantly volatile under the experimental conditions and that we will be working at room temperature. Finally, it is highly desirable to make the solutions reasonably dilute so that we can expect "ideal" behavior of the solute. Concentrations in the range 1 to 0.1 mg/ml or less are likely to be satisfactory for the simple measurement of C. The estimation of C for a ^{14}C-labeled drug (metabolite) in crude extracts of urine or tissue, where it is present in unknown concentration with other ^{14}C-labeled substances, can be done by SME or CCD (see following).

Solvent Systems

The choice of solvent systems to be studied will usually involve an aqueous phase, suitably buffered if our substance is an acid or a base. If the pK' is known, the pH of the buffer can be chosen to give a value of C for the unionized species; or, two values for C_0's and the C calculated via Eq. (5). If the pK' is unknown, a value for it can be estimated from several C's at different pH's. A buffer concentration of 0.1 M should be satisfactory, although we find 1 M, 2 M, etc., to be advantageous in special cases.

Organic Solvents

The organic solvents that have been found useful have been listed in large numbers.[1-4] Here I will mention a selected few, in order of increasing polarity: isooctane (2,2,4-trimethylpentane), cyclopentane, 1-chlorobutane, benzene, diisopropyl ether, dichloromethane, diethyl ether, ethyl acetate, *tert*-pentanol. I have chosen these few because they cover the range of polarities useful for most drug metabolism studies; also the pure compounds are available commercially and are readily volatile. This latter property is important in those many instances where it is necessary to concentrate or completely remove the solvent from a very dilute extract. They all have a more or less limited solubility in water, of course. If a less volatile solvent is desired, we could choose cyclohexane, toluene, di-*n*-butyl ether, 1,2-dichloroethane, isopropyl or *n*-butyl acetate, or *n*-butanol. The first solvent to be tried will be decided by a knowledge of the properties (chemical structure) of the substance to be partitioned, or if such knowledge is not available, a series of measurements must be made with several solvents of increasing polarity until useful C's are found. As a rule, it is economical first to try a solvent of intermediate polarity, such as dichloromethane or diethyl ether. C's for a great many drugs and especially their metabolites are smaller in the less polar solvents. Two-phase solvent systems without water can sometimes be useful for further fractionation of drugs (metabolites) that have already been recovered by extraction from biological material and after removal of the solvent along with most of the usual small amount of water. The use of octane/acetonitrile in such a manner will be illustrated later. Titus[4] discusses some uses of similar solvent systems. Aqueous phases containing some water-miscible organic solvent (ethanol, acetonitrile, acetic acid) give two-phase systems with solvents of very low polarity (isooctane, cyclohexane) that may be useful.

Apparatus

The most satisfactory and commonly used apparatus for carrying out a "single-shake" determination of C is the glass-stoppered or screw-capped (Teflon-lined) centrifuge tube. Conical, preferably graduated, tubes of 12- to 15-ml, 45- to 50-ml, and 140-ml capacity are readily available commercially. I much prefer a screw-capped tube.[15] Before measuring reagents into the tube, it is important to prepare a stock of the preequilibrated solvents, say 100 ml or so of each in a glass-stoppered bottle. Portions such as 5 ml or 20 ml of each are pipetted into the centrifuge tube that has already been charged with the weighed solute to be distributed or an accurately measured small portion of a stock solution of it. The closed tube is then tilted back and forth or shaken in such a way as thoroughly to mix the two phases, for a period of 2 min; 60 sec is usually enough to achieve equilibrium, i.e., the point at which further shaking does not result in any change in the concentrations of the distributed solute.[15] A screw-capped tube need not be vented[15] at this point and is allowed to stand for 0.5 hr to allow the phases to separate thoroughly; by the Tyndall effect it can often be seen that very fine drops of the heavy liquid remain suspended in the otherwise clear upper phase. Alternatively, the suspension is centrifuged (horizontally) for a few minutes at moderate speed. Samples of the two phases are then measured out for assay. It is generally better to estimate a C by a multiple extraction, even for a solute purported to be pure.

[15] On shaking a volatile solvent with an aqueous phase, considerable positive pressure may develop within the tube. Glass stoppers may be blown out but the screw-caps hold. Even with the presently available high-quality commercial glassware, leakage is sometimes encountered. Because this is intolerable, we now use only screw-cap tubes, preferably fabricated from threaded glass blanks that permit the use of open-top caps and replaceable Silicone-Teflon diaphragms, as we have described.[16] Care must be taken that the dimensions of these tubes permit their use in the available centrifuges. Shaking longer than 2 min may be necessary to reach equilibrium in the extraction of some drugs from tissue homogenate. Too vigorous shaking may unnecessarily give stable emulsions.

[16] M. T. Bush and E. Sanders-Bush, *Anal. Biochem.* **106**, 351 (1980).

For carrying out an SME of a few stages as illustrated in Figs. 1 and 2, a bench-top manual apparatus[17] is very helpful, although conical centrifuge tubes can be satisfactory. Transfer of the lower layer from one tube to the next is much easier than transfer of the upper. A special pipet[18] or syringe with a long needle is preferred, although a siphon device has been described.[11] If it is desirable to handle large volumes, separatory funnels can be used, preferably with Teflon stopcock plugs. In the bench-top apparatus[17] (4-tube, 6-tube, or 10-tube), the upper phases are decanted; these machines are still available commercially.[20]

Assay Methods

Only the two most general methods of assay will be discussed here: weight analysis and determinations of radiolabeled substances by liquid scintillation counting (of ^{14}C).

Weighing the residue (one or a few milligrams) from evaporation of a portion of the solvent extract gives a datum from which (together with the weight of the original substance) a value of C can be calculated.[9] If the solute and the solvents taken at the beginning were *pure,* this C should be accurate within the relative accuracy of the weighing. Correction must be

[17] M. T. Bush and O. Post, *Anal. Biochem.* **32,** 145 (1969).

[18] For transfer or sampling of the lower layer, the author has used long pipets with fine glass tips (ca. 100 mm long, ca 3 mm o.d., and 1–1.5 mm i.d.) drawn from broken-tip volumetric or graduated pipets. These provide good visibility but are fragile. It is more practical to provide long stainless steel needles with Kel-F hubs and to attach these to glass syringes or to glass male Luer joints sealed to such pipets. The needles should be 4- or 6-in. long and of 16, 18, or 20 gauge (Hamilton Co., Reno, NV) depending on the viscosity of the solution to be handled. To control the flow of liquid into and out of pipets, a one-arm technique can be used, with a rubber tube attached to the pipet and wrapped around the neck and controlled by the mouth—interposing a small glass double-trap. For most workers it is better to use an irrigation or other large glass syringe (clamped on a ring-stand) with greased plunger and attached to the pipet by a $\frac{1}{4} \times \frac{1}{8}$ in. rubber tube.[19] Of course, rubber bulbs especially designed for this purpose are commonly available. Either manual or decantation transfer of the upper phase involves larger mechanical errors. Some loss of a volatile solvent is difficult to avoid by either technique. The same pipet or syringe can be used for successive transfers, without washing, but draining thoroughly. The smaller the volume the greater will be the error due to the holdup.

[19] J. Axelrod, *in* "Toxicology" (C. P. Stewart and A. Stolman, eds.), Vol. 1, p. 714. Academic Press, New York, 1960.

[20] These small machines can be obtained on special order from Spectrum Medical Industries, Inc. (Mr. Kurt Schuerger) 430 Middle Village Station, New York, NY 11379. (212) 894-2200. The writer has a dozen-odd of these machines and would be willing to lend one or another. Ten-tube and larger manual machines of a different design have been available from E-C Apparatus Corp., 5000 Park St. N., St. Petersburg, FL 33733, (813) 544-8875. Many CCD machines with large numbers of tubes, some with robot controls, are probably in storage in various laboratories. Mr. Schuerger may have information about some of these.

made if the solvent extracts some of the buffer salts.[21] Ideally, the aqueous phase should also be analyzed, but its evaporation is tedious and the residue may also contain buffer salts[21]; thus, other methods for its assay are usually used. The evaporation of the organic solvent can be allowed to occur spontaneously or by impingement in a hood from a tared small beaker, or carried out from a test tube by impingement,[22] or in special light-weight vessels for greatest accuracy.[23] (A number of elegant laboratory evaporators have become available commercially, mostly for other purposes.) In any case, constancy of weight should be carefully checked, preferably using a suitable tare (beaker or test tube plus necessary small weights). Unless the weighings can be made under conditions of constant temperature and humidity, the one-pan balance is disadvantageous. The pure solute should be checked for hygroscopicity and volatility and appropriate steps taken to avoid error from these sources. Weighing residues is not appropriate for assaying crude extracts of biological material, although it can be applied after these have been subjected to SME[24] or CCD.[23] In a multiple extraction procedure, each weighed residue can be assayed by other appropriate methods to show constancy or variation of its properties.

Assay by means of *radiolabeling*, e.g., with ^{14}C, is certainly much more convenient than most other methods and can be made as accurate. It also has special advantages over all other methods. It can be applied directly to small amounts of crude extracts of biological material, even directly to urine and bile. If the pure ^{14}C-labeled drug is available, its C's in several solvent-aqueous systems can be rather quickly determined by assay of both phases, even if the aqueous phase contains salts. Assay of both phases can be critical if adsorption of the solute on the glass occurs; this is not uncommon with solvents of low polarity. If the solutions are very dilute (<1 μg/ml), the loss may be relatively large.[25] The use of [14]C

---

[21] Evaporation of the solvent or aqueous phase to dryness and extraction of the residue with dry methanol usually recovers all the ^{14}C-labeled material while leaving the salt behind. The less polar solvents extract practically no inorganic salts but solvents that extract significant amounts of water (diethyl ether, ethyl acetate, n-butanol) take out salts roughly in proportion to the water that is extracted. From solutions 2 M or more in phosphate buffer, practically none of the salt is taken out by diethyl ether.

[22] M. T. Bush, *Microchem. J.* **1**, 109 (1957).

[23] L. C. Craig, *Anal. Chem.* **23**, 1326 (1951).

[24] M. T. Bush, O. Touster, and J. E. Brockman, *J. Biol. Chem.* **188**, 685 (1951).

[25] Adsorption of certain basic drugs onto glass from heptane extracts is discussed in some detail by Brodie.[7] His group discovered that adding 1.5% of isoamyl alcohol to the heptane before extraction prevented this loss. The polarity of the solvent is thus increased, of course, and the use of a more polar solvent, e.g., 1,2-dichloroethane, in the first place avoids the adsorption. Increasingly polar mixtures can be made by adding more and more isoamyl alcohol to heptane, but whether this is advantageous over single solvents of appropriate polarity must be determined for each case.[19]

methodology permits the easy monitoring of such small amounts of substance and many an extract of biological material is even more dilute. The accuracy of assay of ^{14}C by current liquid scintillation counting techniques can be made adequate by the use of internal (or external) standards. It is necessary thus to check the efficiency of counting for each solvent extract and each aqueous phase, under the conditions of the experiment.

Sources of Error

The error of determination of a C, which is a ratio, is likely to be somewhat greater than the error of a single analysis. Failure to control adequately any of a host of factors may lead to more or less variable results.[1] Given reproduced conditions [such as temperature, equilibration, solvent composition, pH, adequate amounts of solute in each phase to permit accurate analyses and complete separation of the phases after equilibration (centrifugation)], the most likely cause of excessive variability in the estimation is imprecise sampling of the two phases. Ordinary volumetric or graduated pipets, especially with volatile solvents, may give larger errors for one phase than the other. Even so, a reproducibility of $\pm 10\%$ can easily be achieved. For greater accuracy, special pipets[18] with Luer male joints fitted with long needles and recalibrated are desirable for sampling (or transfer of) the aqueous lower phase in a centrifuge tube; Class A volumetric pipets are suitable for sampling the upper organic solvent phase. If the solvent is dichloromethane or chloroform (heavy phase), a syringe with long needle is more suitable. While thrusting the needle through the upper phase to sample the lower phase, air must be extruded to prevent entry of the upper phase. To avoid any evaporation of volatile solvent during sampling or transferring, it is necessary to use a glass syringe. Ordinary syringes must be recalibrated because their plungers often do not seat at zero. The use of precision syringes (Hamilton Co., Reno, NV; Glenco Scientific, Houston, TX) with cemented needles is undoubtedly the most accurate method of sampling the phases. For the lower phase, the syringe can be wet and the needle filled with water, buffer or the pure heavy solvent. The needle can then be thrust through the upper layer with no danger of the liquid entering. The sample and needle are withdrawn, and the outside of the needle washed and blotted off before the sample is delivered. The syringe and needle are then washed out with fresh buffer or solvent, and the washings added to the sample. If necessary this washing may be repeated. Alternatively, a suitable volume of back-up pure solvent can be adjusted in the syringe before sampling, then extruded behind the sample.

It is difficult to match the $\pm 1\%$ achieved by Lyman Craig,[1] but not the

±5% that is adequate for the present purposes. C's that are close to 1 can be measured with greater accuracy than those far removed. In estimating C from an SME or a CCD, an error will be introduced if the volume ratio is permitted to change.

Because it is unsafe to assume that a ^{14}C-labeled drug is "pure," its examination by at least one fractionation procedure is advisable. Not only should it be examined for radiochemical purity but also for chemical identity and purity. Comparisons with an authentic sample of the unlabeled substance in thin-layer chromatography (TLC) systems and by analytical solvent extractions should suffice. If the specific activity is not too high, i.e., about 1–5 mCi/mmol, it should be possible to check chemical purity and identity by weight analysis combined with nonradio methods of assaying the fractions; radioassay of the fractions will characterize only the labeled substance(s) present. Small amounts, sometimes large amounts of substances with incorrect R_f's and C's are found. The *routine* application of TLC or of analytical extraction with a small number of stages may not reveal the presence of closely related impurities, especially if inappropriate solvent systems are used. However, the simplest of SME with an appropriate solvent system can disclose both the presence and the amount of an impurity with a substantially different C.

Bioassay

Besides the various physicochemical methods, bioassay can often be used to estimate C's of unknown (impure) substances extracted from biological material. As an example, the quantitative separations by SME of two very different crude toxic fractions from the relatively nontoxic antibiotic "flavicin" were designed and followed by means of assays of toxicity in mice and *in vitro* inhibitory activity against *Staphylococcus aureus*.[26]

The theoretical number of stages of an SME required to achieve any specified degree of separation of two substances with known C's has been given in graphical form[5] through $n = 10$. For $n = 20$ and $p = 0.4$ and $q = 0.6$, the separation is 89%. This has been discussed by Titus[4] for CCD. The concentrations of the solutes diminish during a CCD or an SME, and such calculations involve the assumptions that a C remains constant during the process and that two or more C's are independent.

Practical Applications of Solvent Extraction

In order to simplify our discussion, let us assume that the drug or compound to be studied is of known structure and is available in ample

[26] M. T. Bush, A. Goth, and H. L. Dickison, *J. Pharmacol. Exp. Ther.* **84**, 262 (1945).

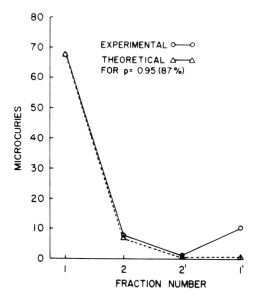

FIG. 3. Fractionation of commercial [^{14}C]pentobarbital. [^{14}C]Pentobarbital (3.8 mg, ostensibly 100 μCi) was partitioned between 1-chlorobutane (40 ml) and 0.1 M phosphate buffer, pH 6.8 (5 ml), in a 2 × 2 SME in centrifuge tubes. The total ^{14}C in all fractions was 87 μCi, and 76 μCi was sharply separated from 11 μCi of a very polar impurity. See text for calculations of partition coefficients and the theoretical distribution for the 76 μCi. In a mathematical sense these "curves" are not continuous, the points being connected only to make their relationships easily visible.

amount in pure unlabeled form and that a sample of the same compound labeled with ^{14}C in a "metabolically stable" position is provided for the actual metabolism studies.

Identity and Purity of the Labeled Material

We will check the identity and purity of the labeled material by comparing it with the unlabeled reference substance as to various physicochemical properties and as to its behavior in several suitable fractionation procedures. The application of the simplest SME to such a problem is illustrated in Fig. 3. Here our (unpublished) data show that this batch of commercial [2-^{14}C]pentobarbital contained an unacceptable amount (12.5%) of a very polar impurity. The partition coefficient of the material in fractions 1 and 2 can be calculated as follows: $p^2/2p^2q =$ 68.0/7.5 from which $q = 0.055$, and by Eq. (2), $p = 0.945$; by Eq. (3') $C = (0.945/0.055)(1/8) = 2.15$, a value close to that obtained with authentic pentobarbital. Further comparisons indicated that the separated labeled pentobarbital was acceptably pure (98+%). We calculate the theoretical amounts of the pentobarbital in fractions 2' and 1' as follows: (1) the total original amount of the substance having $p = 0.945$ is $68.0/(0.945)^2 = 76.2$ μCi; (2) the amount of this that should be in 2' is $(2pq^2)(76.2) = (2)(0.945)(0.055)^2(76.2) = 0.43$ μCi; (3) for fraction 1' the amount is $76q^2 = 0.23$ μCi. The actual amounts of ^{14}C in these fractions were 1.0 μCi and 10.5 μCi, respectively. For this polar material, $C = 0.003$ is a rough estimate. We have found several other commercial ^{14}C-labeled products to have unacceptable amounts of readily separable labeled impurities, sometimes shockingly large amounts. Of course, many labeled compounds are somewhat unstable, but probably not pentobarbital and certainly not the closely related hexobarbital, which showed no change in its partition behavior after storage for 3 years at 25°.[27]

The general problems of identification and estimation of chemical and radiochemical purity and stability have been reviewed lucidly and in some detail.[1,28]

The reasons why we have learned to choose, when feasible, solvent systems in which most of the sought substance accumulates in the organic solvent fractions, as in the example of Fig. 3, are that this usually gives better separations from polar impurities, weight analysis is facilitated and recovery of the substance (frequently in a higher state of purity) is simplified. Obviously, these conditions are not optimum if we wish to test our substance for the presence of *less* polar impurities. If we feel that our tests of other physicochemical properties are not sufficiently sensitive to demonstrate adequate purity, we can substantially improve this by "reversing" the conditions used before: that is, choosing volume ratio and/or pH and/or solvent composition so that the sought substance accumulates in the aqueous phases and less polar substances will be concentrated in the organic solvent fractions. For example, in the purification of our newly synthesized [2-^{14}C]hexobarbital we thereby removed an impurity (in 30% yield!) that had a partition coefficient of 8 in the solvent system butyl chloride/0.02 M KOH (equilibrium pH 11.5) in which C for hexobarbital is ca. 0.01.[27] The extractions were carried out as a 3 × 2 SME in relatively large volumes in 1-liter separatory funnels in order to dissolve 0.8 g of the impure drug. The purified substance was subsequently promptly recovered by acidification, extraction, and crystallization.

It should be emphasized that if the partition coefficients of two substances are rather close, i.e., about 2-fold different or less, a large number of extractions is required to give a useful degree of separation. It is,

[27] M. T. Bush and W. L. Weller, *Drug Metab. Rev.* 1, 249 (1972).
[28] J. R. Catch, *in* "International Conference on Radioactive Isotopes in Pharmacology" (P. G. Waser and B. Glasson, eds.), p. 19. Wiley (Interscience), New York, 1969.

therefore, difficult to detect small amounts of closely related impurities. For example, in a 4 × 4 SME where the C's differ by 10-fold, 10% of an impurity can barely be detected. If the C's differ by 20-fold or more, as is frequently the case, the separation is clear cut. If the C's differ by 2.25, the 10% impurity can readily be detected and partly separated by a 10 × 10 SME.[5] Of course, other methods must also be used when two or more C's are not very different.

Extraction from Biological Material

Extraction from biological material will usually involve a number of considerations that have not yet been discussed. Intractable emulsions are often formed when urine, or diluted blood or plasma, or homogenized tissue is shaken with an immiscible organic solvent. The less polar solvents are less prone to form such emulsions. Chloroform is said to be especially bad although it is often used. Even the more polar solvents produce less emulsion if they are used in relatively large volume. Addition of high concentrations of various salts to the aqueous phase, especially NaH_2PO_4, K_2HPO_4, or K_3PO_4, or mixtures of these, are often effective in preventing undue emulsification. These particular salts are very soluble and also serve as buffers. Smaller quantities of such salts are extracted by less polar solvents.[21] An important additional advantage of using less polar solvents is that they usually give greater differences between the C's of drug and its metabolites. The following are several examples from our work (compound, least polar known metabolite, solvent system, respective C's): N-methylalurate, alurate, benzene/water, 35 and 0.8[9]; hexobarbital, norhexobarbital, butyl chloride/0.1 M pH 6.8 phosphate, 9 and 0.15[27]; phenobarbital, 4'-hydroxy-phenobarbital, benzene/2 M pH 6.8 phosphate, 3.7 and 0.03[29]; N-n-butylbarbital, 2'-keto derivative, octane/0.1 M pH 6.8 phosphate, 8.3 and 0.1.[30] Thus to recover unchanged drug from urine, blood, tissue homogenate, etc., it will always be advantageous to choose the least polar solvent with which we expect to have a reasonable recovery of the drug (so as to reduce the amount of metabolite extracted) and to use it in a volume ratio of at least 3 (to reduce emulsification). For $C = 3.7$ and volume of solvent 3 times that of the aqueous phase, $p = 0.92$, 92% of the drug should be extracted. For $C = 0.03$ (metabolite), $p = 0.09$ and 9% of the metabolite is extracted. However, our experience is that in the presence of biological material, C's are somewhat lower than with water or dilute

[29] J. Alvin, T. McHorse, A. Hoyumpa, M. Bush, and S. Schenker, *J. Pharmacol. Exp. Ther.* **192**, 224 (1975).

[30] M. Vore, B. J. Sweetman, and M. T. Bush, *J. Pharmacol. Exp. Ther.* **190**, 384 (1974).

buffer.[27] Recoveries must therefore be monitored and C's checked for the specific conditions of the extraction. If the recoveries are consistent over the concentration range to be studied, a simple correction can be made. If recoveries are unexpectedly low or unsatisfactory, a second extraction can be made, or a more polar solvent used. In any case, the extract must be examined for the presence of metabolites. In the case cited earlier, the amount of the known metabolite can be substantially reduced by a single "buffer wash"—equilibration of the extract with one-third volume of the buffer. This by itself does not insure that the ^{14}C remaining in the extract actually represents unchanged drug. This must be shown by use of suitable analytical SME and other identification procedures, as pointed out previously.

An acidic drug that has been extracted at a pH where it is not very much ionized can usually be separated to a considerable degree from neutral organic impurities by a simple "back-extraction" into a small volume of an aqueous phase of high pH. Here it might be quantifiable by ultraviolet spectrophotometry or after another back-extraction from the acidified aqueous solution, by ^{14}C assays. Basic drugs can be handled in a converse manner.

We must not lose sight of the fact that theoretical recoveries based on C's are given as percentages of the amount of a substance present in the original aqueous, biological material. When the disappearance of a drug is being studied in a biological system, analysis for the unchanged substance may be complicated by the presence of metabolites if these become preponderant. This is likely to be the case in late stages of an *in vitro* study, or *in vivo* when a drug does not appear in bile or urine in more than trace amounts while its metabolites are secreted or excreted in large amounts. In such a situation, a small percentage of the large amount of metabolite may exceed a large percentage of the small amount of the drug. These and other problems have required the development of methodologies especially suited for quantitating metabolites.

Extraction and Separation of Drug Metabolites from Biological Material

Extraction and separation of drug metabolites from biological material is generally a much more complicated matter than dealing with the unchanged drug. Not only are the metabolites relatively polar but there are usually several, sometimes a large number of them. The difficulties associated with their recovery and separation from the usually huge amounts of extraneous material, and from each other, are greater or lesser in relation to whether they are more or less polar and more or less closely related in physicochemical properties. The more polar solvents that may be necessary to extract them also extract many other polar compounds, often in quantities far exceeding those of the sought substances. Partial or even adequate separations can sometimes be achieved by judicious use of solvent extractions alone. More often, however, these are used as "clean-up" procedures preliminary to the use of more powerful methods of separation.

Just as for extracting the drug itself, it may be advantageous to find the least polar solvent that will extract the metabolite(s). Often, the addition of large amounts of sodium chloride or phosphate to the aqueous phase will substantially increase partition coefficients of both unchanged drug and its polar metabolites while maintaining or augmenting useful differences between the C's. An excellent example of this is given by Kuntzman et al.[31] in their study of the rates of formation of metabolites of [^{14}C]pentobarbital by rat liver microsomes. The method was designed to measure small amounts of specific metabolites in the presence of large amounts of unchanged drug. The latter was removed from the aqueous suspension at pH 5 in the presence of excess sodium chloride by extracting three times with 7.5 volumes of petroleum ether containing 1.5% isopentanol (our values are, for pentobarbital $C = 1.6$ and for the hydroxy metabolite $C \leq 0.01$). Most of the polar metabolites remaining in the aqueous phase were then extracted by 5 volumes of ethyl acetate. This extract was concentrated and the several metabolites separated by paper chromatography. An analogous procedure was designed by Alvin et al.[29] for the separation of phenobarbital (PB) and its 4'-hydroxy metabolite (HO-PB) from each other and from the much more polar labeled substances also present in human urine after administation of [^{14}C]PB. A 5-ml portion of the urine was diluted with 5 ml of 4 M phosphate buffer (1 : 1 mixture of NaH_2PO_4 and K_2HPO_4) and extracted with 30 ml of diethyl ether. This removed ca. 99% of both PB and HO-PB and practically none of the conjugates. The two compounds could then be separated almost quantitatively (ca. 97%) after removal of the ether, by a 2 × 2 SME between benzene (21 ml) and 2 M phosphate, pH 6.8 (7 ml), in which C for PB is 3.7 and for HO-PB is 0.03.

In the first of these examples, no effort was made to measure the ^{14}C substances not extracted by ethyl acetate. In the second example, the glucuronide of HO-PB was hydrolyzed by Glusulase, then extracted and quantitated as free HO-PB. Thus, some 85% of the ^{14}C substances in the 5-day urine (66% of the dose) were identified and quantitated. Because of the availability of the ^{14}C materials and methods, this procedure for quantitating PB and these metabolites could be made much simpler than the 30-tube CCD method Butler had to devise because the only method of

[31] R. Kuntzman, M. Ikeda, M. Jacobson, and A. H. Conney, J. Pharmacol. Exp. Ther. **157**, 220 (1967).

assay available to him (ultraviolet spectrophotometry) required that the compounds be separated from interfering substances.[32]

The metabolism of the highly lipid-soluble drug, N-n-butylbarbital (nBB), is somewhat more complicated.[30] This drug is metabolized rapidly and almost completely by the rat. In a typical experiment, some 90% of the ^{14}C from an iv dose of 3 μCi of [2-^{14}C]nBB (10 mg) was excreted in the 144-hr urine. A portion of the pooled urine was extracted by 2 volumes of isooctane, and this extract was found to contain practically no nBB ($C =$ 8). The aqueous phase was made 2 M with phosphate buffer at pH 6.8 and exhaustively extracted with ether (4 times with 2-volume portions; the fourth extract contained <5% as much ^{14}C as the first). The combined extracts contained 30% of the ^{14}C that was in the urine. Almost all of this ^{14}C was accounted for by five metabolites, which were separated by TLC, recovered, and characterized; three were identified by UV and gas chromatography-mass spectroscopy (GC/MS). The remaining 70% of the ^{14}C that was not extracted by ether was then quantitatively extracted by two volumes of acetone. A portion of the acetone extract was concentrated to 2–3 ml and put into a 6 × 6 SME between tert-pentanol (9 ml) and 4 M phosphate (pH 6.8, 7 ml). Some 95% of the ^{14}C distributed as a single substance ($p = 0.73$, $C = 2.1$ and 93% of it was in the tert-pentanol fractions). Either acid or Glusulase hydrolysis of this material gave an 85% yield of 3'-hydroxy-nBB.

The effectiveness of acetone in extracting drugs and their polar metabolites under such conditions is undoubtedly related to its ability to extract water. Evaporation of the acetone in the extract leaves an aqueous slurry of salt crystals and organic matter from which water can be removed conveniently by codistillation under reduced pressure with alcohol or acetonitrile, to yield finally a concentrated solution of practically all the ^{14}C-labeled substances and containing very little water or salt. Portions of this solution can then be examined by SME and other methods.

If a fatty tissue is extracted with dry acetone, the extract will contain a good deal of extraneous organic matter. It may be possible to effect clean-up by a simple partition or SME between isooctane and acetonitrile, in which system fats and higher fatty acids have large C's ($C = 3$ for myristic acid) and even highly lipid-soluble drugs have low C's ($C = 0.02$ for thiopental); thus this drug and its metabolites would be recovered in the acetonitrile fraction(s). Although homogenization of whole mice in dry acetone gives extracts containing all of the [^{14}C]hexobarbital administered to the animal,[33] the large amount of fatty substances present in the extract (about 2 grams) precludes this simple method of clean-up, because of the

[32] T. C. Butler, J. Pharmacol. Exp. Ther. 116, 326 (1956).

[33] N. Gerber, R. Lynn, R. Holcomb, W. L. Weller, and M. T. Bush, J. Pharmacol. Exp. Ther. 177, 234 (1971).

formation of intractable emulsions. It is undoubtedly better, as a rule, to homogenize tissues or whole animals (rats[34]) in aqueous media and subject the homogenate to extraction with suitable solvent(s) before applying the acetone procedure. Even 1-g portions of fat have been satisfactorily analyzed for nBB and its metabolites by homogenizing in 10-ml portions of 4 M phosphate at pH 6.8, extracting with acetone, concentrating the extract, and partitioning the residue between isooctane and 0.1 M buffer at the same pH.[35]

Polar metabolites of drugs have often been extracted from urine by n-butanol. This solvent will form a 2-phase system but also extracts a considerable amount of water. The addition of much salt to the aqueous phase in this case may even decrease the C for the sought metabolite. For example, we have found the C for [^{14}C]urea between butanol and water to be 0.16, whereas between butanol and 2 M NaH$_2$PO$_4$, the C is 0.13! If one is to extract a batch of urine with many portions of butanol, it is important to saturate the alcohol beforehand with water if we wish to prevent the eventual complete dissolution of the urine. Butanol, like acetone, extracts considerable amounts of the normal urinary substances along with the sought drug metabolites. Because of its lower boiling point, it may be advantageous to use *tert*-pentanol instead of n-butanol. Either of these solvents can be evaporated under reduced pressure and the extract residue treated like that from acetone extracts.

Laboratory Apparatus

Laboratory apparatus for the extraction of aqueous solutions by solvents such as ether or dichloromethane have been reviewed[1] and other improved setups described.[36] The main difficulty with urine is emulsification. Small volumes are easily handled by the procedures I have already described, but large volumes present additional difficulties in the laboratory. For the recovery of solutes that are not too polar, a liter of urine can conveniently be divided into several small portions of about 200 ml and these extracted *successively* with several large portions, 600 to 700 ml, of solvent in one or more 1-liter separatory funnels. This permits a large volume ratio, thereby reducing emulsification, without the need for a 6-liter separatory funnel; a smaller total amount of solvent is required for a stipulated recovery of solute. Several liters of urine per hour can be handled by a spray-column device, which pretty well avoids emulsions,[37] although recoveries may be unacceptable if the C is small ($<$ about 1). For extracting very polar metabolites from a large volume of urine, ion-

[34] R. Holcomb, N. Gerber, and M. T. Bush, *J. Pharmacol. Exp. Ther.* **188**, 15 (1974).
[35] M. Vore and M. T. Bush, *J. Pharmacol. Exp. Ther.* **212**, 103 (1980).
[36] M. T. Bush, *Microchem. J.* **1**, 269 (1957).
[37] M. T. Bush and A. Goth, *Ind. Eng. Chem., Anal. Ed.* **16**, 528 (1944).

exchange liquid chromatography[38] or adsorption chromatography[39] undoubtedly are more efficient and convenient than solvent extraction.

It should be mentioned that small amounts of compounds of almost any degree of polarity can often be detected or isolated by TLC of urine directly, perhaps tentatively identified by comparison with pure reference substances. TLC bands of unknown labeled substances can be eluted and C values determined, to guide the recovery of larger quantities by extraction. We have often recovered the TLC "band" of labeled substance from the scintillation fluid in appropriate counting vials and subjected it to limited analytical solvent extraction to estimate C's and pK's. In a very instructive study of the metabolic fate of [^{14}C]pentobarbital in the dog, Titus and Weiss[40] used analytical and preparative paper chromatography of the urine together with butanol extraction of large volumes of the urine to separate and quantitate a trace of unchanged drug and nine metabolites, six of which were identified and several of which were isolated by CCD in sufficient amount to provide the pure labeled compounds for isotope dilution studies.

Ion-Pair Extraction

Ion-pair extraction of the very polar glucuronic acid, sulfuric acid, and other conjugates of hydroxy and carboxy metabolites of many drugs is a powerful relatively new technique. The extraction depends on the fact that these conjugates can form ion-pairs with hydrophobic symmetrical tetraalkylammonium ions and that these are more or less readily extractable by such a solvent as chloroform.[41,42]

Many drugs are tetraalkylammonium compounds not extractable by organic solvents. By reaction with such a dye as bromphenol blue in properly prepared biological material, many of these drugs can be converted to extractable salts or ion-pairs and quantified by colorimetric measurement of the dye in the extract.[19] Additional studies of ion-pair extractions of this type have been described.[43]

Derivatization in Situ

Derivatization *in situ* may sometimes be applicable to diminish the polarity and perhaps increase the stability of a metabolite. Very polar carboxylic acids in dilute aqueous solution can be esterified with

[38] J. E. Mrochek and W. T. Rainey, Jr., *Anal. Biochem.* **57**, 173 (1974).
[39] B. K. Tang, W. Kalow, and A. A. Grey, *Drug Metab. Dispos.* **7**, 315 (1979).
[40] E. Titus and H. Weiss, *J. Biol. Chem.* **214**, 807 (1955).
[41] B. Fransson and G. Schill, *Acta Pharm. Suec.* **12**, 107 (1975).
[42] G. Schill, R. Modin, K. O. Borg, and B.-A. Persson, *in* "Handbook of Derivatives for Chromatography" (K. Blau and G. S. King, eds.), p. 500. Heyden, London, 1978.
[43] H. M. Stevens and A. C. Moffat, *J. Forensic Sci. Soc.* **14**, 141 (1974).

diazomethane[44]: the aqueous solution is adjusted to pH 3–4 with HCl, cooled in ice, and with good stirring treated with diazomethane in ether until an excess (yellow color) is present. The ester may then be extracted by the ether. We have not tried this with glucuronides but the carboxyl derivatives should thus be esterified and probably made more readily extractable. Phenolic hydroxyl groups can be acetylated in dilute aqueous solution simply by alkalinization and shaking with an excess of acetic anhydride. The very polar catechol amines are thus made readily extractable.[45] This procedure should be useful also for phenolic carboxylic acids in urine.

Tetrabutylammonium hydroxide and the like are widely used catalysts for the alkylation of carboxyl, phenolic, or imide groups by methyl iodide and other such reagents under ion-pair extraction conditions.[46]

Isotope Dilution Analysis

In an isotope dilution analysis, analytical solvent extraction is an excellent method for determining radiochemical and chemical purity "after the substance has been recrystallized to constant specific activity."[47] In fact, when a relatively large amount of carrier has been added, the specific activity of the recrystallized product may be so low that checks on its purity by any other method would be much more difficult if not impossible. The ordinary criteria of the organic chemist are far too insensitive. Solvent partitioning is particularly advantageous because several milligrams of the recrystallized substance containing only 1 or 2 nCi can be fractionated and partition coefficients determined at different pH values by weight analysis as well as by radioassay and other methods.

Acknowledgment

The writer has been supported by Research Grant NS-03534 from the National Institute of Neurological and Communicative Disorders and Stroke.

[44] E. J. Eisenbraun, R. N. Morris, and G. Adophen, *J. Chem. Educ.* **47**, 710 (1970).
[45] M. Goldstein, A. J. Friedhoff, and C. Simmons, *Experientia* **15**, 80 (1959).
[46] R. A. Jones, *Aldrichimica Acta* **9**, 35 (1976).
[47] M. Calvin, C. Heidelberger, J. Reid, B. Tolbert, and P. Yankwich, "Isotopic Carbon," pp. 240–247, esp. pp. 244–246. Wiley, New York, 1949.

[48] Assay of Glutathione, Glutathione Disulfide, and Glutathione Mixed Disulfides in Biological Samples

By THEODORUS P. M. AKERBOOM and HELMUT SIES

The tripeptide glutathione (γ-glutamylcysteinylglycine) is the major free thiol in most living cells and participates in such diverse biological processes as the detoxication of xenobiotics, removal of hydroperoxides, protection against effects of ionizing radiation, maintenance of the sulfhydryl status of proteins, and modulation of enzyme activity by disulfide interchange.[1-5]

In cells, oxidation of glutathione (GSH) leads to the formation of glutathione disulfide (GSSG) and, by disulfide interchange catalyzed by thioltransferases, glutathione can be incorporated into mixed disulfides (GSSR). The glutathione status, particularly the proportion present as GSSR, undergoes marked diurnal variation and probably is subject to hormonal and nutritional regulation.[6-10] Intracellular glutathione is effectively maintained in the reduced state by GSSG reductase linked to the NADPH/NADP+ system.

Apart from (incorrectly called) "total glutathione," comprising GSH and GSSG, interest has been focused on the redox state (GSSG/GSH)[11] and the presence of mixed disulfides.[12,13] Whereas the available assays per se are both sensitive and specific due to the availability of highly purified

[1] A. Meister, *in* "Metabolic Pathways" (D. M. Greenberg, ed.), 3rd ed., p. 101. Academic, New York, 1975.

[2] I. M. Arias and W. B. Jakoby, "Glutathione: Metabolism and Function." Raven, New York, 1976.

[3] H. Sies and A. Wendel, eds., "Functions of Glutathione in Liver and Kidney." Springer-Verlag, Berlin and New York, 1978.

[4] N. S. Kosower and E. M. Kosower, *Int. Rev. Cytol.* **54**, 109 (1978).

[5] D. J. Reed and P. W. Beatty, *Rev. Biochem. Toxicol.* **2**, 213 (1980).

[6] N. Tateishi, T. Higashi, S. Shinya, A. Naruse, and Y. Sakamoto, *J. Biochem. (Tokyo)* **75**, 93 (1974).

[7] T. Higashi, N. Tateishi, A. Naruse, and Y. Sakamoto, *Biochem. Biophys. Res. Commun.* **68**, 1280 (1976).

[8] J. Isaacs and F. Binkley, *Biochim. Biophys. Acta* **497**, 192 (1977).

[9] J. Isaacs and F. Binkley, *Biochim. Biophys. Acta* **498**, 29 (1977).

[10] G. Harisch and M. F. Mahmoud, *Hoppe-Seyler's Z. Physiol. Chem.* **361**, 1859 (1980).

[11] H. Sies, A. Wahlländer, C. Waydhas, S. Soboll, and D. Häberle, *Adv. Enzyme Regul.* **18**, 303 (1980).

[12] G. Harisch and J. Schole, *Z. Naturforsch. C* **29C**, 261 (1974).

[13] B. Mannervik and K. Axelsson, *Biochem. J.*, **190**, 125 (1980).

METHODS IN ENZYMOLOGY, VOL. 77

enzymes, sample preparation remains a problem, particularly with respect to GSSG. Sample preparation will, therefore, be emphasized here.

Preparation of Samples

Quenching [14-16]

The metabolic reactions in tissue material are quenched by means of the freeze-stop technique, and the frozen sample is pulverized, weighed, and treated with acid. To an aliquot of powdered tissue (0.3–0.5 g), four volumes of 1 M $HClO_4$, 2 mM in EDTA, is added, mixed quickly with a glass rod, and homogenized. Biological material in suspension (blood, plasma, cultured or freshly isolated cells or cell organelles) is treated by addition of an equal volume of 2 M $HClO_4$, 4 mM in EDTA, or, for particulate material, the suspending medium is removed by centrifuging through a silicone oil layer into 1 M $HClO_4$ containing 2 mM EDTA.

The use of perchloric acid is advantageous compared to acids such as trichloroacetic, metaphosphoric, picric, or sulfosalicylic because it can be removed simply by precipitation at neutral pH as the potassium salt. The presence of the acids in the samples may disturb such subsequent treatment as chromatographic or enzymatic steps.

The acid extracts are centrifuged at 5000 g for 5 min to remove the protein. The supernatant liquid contains the soluble components (GSH, GSSG, and acid-soluble mixed disulfides), whereas the pellet contains acid-insoluble mixed disulfides. Although autoxidation in acid extracts is minimal (0.1–0.2% GSH per hour as measured by the formation of GSSG), the low amount of GSSG (often less than 1% of GSH) necessitates quick processing.

Sampling for GSH Assay

An aliquot of the deproteinized extract is neutralized with a solution containing 2 M KOH and 0.3 M N-morpholinopropanesulfonic acid (MOPS) and assayed immediately with Methods II or III (see following sections). The neutralized sample may also be used for assay of the sum of GSH and GSSG (Method I). After neutralization of a deproteinized liver extract, oxidation of GSH does occur at 0° as indicated by a time-dependent increase in GSSG levels. A rapid increase is observed during

[14] B. Hess and K. Brand, in "Methods in Enzymatic Analysis (H. U. Bergmeyer, ed.), 2nd ed. (Engl.), p. 396. Verlag Chemie, Weinheim, 1974.
[15] J. R. Williamson and B. E. Corkey, this series, Vol. 13, p. 434.
[16] P. F. Zuurendonk, M. E. Tischler, T. P. M. Akerboom, R. Van der Meer, J. R. Williamson, and J. M. Tager, this series, Vol 55, p. 207.

the first 5 min, followed by a slower but steady increase, amounting to a total of about 5% GSH after 1 hr. The addition of EDTA[17] does not completely prevent the oxidation.

Sampling for GSSG Assay

The problem of autoxidation of GSH in neutralized extracts is circumvented by trapping of GSH with suitable reagents such as N-ethylmaleimide.[18-20] Subsequently, excess reagent must be removed if there is the possibility of interference with the assay. Because these steps are critical, a detailed description is presented.

Addition of NEM. N-ethylmaleimide (NEM) is added to an aliquot of the acid extract to a final concentration of 50 mM. This can be done either from a stock solution e.g., 0.2 M NEM, or else NEM may be added directly with the perchloric acid solution because the compound is relatively stable at low pH. The mixture is then carefully neutralized to pH 6.2 with a solution containing 2 M KOH and 0.3 M N-morpholinopropanesulfonic acid. The formation of N-ethyl succinylimide-S-glutathione is effective in the range of pH 4–6. Because autoxidation is a competing reaction at higher pH values, care must be taken to avoid local alkalinization during neutralization; this is achieved by the addition of a buffer such as N-morpholinopropanesulfonic acid or triethanolamine, or with K_2CO_3. Optimal precipitation of $KClO_4$ is obtained by freezing and thawing the neutralized sample.

In some methods,[18-21] NEM is added to the tissue first and quenching with acid is carried out after an additional period of time. Metabolic changes may occur during the time of reaction with NEM; initial acidification avoids this complication.

Removal of Excess NEM. NEM can be removed by extraction with organic solvents such as diethyl ether, although this is a laborious step.[20] The sample is extracted ten times with an equal volume of ether, and traces of ether are removed by vigorous shaking under reduced pressure (suction pump) or by bubbling nitrogen through the sample for 10–20 min.[19]

Alternatively, several chromatographic procedures have been described using Sephadex G-10[18] or, depending on sample pH, cationic or anionic exchange resins.[21-23] We present here a simplified and modified

[17] J. Vina, R. Hems, and H. A. Krebs, *Biochem. J.* 170, 627 (1978).
[18] H. Güntherberg and S. Rapoport, *Acta Biol. Med. Ger.* 20, 559 (1968).
[19] S. K. Srivastava and E. Beutler, *Anal. Biochem.* 25, 70 (1968).
[20] F. Tietze, *Anal. Biochem.* 27, 502 (1969).
[21] P. L. Wendell, *Biochem. J.,* 117, 661 (1970).
[22] E. Beutler and C. West, *Anal. Biochem.* 81, 458 (1977).
[23] M. Pace and J. E. Dixon, *J. Chromatogr.* 147, 521 (1978).

GSH, GSSG, AND GLUTATHIONE MIXED DISULFIDES IN BIOLOGICAL SAMPLES[*]

Source	GSH+GSSG (GSH-eq)	GSH	GSSG	GSSR
Liver[†]				
Human[a]	1.1			
Rat				
Fed, perfused[b]	5.6	5.5	0.018	
Fed[c]		7.0		
Fasted[c]		4.4		
Fed, 10 AM[d]		3.1	<0.15	1.6
Fed, 6 PM[d]		2.0	<0.15	2.6
Fed[e]		6.5	0.15	2.1
Isolated hepatocytes[b]		3.1	0.007	
Mouse, Fed[h]		8.6	0.084	
Kidney[†]				
Rat[i]	2.8		0.060	
Isolated renal cells[j]		2.9		
Mouse[h]		3.3	0.004	
Heart[†]				
Rat, perfused[k]		1.1	0.014	
Mouse[h]		0.9	0.006	
Nervous tissue, Rat[†]				
Cerebral cortex[f]		2.0	0.014	
Brain[g]	1.8		0.002	
Adipocytes				
Rat[gg]	3.0		0.001	
Erythrocytes[†]				
Human[l]	2.3		0.004	
Whole blood[‡]				
Rat[i]	1.2		0.002	
Plasma[‡]				
Human[m]	0.0003			
Rat[i]	0.005			
Mouse[n]	0.028			
Bile[‡]				
Rat[o]	2.2	1.4	0.38	

[*] Selected literature data; values are useful as a guideline and are subject to diurnal, nutritional, and hormonal variation as is exemplified for rat liver. [†] Values expressed in micromoles per gram wet weight. [‡] Values expressed in micromoles per milliliter. [a] K. H. Konz, *J. Clin. Chem. Clin. Biochem.* **17**, 353 (1979). [b] T. P. M. Akerboom, M. Gärtner, and H. Sies, *Bull. Eur. Physiopathol. Respir.* **17**, 221 (1981). [c] N. Tateishi, T. Higashi, S. Shinya, A. Naruse, and Y. Sakamoto, *J. Biochem.* (*Tokyo*) **75**, 93 (1974). [d] J. Isaacs and F. Binkley, *Biochim. Biophys. Acta* **497**, 192 (1977); a protein content of 0.15 g/g wet wt. was assumed. [e] G. Harisch and J. Schole, *Z. Naturforsch.* **29C**, 261 (1974). [f] J. Folbergrova, S. Rehncrona and B. K. Siesjö, *J. Neurochem.* **32**, 1621 (1979). [g] A. J. L. Cooper, W. A. Pulsinelli and T. E. Duffy, *J. Neurochem.* **35**, 1242 (1980). [gg] J. M. May, *Arch. Biochem. Biophys.* **207**, 117 (1981); micromoles/10^8 cells.

elution from QAE-25 Sephadex.[23] In two steps, NEM and GSSG are resolved in neutralized (pH 6.2) extracts at 4°.

QAE-25 Sephadex is swollen in 0.01 M potassium phosphate at pH 6.2, and 1 ml of the gel is packed into a column (4.0 × 0.55 cm) consisting of a Pasteur pipet, fritted with glass wool, and is equilibrated with 10 ml of the same buffer. A portion of the neutralized sample (maximally 2.5 ml) is applied to the column. A 15-ml reservoir (e.g., polypropylene syringe cylinder) is put on top of the pipet, permitting elution of NEM with 12 ml 0.01 M potassium phosphate, pH 6.2. GSSG is subsequently eluted with 4.0 ml of 0.25 M potassium phosphate at pH 6.2. The pooled eluate (4.0 ml) is adjusted to a pH of about 7.2 with 0.15 ml of a solution containing 2 M KOH and 0.3 M N-morpholinopropanesulfonic acid and is assayed for GSSG (Methods I or IV), thereby adding little further sample dilution.

After the elution step with 0.01 M potassium phosphate, the amount of NEM in the eluate is less than 3 μM. GSSG is recovered completely in the 0.25 M potassium phosphate eluate. Measurement of GSH plus GSSG (Method I) and of GSSG alone (Method IV) in the eluate fractions will reveal no differences, indicating that all the GSH is completely removed by conjugation with NEM.

Comments. The very low amount of GSSG in rat liver found with this procedure (18 nmol/g wet wt) is in accord with other reports in which precautions against autoxidation were taken[18,22] (see also the table). This amount was not significantly influenced by the addition of 5 μmol GSH (made virtually free of GSSG) per gram tissue powder, indicating the lack of autoxidation during tissue processing.

A method for measurement of GSSG in the presence of NEM using o-phthaldialdehyde at high pH[24] was found inadequate for tissue analysis.[22] The drawback of the requirement for removal of NEM has led to a search for alternative trapping agents that do not interfere with the assays. 2-Vinylpyridine has been proposed as such an alternative agent.[25] However, it derivatizes GSH at about neutral pH only within 20–60 min,

[24] P. J. Hissin and R. Hilf, *Anal. Biochem.* **74**, 214 (1976).
[25] O. W. Griffith, *Anal. Biochem.* **106**, 207 (1980).

[h] G. A. Hazelton and C. A. Lang, *Biochem. J.* **188**, 25 (1980)
[i] F. Tietze, *Anal. Biochem.* **27**, 502 (1969)
[j] D. P. Jones, G.-B. Sundby, K. Ormstad, and S. Orrenius, *Biochem. Pharmacol.* **28**, 929 (1979); assuming 10^8 cells/g wet wt.
[k] P. L. Wendell, *Biochem. J.* **117**, 661 (1970)
[l] S. K. Srivastava and E. Beutler, *Anal. Biochem.* **25**, 70 (1968)
[m] A. Wendel and P. Cikryt, *FEBS Lett.* **120**, 209 (1980)
[n] O. W. Griffith and A. Meister, *Proc. Natl. Acad. Sci. U.S.A.* **76**, 5606 (1979)
[o] H. Sies, O. R. Koch, E. Martino, and A. Boveris, *FEBS Lett.* **103**, 287 (1979)

leading to artificially high GSSG values.[25] In agreement, we observed similar results whether or not 2-vinylpyridine was used in preparing the rat liver extract (T. P. M. Akerboom and H. Sies, unpublished).

Other derivatives that have been investigated include S-carboxymethylglutathione, formed with iodoacetic acid,[26,27] and subsequent chromatography. Here, too, a minimum reaction time of 15 min at room temperature is required. Thus, the use of NEM appears preferable due to its rapid rate of reaction, i.e., completion within 1 min under the conditions given earlier.

Sample Preparation for Assay of Mixed Disulfides (GSSR)

Glutathione mixed disulfides (GSSR) can be either protein-bound (cysteinyl-S-glutathione) or freely soluble (cysteinyl or other thiols such as CoA). GSH is released from GSSR by reduction with 2% sodium borohydride (30 min incubation in Tris-HCl, pH 7.4, at 40°).[28] The reduction of protein-bound glutathione is usually carried out in the presence of 8 M urea[28,29] or 8 M guanidine[30,31] to ensure unfolding of the protein. Modig[28] observed however that the amount of GSH released is almost independent of urea concentration. Borohydride was more effective as a reducing agent than dithiothreitol or 2-mercaptoethanol.[30] Excess borohydride should be removed by acidification because it interferes with the assay. The neutralized extracts are assayed for glutathione by Method III (however, see Comments on p. 381).

Enzymatic and Nonenzymatic Assays of Glutathione

Reduced Plus Oxidized Glutathione (Method I)[20,32]

Principle. The sum of the reduced and oxidized forms of glutathione can be determined using a kinetic assay in which catalytic amounts of GSH or GSSG and glutathione reductase bring about the continuous reduction of 5,5'-dithiobis(2-nitrobenzoic acid) (abbreviated DTNB) by NADPH according to the following reactions:

[26] D. J. Reed, J. R. Babson, P. W. Beatty, A. E. Brodie, W. W. Ellis, and D. W. Potter, *Anal. Biochem.* **106,** 55 (1980).
[27] C. W. Tabor and H. Tabor, *Anal. Biochem.* **78,** 543 (1977).
[28] H. Modig, *Biochem. Pharmacol.* **17,** 177 (1968).
[29] K. R. Harrap, R. C. Jackson, P. G. Riches, C. A. Smith, and B. T. Hill, *Biochim. Biophys. Acta* **310,** 104 (1973).
[30] A. F. S. A. Habeeb, *Anal. Biochem.* **56,** 60 (1973).
[31] G. Harisch, J. Eikemeyer, and J. Schole, *Experientia* **35,** 719 (1979).
[32] C. W. I. Owens and R. V. Belcher, *Biochem. J.,* **94,** 705 (1965).

$$2 \text{ GSH} + \text{DTNB} \xrightarrow{\text{nonenzymic}} \text{GSSG} + 2 \text{ TNB}$$

$$\text{GSSG} + \text{NADPH} + \text{H}^+ \xrightarrow{\text{GSSG reductase}} 2 \text{ GSH} + \text{NADP}^+$$

$$\text{NADPH} + \text{H}^+ + \text{DTNB} \xrightarrow[\text{GSSG reductase}]{\text{GSH/GSSG}} 2 \text{ TNB} + \text{NADP}^+$$

The reaction rate is proportional to the concentration of glutathione at values up to about 2 μM. The formation of 5-thio-2-nitrobenzoate (TNB)[33] is followed spectrophotometrically at 412 nm (or at 405 nm with Hg-line photometers). The sensitivity of the assay may be enhanced by measuring NADPH fluorometrically.[34]

Reagents

Buffer: 0.1 M Potassium phosphate and 0.001 M EDTA at pH 7.0; prepare daily from stock solutions of 0.1 M KH$_2$PO$_4$ and 0.1 M K$_2$HPO$_4$, each containing 1 mM EDTA

5,5'-Dithiobis(2-nitrobenzoic acid) (DTNB), 1.5 mg/ml: Prepare fresh each day; dissolve DTNB in 0.5% NaHCO$_3$ and store in the dark

NADPH, 4 mg/ml: Dissolve NADPH in 0.5% NaHCO$_3$ and store at 4°

Glutathione reductase, 6 units/ml: Dilute commercial enzyme in buffer each day

Glutathione disulfide (GSSG), 10 μM: Prepare standard daily from stock solution (1 mM)

Procedure. Pipet into cuvette: 1.0 ml of buffer, sample, 100 μl, containing 0.5–2 nmol glutathione, 50 μl NADPH, 20 μl DTNB, 20 μl glutathione reductase. After mixing the contents of the cuvette, record the linear increase in absorbance at 412 nm. A blank assay, without glutathione, is run separately. For calibration, the procedure is repeated using 100 μl GSSG (10 μM, analyzed by method IV) instead of sample. Temperature is controlled at 25°.

Comments. Because the assay is calibrated with a standard, the amount in the sample must be adjusted to within the range in which the reaction rate is linearly related to the standard GSSG concentration.

The reaction rate is dependent not only on GSSG concentration but also on the glutathione reductase activity in the assay. Thus, physiological or artificial effectors of the enzyme (e.g., high concentrations of quenching material present in the extract) may give rise to nonlinear responses and underestimated values. The use of GSSG standards added to the assay cuvette will greatly reduce or eliminate such problems that are also encountered with perfusate and blood samples.[35,36]

[33] G. L. Ellman, *Arch. Biochem. Biophys.* **82**, 70 (1959).
[34] J. E. Brehe and H. B. Burch, *Anal. Biochem.* **74**, 189 (1976).
[35] N. Oshino and B. Chance, *Biochem. J.*, **162**, 509 (1977).
[36] G. M. Bartoli, D. Häberle, and H. Sies, *in* "Functions of Glutathione in Liver and

Because both the reduced and oxidized forms of glutathione are measured by this method, the data are usually expressed in "GSH equivalents": GSH + 2 GSSG.

The specificity of the method is governed by that of glutathione reductase and the assumption that DTNB leads to the quantitative formation of GSSG without interference of other thiols. A 100-fold excess of cysteine does not influence the assay significantly.[20]

Reduced Glutathione (GSH)

GLYOXALASE I METHOD (METHOD II)[37,38]

Principle. Reduced glutathione is specifically converted by glyoxalase I according to the reaction

$$\text{GSH} + \text{methylglyoxal} \xrightarrow{\text{glyoxalase I}} S\text{-lactoyl-GSH}$$

The formation of the reaction product can be monitored directly at 240 nm ($\epsilon_{240} = 3.37 \times 10^6$ cm^2/mol) or, at very low levels of GSH, a dual-wavelength spectrophotometer can be used at the wavelength-pair 240–270 nm with the same extinction coefficient.

Reagents

Buffer: 0.05 M Potassium phosphate, pH 7.0; prepare daily from stock solutions of KH_2PO_4 (0.05 M) and K_2HPO_4 (0.05 M)

Methylglyoxal, 110 mM: Dilute 20 μl methylglyoxal in 1 ml buffer; prepare daily

Glyoxalase I (lactoyl-glutathione lyase), 1000 units/ml: Commercially available

Procedure. Pipet into quartz cuvette: 1.0 ml of buffer, sample at volume containing 2 to 100 nmol GSH and 2 μl glyoxalase I. After mixing, the absorbance at 240 nm or at 240–270 nm is recorded; the reaction is started by the addition of 20 μl methylglyoxal. After the reaction is completed, methylglyoxal is again added for assessing methylglyoxal blank absorbance, of importance with samples containing low concentrations of glutathione.

o-PHTHALDIALDEHYDE METHOD (METHOD III)[39]

Principle. GSH forms a fluorescent complex with o-phthaldialdehyde(OPT), which shows an excitation maximum at 350 nm and an emis-

---

Kidney" (H. Sies and A. Wendel, eds.), p. 27. Springer-Verlag, Berlin and New York, 1978.

[37] E. Racker, *J. Biol. Chem.* **190**, 685 (1951).

[38] E. Bernt and H. U. Bergmeyer, *in* "Methods in Enzymatic Analysis" (H. U. Bergmeyer, ed.), 2nd ed. (Engl.), p. 1643. Verlag Chemie, Weinheim, 1974.

[39] V. H. Cohn and J. Lyle, *Anal. Biochem.* **14**, 434 (1966).

sion maximum at 420 nm. Under the conditions that are presented, the fluorescence intensity is proportional to the amount of GSH present.

Reagents

Buffer: 0.1 *M* Potassium phosphate and 0.005 *M* EDTA at pH 8.0; prepare daily

o-Phthaldialdehyde, 1 mg/ml: Dissolve OPT in methanol and store in the dark; prepare daily

GSH, 0.1 m*M*: Prepare standard solution daily

Procedure. Pipet into quartz cuvette: 1.0 ml buffer and sample at volume containing 0.2 to 20 nmol GSH. Record the fluorescence baseline. When temperature equilibration (25°) is complete, add 50 μl *o*-phthaldialdehyde. The increase in fluorescence intensity will end after about 15 min. The assay is calibrated by the addition of an approximately equal amount of GSH standardized according to Method II (external standard).

Comments. As with other nonenzymatic assay systems for biological samples, a disadvantage here is that a high specificity is not guaranteed. Indeed, several other biological thiols show also a fluorescence intensity when reacting with *o*-phthaldialdehyde.[40] However, this limitation may not invalidate the method if GSH is the major thiol present in the sample.

Glutathione levels in freeze-stopped samples of livers from fed rats were found to be identical within experimental error for Methods I, II, and III.

Other nonenzymatic methods for the determination of GSH have been compiled by Jocelyn.[41]

Glutathione Disulfide, "Oxidized Glutathione" (GSSG) (Method IV)[42]

Principle. GSSG is determined specifically with glutathione reductase according to the reaction:

$$\text{GSSG} + \text{NADPH} + \text{H}^+ \xrightarrow{\text{GSSG reductase}} 2 \text{ GSH} + \text{NADP}^+$$

The reaction is monitored directly by measuring the stoichiometric conversion of NADPH either spectrophotometrically (at 340–400 nm by means of a dual-wavelength spectrophotometer) or fluorometrically (at 366 nm excitation and 400–3000 nm emission) to levels as low as 0.1 μ*M*.[43]

Reagents

Buffer: 0.1 *M* Potassium phosphate, pH 7.0; prepare daily
EDTA, 100 m*M*

[40] P. C. Jocelyn and A. Kamminga, *Anal. Biochem.* 37, 417 (1970).
[41] P. C. Jocelyn, "Biochemistry of the SH Group." Academic Press, New York, 1972.
[42] T. W. Rall and A. L. Lehninger, *J. Biol. Chem.* 194, 119 (1952).
[43] H. Sies, G. M. Bartoli, R. F. Burk, and C. Waydhas, *Eur. J. Biochem.* 89, 113 (1978).

NADPH, 1 mM: Dissolve NADPH in 0.5% NaHCO$_3$

Glutathione reductase, 20 units/ml: Dilute the commercial enzyme in buffer containing 10 μM NADPH

Procedure. Pipet into cuvette: 1.0 ml of sample (the eluate in 0.25 M potassium phosphate at pH 7.2, obtained following NEM removal as described earlier), 10 μl of EDTA and 10 μl of NADPH. Mix and record the baseline level of NADPH absorbance (or fluorescence). When temperature equilibration (25°) is complete, the reaction is started by the addition of 5 μl glutathione reductase. For the measurement of standard GSSG solution, 1.0 ml sample is replaced by 1.0 ml buffer and standard GSSG at volume containing 5 nmol.

Chromatography of Glutathione and Related Compounds

Chromatography with an amino acid analyzer and detection of ninhydrin-positive material[26] or with HPLC of DNP-derivatives[27] have yielded useful patterns with quantitation available in the nanomole region. Recently, separation of thiol compounds after derivatization with the fluorescent labeling agent monobromobimane proved to be useful and sensitive.[45]

Concluding Remarks

From the large variety and number of published methods, we have selected Methods I–IV as representing the most suitable assay systems. Sample preparation for accurate determination of GSSG and GSSR is an equally crucial step. Some results from the literature are collected in the table (p. 376); the data are presented as tissue *contents,* in μmol/g wet wt.; tissue *concentrations,* referring to intracellular water spaces, may be roughly calculated by assuming 0.5 ml of solvent space per gram of wet tissue. Hepatic cytosol and mitochondrial matrix of the rat were found to contain approximately 10 mM GSH + GSSG (Method I).[44]

Acknowledgments

Excellent technical assistance by Maria Gärtner is gratefully acknowledged. Experimental work from this laboratory was supported by Deutsche Forschungsgemeinschaft, Schwerpunkt "Mechanismen toxischer Wirkungen von Fremdstoffen," and by Ministerium für Wissenschaft und Forschung, Nordrhein-Westfalen.

[44] A. Wahlländer, S. Soboll, and H. Sies, *FEBS Lett.* **97,** 138 (1979).
[45] R. C. Fahey, G. L. Newton, R. Dorian, and E. M. Kosower, *Anal. Biochem.* **111,** 357 (1981).

[49] Assays for UDPglucuronyltransferase Activities

By G. J. Dutton, J. E. A. Leakey, and M. R. Pollard

UDPglucuronic acid + ROH $\rightleftharpoons$ UDP + R-*O*-glucuronic acid

The displayed reaction is the major biosynthetic route for simple glucuronides.[1] The enzyme should be termed UDPglucuronyl transferase (or UDPglucuronosyltransferase) but not "glucuronyltransferase"; the last name belongs to the few catalytic entities so far known that do not require UDPglucuronic acid (UDPGlcUA) for glucuronidation.[2]

UDPglucuronyltransferase is best classified under EC 2.4.1.17 (UDPglucuronosyltransferase); specificity of other forms listed is premature. Its important physiological and pharmacological roles in vertebrate tissues have been comprehensively reviewed and some of its many substrates listed.[1,2] Recent work suggests physical and "functional" heterogeneity.[2] We will briefly outline precautions required before and during assay, and then describe four simple assays that we find reliable and that illustrate principles applicable to other assays with both broken- and intact-cell preparations. The substrates chosen form *O*-glucuronides, much the commonest studied. Transferase activity in forming *S*-glucuronides has been assayed[2]; that forming *N*-glucuronides is controversial[2]; and its role in forming *C*-glucuronides is unknown.[2]

The enzyme is both activatable and inducible. Activation is essentially an *in vitro* phenomenon, controllable as such (see Latency). Induction, by contrast, occurs in the living animal or cultured tissue and requires control tissue. Hormonal induction or induction by administered xenobiotics is readily checked, but accidental induction (e.g., due to environmental factors such as changed diet, bedding, or antiseptics) can be confusing. Because both activation and induction may vary with species, tissue, age, substrate, and similar factors,[2] transferase activity should be related to these parameters when assay conditions or rates are quoted.

Assay in Broken-Cell Preparations

Donor Substrate

The donor substrate, UDPGlcUA, must be added to the assay system, and in considerable molar excess because of breakdown there. Control

[1] G. J. Dutton, *in* "Glucuronic Acid" (G. J. Dutton, ed.), p. 185. Academic Press, New York, 1966.
[2] See G. J. Dutton, "Glucuronidation of Drugs and Other Compounds." CRC Press, Boca Raton, Florida, 1980.

METHODS IN ENZYMOLOGY, VOL. 77

mixtures must contain acceptor substrate (termed "substrate" hereafter) and enzyme, but no exogenous UDPGlcUA.[3] Quoted activity is understood as due to added UDPGlcUA. Control values are usually 0–10%.

Latency

Assay of this enzyme is complicated by its often-high latency, associated with constraint at or within the microsomal membrane. The transferase can be activated by its known endogenous activator, UDP-*N*-acetylglucosamine (UDPGlcNAc), or nonspecifically by membrane perturbation. Activations are generally not additive, and increasing perturbation leads first to optimal activation and then to inactivation. As storage and manipulation of tissue or enzyme alter activation characteristics, preparations of "latent" and "optimally activated" enzyme are required. "Latent" transferase of fresh homogenates, or of microsomes rapidly prepared in 0.25 M sucrose, is activated by adding UDPGlcNAc or by a wide variety of perturbation procedures.[2,4] Each investigator must ascertain the conditions best for his investigation. An example is given in Assay A below.

Specimen Assays in Broken-Cell Preparations

Liver being the most active tissue for glucuronidation and rat the commonest species employed, assays are described for liver preparations from young male Wistar rats (30–50 days old; 125–200 g). The basic procedures given ensure linearity for the period specified; kinetic studies may be developed from them as required. Sensitivities quoted are based on the minimum significant conversion to glucuronide detectable above the controls for procedures indicated.

2-Aminophenol as Substrate[5] *(Assay A)*

Principle. 2-Aminophenylglucuronide is determined colorimetrically after diazotization and coupling with a secondary arylamine to produce an azo dye. Selective conditions of colorimetric incubation eliminate interference from unconjugated substrate.

Comment. The method directly measures glucuronide, without inter-

---

[3] With *N*-glucuronidation, nonenzymatic synthesis of glucuronide from UDPGlcUA should also be controlled.

[4] Activation by UDPGlcNAc and by perturbation confer different kinetic properties on GT; references are listed in Ref. 2.

[5] More detailed comments are given by G. J. Dutton and I. D. E. Storey, this series, Vol. 5, p. 159, and in Refs. 1 and 2.

ference from the sulfate conjugate. Other glycoside conjugates of the substrate interfere, but their formation in vertebrate broken-cell preparations in presence of UDPGlcUA is negligible. As colorimetric conditions are critical, the assay must be checked periodically with authentic 2-aminophenylglucuronide (Koch-Light) added to the complete nonincubated reaction mixture. Procedure and precautions have been detailed earlier in this series[5] and elsewhere.[2] We summarize our current modification.

Reagents

UDPGlcUA, 40 mM, neutralized; stable at $-20°$ for months

Substrate solution, containing 1 mM 2-aminophenol (resublimed[5]), 2 mM ascorbic acid, 20 mM MgCl$_2$, 150 mM maleate, and 150 mM Tris, at pH 7.4; stable at $-20°$ under N$_2$ in the dark for some months; discard when color becomes appreciably yellow

Trichloroacetic acid, 1 mM

Sodium phosphate, 1.2 M, pH 2.35–2.45 (pH is critical)

Sodium nitrite, 0.2% (w/v)

Ammonium sulfamate, 1.0% (w/v)

N-1-Naphthylethylenediamine dihydrochloride, 0.2% (w/v)

Procedure. Incubation mixtures contain 62.5 μl of substrate solution, 12.5 μl of UDPGlcUA solution, enzyme, and any additions, in a total 125 μl. Begin reaction by adding 25 μl of the enzyme preparation (0.5–1 mg homogenate protein or 0.1–0.2 mg microsomal protein) and incubate at 37° for 20 min, shaking to ensure continuous mixing. Add 0.125 ml 1 M trichloroacetic acid, mix, centrifuge, and transfer 0.2 ml of the supernatant solution to 0.5 ml of the phosphate buffer (final pH should now be 2.20–2.25). Add, at 3-min intervals, with *thorough* shaking after each addition (excess nitrite must be destroyed by the sulfamate), 0.1 ml each of NaNO$_2$, ammonium sulfamate, and N-1-naphthylethylenediamine dihydrochloride. Leave in dark at room temperature for 90 min, read absorbance of pink-purple color at 550 nm ($\epsilon_{550} = 3.9 \times 10^4$ mol^{-1} liter cm^{-1}. Sensitivity about 0.7 nmol.

Typical Activation Characteristics with 2-Aminophenol and Rat Liver

For activation of transferase activity (units are nanomoles 2-aminophenyl glucuronide formed per minute per milligram protein) with a membrane perturbant, we use digitonin. Depending on age, optimal activation with homogenates can be up to 15-fold (up to 6-fold for microsomes), with latent activities of up to 0.2 units (1.5 units for microsomes) and maximally activated activities of up to 1.5 units (about 6 units for microsomes). Activation depends both on digitonin concentration and on the ratio of digitonin to protein; as the latter is more critical, activations should be referred to the (w/w) ratio; optimal activation seems largely independent of total protein present. The activation profile shows three distinct phases: at low digitonin:protein ratios, latency persists; with progressively higher ratios, there is a rapid rise in transferase activity to the maximum, which usually occurs at a ratio of 0.075–0.25 for homogenates (0.5–1.0 for microsomes): at still higher ratios there is a plateau region or a slow decline in activity. High detergent concentrations cause slight, controllable, colorimetric interference. We use as stock, fine aqueous suspensions of digitonin over the range 0.4–4.0% (w/v), prepared by homogenization. When we use the endogenous activator, UDPGlcNAc, we add it as 4 mM; it possesses a $K_{1/2}$ value of about 0.2–0.4 mM for rat liver UDPglucuronyltransferase.

1-[^{14}C]Naphthol as Substrate[6] (Assay B)

Principle. After reaction, unconjugated substrate is extracted by an organic solvent and the label remaining in the aqueous phase is counted.

Comment. A simple reliable extraction procedure leaving 1% or less of the substrate unextracted. Obviously extraction is not specific for glucuronide, as any water-soluble conjugate or metabolite will remain in the aqueous phase.

Reagents

UDPGlcUA as for Assay A

[1-^{14}C]Naphthol, 1 mM stock, made by evaporating to dryness 0.1 ml of a stock 0.1 M [1-^{14}C]naphthol solution in ethanol and dissolving the residue in 10 ml of the Tris/maleate/MgCl$_2$ buffer of Assay A; [1-^{14}C]naphthol should be present at ca. 8×10^5 cpm ml^{-1}

Trichloroacetic acid, 0.5 M, and glycine, 0.8 M at pH 2.2, mixed 1:1 (v/v)

Chloroform

Procedure. Incubates contain 4 mM UDPGlcUA, 10 mM MgCl$_2$, Tris/maleate buffer at pH 7.4 all as for Assay A, with 0.5 mM labeled substrate (50,000 cpm), in a total 0.125 ml. As GT activity toward 1-naphthol is high in rat liver, we recommend about 0.3 mg homogenate protein or about 0.05 mg microsomal protein, and incubation for 10–15 min. After incubation, add 0.375 ml of the trichloroacetic acid–glycine solution, mix with 2 ml CHCl$_3$, centrifuge, and count a 0.25 ml aliquot of the (upper) aqueous phase. Sensitivity: about 0.4 nmol.

---

[6] G. Otani, M. M. Abou-El-Makarem, and K. W. Bock, *Biochem. Pharmacol.* **25**, 1293 (1976).

[^{14}C]Testosterone as Substrate[7] (Assay C)

Principle. As for Assay B.

Comment. As for Assay B. An example of how a wholly water-insoluble endogenous steroid substrate can be put through this procedure with only slight modifications.

Reagents

UDPGlcUA, as for Assay A
[^{14}C]Testosterone, 50 mM, in ethanol
Dichloromethane

Procedure. Prepare incubation mixtures as for Assay B, except the substrate is [^{14}C]testosterone, final concentration 1 mM (50,000 cpm), added in 2.5 μl ethanol with a microsyringe. Suggested protein weight and incubation time are the same as for Assay A. After incubation, add 2 ml cold dichloromethane and 0.125 ml water. Mix, centrifuge, and count a 0.125-ml aliquot of the (upper) aqueous phase. Sensitivity: about 0.7 nmol.

Umbelliferone as Substrate[8] (Assay D)

Principle. A continuous assay following the disappearance of substrate fluorescence upon conjugation.

Comment. Rates are constantly displayed and kinetic studies facilitated. Calibration of each sample compensates for quenching. Degree of fluorescence is pH-dependent, requiring control in pH studies.

Reagents

UDPGlcUA, 80 mM, neutralized; stable at $-20°$ for months
Umbelliferone, 0.2 mM: Dilute 50 mM umbelliferone, thoroughly dissolved in ethanol, 250-fold with 0.1 M sodium phosphate at pH 7.4 and 37°; prepare fresh daily

Procedure. Set fluorimeter for excitation at 376 nm, emission at 460 nm, and couple directly to a chart recorder. To 0.8 ml 0.2 mM umbel-

---

[7] G. S. Rao, G. Haueter, M. L. Rao, and H. Breuer, *Anal. Biochem.* **74,** 35 (1976).

[8] We find the method also suitable for 4-methylumbelliferone [see I. M. Arias, *J. Clin. Invest.* **41,** 2233 (1962), and other sources in Refs. 1 and 2] and for resorufin (buffered substrate concentration is 0.005 mM and excitation and emission are set at 550 and 585 nm, respectively); sensitivities are 15 pmol min^{-1} and 580 pmol min^{-1}, respectively. Resorufin is labile to daylight but not to light at the excitation wavelength. The method is suitable for coupling with phase 1 reactions: umbelliferone is formed by monooxygenase action from 7-ethoxycoumarin, and resorufin from 7-ethoxyresorufin. For an example of the former coupling using hepatocytes, see P. Moldéus, J. Hogberg, and S. Orrenius, this series, Vol. 52, p. 60. Fluorescence assay of the transferase is very sensitive. With 3-hydroxybenzo[a]pyrene, picograms of glucuronide can be detected: J. Singh and F. J. Wiebel, *Anal. Biochem.* **98,** 394 (1979). Other examples are referred to in Ref. 2.

liferone and 0.1 ml water (or activator, etc.) in the cuvette at 37°, add 25 μl microsomal preparation (50 to 100 μg protein). Remove any bubbles. When base line becomes straight, add 25 μl umbelliferone (5 nmol); this calibrates the trace and allows for any quenching caused, for example, by digitonin used as activator. Add 50 μl UDPGlcUA and record the slope. Sensitivity, quoted as a rate for this procedure, is 15 pmol min^{-1}.

Assays with Other Substrates in Broken-Cell Preparations

Detailed parallel procedures and comments on the following other transferase assays can be found under the review references indicated: with estradiol,[2] harmol and harmalol,[2] menthol,[1] morphine,[2] 4-nitro-thiophenol,[2] phenolphthalein.[1,9] Assays with bilirubin (Article [22]), estrone (Article [21]), and 4-nitrophenol (Articles [20] and [21]) are described elsewhere in this volume. Bilirubin forms "ester" (acyl-linked) glucuronides; 4-aminobenzoate also forms an ester glucuronide but the assay[9,10] suffers in rat liver from the low enzyme activity that is present. Ester links are labile, and the aglycone can migrate to positions on the sugar possibly unsuitable for assay or characterization.[2]

References to assays with many other substrates have been listed.[2] Assay of the reverse reaction has been described[2,9]; with 4-nitro-phenylglucuronide as substrate, it measures appearance of 4-nitro-phenol, being thus more sensitive than the forward reaction with this phenol.[9]

Until high performance liquid chromatography is more widely available, there is no single reliable procedure assaying a wide substrate range. Simple phenols are covered by one colorimetric method of varying sensitivity[11]; an apparently more flexible coupled assay[12] possesses limitations.[13]

Assay in Intact-Cell Preparations

Because of the unknown degree of latency of UDPglucuronyl-transferase in the intact cell, assay of the enzyme in broken-cell preparations is difficult to relate to its activity *in vivo*. With broken cells one can vary the degree of activation and employ quasi-physiological concentrations of UDPGlcUA and UDPGlcNAc, and thereby approximate the rate

[9] D. Zakim and D. A. Vessey, *Methods Biochem. Anal.* 21, 1 (1973).
[10] Y. Shirai and T. Ohkubo, *J. Biochem. (Tokyo)* 41, 337 (1954); for later modifications see Refs. 1, 2, and 9.
[11] G. J. Wishart and M. T. Campbell, *Biochem. J.* 178, 443 (1979).
[12] G. J. Mulder and A. B. D. van Doorn, *Biochem. J.* 151, 131 (1975).
[13] S. A. E. Finch, T. Slater, and A. Stier, *Biochem. J.* 177, 925 (1979).

of glucuronidation seen with intact-cell preparations; however, the latter remain the point of reference. Assay of glucuronidation in intact-cell preparations is therefore included here. In nonstarved liver, it effectively measures UDPglucuronyltransferase activity.[2]

Advantages and limitations of the various methods of assay of glucuronidation in intact cells, i.e., with perfused organ, cultured tissue, sliced tissue, or cell suspensions, have been reviewed.[2] We describe herein a process we find simpler, quicker, and more reproducible than these.

Principle. Intact-cell preparations do not (unless leaky) respond to added UDPGlcUA, and must make their own. They require energy, a carbohydrate source, and oxygen. This procedure aerobically incubates snipped pieces of liver with substrate. The assay then continues as for broken cells, economizing in time and material when the two preparations are compared. As a ±UDPGlcUA control is not possible, the reaction product must be identified as a β-glucuronide of the substrate. We quote the convenient direct assay with 2-aminophenol in rat liver.

Reagents

> Krebs bicarbonate-Ringer solution containing equivalent $MgCl_2$ in place of $MgSO_4$[14]
> 2-Aminophenol, 20 mM, in 40 mM ascorbic acid (stability, see Assay A)
> Trichloroacetic acid, phosphate buffer, $NaNO_2$, ammonium sulfamate, and N-naphthylethylenediamine dihydrochloride as for Assay A.

Procedure. Place freshly excised, unchilled liver in a Petri dish of warm (30 –37°) Ringer solution. Hold liver gently but firmly with forceps and cut into constituent lobes or pieces of similar size. As cell integrity is paramount, minimize damage from forceps. Hold the lobe in another dish of warm Ringer solution and snip round its edge to produce pieces that are as uniform as possible and no more than 4 mm long × 1 mm wide. With practice this operation is rapid and simple. We find iridectomy scissors (any surgical supplier) ideal. Drain off the Ringer solution and transfer some 30 snips (10 mg protein) to each 25 ml flask containing 4 ml of the same modified Ringer solution. Add 100 μl fresh 2-aminophenol/ascorbic acid solution to give a final concentration of 0.5 mM 2-aminophenol and 1 mM ascorbic acid.[15] Gas rapidly with 5% CO_2 : 95% O_2 for approximately 30 sec. Stopper tightly and place in a water bath at 37°; after 2 min, loosen

---

[14] G. A. Levvy and I. D. E. Storey, *Biochem. J.* **44**, 295 (1949).
[15] Substrate concentration must often be lower with intact cells than with broken cells, because of cytotoxicity.

the stopper momentarily to equilibrate; retighten and incubate at 37° for 100 min, shaking continuously. At 25-min intervals remove 125 μl medium, add to 125 μl trichloroacetic acid on ice, and regas the flask. At the end of incubation, homogenize flask contents for protein determination and measure conjugate in the 250-μl portions as for Assay A. A control, incubated without tissue, is not possible with this readily oxidizable substrate; suitable controls are zero time, plus a complete incubation mixture with substrate added only after protein precipitation.

By employing six flasks, we find that, for one person, the time from excision of the liver to start of incubation is 30 min or less. After an initial lag (20 min), the rate of appearance of conjugate is linear for approximately 2 hr and is identical in medium and medium plus tissue pieces. The rate is proportional to enzyme concentration up to 15 mg protein per flask. With male Wistar rats (150 g), rates are 0.33–0.43 nmol 2-aminophenylglucuronide formed per minute per milligram liver protein. We obtain comparable rates in tissue pieces and in homogenates treated with quasi-physiological concentrations of UDPGlcUA (0.3 mM), UDPGlcNAc (0.3 mM), and 0.5 mM substrate. Standard deviation is $\pm 10\%$ for multiple determinations of a single experiment.[16] With tissues other than unstarved liver, 20 mM glucose in the medium may increase the glucuronidation rate, presumably by assisting UDPGlcUA synthesis.[17] If required, conjugate present in, but not excreted from, the tissue is readily measured following homogenization of the washed tissue pieces in cold trichloroacetic acid.

Identification of Conjugate as a β-Glucuronide of the Substrate

This is essential whenever a new substrate or tissue is employed. The conjugate is hydrolyzed with β-glucuronidase from rat preputial gland[18] at pH 4.5, and this hydrolysis must be 80–100% inhibited by 10–20 mM glucarolactone.[19] The liberated aglycone should be identified chromatographically; the sugar moiety also can be distinguished this way from galacturonic acid.[9,20] High performance liquid chromatography should facilitate identification.

[16] We have as yet come across no clear evidence suggesting differences between tissue pieces from different lobes, or from lobe edges and lobe interiors; this technique is better suited than slices or hepatocytes for future studies on the problem.

[17] I. H. Stevenson and G. J. Dutton, *Biochem. J.* **82**, 330 (1962).

[18] Prepared as far as ammonium sulfate precipitation: see G. A. Levvy, A. McAllan, and C. A. Marsh, *Biochem. J.* **69**, 22 (1958). β-Glucuronidase from this source is practically free from sulfatase. We find it hydrolyzes a wider range of glucuronides than bacterial β-glucuronidase which, with its higher optimal pH, is difficult to inhibit with glucarolactone.

[19] G. A. Levvy, *Biochem. J.* **58**, 462 (1954).

[20] I. D. E. Storey and G. J. Dutton, *Biochem. J.* **59**, 279 (1955).

Assay Utilizing High-Performance Liquid Chromatography

Reversed-phase liquid chromatography has shown great potential for glucuronide analysis. Its advantages include high sensitivity, minimum sample manipulation, simultaneous monitoring of both substrate and product(s), wide substrate coverage, and an ability to handle simultaneously complex mixtures of conjugates, not just glucuronides.[2] It has already been applied to routine assay of glucuronidation in broken-[21] or intact-cell[22] preparations.

[21] M. Matsui and F. Nagai, *Anal. Biochem.* **105**, 141 (1980).
[22] P. Moldéus, *Biochem. Pharmacol.* **27**, 2859 (1978).

[50] Analysis of Bilirubin Conjugates

By KAREL P. M. HEIRWEGH and NORBERT BLANCKAERT

Assays of enzymic formation of bilirubin[1] monoglucuronides from bilirubin (Article [22], this volume), and bilirubin diglucuronide from bilirubin monoglucuronides (Article [23], this volume) are technically difficult because substrates and reaction products are labile, and little or no adequate methodology was available in the past for isolation and specific measurement of bilirubin conjugates. Contaminants present in substrates or formed during storage, incubation, and handling obviously will complicate interpretation of results of enzymic assays and should be avoided. In this chapter we will discuss measures for inhibiting artificial changes and breakdown of rubins, and simple techniques for structure elucidation, measurement, and assessment of purity of conjugated bilirubins. Bilirubin from commercial sources is usually employed as substrate for assay of UDPglucuronyltransferase activity and synthesis of bilirubin glucuronides. As the composition of such bilirubin preparations may widely vary, it is mandatory to purify impure preparations by recrystallization.[2]

Lability of Bilirubin and Bilirubin Glucuronides in Aqueous Solution

A number of alterations that frequently occur in conjugated bilirubins are listed in the table. Chemical changes of the tetrapyrrole skeleton have

[1] In agreement with the recommendations of the IUPAC/IUB Commission on Biochemical Nomenclature, the name bilirubin represents bilirubin-IXα.
[2] A. F. McDonagh, *in* "The Porphyrins" (D. Dolphin, ed.), Vol. 6, Part A, p. 293. Academic Press, New York, 1979.

CHEMICAL CHANGES OF BILIRUBIN AND BILIRUBIN CONJUGATES

	Chemical changes
Tetrapyrrole moiety	Decomposition (i.e., autoxidation)
	Disproportionation, resulting in formation of IIIα and XIIIα isomers
	Formation of geometric isomers (e.g., 4Z, 15E)
Ester linkage	Hydrolysis
	Transesterification (alcoholysis)
	Positional isomerization (acyl-shifting)

been investigated almost exclusively with unconjugated bilirubin[2] but undoubtedly also readily occur with conjugated bilirubins. The principal initiating factors in these reactions are light and molecular oxygen.[2,3] Even in the absence of oxygen, irradiation with visible, particularly blue, light causes decomposition, disproportionation, and geometric isomerization. Therefore, all manipulations should be done in subdued, preferably red, light. Even in darkness, trace amounts of molecular oxygen may suffice to cause autoxidation (catalyzed by trace amounts of metal ions) and disproportionation.[3] Autoxidation is inhibited but not stopped by EDTA (1–5 mM). Disproportionation does not occur in strongly alkaline medium (e.g., 0.1 M NaOH), and both autoxidation and disproportionation are inhibited by a molar excess of albumin and by autoxidants such as ascorbic acid (1–5 mM). Similarly, in aqueous solution at pH 7.4, ascorbic acid (1 mM) effectively inhibits disproportionation and autoxidation of bilirubin glucuronides.[4] Consequently, it is recommended that ultrapure reagents and solvents be used and that the solvents be deoxygenated. Whenever possible, both the dry pigments and pigment solutions should be maintained in an inert atmosphere. It should be noted, however, that deoxygenation by bubbling of gases is likely to facilitate aggregation of supersaturated solutions of unconjugated bilirubin.[3]

Nonenzymic hydrolysis of bilirubin ester conjugates[5,6] and shifting of bilirubin acyl groups on the conjugating sugar,[7] which occur at pH values above 6, can be avoided by buffering the solutions below pH 6.[4,7] Hydrolysis by β-glucuronidase can be inhibited with glucaro-1,4-lactone.[6,8]

[3] D. A. Lightner, A. Cu, A. F. McDonagh, and L. A. Palma, *Biochem. Biophys. Res. Commun.* **69**, 648 (1976).

[4] A. Sieg and K. P. M. Heirwegh, unpublished work.

[5] B. H. Billing, P. G. Cole, and G. H. Lathe, *Biochem. J.* **65**, 774 (1957).

[6] K. P. M. Heirwegh and F. Compernolle, *Biochem. Pharmacol.* **28**, 2109 (1979).

[7] N. Blanckaert, F. Compernolle, P. Leroy, R. Van Houtte, J. Fevery, and K. P. M. Heirwegh, *Biochem. J.* **171**, 203 (1978).

[8] G. A. Levvy and J. Conchie, *in* "Glucuronic Acid: Free and Combined, Chemistry,

Finally, base-catalyzed alcoholysis of bilirubin conjugates can occur, particularly in solutions containing methanol.[6,9] Therefore, alkaline conditions in alcohol-containing solutions of bilirubin conjugates must be avoided.

Selective Determination of Bilirubin Conjugates with Diazotized Ethyl Anthranilate

Principle. At 20°–25° and pH 2.6–2.7, both bilirubin mono- and diglucuronides are quantitatively converted to dipyrrolic azo derivatives by reaction for 30 min with the diazonium salt of ethyl anthranilate.[10,11] Excess reagent is destroyed with ascorbic acid, and azo color is measured at 530 nm, after extraction in organic solvent. With purified preparations of bilirubin mono- and diglucuronides in protein-free aqueous solutions, prolongation of reaction time up to 3 hr does not change color yields or the composition of the reaction products.[4] Adequate buffering of the reaction medium at pH 2.6–2.7 is mandatory in preserving reaction selectivity for bilirubin conjugates, since "parasitic" reaction of unconjugated bilirubin dramatically increases above pH 3.[10] For unknown reasons, diazo reaction of unconjugated bilirubin increases with increasing liver tissue concentration and correction with appropriate controls is required.[12] Bile samples must be diluted at least 20-fold to prevent diazo cleavage of unconjugated bilirubin, which may be promoted by "accelerators," such as bile salts, present in the samples.

Reagents

Buffer: 0.4 M HCl adjusted to pH 2.7 with solid glycine

$NaNO_2$, 5 mg/ml, freshly prepared from stock solution, 100 mg/ml, which can be stored for up to 1 week in a dark bottle at 4°

Ammonium sulfamate solution, 10 mg/ml; store at 4°

Diazo reagent: A mixture of 10 ml 0.15 M HCl and 0.1 ml ethyl anthranilate is shaken vigorously at room temperature to prepare a suspension that is then mixed with 0.3 ml $NaNO_2$ solution. After 5 min, 0.1 ml ammonium sulfamate solution is added and the mixture is used after 3 min

Biochemistry, Pharmacology and Medicine" (G. J. Dutton, ed.), p. 301. Academic Press, New York, 1966.

[9] N. Blanckaert, *Biochem. J.* **185**, 115 (1980).

[10] F. P. Van Roy and K. P. M. Heirwegh, *Biochem. J.* **107**, 507 (1968).

[11] K. P. M. Heirwegh, J. Fevery, J. A. T. P. Meuwissen, J. De Groote, F. Compernolle, V. Desmet, and F. P. Van Roy, *Methods Biochem. Anal.* **22**, 205 (1974).

[12] K. P. M. Heirwegh, M. Van de Vijver, and J. Fevery, *Biochem. J.* **129**, 605 (1972).

　　　　Diazo reagent blank: Prepare as diazo reagent but omit $NaNO_2$
　　　　Ascorbic acid, 10 mg/ml: Freshly prepared before use
　　　　Extraction solvent: Pentan-2-one; dried with $CaSO_4$ and redistilled
　　　　Standard Procedure. Sample (1 ml; 60 nmol of conjugated bilirubin or
less) is mixed with 1 ml buffer. After addition (with immediate mixing) of 1
ml diazo reagent, the reaction mixture is kept for 30 min at 20–25°. After
addition of 1 ml of ascorbic acid solution, azo color is extracted 3 min later
by vigorous shaking with 2 or 3 ml of pentan-2-one. The organic phase,
obtained by centrifugation for 10 min at 3000 g, is measured at 530 nm,
using pentan-2-one as reference. A blank is obtained by treating, in parallel, a duplicate sample with buffer and diazo reagent blank. The total
concentration of conjugated bilirubins in the sample submitted to the
assay is calculated from the value ΔA_{530}, equal to $(A_{test}^{530} - A_{blank}^{530})$ and ϵ_{530}
$44.4 \times 10^3 \, M^{-1} \, cm^{-1}$ as follows:

$$\text{Concentration} \, (\mu M) = \Delta AF/0.0444 \, d$$

with d equal to the optical path length in centimeters and F equal to 1.82
and 2.87 for 2 and 3 ml of added pentan-2-one, respectively. To obtain
the total concentration of dipyrrolic azo derivatives, use ϵ_{530} 22.2×10^3
$M^{-1} \, cm^{-1}$.

　　　　Comment. Difficulties are frequently experienced with extraction of
azo pigments from protein-rich samples. For serum, extraction yields are
improved by diluting the samples at least 4-fold and by adding 0.2 ml of
trichloracetic acid solution, 30% (w/v), following the destruction of diazo
reagent with ascorbic acid. Extraction is facilitated further by shaking the
aqueous medium with solvent, then consecutively freezing and thawing
the mixtures, followed by shaking again before centrifugation.

Separation of Azo Derivatives

　　　　Extracts containing TCA should be washed 3 times with an equal
volume of glycine-HCl at pH 2.7 before application to a thin-layer plate.
The azo derivatives are conveniently separated by tlc using silica gel-
coated glass plates and the following development sequence: (a) toluene/
ethyl acetate (17:3, v/v) over 18 cm[13]; (b) chloroform/methanol/water
(65:25:3, by volume) over 15 cm; (c) chloroform/methanol (9:1, v/v)
over 15 cm. Reference unconjugated azopyrromethene $(\alpha_0)^{[10,14]}$ and
azopyrromethene glucuronide $(\delta)^{[7]}$ are prepared, respectively, from pure
bilirubin IXα and from rat bile, freshly collected at 0°C in sodium citrate at
pH 6 (0.1 M citric acid adjusted with 0.2 M Na_2HPO_4 to pH 6. Reference

---

[13] Benzene, previously used in solvent systems, can be replaced by the less toxic compound,
　　toluene.
[14] N. Blanckaert, K. P. M. Heirwegh, and F. Compernolle, *Biochem. J.* **155**, 405 (1976).

azopyrromethene xyloside (α_2) and glucoside (α_3) are obtained from dog bile.[11] The Greek letter symbols in parentheses indicate the denotation of chromatographic fractions.[11] It is important to realize that chromatographic identity of unknowns and reference pigments does not necessarily imply structural identity. Therefore, it is mandatory to verify the structures of the isolated fractions, particularly when the sample is from a source not previously investigated.

The pigment fractions are quantified by either densitometry or elution with methanol followed by reading A_{530} of the eluates. From R, the ratio unconjugated azopyrromethene/(unconjugated azopyrromethene + conjugated azopyrromethenes), one calculates $2R/(1 + R)$, which equals the ratio of monoconjugated/total conjugated bilirubin. The separated azo pigment fractions comprise two azopyrromethene isomers (so-called endo- and exovinyl isomers), which are derived from the two different dipyrrolic halves of the dissymmetric bilirubin IXα skeleton.

Structure Verification of Dipyrrolic Azo Derivatives

Overall confirmation of azo pigment structure[6,11] is obtained by treating pentan-2-one/formamide (9 : 1, v/v) eluates (washed with glycine-HCl at pH 2.7 to remove formamide) with ethereal diazomethane (30–60 sec),[7] followed by tlc against similarly treated reference standards. Exo- and endovinyl isomers of azopyrromethene methyl ester are separated by development with chloroform/methanol (9 : 1, v/v) over 1 to 2 cm followed by toluene/ethyl acetate (9 : 1, v/v) over 18 cm. For the methyl ester of azopyrromethene glucuronide, chloroform/methanol (9 : 1, v/v)[7] or (17 : 3, v/v)[11] is used. After acetylation in pyridine/acetic acid anhydride (2 : 1, v/v), the exo- and endovinyl isomers of the glucuronide can be separated by development with toluene/ethyl acetate (17 : 3, v/v).[7,15] Verification of the azopyrromethene xyloside and glucoside structures of α_2 and α_3 azo pigment fractions, respectively, is conveniently achieved by acetylation in pyridine/acetic acid anhydride (2 : 1, v/v); the acetylated derivatives are chromatographed as described for the azopyrromethene glucuronide methyl ester derivative.

The aglycone structure of azopyrromethene glucuronide and the presence of an ester linkage can be confirmed by treating a methanol solution (30–60 sec) of the free acid[6,15] or an ethanol solution of the fully acetylated methyl ester (1–2 min),[16] at room temperature, with an equal volume of

[15] K. P. M. Heirwegh, J. Fevery, R. Michiels, G. P. Van Hees, and J. Fevery, *Biochem. J.* **145**, 185 (1975).
[16] N. Blanckaert, J. Fevery, K. P. M. Heirwegh, and F. Compernolle, *Biochem. J.* **164**, 237 (1977).

NaOH (10 mg/ml) in the corresponding alcohol. After addition of 2–8 volumes of glycine-HCl at pH 2.7, the formed azopyrromethene alkyl esters are separated as indicated earlier for the methyl esters. Ethanolysis is the preferred approach, since partial alcoholysis of the ethyl anthranilate moiety also occurs, leading to formation of methyl anthranilate side products during methanolysis.

For confirmation of the ester linkage and the sugar structure, unknown and reference glucuronide, applied to a thin-layer plate, are kept in contact with the vapor produced by concentrated ammonia (27%, w/v) for 2–18 hr.[11] After removal of excess ammonia and application of reference sugars, the plate is developed with chloroform/methanol/water (65 : 25 : 3, by volume) for removal of azo pigment reaction products, followed by propan-1-ol/water (17 : 3, v/v) for sugar separation. If desirable, the plates are further cleaned by another development with chloroform/methanol/water mixture. Sugars can be detected by spraying with, for example, naphthoresorcinol/sulfuric acid reagent.

Differential Assay of Bilirubin and Bilirubin Mono- and Diglucuronides[9,17]

Purified preparations of bilirubin mono- or diglucuronides are often contaminated by unconjugated bilirubin and, respectively, the di- or monoglucuronide. These contaminants easily arise by disproportionation of bilirubin monoglucuronide or by hydrolysis of glucuronides. Accurate differential assay of bilirubin and its mono- and diglucuronides therefore is of key importance in studies with these rubins.

Principle. Mono- and diester conjugates of bilirubin are quantitatively converted to the corresponding mono- and dimethyl esters by brief base-catalyzed methanolysis in the presence of ascorbate and disodium-EDTA as protective agents. Unconjugated bilirubin is not esterified under the applied reaction conditions.[9] The tetrapyrrolic reaction products are separated and quantified by tlc[9] or HPLC.[17] Addition of an internal standard to the sample prior to alkaline methanolysis permits direct measurement of the absolute concentrations of the various pigment fractions.

Reagents

KOH, 2% (w/v), in methanol
Buffer: 0.4 M HCl adjusted to pH 2.7 at room temperature with solid glycine
Chloroform (containing 0.6 to 1% ethanol)

[17] N. Blanckaert, P. M. Kabra, F. A. Farina, B. E. Stafford, L. J. Marton, and R. Schmid, *J. Lab. Clin. Med.* **96**, 198 (1980).

Xanthobilirubic acid methyl ester (ϵ_{410} 34 × 10^3 M^{-1} cm^{-1}, in methanol),[18] 4 to 5 mg/ml, in methanol (only needed for HPLC procedure)

Alkaline Methanolysis and Determination of Relative Amounts by Chromatography[9]

Methanol (2 ml), about 20 mg sodium ascorbate, and a trace disodium-EDTA are added to 0.2 ml sample (aqueous solution, bile, serum, or subcellular fraction) containing up to 120 nmol bile pigment. The mixture is vortex-mixed with 2 ml KOH reagent and left for 60–90 sec at 20–25°. After subsequent addition of 2 ml chloroform and 4 ml glycine-HCl buffer, the mixture is shaken and briefly centrifuged. The organic phase is removed, evaporated under N_2 at 30°, and stored under argon at −15° (up to 1 week).

Chromatography (tlc) with chloroform/methanol/acetic acid (97:2:1, by vol.) on silica gel plates separates bands of unconjugated bilirubin (IIIα, IXα, XIIIα), bilirubin IIIα monomethyl ester, bilirubin IXα C-8 monomethyl ester, bilirubin IXα C-12 monomethyl ester and bilirubin XIIIα monomethyl ester, and bilirubin dimethyl ester (IIIα, IXα, XIIIα), in that order. Pigment bands are scraped from the plates, eluted with a known volume of chloroform/methanol (1:1, v/v), and measured at 450 nm. The following ϵ_m^{450} values (M^{-1} cm^{-1}) are used for calculation: bilirubin IXα, 61.5 × 10^3; bilirubin IXα C-8 monomethyl ester, 58.4 × 10^3; bilirubin IXα C-12 monomethyl ester, 57.0 × 10^3; bilirubin IXα dimethyl ester, 60.8 × 10^3. To avoid breakdown of rubins, which is particularly rapid when the pigments are adsorbed to dry silica gel, only one sample is applied to a plate (5 × 20 cm), and developed and quantified without delay. Reference isomers of bilirubin (IIIα, IXα, and XIIIα) and of the methyl esters are readily synthesized.

Determination of the Absolute Amounts of Bilirubin and Methyl Esters by Chromatography[17]

For preparation and storage of reaction products, the preceding procedure is used, except for inclusion in the methanol diluent (2 ml) of xanthobilirubic acid methyl ester as an internal standard. Chromatography (HPLC) by gradient-elution on a silica gel column (separation time, 14 min; reequilibration time, 10 min) separates a peak of bilirubin (IIIα, IXα, and XIIIα isomers, if present) and the various monomethyl ester isomers

[18] J. O. Grunewald, R. Cullen, J. Bredtfeld, and E. R. Strope, *Org. Prep. Proced. Int.* 7, 103 (1975).

in the same order as with tlc. More slowly moving peaks consist of xan-thobilirubic acid methyl ester, and of individual peaks of the dimethyl esters of bilirubin IIIα, IXα, and XIIIα. The limit of detection for serum samples is at least 0.6 μM for bilirubin and 0.3 to 0.6 μM for monoconju-gates. A 3-fold more sensitive variant of the procedure employs 0.6 ml instead of 0.2 ml of sample.

Comment. Heme and non-α isomers of bilirubin are immobile in the chromatographic systems used. The methods permit accurate determina-tion of bilirubin-IXα and its mono- and diester conjugates, and of isomeric rubins that may arise from any of these natural rubins by disproportiona-tion. Verification of the identity of the separated peaks is conveniently done (1) by comparison of the retention times with those of reference compounds, and (2) for the methyl esters, by running a duplicate sample through the same procedure except for replacement of alkaline methanol by methanol. In this reagent blank, peaks corresponding to methyl esters derived from sugar conjugates do not appear in the chromatogram.

General Comments on Structure Elucidation of Bilirubin Conjugates

The ethyl anthranilate diazo method and the alkaline methanolysis procedures described earlier permit the assessment of several structural features. These include (1) the nature of the tetrapyrrole skeleton (IIIα, IXα, or XIIIα) (alkaline methanolysis), (2) the ratio of mono- to diester conjugates (both methods) and point of attachment of the sugar residue in monoester conjugates, C-8 or C-12 (alkaline methanolysis), (3) the nature of the sugar residue (diazo method), and (4) the nature of the bond be-tween bilirubin aglycone and sugar, involving characterization of config-uration (α-D versus β-D) with β-glucuronidase (diazo method),[6] the ratio of glycosidic (C-1) versus acyl-shifted (non-C-1) conjugates (diazo method),[7] and the type of linkage (ester-linkage or not) (both methods). Except for the methods referred to in this paragraph, all procedures have been previously outlined.

[51] Assays for Differentiation of Glutathione S-Transferases

By WILLIAM H. HABIG and WILLIAM B. JAKOBY

The glutathione S-transferases catalyze that wide range of reactions that can be described as accompanying the many varieties of catalysis in which the glutathione thiolate anion participates as a nucleophile. The

METHODS IN ENZYMOLOGY, VOL. 77

scope[1,2] and limitations[3] of this view have been presented in detail. This chapter is to serve as a compilation of assay methods for this group of enzymes. The assays include tests of thioether and thioester formation, i.e., nucleophilic attack at carbon, as well as an example of attack at electrophilic sulfur (an organic thiocyanate as substrate) and nitrogen (an organic nitrate ester as substrate). Because there are frequently several species of glutathione S-transferase in a single tissue (see Article [27]), a brief discussion is appended concerning the differentiation of such transferases.

Thioether Formation

Spectrophotometric Assays

$$O_2N-\text{\textbenzene}-Cl + GSH \longrightarrow O_2N-\text{\textbenzene}-SG + HCl \qquad (1)$$

Principle. An outstanding feature of these assays for thioether formation is their convenience; they depend upon a direct change in the absorbance of the substrate when it is conjugated with glutathione (GSH). Because each of the reactions is catalyzed at a finite rate in the absence of enzyme, care is taken to reduce nonenzymatic catalysis by minimizing substrate concentrations and by decreasing pH wherever necessary.[4] For each assay, the conditions have been selected to take into account the often limited solubility of substrates and the nonenzymatic rate of reaction under conditions that nevertheless allow for enzymatic catalysis.

Reagents. All of the spectral assays shown in Table I are carried out in 0.1 M potassium phosphate of the indicated pH and 5 mM GSH *except as indicated.* Substrates with limited water solubility are prepared as stock solutions in ethanol; the concentration of ethanol in the final assay is maintained below 5%.[5]

Procedure. The conditions for the spectral assay of a variety of substrates are shown in Table I. A good quality spectrophotometer is essential because of the high absorbance of some of the assay solutions. Assays

[1] W. B. Jakoby, *Adv. Enzymol.* **46**, 383 (1978).

[2] W. B. Jakoby and W. H. Habig, *in* "Enzymatic Basis of Detoxication" (W. B. Jakoby, ed.), Vol. 2, p. 63. Academic Press, New York, 1980.

[3] J. H. Keen and W. B. Jakoby, *J. Biol. Chem.* **253**, 5654 (1978).

[4] W. H. Habig, M. J. Pabst, and W. B. Jakoby, *J. Biol. Chem.* **249**, 7130 (1974).

[5] Because ethanol can be an inhibitor of the activity of the transferases, the lowest concentration allowing solubility of the substrate should be used.

TABLE I
CONDITIONS FOR SPECTROPHOTOMETRIC ASSAYS[a]

Substrate	[Substrate] (mM)	[GSH] (mM)	pH	Wave length (nm)	$\Delta\epsilon$ (nM^{-1} cm^{-1})
1,2-Dichloro-4-nitrobenzene[b]	1.0	5.0	7.5	345	8.5
1-Chloro-2,4-nitrobenzene	1.0	1.0	6.5	340	9.6
p-Nitrophenethyl bromide	0.1	5.0	6.5	310	1.2
p-Nitrobenzyl chloride	1.0	5.0	6.5	310	1.9
1,2-Epoxy-3-(p-nitrophenoxy)propane	0.5	5.0	6.5	360	0.5
1-Menaphthyl sulfate[c]	0.5	5.0	7.5	298	2.5
trans-4-Phenyl-3-buten-2-one	0.05	0.25	6.5	290	−24.8
Ethacrynic acid	0.2	0.25	6.5	270	5.0
Δ^5-Androstene-3,17-dione[d]	0.068	0.1	8.5	248	16.3

[a] Unless otherwise noted, data are from W. H. Habig, M. J. Pabst, and W. B. Jakoby, *J. Biol. Chem.* **249**, 7130 (1974).

[b] A modification of an assay devised by J. Booth, E. Boyland, and P. Sims, *Biochem. J.* **79**, 516 (1961).

[c] Assay devised by B. Gillham, *Biochem. J.* **121**, 667 (1971).

[d] Assay described by A. M. Benson, P. Talalay, J. H. Keen, and W. B. Jakoby, *Proc. Natl. Acad. Sci. U.S.A.* **74**, 158 (1972).

are conducted in a thermostatted cell compartment at 25°. A complete assay mixture without enzyme serves as the control. For each substrate presented here, the change in absorbance is a linear function of enzyme concentration and of time for at least 3 min when the rate of absorbance change is limited to less than 0.05 per min.

Titrimetric Assay

$$GSH + CH_3I \rightarrow GSCH_3 + HI \qquad (2)$$

Principle. The conjugation of alkyl halides with GSH can be measured titrimetrically. Although acid production accompanies many of the transferase catalyzed reactions in which thioethers are formed, titrimetry is only used where more convenient assays are not available. In the case of methyl iodide, other, equally tedious, assays have been described which quantitate I^- or nonvolatile S-methylglutathione formation from radiolabeled substrate.[6] Methyl iodide is the alkyl halide most often used because the relative rate of the enzymatic reaction decreases with increasing chain length; halide derivatives are reactive with the transferases in the expected order of I > Br > Cl.[4]

[6] M. K. Johnson, *Biochem. J.* **98**, 44 (1966).

Reagents

GSH, 20 mM, neutralized with NaOH
Methyl iodide, 10 mM, prepared daily
NaOH, 5 mM

Procedure. The standard assay mixture consists of 3.0 ml of 10 mM methyl iodide and 0.25 ml of 20 mM GSH. The rate of addition of 5 mM NaOH required to maintain the pH at 7.2 is measured in a stirred reaction vessel using a recording titrimeter. Full scale on the recorder is set to represent the addition of 2.5 μmol of base. The nonenzymatic rate is measured for several minutes prior to addition of enzyme. Samples of low activity or high buffering capacity may require either concentration by ultrafiltration or dialysis prior to assay. Enzyme activity is linear in the range of 0.01 to 0.4 μmol of product formed per minute; less than 1 μmol should be produced.

Nitrite Assay

$$RCH_2NO_2 + GSH \rightarrow RCH_2SG + HNO_2 \tag{3}$$
$$RCH_2ONO_2 + GSH \rightarrow RCH_2OH + HNO_2 + GSSG \tag{4}$$

Principle. Nitrite is released when GSH reacts with nitroalkanes[4] or with organic nitrate esters.[7-9] The nitrite is assayed as the limiting factor in a diazotization reaction with sulfanilamide that produces a readily quantitatable pink dye.[10]

The reaction with nitroalkanes involves attack of an electrophilic carbon atom leading to formation of a thioester (Reaction 3). In the case of nitrate esters, the attack is on an electrophilic nitrogen (Reaction 4). Experimental evidence[11] supports the glutathione sulfenyl nitrite intermediate (Reaction 5), which is attacked nonenzymatically by a second molecule of thiol (Reaction 6). This enzymatic reaction can also be assayed with glutathione reductase (see Article [48]).

$$RCH_2ONO_2 + GSH \rightarrow RCH_2OH + [GSNO_2] \tag{5}$$
$$[GSNO_2] + GSH \rightarrow GSSG + HNO_2 \tag{6}$$

Reagents

Potassium phosphate, 0.1 M, pH 7.5
GSH, 50 mM

---

[7] L. A. Heppel and R. J. Hilmoe, *J. Biol. Chem.* **183**, 129 (1950).

[8] W. H. Habig, J. H. Keen, and W. B. Jakoby, *Biochem. Biophys. Res. Commun.* **64**, 501 (1975).

[9] Reaction with this class of substrates had been attributed to "glutathione organic nitrate reductase," an activity now known to reside in the glutathione transferases.

[10] F. D. Snell and C. T. Snell, "Colorimetric Methods of Analysis," 3rd ed., Vol. 2, p. 804. Van Nostrand/Reinhold, Princeton, New Jersey, 1949.

[11] J. H. Keen, W. H. Habig, and W. B. Jakoby, *J. Biol. Chem.* **251**, 6183 (1976).

Sulfanilamide, 1% (w/v) in 20% HCl

N-(1-Naphthyl)ethylenediamine dihydrochloride, 0.02% (w/v)

Second substrate: The nitrate esters are normally available as a mixture with lactose; such powders are extracted with 95% ethanol to prepare 100 mM stock solutions of nitroglycerin or erythrityl tetranitrate. The nitroalkanes (1- or 2-nitropropane) are soluble in water, whereas 2,3,5,6-tetrachloronitrobenzene is prepared as a 20 mM stock solution in ethanol

Procedure. Reaction mixtures normally consist of 2.0 ml of buffer containing the nitro compound and 0.1 ml of the GSH solution. The second substrate concentrations used are 1 mM for nitrate esters, 20 mM for nitropropanes, and 0.05 mM for tetrachloronitrobenzene. The control consists of the same reagents without enzyme. After addition of enzyme, aliquots of 0.4 ml are removed at intervals, usually 0, 5, 15, and 30 min, and are added directly to 2.0 ml of the 1% sulfanilamide reagent. N-(1-Naphthyl)ethylenediamine dihydrochloride, 2 ml, is added and the mixture shaken. After 20 min, color development is measured in a Klett–Summerson colorimeter with a 540 nm filter. Assays are linear up to 10 nmol of nitrite formed per minute. One nanomole of nitrite equals approximately 6 Klett Units. Higher concentrations of GSH interfere with color development.

Cyanide Assay

$$RSCN + GSH \rightarrow RSSG + HCN \qquad (7)$$

Principle. The glutathione transferases catalyze the attack of the glutathione thiolate ion on the electrophilic sulfur atom of several organic thiocyanates, resulting in the formation of an asymmetric glutathionyl disulfide and cyanide[11] (Reaction 7). Cyanide is readily quantitated by a colorimetric method.[12]

Reagents

Potassium phosphate, 0.2 M, pH 7.5

Thiocyanate (ethyl, octyl, or benzylthiocyanate), 30 mM, in ethanol

GSH, 20 mM

N-Ethylmaleimide, 0.3 mM, in 0.1 M potassium phosphate, pH 7.5

Chloramine-T, 0.1% (w/v)

Pyrazolone reagent: Two volumes of a saturated, filtered solution of 1-phenyl-3-methyl-5-pyrazolone in distilled water are mixed with one volume of a fresh solution of 0.1% (w/v) 3,3'-dimethyl-1,1'-diphenyl (4,4'-bi-2-pyrazoline)-5,5'-dione in pyridine

Procedure. The assay is conducted in narrow glass tubes (10 × 65 mm)

---

[12] A modification of the method of J. Epstein, *Anal. Chem.* **19**, 272 (1947).

in a total volume of 0.1 ml containing 0.1 M phosphate buffer, 1 mM GSH, 0.75 mM thiocyanate, and enzyme. After 10 min, the reaction is terminated by adding 0.5 ml of 0.3 M N-ethylmaleimide. One minute later, 10 μl of 0.1% chloramine-T is added and allowed to incubate for 3 min. Finally, 0.4 ml of pyrazolone reagent is added and the absorbance at 618 nm is determined after 20 min.

The loss of volatile cyanide under the assay conditions described is less than 5%. The organic thiocyanates at 0.8 mM do not interfere with the assay for cyanide. Inorganic thiocyanate, however, produces about 80% of the color value of cyanide. The assay is linear in the range of 0.4 to 10 nmol of cyanide.

Thiolysis

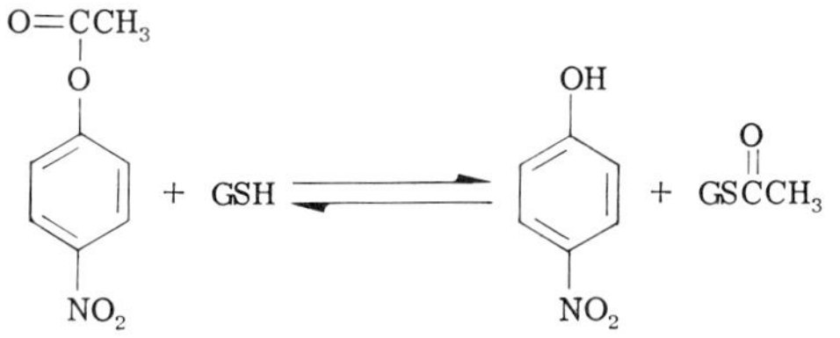

$$\text{(8)}$$

Principle. p-Nitrophenyl acetate will react with thiols to produce p-nitrophenol and an acylated thiol (Reaction 8). Each of the glutathione transferases that have been tested catalyze this reaction yielding p-nitrophenol and acetyl-S-glutathione.[3] p-Nitrophenol is quantitated directly by its absorbance at 400 nm.

Reagents

p-Nitrophenyl acetate, 0.2 mM, in 0.1 M potassium phosphate, pH 7.0
GSH, 50 mM

Procedure. The assay solution includes 1 ml of the phosphate buffer containing p-nitrophenyl acetate, 10 μl of the GSH solution, and enzyme. The rate of change of the absorbance at 400 nm is recorded. A correction is applied for the nonenzymatic rate; there is no detectable reaction in the absence of added GSH. The extinction coefficient for p-nitrophenolate at pH 7.0 is taken as 8.79 mM^{-1} cm^{-1} at 400 nm.[13]

p-Nitrophenyl trimethylacetate also serves as a substrate.[3]

Differentiation of Glutathione Transferases

When originally isolated from rat liver,[4] these enzymes were described as having an overlapping specificity. This is obvious from Table II, the

[13] F. J. Kezdy and M. L. Bender, *Biochemistry* **1**, 1097 (1962).

TABLE II. SPECIFIC ACTIVITIES[a] OF THE GLUTATHIONE TRANSFERASES WITH SELECTED SUBSTRATES

Substrate	Rat transferases						Human transferases			References
	A	AA	B	C	E	M	β	δ	ρ	
1-Chloro-2,4-dinitrobenzene	62	14	11	10	0.01		16	37	66	d–g
1,2-Dichloro-4-nitrobenzene	4.3	0.008	0.003	2.0	0	0.004	0.065	0.050	0.025	d–h
1,2-Epoxy-3-(p-nitrophenoxy)propane	0.1		0[b]	0[b]	6.7		0[b]	0[b]		d,g
Ethacrynic acid	0[b]	0.3	0.26	0.11	0[b]		0.21	0.28	2.9	d–g
Iodomethane	0[b]	1.4	0.59	0[b]	8.9		1.7	4.2		d,e,g,h
Menaphthyl sulfate	0		0.004	0[b]	0[b]	0.1				h
p-Nitrobenzyl chloride	11.4	0.09	0.1	10.2	4.1	0.5	0.22	0.20		d,e,g,h
trans-4-Phenyl-3-buten-2-one	0.02		0.001	0.40	0[b]		0.001	0.001		d,g
2,3,4,6-Tetrachloronitrobenzene	3.9		0[b]	0.001	0.001	0[b]				h
Ethyl thiocyanate	0.076	0.20	0.019	0.057			0.28	0.13		i,j
Benzyl thiocyanate	0.59	0.56	0.084	1.1			0.28	0.74		i,j
2-Nitropropane	0.012	0.01	0.008	0.014	0[b]					d
Nitroglycerin	0.06	0.14	0.09	0.37			0.05	0.32	0[b]	g,i,j
Erythrityl tetranitrate	0.76	0.28	0.36	0.15			0.27	0.89	0[b]	g,i,j
p-Nitrophenyl acetate[c]	0.56	0.27	0.33	0.36				15.0	0.2	f,k
Δ^5-Androstene-3,17-dione[c]	0.01	0.001	1.87	0.005				10.2	0[b]	f,l

[a] Specific activities in μmol min^{-1} mg^{-1} protein determined under the standard assay conditions described in this chapter.

[b] No activity detected at the greatest enzyme concentration tested.

[c] In a previous review [W. B. Jakoby and W. H. Habig, in "Enzymatic Basis of Detoxication" (W. B. Jakoby, ed.), Vol. 2, p. 63. Academic Press, New York, 1980], these activities were incorrectly presented as nmol min^{-1} instead of μmol min^{-1} as intended.

[d] W. H. Habig, M. J. Pabst, and W. B. Jakoby, J. Biol. Chem. 249, 7130 (1974).

[e] W. H. Habig, M. J. Pabst, and W. B. Jakoby, Arch. Biochem. Biophys. 175, 710 (1976).

[f] C. J. Marcus, W. H. Habig, and W. B. Jakoby, Arch. Biochem. Biophys. 188, 287 (1978).

[g] K. Kamisaka, W. H. Habig, J. N. Ketley, I. M. Arias, and W. B. Jakoby, Eur. J. Biochem. 60, 153 (1975).

[h] M. J. Pabst, W. H. Habig, and W. B. Jakoby, Biochem. Biophys. Res. Commun. 52, 1123 (1973).

[i] W. H. Habig, J. H. Keen, and W. B. Jakoby, Biochem. Biophys. Res. Commun. 64, 501 (1975).

[j] J. H. Keen, W. H. Habig, and W. B. Jakoby, J. Biol. Chem. 251, 6183 (1976).

[k] J. H. Keen and W. B. Jakoby, J. Biol. Chem. 253, 5654 (1978).

[l] A. M. Benson, P. Talalay, J. H. Keen, and W. B. Jakoby, Proc. Natl. Acad. Sci. U.S.A. 74, 158 (1977).

major purpose of which is to allow differentiation of the known glutathione transferases, at least from the rat, and to point to such differences in specificity as exist.

It will be evident that the purified enzymes from rat can be differentiated, in large part by their activity with 1-chloro-2,4-dinitrobenzene, 1,2-dichloro-4-nitrobenzene, 1,2-epoxy-3-(p-nitrophenoxy)propane, ethacrynic acid, iodomethane, and menaphthyl sulfate. Rat transferases AA and B can be distinguished from each other by the greater inhibition of transferase B ($K_i = 3 \mu M$) as compared to transferase AA ($K_i = 100 \mu M$) by indocyanine green.[14] It is worth emphasizing that attempts at differentiation of the rat liver transferases by the pattern of substrate activity in *crude* organ extracts are usually unsound.

The transferases from human liver, apparently all products of a single gene,[15] are quite similar in their substrate spectrum, varying within narrow quantitative limits.

[14] J. N. Ketley, W. H. Habig, and W. B. Jakoby, *J. Biol. Chem.* **250**, 8670 (1975).
[15] K. Kamisaka, W. H. Habig, J. N. Ketley, I. M. Arias, and W. B. Jakoby, *Eur. J. Biochem.* **60**, 153 (1975).

[52] Hydrolysis of Aromatic Amides as Assay for Carboxylesterases-Amidases

By EBERHARD HEYMANN, ROLF MENTLEIN, and HELLA RIX

Most carboxylic ester hydrolases (EC 3.1.1.1) also hydrolyze aromatic amides.[1] Normally, it is not possible to assay the hydrolysis of these amides by titration of the released acid,[2] because the released amine will itself serve as titrant. Actually the rate of cleavage of amides by carboxylesterases-amidases is usually low, requiring more sensitive methods. A number of assays are described here that are useful for the determination of the hydrolysis of aromatic amides both by cell fractions and by purified enzymes.

Assay Involving Diazotation of Aniline or Aniline Derivatives[3]

General. This procedure can be used with substrates that yield aniline or p-phenetidine. It cannot be applied to substrates producing 2,6-

[1] E. Heymann, *in* "Enzymatic Basis of Detoxication" (W. B. Jakoby, ed.), Vol. 2, p. 291. Academic Press, New York, 1980
[2] Article [45] of this volume.
[3] Adapted from K. Krisch, *Biochem. Z.* **337**, 531 (1963).

dimethylaniline or 2-chloro-6-methylaniline. The assay is well suited for weakly active cell fractions and homogenates. It is linear with the amount of enzyme up to absorbance differences of 1.2. All solutions are kept refrigerated.

Solutions

Buffer: 0.1 M Tris-HCl pH 8.6

Substrate A: 20 mM Acetanilide. Dissolve 270 mg acetanilide in 90 ml 0.1 M Tris-HCl, pH 8.6, at 80°. Cool to room temperature and adjust to 100 ml with buffer; prepare weekly

Substrate B: 10 mM phenacetin (179 mg/100 ml), dissolved as for Substrate A

NaNO$_2$, 2.5 g, in 100 ml water

Ammonium sulfamate, 2.5 g, in 50 ml water

Anhydrous sodium acetate, 30 g, dissolved in water to 100 ml

N-(1-Naphthyl)ethylenediamine dihydrochloride, 0.2 g, in 100 ml water; keep in a dark bottle and prepare weekly

Isoamyl alcohol, 3 ml, in 200 ml toluene

Trichloroacetic acid, 25 g, made up to 100 ml with 1,2-dichloroethane

HCl, Half concentrated, 5–6 M

NaOH, 5 M

Standard A: 130 mg of recrystallized aniline hydrochloride in 1000 ml water (1 mM); prepare weekly

Standard B: 137 mg p-phenetidine in 1000 ml water (1 mM); prepare weekly

Procedure. Pipet into glass test tubes bearing Teflon-sealed screwcaps: the enzyme solution ($\leq$0.5 ml); sufficient buffer to bring the volume to 0.5 ml; and 0.5 ml of Substrate A or B. Shake in a water bath of 30° or 37° for 10–60 min, according to the expected activities. Then add 0.1 ml 5 M HCl and shake well; add 20 μl 2.5% NaNO$_2$ and wait 10 min; add 50 μl 5% ammonium sulfamate and wait 3 min; add 0.5 ml 30% sodium acetate and 0.25 ml 0.2% naphthylethylenediamine and wait 20 min; add 0.25 ml 5 M NaOH. Extract the yellow dye with 3 ml isoamyl alcohol/toluene. Mix 2 ml of the organic extract with 0.3 ml of the trichloroacetic acid solution and estimate the resulting violet color at 578 nm in a cuvette of 1-cm optical path length. Use a blank without enzyme for each series of assays.

Standardization. Standard values are obtained with 0.1 ml standard solution A or B, 0.9 ml buffer, and the preceding procedure, except without substrate. These values are compared with a reagent blank obtained without both substrate and standard solution.

TABLE I
ASSAY CONDITIONS FOR DIRECT PHOTOMETRIC ASSAYS

Substrate	Substrate concentration[a] (stock solution, mM)	Wavelength (nm)	Molar absorbance of the produced amine, $\Delta\epsilon$, at pH 8.6 (liters $\times$ mol^{-1} cm^{-1})
Butanilicaine[b,c]	10	285	2080
Acetanilide[d]	20	286	1200
4-Nitroacet-anilide[e]	1	405	9350

[a] All substrates are dissolved in 0.1 M Tris-HCl buffer, pH 8.6.
[b] N-Butylaminoacetyl-2-chloro-6-methylanilide.
[c] Eckert et al.[4]
[d] Franz and Krisch.[5]
[e] Alt et al.[6]

Direct Photometric Assay for the Hydrolysis of Butanilicaine,[4]
Acetanilide,[5] or 4-Nitroacetanilide[6]

General. Except for butanilicaine, which is a very good substrate for liver carboxylesterases-amidases,[1] this type of assay should only be used for particulate-free enzyme solutions. Mammalian carboxylesterases-amidases normally are not active toward 4-nitroacetanilide, which is a good substrate for some bacterial enzymes.[6]

The aromatic amines produced by hydrolysis of the amides listed in Table I have spectral properties that allow a distinction from their parent amides at the wavelengths given in Table I. The increase in absorbance is recorded using a spectrophotometer equipped with a thermostatted cell holder. These assays are well suited for on-line computation with digital voltmeter and a microcomputer.

Procedure. Pipet the enzyme sample into the thermostatted quartz cuvette of 1-cm optical path length. Add sufficient 0.1 M Tris-HCl at pH 8.6 to bring the volume to 1 ml. Start the enzyme reaction with 1 ml of prewarmed substrate solution (Table I). Record the increase in absorbance at the wavelength given in Table I for at least 2 min. Under the conditions described here, corrections for spontaneous hydrolysis are not necessary.

[4] Adapted from T. Eckert, J. Reimann, and K. Krisch, *Drug Res.* **20**, 487 (1970).
[5] W. Franz and K. Krisch, *Hoppe-Seyler's Z. Physiol. Chem.* **349**, 1413 (1968).
[6] J. Alt, K. Krisch, and P. Hirsch, *J. Gen. Microbiol.* **87**, 260 (1975).

TABLE II
CONDITIONS FOR FLUOROMETRIC ASSAYS

| Substrate | Substrate concentration[a] (stock solution, mM) | Characteristics of the released amine[b] | | |
		Excitation wavelength (nm)	Emission wavelength (nm)	Relative fluorescence (aniline = 1)
Acetanilide	20	282	348	1
Phenacetin	10	305	372	0.94
Paracetamol	20	304	378	0.78

[a] All substrates are dissolved in 0.1 M Tris-HCl buffer, pH 8.6.
[b] At pH 8.6.

Fluorometric Assay[7] for the Hydrolysis of Acetanilide, Phenacetin, or Paracetamol

General. The direct fluorometric method, Procedure A, can be used for clear enzyme solutions. With particle-containing cell fractions, Procedure B, is preferred. A fluorometer with two monochromators is used so that both exciting and fluorescent light may be collimated; it is not known whether these assays can be adapted to filter fluorometers. The direct method, Procedure A, is well suited for on-line computation (see earlier).

Procedure A. Pipet the enzyme sample into a thermostatted fluorometric quartz cuvette. Add sufficient water to bring the volume to 1 ml. Select the excitation wavelength (Table II). Start the enzymatic reaction with 1 ml of prewarmed substrate solution (Table II). Record the increase in emission at the wavelength given in Table II for at least 2 min. If absolute activities are to be calculated, the fluorometer must be standardized using 2 ml of 50 μM solutions of aniline, 4-hydroxyaniline, or p-phenetidine in 0.1 M Tris-HCl at pH 8.6. Under the conditions described, corrections for spontaneous hydrolysis of the substrates are not necessary.

Procedure B. Pipet into a capped 1.5-ml test tube the enzyme sample, add sufficient 0.1 M Tris-HCl, pH 8.6, to bring the volume to 500 μl and start with 500 μl substrate solution (Table II). Shake for 30 min in a water bath at 30°. Stop the enzymatic reaction by vigorous shaking with 200 μl 3 M trichloroacetic acid. Centrifuge for 2 min at 10,000g with, for example, an Eppendorf centrifuge 3200. Mix 1 ml of the supernatant liquid with 1 ml 1

[7] R. Mentlein and H. Rix, unpublished.

M NaOH and estimate the fluorescence as described in Table II. Run a blank containing 500 μl buffer and 500 μl substrate solution under the same conditions. Standardize by using 120 μl of 1 m*M* aniline, (phenetidine or 4-hydroxyaniline, respectively) instead of the enzyme sample. Corrections for spontaneous hydrolysis of the substrates are not necessary.

Section IV

Synthesis

[53] Adenosine 3'-Phosphate 5'-Phosphosulfate

By RONALD D. SEKURA

Research in the area of sulfate ester biosynthesis has been somewhat hindered by the unavailability of satisfactory methods for the large scale preparation of adenosine 3'-phosphate 5'-phosphosulfate (PAPS).[1] Although several methods have been described using either chemical[3,4] or biological[5] approaches, the techniques are difficult to implement and often provide product in low yield. The synthesis described in this chapter is a composite of elements from previous work with some new innovations. It provides PAPS at high levels of purity and in quantities greater than 400 mg.[6] A similar synthetic approach also has been reported.[7]

The initial step is the preparation of adenosine 2',3'-cyclic phosphate 5'-phosphate from a mixture of adenosine 2',5'- and 3',5'-bisphosphates by treatment with dicyclohexylcarbodiimide.[8] The product is then sulfated with triethylamine-N-sulfonic acid.[4] Treatment of the sulfate ester thus formed with immobilized ribonuclease T_2 specifically cleaves the cyclic phosphate ester to the 3'-phosphate. PAPS generated in this manner is purified by chromatography on DEAE-cellulose.

Immobilized Ribonuclease T_2

Cyanogen bromide-activated Sepharose 4B (Pharmacia) is swelled and washed according to the manufacturer's suggestions. To a chilled suspension prepared from 1 g of the dry resin, a 3-ml solution containing 1000 units ribonuclease T_2, in 1.3 M NaCl and 0.3 M NaHCO$_3$, is added. The total volume is adjusted to 8 ml and the suspension is agitated for 14 hr at 4°. Buffer is removed by filtration and the resin is suspended in a solution of 1 M ethanolamine acetate at pH 8.0. After 2.5 hr at 25°, the resin is washed with 200 ml of a solution containing 1 M NaCl and 0.1 M

---

[1] The approved name for the compound is adenosine 3'-phosphate 5'-phosphosulfate. However, it was originally designated as 3'-phosphoadenosine 5'-phosphosulfate and abbreviated as PAPS.[2] Because of common usage, PAPS is retained here as the abbreviation.

[2] P. W. Robbins and F. Lipmann, *J. Biol. Chem.* **229**, 837 (1957).

[3] J. Baddiley, J. G. Buchannan, R. Letters, and A. R. Sanderson, *J. Chem. Soc.* p. 1731 (1959).

[4] R. Cherniak and E. A. Davidson, *J. Biol. Chem.* **239**, 2986 (1964).

[5] M. L. Tsang, J. Lemeiux, and J. A. Schiff, *Anal. Biochem.* **74**, 623 (1976).

[6] R. D. Sekura and W. B. Jakoby, *J. Biol. Chem.* **254**, 5658 (1979).

[7] J. P. Horwitz, J. P. Neenan, R. S. Mesia, J. Rozhin, A. Huo, and K. D. Philips, *Biochim. Biophys. Acta* **480**, 376 (1977).

[8] J. G. Moffatt and H. G. Khorana, *J. Am. Chem. Soc.* **33**, 663 (1961).

NaHCO$_3$, and then with 100 ml water. The resin is suspended in 0.1 M imidazole hydrochloride at pH 6.5 and stored at 0 –4°. After use the immobilized enzyme is washed with the imidazole buffer and stored at 0 –4°. Immobilized enzyme treated in this manner has been used for multiple preparations over a period of more than a year.

Triethylamine-N-Sulfonic Acid

To a solution of 35 ml of triethylamine in 80 ml of chloroform, chilled with a salt/ice bath, a solution containing 8 ml of chlorosulfonic acid in 25 ml of chloroform is added dropwise under anhydrous conditions. During addition of the chlorosulfonic acid, the reaction mixture is stirred vigorously, and stirring is continued for 30 min after addition is complete. The chloroform solution is extracted six or seven times with 100 ml of ice-cold water, and the resulting chloroform phase is evaporated with a rotary evaporator at 40°. The off-white solid is collected and dried under reduced pressure over phosphorus pentoxide. This product may be kept for extended periods when stored desiccated over phosphorus pentoxide at room temperature.

Adenosine 3'-Phosphate 5'-Phosphosulfate

The disodium salt of mixed adenosine 2',5'- and 3',5'-bisphosphates[9] is converted to the triethylammonium salt by passing a 5-ml solution containing 2 mmol of nucleotide through a column (2.5 × 6 cm) of Dowex-50 in the triethylammonium form. The column is washed with 30 ml of water and the combined washings are evaporated to dryness with a rotary evaporator at 37°. After dissolving the residue in 20 ml of water, 40 ml *tert*-butanol, 0.84 ml triethylamine, and 2 g of dicyclohexylcarbodiimide are added and the solution is refluxed for 3 hr. A profuse precipitate of dicyclohexylurea forms overnight at 25° and is removed by filtration. The resulting filtrate is evaporated to dryness with a rotary evaporator at 45°. Additional traces of the urea are removed by suspending the residue in 40 ml of water containing 0.5 ml triethylamine and extracting the suspension exhaustively with ether.

The resulting solution containing adenosine-2',3'-cyclic phosphate 5'-phosphate, is evaporated to dryness in a 100-ml round bottom flask, and the product is rendered anhydrous by successive solution in, and evaporation from, three 30-ml portions of absolute ethanol and one 10-ml portion of toluene. To the resulting residue, 17 ml dioxane, 3.4 ml pyridine, 17 ml

[9] The sodium salts of adenosine 2',5'- and 3',5'-bisphosphates can be obtained relatively inexpensively from Sigma Chemical Company.

dimethylformamide,[10] and 2 g triethylamine-N-sulfonic acid are added. The flask is closed with a glass stopper and sealed with Parafilm; the reaction mixture is agitated until solution is complete and allowed to remain overnight at 25°. Solvents are removed with a rotary evaporator at 35°, and the resulting oil is dissolved in 25 ml chloroform. This solution is extracted twice with 20 ml of 20 mM ammonium hydroxide and once with 20 ml of water. The aqueous fractions are pooled and immediately adjusted to pH 7.5 to prevent hydrolysis of the labile sulfate.

The aqueous extract is incubated with 4 ml (packed volume) of immobilized ribonuclease T_2 for 2 hr at 30°. The reaction mixture is gently and frequently agitated during this period, and the pH is maintained at 7.4 by the addition of dilute ammonium hydroxide. Hydrolysis of the cyclic phosphate to PAPS can be monitored by thin-layer chromatography on PEI cellulose plates (Brinkmann) developed with 1.2 M LiCl. Immobilized enzyme is recovered by passing the reaction mixture through a sintered glass filter and washing the resin with 100 ml of water.

The combined washings are applied to a column (2.5 × 45 cm) of DEAE-cellulose (Whatman DE-52), equilibrated with 0.1 M triethylamine carbonate at pH 7.6.[11] The column is eluted with a 2-liter linear gradient established between equal volumes of equilibrating buffer and the 1 M stock buffer at pH 7.6. When monitored at 260 nm, two peaks are eluted from the column; the second peak is PAPS. Fractions containing PAPS are pooled and evaporated with a rotary evaporator at 37°. The residue is repeatedly dissolved in water and evaporated until traces of triethylamine are removed. The residue is dissolved in water to give a 100 mM PAPS solution, which is stored in small aliquots at −20°. PAPS thus stored is stable for more than a year.

The final product is analyzed by chromatography on PEI cellulose plates as previously described. In this sytem PAPS and adenosine 3′,5′-bisphosphate migrate with respective R_f's of 0.17 and 0.33. On the basis of ultraviolet absorption, PAPS prepared by this method is more than 95% pure with the major contaminant being the diphosphate.

---

[10] Solvents used for this synthesis must be anhydrous. Dimethylformamide is dried by placing 200 ml of the solvent over 20 g of sodium hydroxide for at least 24 hr. Pyridine is dried by passing the solvent over a column of a molecular sieve (Linde, Type 3A) immediately before use. Dioxane is always taken from a fresh bottle of spectroscopic grade reagent.

[11] A concentrated 1 M stock solution of triethylamine carbonate buffer is prepared by bubbling carbon dioxide into a mixture of 150 ml of triethylamine and 900 ml of water until the desired pH is reached. The carbon dioxide may be generated from dry ice.

[54] Mono- and Di-β-Glucuronosides of Bilirubin

By FRANS COMPERNOLLE

Bilirubin has its two carboxyl groups shielded by multiple intramolecular hydrogen bonds and is poorly soluble in water near physiological pH.[1] It is excreted by mammalian liver cells as β-glycosidic esters, i.e., mainly the mono- and di-β-glucuronosides (I) and (II).[2] The chemical synthesis of (I) and (II) involves preparation of an acetal-protected derivative of glucuronic acid with free HO-1 group (VI).[3] Direct esterification of (VI) with bilirubin diimidazole (having the two carboxyl groups in activated form) gives rise to an anomeric mixture of bilirubin glucuronosides.[4] Selective formation of the β-anomers is accomplished by converting (VI) to a 1-mesylate (VII) and displacement of the mesylate group with the carboxyl dianion of biliverdin.[5]

Outline of Synthesis

Saponification of methyl (benzyl 2,3,4-tri-O-acetyl-β-D-glucopyranosid) uronate (III)[3,6] yields benzyl β-D-glucopyranosiduronic acid (IV). The carboxyl group and HO-2,3,4 of (IV) are protected as the acetals formed with ethyl vinyl ether (IV → V) followed by hydrogenolysis of the benzyl group (V → VI).

$$\text{Glc U(Prot)-OSO}_2\text{Me} + \text{R}^1(\text{CO}_2{}^-\text{Bu}_4\text{N}^+)_2 \rightarrow \text{R}^1[\text{CO}_2\text{Glc U(Prot)}]_2$$
$$\text{(VII)} \qquad \text{R}^1 = \text{biliverdin residue} \qquad \text{(VIII)}$$

$$\text{(VIII)} \xrightarrow{\text{H}^+} \text{R}^1(\text{CO}_2\text{H})_2$$
$$+ \text{R}^1(\text{CO}_2\text{H})(\text{CO}_2\text{Glc U}) \text{ (IX)} \rightarrow \text{(I)}$$
$$+ \text{R}^1(\text{CO}_2\text{Glc U})_2 \qquad \text{(X)} \rightarrow \text{(II)}$$

The free HO-1 group of (VI) is transformed to a mesylate (VI → VII), and the mesylate group displaced with the carboxyl dianion of biliverdin. The resulting biliverdin esters (VIII) are a mixture of true β-glycosidic esters (displacement at C-1) and α-ethyl glycosides (biliverdin esterified at the C-2 acetal function with concomitant rearrangement of the

[1] R. Bonnett, J. E. Davies, and M. B. Hursthouse, *Nature (London)* **262**, 326 (1976).
[2] K. P. M. Heirwegh, J. Fevery, R. Michiels, G. P. Van Hees, and F. Compernolle, *Biochem. J.,* **145**, 185 (1975).
[3] F. Compernolle, *Carbohydr. Res.* **83**, 135 (1980).
[4] F. Compernolle, *FEBS Lett.* **114**, 17 (1980).
[5] F. Compernolle, *Biochem. J.,* **187**, 857 (1980).
[6] N. Pravdic and D. Keglevic, *J. Chem. Soc.* p. 4633 (1964).

METHODS IN ENZYMOLOGY, VOL. 77

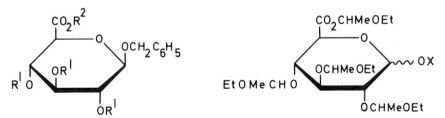

FIG. 1. Bilirubin mono- (I) and di-β-glucuronosides (II) (R^1 = H and β-D-glucopyranuronic acid, respectively).

acetal ethoxy group to C-1).[5] Mild acid treatment of (VIII) removes the protecting groups *and* the biliverdin ester acetals at C-2, yielding biliverdin and biliverdin mono- and di-β-glucuronosides (IX) and (X). The biliverdin glucuronosides are separated by tlc. and reduced with NaCNBH$_3$ to the corresponding bilirubin mono- and di-β-glucuronosides (I) and (II).

Procedure

Benzyl β-D-Glucopyranosiduronic Acid (IV)

A mixture of methyl (benzyl 2,3,4-tri-O-acetyl-β-D-glucopyranosid)-uronate (III)[3,6] (500 mg) with methanol (30 ml), triethylamine (15 ml), and water (15 ml) is stirred at 0° until dissolution is complete (3 hr).

FIG. 2. Intermediates in the synthesis of bilirubin mono- and di-β-glucuronosides. (III) R^1 = Ac, R^2 = Me; (IV) R^1 = R^2 = H; (V) R^1 = R^2 = CHMeOEt; (VI) X = H; (VII) X = SO$_2$Me.

After 3 days at 22°, the solution is evaporated under reduced pressure at 40°–50° and an aqueous solution of the residue is passed through Dowex 50 ion-exchange resin (Fluka, 30 ml wet volume, 50 W–X8, 20–50 mesh, H⁺) packed in water in a 20 × 2 cm column. Elution with water (100 ml collected) followed by evaporation and drying under reduced pressure yields (**III**) (0.31 g) as a colorless syrup, $[\alpha]_D$ −74° (C = 0.5, water), R_f value[7] ca. 0.5.

α-*Ethoxyethyl 2,3,4-Tri-O-(α-Ethoxyethyl)-*D-*Glucopyranuronate* (**VI**)

To a solution (at 0°) of (**IV**) (0.31 g) in dry tetrahydrofuran (30 ml) are added ethyl vinyl ether (5 ml) and trifluoroacetic acid (0.3 ml). After closing the reaction vessel (ground-glass joint and Parafilm), the mixture is kept at 0° for 3 days and then at 22° for 3 hr. Triethylamine (5 ml), dichloromethane (50 ml), and water (50 ml) are added in sequence, and the organic layer is washed with water (4 × 50 ml), dried over molecular sieves (2 hr), and concentrated, to yield syrupy (**V**), $[\alpha]_D$ −20° (C = 1, chloroform). Compound (**V**) is dissolved immediately in ethyl acetate (30 ml); triethylamine (0.2 ml) and 10% palladium-on-carbon (500 mg) are added; and the mixture is shaken with hydrogen (2 atm) at 22° for 3 to 6 hr until complete disappearance of the peak at m/e 499 in the electron-impact mass spectrum.[8] The solution of (**VI**) is filtered and stored moisture-free at 0° until needed for preparation of (**VII**). Evaporation under reduced pressure yields 0.55 g of oily (**VI**).

α-*Ethoxyethyl 2,3,4-Tri-O-(α-Ethoxyethyl)-1-O-Mesyl-* D-*Glucopyranuronate* (**VII**)

Compound (**VI**) (30 mg) is dissolved in dry dichloromethane (5 ml) and dry s-collidine (0.3 ml). After mixing the cooled solution (0°) with methanesulfonyl chloride (50 μl) and reacting at 0° for 10 min, the organic layer is shaken vigorously and repeatedly with a saturated

---

[7] Thin-layer chromatography was conducted on silica gel (E. Merck F254, layer thickness 0.25 mm) with 9:2, v/v, 1-propanol/water. Detection was effected by spraying with a 3:1, v/v, mixture of 0.2% naphthalene-1,3-diol in ethanol and 20% sulfuric acid, followed by heating at 120°.

[8] Alternative procedures A and B for determining the extent of hydrogenolysis of the benzyl group are based on acetylation of HO-1 of (VI) and removal of protecting groups. Analysis of the resulting 1-*O*-acetyl-α,β-D-glucopyranuronic acids reveals the presence of any residual 1-*O*-benzyl compound, detected as (IV). (A) A sample (0.2 ml) of the solution is filtered and concentrated under reduced pressure. The residue is treated with pyridine (0.2 ml) and acetic anhydride (0.1 ml) for 1 hr at 22°. Reagents are evaporated under reduced pressure and the residue is dissolved in acetic acid (0.4 ml) and 0.1 M hydrochloric acid

$NaHCO_3$–Na_2CO_3 buffer solution (pH 9.0, 6 × 5 ml, 22°) to remove excess methanesulfonyl chloride. The dichloromethane layer is washed with water (3 × 5 ml) and dried over molecular sieves (0.4-nm pore diameter) for 2 hr.

Biliverdin Glucuronosides (**IX**) *and* (**X**)

Biliverdin[9] (6.0 mg) is converted into the dianion by addition of 0.1 M tetrabutylammonium hydroxide (Merck, methanol/2-propanol solution, 0.21 ml) and dichloromethane (1 ml). The solution is evaporated to dryness under reduced pressure (60°, rotary pump). The residue is dissolved in dichloromethane (5 ml) containing mesylate (**VII**), prepared from 30 mg of (**VI**). Molecular sieve (pore size 0.4 nm) is added, and after 1 day at 22° the solvent and remaining s-collidine are evaporated under reduced pressure (the temperature is not allowed to exceed 30° by using an efficient rotary pump). Thin-layer chromatography with 9 : 1, v/v, chloroform/methanol reveals formation of mainly diesters (R_f 0.72 and 0.69), and smaller amounts of monoesters (R_f 0.22) and biliverdin (R_f 0.02). The residue is hydrolyzed with acetic acid (1 ml) and aqueous 0.1 M HCl (0.25 ml) at 22° for 30 min, followed by evaporation under reduced pressure (22°, rotary pump). The residue is dissolved in a minimal amount of acetic acid and nine volumes of 2-propanol. This solution is applied to precoated tlc. plates (Merck F254, silica gel, 20 × 20 cm, one-tenth and one-half of total amount, respectively, to plates of 0.25- and 1.0-mm layer thickness). The plates are developed in the dark with 100 : 50 : 10 : 1, v/v, chloroform/methanol-water/acetic acid for about 18 cm. Biliverdin (R_f 0.82), biliverdin mono-β-glucuronoside (R_f 0.33) and biliverdin di-β-glucuronoside (R_f 0.13) are eluted from the adsorbent with 3 : 1, v/v, methanol/acetic acid. Yields are approximately 6 and 1.5% for the mono- and diglucuronoside, respectively, as calculated from values for A_{376} and A_{666} (assuming ϵ = 50,800 and 14,400 liter mol^{-1} cm^{-1}, respectively, as for biliverdin[9]). The eluates are evaporated under reduced pressure in the dark at 22°.

(0.1 ml) for 30 min. The solution is concentrated under reduced pressure at 20°–30° (Residue a). R_f-values[7] are ca. 0.3 for 1-O-acetyl and ca. 0.5 for 1-O-benzyl compounds. (B) Gas-liquid chromatography (1.8 m × 6 mm packed column of 3% OV-101, 150°, 4°/min) is performed on trimethylsilyl derivatives prepared by reaction of Residue a with N,O-bistrimethylsilylacetamide (20 μl) and pyridine (50 μl) for 30 min. The α- and β-1-O-acetyl derivatives are eluted at 172.5 and 174° (ratio 2 : 1) whereas the β-1-O-benzyl compound is eluted at 201°.

[9] A. F. McDonagh, *in* "The Porphyrins" (D. Dolphin, ed.), Vol. 6, p. 293. Academic Press, New York, 1979.

Bilirubin Mono- and Di-β-Glucuronosides (**I**) *and* (**II**)

The purified biliverdin glucuronosides are dissolved in a minimal amount of acetic acid. After dilution with nine volumes of 2-propanol, the solution is shaken with an excess of solid $NaCNBH_3$ for a few minutes until complete disappearance of blue-green color. When kept in the dark at $0°$ in the presence of $NaCNBH_3$, the resulting solutions of bilirubin mono-β-glucuronoside (**I**) and di-β-glucuronoside (**II**) are rather stable with respect to oxidation and disproportionation (exchange of dipyrrole groups). Preparative TLC with $10:5:1$, v/v, chloroform/methanol/water under an atmosphere of argon in the dark (elution with methanol) affords (**I**) (R_f 0.46) and (**II**) (R_f 0.12). To avoid dipyrrole exchange during evaporation ($22°$) of methanol eluates, care should be taken to exclude light and oxygen. As a protective measure, ascorbic acid or $NaCNBH_3$ may be added to the methanol eluates.

[55] Mixed (Unsymmetric) Disulfides:
Coenzyme A–Glutathione Disulfide as an Example

By Bengt Mannervik and Kerstin Larson

Strategy

Mixed disulfides exist in various tissues and cells (see Mannervik and Eriksson[1] for a review). They are easily formed by oxidation of mixtures of thiols, by thiol–disulfide interchange, and by disulfide–disulfide exchange. For preparative procedures it is essential to optimize the yield of mixed disulfide by proper design of synthetic reactions, especially when a component is expensive or difficult to obtain in large quantities. Three types of reactions can be recommended for conversion of a thiol (RSH) to a mixed disulfide (RSSR'):

$$RSH + R'SSR' \rightleftharpoons RSSR' + R'SH \qquad (1)$$
$$RSH + R'SSO_3^- \rightleftharpoons RSSR' + HSO_3^- \qquad (2)$$
$$RSH + R'SSO_2R' \rightarrow RSSR' + R'SO_2^- + H^+ \qquad (3)$$

The first reaction has an equilibrium constant near 1 for many naturally occurring thiols and disulfides, but may be used for almost complete conversion of RSH to RSSR' by use of a large excess of R'SSR' as shown by Eldjarn and Pihl.[2] This reaction may be suitable when R'SH and surplus R'SSR' can easily be separated from the mixed disulfide. Exam-

[1] B. Mannervik and S. A. Eriksson, *in* "Glutathione" (L. Flohé, H. C. Benöhr, H. Sies, H. D. Waller, and A. Wendel, eds.), p. 120. Thieme, Stuttgart, 1974.
[2] L. Eldjarn and A. Pihl, *J. Biol. Chem.* **225**, 499 (1957).

METHODS IN ENZYMOLOGY, VOL. 77

ples include the syntheses of mixed disulfides of glutathione and egg white lysozyme[3] and rat liver proteins,[4] in which the protein derivative can be freed from low-molecular-weight reactants by dialysis or gel filtration. When an aliphatic thiol reacts with an aromatic disulfide, a stoichiometric conversion of the thiol to mixed disulfide may be expected according to Reaction 1. This combination of reactants, which does not require excess disulfide, has been exploited in the preparation of the mixed disulfide of 3-carboxy-4-nitrobenzenethiol and glutathione.[5]

Reaction (2) generally also has an unfavorable equilibrium constant for stoichiometric utilization of RSH. In this case, however, as an alternative to use of a large excess of thiosulfate ester $(R'SSO_3^-)$, strontium ions may be introduced to precipitate sulfite and drive the reaction to completion. Procedures have been published for preparation of the naturally occurring thiosulfate esters S-sulfocysteine[6] and S-sulfoglutathione.[7]

Reaction (3) is practically irreversible and proceeds essentially to completion with stoichiometric amounts of reactants. Here the difficulty may be the preparation of the thiolsulfonate, $R'SSO_2R'$. The procedure has been exploited in the synthesis of anti-radiation drugs[8] and has been used in our laboratory[9] for syntheses of mixed disulfides of coenzyme A and glutathione,[10] of cysteine and glutathione,[11] and of pantetheine and glutathione.[12] The synthesis of the first of these mixed disulfides (CoASSG) is described as an example.

Preparation of Coenzyme A–Glutathione Disulfide

Principle

The thiolsulfonate analog of glutathione disulfide $(GSSO_2G)$ is prepared and allowed to react with the more expensive coenzyme A.[10]

Preparation of $GSSO_2G$

$GSSO_2G$ is obtained from glutathione disulfide (GSSG) by oxidation with H_2O_2/formic acid. GSSG (400 mg) is dissolved in 3 ml of formic acid

[3] K. Axelsson and B. Mannervik, *FEBS Lett.* **53**, 40 (1975).

[4] K. Axelsson and B. Mannervik, *Biochim. Biophys. Acta* **613**, 324 (1980).

[5] B. Mannervik, *Acta Chem. Scand.* **24**, 1847 (1970).

[6] I. H. Segel and M. J. Johnson, *Anal. Biochem.* **5**, 330 (1963).

[7] B. Eriksson and M. Rundfelt, *Acta Chem. Scand.* **22**, 562 (1968).

[8] L. Field, T. C. Owen, R. R. Crenshaw, and A. W. Bryan, *J. Am. Chem. Soc.* **83**, 4414 (1961).

[9] By B. Mannervik (formerly B. Eriksson) and co-workers.

[10] B. Eriksson, *Acta Chem. Scand.* **20**, 1178 (1966).

[11] B. Eriksson and S. A. Eriksson, *Acta Chem. Scand.* **21**, 1304 (1967).

[12] B. Mannervik and G. Nise, *Arch. Biochem. Biophys.* **134**, 90 (1969).

(analytical grade) containing 0.06 ml of concentrated HCl. (Omission of HCl leads to formation of the sulfonic acid, GSO_3H, instead of the thiolsulfonate.) A total of 160 μl of 30% H_2O_2 (analytical grade) is added with efficient stirring in portions of 5 μl over a period of 60 min. Ninety minutes after the start of the oxidation, the reaction mixture is concentrated to a clear syrupy residue on a rotary evaporator. The residue is dissolved in 4 ml of deionized water. At this stage, a 50% yield of $GSSO_2G$ should have been obtained[10]; the remaining material is mostly unreacted GSSG. $GSSO_2G$ is labile and is preferably used in crude form under slightly acidic conditions. The absence of excessive amounts of GSO_3H can be checked by electrophoresis or thin-layer chromatography is acidic systems (see Refs. 7 and 11); $GSSO_2G$ migrates essentially as GSSG in most systems. The presence of $GSSO_2G$ in electropherograms and chromatograms can be verified by its strong reaction with the iodoplatinate reagent.[13] Alternatively, $GSSO_2G$ is converted by alkaline dismutation to the sulfinic acid, GSO_2H [Reaction (4)], before analytical separation

$$3GSSO_2G + 4OH^- \rightarrow 4GSO_2^- + GSSG + 2H_2O \qquad (4)$$

Preparation of CoASSG

The crude $GSSO_2G$ preparation is adjusted to pH 3 with about 4 ml of 1 M NH_3. Too much alkali causes decomposition [Reaction (4)]. A solution of 100 mg coenzyme A in 2 ml of deionized H_2O is added dropwise with stirring. The almost instantaneous disappearance of free sulfhydryl groups (of CoASH) can be verified by dipping the tip of a thin strip of filter paper into the reaction medium and then into a solution of nitroprusside.[13]

Purification of CoASSG

The reaction mixture contains, in addition of CoASSG, components from the $GSSO_2G$ solution, impurities in the coenzyme A preparation, and GSO_2H formed according to Reaction (3). The solution is applied to a column containing 10 g of DEAE-Sephadex A-25, packed in deionized H_2O. The column (2 × 17 cm) is rinsed with deionized H_2O until UV-absorbing material no longer is eluted (about 100 ml). CoASSG is eluted and separated from impurities by a linear concentration gradient of NaCl or KCl. The gradient is formed from 1 liter of deionized H_2O and 1 liter of 1 M salt. CoASSG is the major UV-absorbing peak and its identity can be confirmed by positive ninhydrin and nitroprusside plus KCN[13] reactions of aliquots placed on filter paper. The CoASSG-containing fractions (about 90 ml) are pooled, concentrated on a rotary evaporator, and dis-

[13] G. Toennies and J. J. Kolb, *Anal. Chem.* **23**, 823 (1951).

solved in 6 ml of deionized H_2O. The salt is removed by gel filtration on a column (1.5 × 115 cm) of BioGel P-2 (100–200 mesh), packed and eluted with deionized H_2O. CoASSG, which can be detected by its UV absorption and positive ninhydrin reaction, is recovered in about 30 ml of effluent. Chloride ions, which emerge distinctly after CoASSG, can be demonstrated by formation of a precipitate with $AgNO_3$ in the presence of HNO_3. For storage, the CoASSG solution may be freeze-dried. At 4° the solution is stable for several days and at −20° for several weeks provided that the pH value is neutral or, preferably, slightly acidic. The yield of purified CoASSG is typically (more than 10 preparations) 50%, calculated on the basis of coenzyme A used in the preparation. This corresponds to about 70 μmol (approximately 70 mg) of CoASSG in the scale described here. Quantitative analyses of CoASSG can be made by use of its UV absorbance or by enzymatic methods.[10,14] A different procedure based on Reaction (1) yields about 35% of CoASSG.[14]

Quantitative Analysis of Mixed Disulfides

Every mixed disulfide presents its individual possibilities and problems in analyses. Various chromatographic and electrophoretic methods can be quantitated for known compounds. In some cases, reduction of the disulfide may yield a thiol that can accurately and specifically be determined quantitatively, e.g., by enzymatic analysis. A general method for determination of the titer of disulfide bonds has been based on reduction of the disulfide with GSH [Reaction (5)] monitored spectrophotometrically at 340 nm by coupling with the reaction catalyzed by glutathione reductase [Reaction (6)]

$$2GSH + RSSR' \rightleftharpoons GSSG + RSH + R'SH \qquad (5)$$
$$GSSG + NADPH + H^+ \rightarrow 2GSH + NADP^+ \qquad (6)$$

Reaction (5) is catalyzed by thioltransferase, which has a broad substrate specificity.[15–17]

Reagents

Sodium phosphate, 0.2 M, pH 7.5, containing 1 mM EDTA
NADPH, 3 mM
GSH, 10 mM
Glutathione reductase, commercially available
Thioltransferase, prepared as described in Refs. 15 and 17

[14] R. N. Ondarza, this series, Vol. 18, Article [54], p. 318.
[15] K. Axelsson, S. Eriksson, and B. Mannervik, *Biochemistry* 17, 2978 (1978).
[16] B. Mannervik, in "Enzymatic Basis of Detoxication" (W. B. Jakoby, ed.), Vol. 2, p. 229. Academic Press, New York, 1980.
[17] B. Mannervik, K. Axelsson, and K. Larson, this volume, Article [36].

Procedure. Add 800 μl of buffer, 50 μl of NADPH, 50 μl of GSH, 2 units of glutathione reductase, and 0.1 unit of thioltransferase to a 1-ml cuvette in a spectrophotometer and follow the absorbance at 340 nm. After reduction of any GSSG in the GSH solution, the decrease of A_{340} should be low or nil. Add the unknown sample to the cuvette and record the decrease in A_{340} due to oxidation of NADPH. The amount of disulfide added is calculated by use of ΔA_{340} and the extinction coefficient of NADPH (ϵ_{340} = 6.2 mM^{-1} cm^{-1}). A correction must be made for the dilution caused by addition of the sample. The total amount of disulfide added to the cuvette must not exceed 0.1 μmol. Some disulfides may react sufficiently rapidly in the absence of thioltransferase that the analysis can be made without this enzyme. However, generally the end point of the reaction will be difficult to establish in the absence of thioltransferase; during prolonged incubation periods, the thiols in the cuvette may be reoxidized under aerobic conditions.

[56] Thioesters of Glutathione[1]

By Lasse Uotila

Two widely occurring enzymes, formaldehyde dehydrogenase (EC 1.2.1.1) and glyoxalase I (EC 4.4.1.5 lactoyl-glutathione lyase) catalyze the synthesis of thiol esters of glutathione from aldehydes and GSH.[2,3] At least three separate enzymes specifically catalyze the hydrolysis of glutathione thiol esters[4]; all three, glyoxalase II (EC 3.1.2.6 hydroxyacylglutathione hydrolase), S-formylglutathione hydrolase (EC 3.1.2.12), and S-succinylglutathione hydrolase (EC 3.1.2.13), have been purified from human liver.[5-7]

Enzymatic and chemical methods devised for the synthesis of purified glutathione thiol esters[4] (substrates for the above mentioned enzymes) are described in this chapter.

---

[1] Supported in part by the Sigrid Jusélius Foundation, Finland.
[2] L. Uotila and M. Koivusalo, *J. Biol. Chem.* **249**, 7653 (1974).
[3] E. Racker, *J. Biol. Chem.* **190**, 685 (1951).
[4] L. Uotila, *Biochemistry* **12**, 3938 (1973).
[5] L. Uotila, *Biochemistry* **12**, 3944 (1973).
[6] L. Uotila and M. Koivusalo, *J. Biol. Chem.* **249**, 7664 (1974).
[7] L. Uotila, *J. Biol. Chem.* **254**, 7024 (1979).

S-2-Hydroxyacylglutathione

This group of compounds is prepared most conveniently by the reaction catalyzed by glyoxalase I. The reaction mixture contains 50 mM potassium phosphate, pH 6.6, 50 mM GSH (neutralized with NaOH to pH 7), 50 mM α-ketoaldehyde, and yeast glyoxalase I, 0.2–0.5 units per ml. The pH is checked and adjusted to 6.6 with NaOH if necessary. The volume of the mixture may vary; the author has usually carried out the synthesis with 1 mmol of glutathione, which results in a volume of 20 ml. The progress of the reactions at 25° is followed by the increase in absorbance of the synthesized thioester at 240 nm; an exception occurs in the case of S-mandeloylglutathione, in which decreasing absorbance is followed at 263 nm. The following thiol esters of glutathione have been prepared in this way (the corresponding α-ketoaldehyde is mentioned within parentheses): lactoyl (methylglyoxal, purified by distillation), glycolyl (glyoxal), glyceroyl (hydroxypyruvaldehyde[8]), mandeloyl (phenylglyoxal), and 3-ethoxy-2-hydroxybutyryl[9] (kethoxal). S-Lactoylglutathione is prepared in 90% yield. The yield of the other compounds ranges from 55 to 80% but may be improved if a higher enzyme concentration is used in order to shorten the reaction period. In the case of S-lactoylglutathione, the reaction can proceed overnight at 25° because significant nonenzymatic hydrolysis of this thiol ester does not occur.

For purification, the solution of thiol ester is applied at 4° to a Dowex-1 (X-4, 200–400 mesh) column in the formate form. A column, 2 × 50 cm, is sufficient for at least 10 ml of the reaction mixture (containing 0.5 mmol of thiol ester plus glutathione). The thiol esters are assayed by continuous monitoring of the eluate at 254 nm. Traces of α-ketoaldehydes, GSSG, and enzyme are easily removed, but complete separation of the thiol ester from GSH is achieved only for S-mandeloylglutathione and S-lactoylglutathione. In the purification of S-mandeloylglutathione, GSH and GSSG are first removed by eluting with 600 ml of 0.3 M formic acid; the thiol ester is eluted with 600 ml of 1.0 M formic acid. For S-lactoylglutathione and the other thiol esters, a linear gradient of formic acid (volume 1600 ml, established between 0.05 and 0.35 M formic acid) is used. The column may be eluted at 100 ml per hr. The peak of GSH is eluted first with S-lactoylglutathione soon after it (at 0.20–0.25 M formic acid). Although the elution volumes of these compounds differ by only about 12%, they are completely separated because the peaks are eluted sharply. The other three thiol esters can be only partially separated from

---

[8] The compound is synthesized according to H. C. Reeves and S. J. Ajl, *J. Biol. Chem.* **240**, 569 (1965).

[9] L. Uotila and M. Koivusalo, *Eur. J. Biochem.* **52**, 493 (1975).

GSH. The resulting solutions of thiol ester are concentrated and freed from formic acid by lyophilization.

A modification of this method has been devised[10] in which the mixture of synthesized S-lactoylglutathione is treated with a methanolic solution of diamide before Dowex-1 chromatography in order to oxidize GSH to GSSG. Aside from GSSG, both the oxidized and reduced forms of diamide are readily separable from S-lactoylglutathione by chromatography with Dowex-1 as described earlier. It is probable that pure glycolyl, glyceroyl, and 3-ethoxy-2-hydroxybutyrylthiol esters of glutathione could be achieved with this modification.

S-Acetylglutathione

Glutathione (620 mg, 2 mmol) is dissolved to 3 ml water and the pH of the solution adjusted to about 6 with NaOH. Ethanol is added to 35% (v/v), final concentration. Thiolacetic acid, 0.72 ml (10 mmol), is added in a hood with constant stirring. The pH of the solution is adjusted to 4.5. The reaction mixture is stirred for 4 to 5 hr at 25°. The progress of the reaction may be followed by measuring the increase of A_{240} of small aliquots after extraction with ether. The product, in about 80% yield, is cleared by filtration and concentrated in a rotary evaporator at 30°. The residue is freed from thiolacetic acid by extensive ether extractions or, more conveniently, with a column (2 × 50 cm; load 4 ml) of Sephadex G-10 equilibrated with water. The thiol ester is eluted much before thiolacetic acid and is virtually free from glutathione. The purified product is concentrated by lyophilization.

S-Propionylglutathione

A method analogous to that for S-acetylglutathione is used in which thiolpropionic acid is the acylating agent. The ethanol concentration is increased to 65%, 7 mmol of the thiol acid are used for 1 mmol of GSH, and the reaction time is longer (6–7 hr). The yield of S-propionylglutathione is about 45%. Purification is the same as for S-acetylglutathione. Some GSH, 5% of the thiol ester, will remain as an impurity. This may be removed by Dowex-1 (formate) ion-exchange chromatography as described for S-lactoylglutathione.

S-Butyrylglutathione and S-benzoylglutathione can also be synthesized by acylating GSH with the corresponding thiol acids.

[10] J. C. Ball and D. L. Vander Jagt, *Anal. Biochem.* **98**, 472 (1979).

S-Acetoacetylglutathione

This is synthesized analogously to the procedure of Lynen et al.[11] for acetoacetyl-CoA. GSH (310 mg, 1 mmol) is dissolved in water (1 to 2 ml) and the pH adjusted to 7.0 with 1 M NaOH. Diketene (100 μl, 1.27 mmol) is added and the pH adjusted to 7.0 with a few drops of 1 M NaOH. If necessary, more water is added until diketene has dissolved. The mixture is stirred at 0° for 2.5 hr and then frozen. Residual diketene will be destroyed in a few days at −20°. The product is freed from acetoacetate by extracting 10 times with two volumes of ether at pH 2. The yield is 95% with only a small amount of contaminating GSH.

S-Succinylglutathione

GSH (310 mg, 1 mmol) is dissolved and made up to 2 ml of water, and the pH is adjusted to 6.6 with 1 M NaOH. A freshly prepared solution of 0.1 M succinic anhydride in ethanol (15 ml, 1.5 equivalent) is added and the pH readjusted to 6.6 with NaOH. If GSH precipitates, just enough water is added to retain GSH in solution. The reaction is allowed to proceed with mixing at 0° for 90 min. The yield is about 75%; most of the rest is present as GSH. A control without GSH may be included and stored similarly; from this, one can determine when residual succinic anhydride has been hydrolyzed with the hydroxamate assay.[12] The product is freed from succinate and GSH by Dowex-1 formate chromatography as described earlier. S-Succinylglutathione is eluted much later than GSH (thiol ester at 0.5 M and GSH at 0.2 M formic acid). However, more than half of the S-succinylglutathione is hydrolyzed during this procedure; the GSH-free preparation of the thiol ester is very unstable at pH 7 and 25°. The product, containing some free GSH, is more stable and, therefore, easier to use in enzyme assays.

S-Formylglutathione

Equivalent amounts of formic acid (98 to 100%) and acetic anhydride (over 99%) are combined (e.g., 2.90 ml, 75 mmol formic acid and 7.20 ml, 75 mmol acetic anhydride), and the mixture stirred at 23° for 60 min to prepare acetic–formic anhydride. Thioglycolic acid (1.1 ml, 15 mmol) is added and the stirring continued at 23° for another 60 min. Water (10 ml) is

[11] F. Lynen, U. Henning, C. Bublitz, B. Sörbo, and L. Kröplin-Rueff, Biochem. Z. 330, 269 (1958).

[12] F. Lipmann and L. C. Tuttle, J. Biol. Chem. 159, 21 (1945); see also E. R. Stadtman, this series, Vol. 3, Article [39].

added with stirring and cooling to hydrolyze the remaining anhydride. The mixture is concentrated to dryness with a rotary evaporator at 27°. The yield of acylthioglycolate is about 95%; most of it is formylthioglycolate, although some acetylthioglycolate is also formed.

The concentrated solution of formylthioglycolate in water (14 mmol; about 2 ml) is adjusted to pH 5.0 with continuous stirring and cooling. To avoid dilution, a strong base (6 M NaOH) is used. GSH (434 mg, 1.4 mmol) is dissolved in the partially neutralized solution and the pH again set to 5.0 with a few drops of 6 M NaOH. The mixture is stirred for 60 min at 23° and pH 5.0, and then cooled to 0°. The pH is adjusted to 2.5–3.0 with 12 M HCl and some of the formylthioglycolate is removed by extracting two or three times with one volume of ether. The product is further purified by gel chromatography on Sephadex G-10 equilibrated with water. A column of 2 × 50 cm is sufficient for a 5-ml load. The first UV-absorbing peak contains S-formylglutathione and GSH in the approximate ratio of 1 : 4. The fractions containing this peak are concentrated by lyophilization. The yield in the synthesis is about 25% but decreases to about 15% during purification. No S-acetylglutathione is formed. Because of the instability of S-formylglutathione, attempts to remove excess GSH by ion-exchange or paper chromatography have been unsuccessful.

S-Pyruvylglutathione

Pyruvic acid (4 mmol) and thioglycolic acid (4 mmol) are combined in a vessel maintained at 0° and equipped with a drying tube. With continuous stirring, a freshly prepared 0.5 M solution of N,N-dicyclohexyl-carbodiimide (20 ml, 10 mmol) in N,N-dimethylformamide is added dropwise during 90 min. The mixture is stirred for 2.5 hr at 0°. After the addition of water (10 ml), the mixture is filtered and the filter washed with water (40 ml). The filtrate is concentrated under reduced pressure and the pH adjusted to 3.0. The solution is extracted seven or eight times with one volume of ether for each extraction. The ether phases are combined, the solvent is evaporated, and 10 ml of water are added. The mixture is clarified by filtration, the filter is washed with water, and the filtrate is concentrated under reduced pressure. The product contains S-pyruvylthioglycolate in 50% yield.

GSH is shaken with a 5-fold excess of pyruvylthioglycolate for 8 min at pH 7.5 and 25°. The mixture is cooled to 0° and the pH adjusted to 3.0. Purification of S-pyruvylglutathione is carried out as described for S-formylglutathione, i.e., ether extractions, Sephadex G-10 gel chromatography, and lyophilization. The yield of S-pyruvylglutathione is 30%.

Other Methods for S-Acylglutathione Synthesis

Wieland et al.[13,14] have devised a general method by which a mixed anhydride is first formed from the acid and ethyl chloroformate. Subsequently, thiophenol is acylated with the mixed anhydride and the acyl group transferred to GSH (for details, see also Stadtman[15]). Carefully dried and freshly distilled solutions are usually required because of the reactivity of the mixed anhydrides with water. The principles described here for S-pyruvylglutathione synthesis can apparently also be used for incorporation of other acyl groups. Wieland and Köppe[16] have described a method for the synthesis of several acid chlorides as well as for the corresponding thiol esters. However, the use of a potentially hazardous reagent is required. Wieland and Schäfer[17] and Sachs and Waelsch[18] have described methods for S-aminoacylglutathione synthesis.

Assay of Purified Thiol Esters of Glutathione

The hydroxamate method[12] is accurate only when the same hydroxamate is available as standard or can be formed from a standard compound. For example, succinic anhydride may be used as a standard for the assay of S-succinylglutathione. Otherwise S-acylglutathione content can be determined[4] by completely hydrolyzing the thiol ester by glyoxalase II or neutral hydroxylamine and assaying the amount of GSH formed with 5,5'-dithiobis(2-nitrobenzoate),[19] 2,2'-dithiodipyridine, or 4,4'-dithiodipyridine.[20] The hydrolysis time must be short in order to prevent significant oxidation of thiols. The disulfide chosen should be included directly in the hydrolysis mixture. By these techniques, the following $\Delta\epsilon$ values (M^{-1} cm^{-1}) have been determined[4,9] for the thiol ester bond absorption at 240 nm of various glutathione thiol esters: lactoyl, 3310; glyceroyl, 3370; glycolyl, 3260; mandeloyl, 4200; 3-ethoxy-2-hydroxybutyryl, 4100; acetoacetyl, 3400; succinyl, 3250; formyl, 3300; acetyl, 2980; and propionyl, 3070.

[13] T. Wieland and H. Köppe, Justus Liebigs Ann. Chem. 581, 1 (1953).
[14] T. Wieland and L. Rueff, Angew. Chem. 65, 186 (1953).
[15] E. R. Stadtman, this series. Vol. 3, Article [137].
[16] T. Wieland and H. Köppe, Justus Liebigs Ann. Chem. 588, 15 (1954).
[17] T. Wieland and W. Schäfer, Justus Liebigs Ann. Chem. 576, 104 (1952).
[18] H. Sachs and H. Waelsch, J. Am. Chem. Soc. 77, 6600 (1955).
[19] G. L. Ellman, Arch. Biochem. Biophys. 82, 70 (1959).
[20] D. R. Grassetti and J. F. Murray, Jr., Arch. Biochem. Biophys. 119, 41 (1967).

Stability and Storage

S-Lactoylglutathione, S-acetylglutathione, and S-propionyl-glutathione, stored in solution at pH 3 to 6, are stable for several months at $-20°$. S-Formylglutathione and S-succinylglutathione are unstable at $-20°$ but stable for months when stored in small aliquots at $-70°$. The other thiol esters have intermediate stability and should preferably be stored at $-70°$ or lower.

Comment

When using a new chemical method for the synthesis of glutathione thiol esters, it is necessary to ensure by amino group analysis[21] that the potentially reactive, free amino group of GSH is not acylated. This requirement is met in all the syntheses detailed in this chapter.

[21] S. Moore and W. H. Stein, *J. Biol. Chem.* **176**, 367 (1948).

[57] Coenzyme A Thioesters of Benzoic, Hydroxybenzoic, Phenylacetic, and Bile Acids

By LESLIE T. WEBSTER, JR. and PAUL G. KILLENBERG

Substrate quantities of coenzyme A thioesters of benzoic, hydroxy-benzoic, phenylacetic, and bile acids are used for the *in vitro* study of reactions and enzymes involved in either acyl-CoA formation or acyl transfer with resulting peptide bond formation. Chemical synthesis of the CoA derivatives can be achieved by reacting the acid chloride or acid anhydride of the selected acid with reduced coenzyme A in an aqueous solution under mildly alkaline conditions. The resulting CoA thioesters are partially purified by extracting some of the contaminants into organic solvents under acidic conditions; complete purification of the acyl-CoA is achieved by column chromatography on Sephadex LH-20. Solvent extraction and Sephadex LH-20 chromatography may also be used to obtain homogeneous acyl-CoA compounds from impure preparations or from enzymatic reaction mixtures containing labeled substrate.

Benzoyl-CoA

Reagents

Benzoic anhydride: Pulverize with mortar and pestle to increase surface area

CoASH
Sodium bicarbonate

Procedure.[1] The CoA thioester of benzoic acid is synthesized from benzoic anhydride by a modification of the method of Schachter and Taggart.[2] CoASH, 200 mg, is dissolved in 5 ml water, and the pH adjusted to 8.0 by addition of solid $NaHCO_3$. Benzoic anhydride, 70 mg, is added to this solution, and the vessel is flushed with nitrogen. The solution is stirred in a water bath at 38° for 2 hr; pH is maintained between 7 and 8 by occasional addition of $NaHCO_3$ to prevent formation of benzoic acid. When no free thiol is detected by the nitroprusside reaction,[3] the solution is brought to pH 3.5 with concentrated HCl and extracted four times with ether. The final aqueous phase is further purified by chromatography on Sephadex LH-20 as noted in a later section.

Hydroxybenzoyl-CoA

This procedure permits synthesis of the CoA thioesters of *o-*, *m-;* and *p*-hydroxybenzoic acids. The synthesis of *o*-hydroxybenzoyl-CoA (salicyl-CoA) is described.

Reagents

Oxalyl chloride, freshly distilled
o-Hydroxybenzoic acid (salicylic acid)
CoASH
Lithium hydroxide, 2 *M*

Procedure.[4] The acyl chloride of salicylic acid is prepared by refluxing 25 g of oxalyl chloride and 10 g of salicylic acid in 250 ml benzene for 2 hr. Salicyl chloride is obtained by fractional distillation under reduced pressure (vacuum pump) at about 80°. An estimated 2- to 3-fold molar excess by weight of the freshly synthesized salicyl chloride is immediately added dropwise to a rapidly stirred solution of 200 mg CoASH in 4 ml water maintained at pH 8.0 by addition of 2 *M* LiOH. The reaction is allowed to procede until no free thiol is detected by the nitroprusside reaction.[3] The speed of the reaction varies with the nature of the acyl chloride; acylation of CoASH in the presence of salicyl chloride requires less than 3 min for completion. Other acyl chlorides react more slowly. After the reaction is complete, the pH is brought to 3.0 with 5 *N* HCl, and the precipitate is removed by centrifugation. The aqueous supernatant

---

[1] W. B. Forman, E. D. Davidson, and L. T. Webster, Jr., *Mol. Pharmacol.* 7, 247 (1971).
[2] D. Schachter and J. V. Taggart, *J. Biol. Chem.* 203, 925 (1953).
[3] G. L. Ellman, *Arch. Biochem. Biophys.* 82, 70 (1959).
[4] J. J. Mieyal, L. T. Webster, Jr., and U. A. Siddiqui, *J. Biol. Chem.* 249, 2633 (1974).

liquid is reduced almost to dryness with a rotary evaporator. The residue is extracted 8 to 10 times with 8 ml acetone: methanol (10 : 1, v/v) or until chloride (precipitate) no longer is detectable in the extract upon addition of several drops of 1% $AgNO_3$. The chloride-free residue that remains following extraction is dissolved in approximately 2 ml water and purified by chromatography as described later.

Phenylacetyl-CoA

Reagents

Phenylacetyl chloride: Pulvarize with mortar and pestle to increase surface area
CoASH
Lithium hydroxide, 2 *M*

Procedure. Synthesis of phenylacetyl-CoA is accomplished by a procedure similar to that for the hydroxybenzoyl-CoA's. Finely pulvarized phenylacetyl chloride, 300 to 400 mg, is added slowly to a rapidly stirred solution of 200 mg CoASH in 5 ml water at 4°. The pH is maintained at 8.0 during mixing by addition of 2 *M* LiOH. When synthesis is completed as determined by the nitroprusside reaction,[3] the solution is acidified to pH 3.5 with 5 *N* HCl and clarified by centrifugation. The supernatant liquid is reduced to near-dryness on a rotary evaporator and extracted with acetone: methanol (10 : 1, v/v) as described in the preceding section. The chloride-free residue is dissolved in 2 ml water and purified by chromatography on Sephadex LH-20.

Bile Acid-CoA

The following procedure is an adaptation of one described by Shah and Staple.[5] It may be applied to the synthesis of the CoA thioesters of cholic, deoxycholic, chenodeoxycholic, lithocholic, and 3β-hydroxy-5-cholenoic acids; synthesis of choloyl-CoA will be described.

Reagents

Cholic acid, recrystallized
2,4,6-Trimethylpyridine, freshly distilled and stored over BaOH
Tetrahydrofuran, redistilled in the presence of $FeSO_4$
Ethylchloroformate
CoASH
Lithium hydroxide, 1 *M*
Perchloric acid, 12% (v/v)

[5] P. P. Shah and E. Staple, *Steroids* **12**, 571 (1967).

Procedure.[6] The mixed anhydride of cholic acid is formed by dissolving 50 mg of cholic acid into 12 ml anhydrous methylene chloride (over calcium chloride) containing 0.04 ml 2,4,6-trimethylpyridine, 30°. To this is added 4 ml of methylene chloride containing 25 μl ethylchloroformate. After stirring for 2 hr at room temperature, the reaction mixture is taken to dryness with a rotary evaporator at 30°. The residue is dissolved in 10 ml tetrahydrofuran and added dropwise to 10 ml of nitrogen-equilibrated water[7] containing 105 mg CoASH. This solution is stirred, maintaining pH at 8.0 with 1 *M* LiOH. After 5 min, tetrahydrofuran is removed with a rotary evaporator at 30° and the aqueous residue is lyophilized. The lyophilized powder is dissolved in 2 ml water and extracted twice with 3 ml ethyl ether. Residual ether is removed by suffusion with nitrogen at 5°. The ether-free solution is acidified by addition of 6 drops 12% perchloric acid and the precipitate harvested by centrifugation. The precipitate is redissolved in 2 ml of methanol : water (40 : 60, v/v) and purified by Sephadex LH-20 chromatography.

Purification by Chromatography with Sephadex LH-20[4,6]

Approximately 2 ml of the synthetic product is applied to a column (2.5 × 30 cm) of Sephadex LH-20 previously equilibrated at 4° with either ethanol : water, (40 : 60, v/v) for purification of benzoyl-CoA, hydroxybenzoyl-CoA, and phenylacetyl-CoA; or methanol : water, (40 : 60, v/v) for bile acid-CoA. Products are eluted from the column at 4° with the equilibrating solvent; fractions of about 2.3 ml are collected at a flow rate of 9 to 18 ml per minute. Effluent absorbance is recorded continuously at 260 nm. The positions of ultraviolet-absorbing components in the eluate are localized more precisely by determining the $A_{260 \text{ nm}}$ of individual fractions.

The order of elution of the components in the solvent-extracted crude reaction mixtures is the same regardless of the elution solvent used. Thus, with the ethanol : water system used for benzoyl-CoA, hydroxybenzoyl-CoA, and phenylacetyl-CoA, oxidized coenzyme A appears first ($\sim$ 74 ml) followed in order by CoASH ($\sim$ 98 ml), acyl-CoA ($\sim$ 120 ml) and uncharacterized compounds with a stoichiometry of acid to CoA of 2 : 1 ($\sim$ 180 ml). If the pH of the crude acyl-CoA mixture applied to the column is sufficiently low, a second acyl-CoA peak may appear in the elution profile ($\sim$ 165 ml). Although the majority of the aromatic acids are removed from the crude reaction mixture by extraction into organic solvents at low pH, the traces that remain elute well after the acyl-CoA product under the preceding chromatographic conditions.

[6] P. G. Killenberg and D. F. Dukes, *J. Lipid Res.* **17**, 451 (1976).
[7] Water that has been boiled and allowed to cool in a stream of nitrogen.

In the case of the methanol : water system used to purify bile acid CoA thioesters, oxidized coenzyme A elutes at about 100 ml, followed in order by CoASH ($\sim$ 120 ml), bile acid-CoA ($\sim$ 150 ml), uncharacterized compounds with a bile acid to coenzyme A ratio of 2 ($\sim$ 200 ml), and any remaining bile acid ($\sim$ 230 ml). The uncharacterized compounds eluting at 200 ml appear when the ratio of bile acid to CoASH in the original synthesis exceeds 1.2; these compounds are inactive as substrates.

Chromatography on Sephadex LH-20 readily separates acyl-CoA from all other reaction mixture compounds except CoASH. Separation can be improved by oxidizing CoASH in the crude reaction mixture, e.g., with H_2O_2 before chromatography. Benzoyl-CoA, hydroxybenzoyl-CoA, and phenylacetyl-CoA fractions free of CoASH can be identified by thin-layer chromatography.[8] For bile acid-CoA thioesters, fractions on the leading edge of the acyl-CoA peak are assayed for CoASH[9] and bile acid.[10] Those fractions free of CoASH are pooled, lyophilized, redissolved in water, and stored at $-70°$. The overall yield of the various acyl-CoA products ranges from 25 to 60%, based on the initial amount of reduced coenzyme A in the synthesis.

Purity and Properties of Benzoyl-CoA, Hydroxybenzoyl-CoA's, and Phenylacetyl-CoA

The purified acyl-CoA's appear homogeneous when examined by several thin-layer chromatography systems[1] and upon rechromatography on Sephadex LH-20. The predicted stoichiometry is observed following elemental analysis (C,N,P,S,Li,H)[4] and when enzymatic release of CoASH is compared to dry weight.[4] Finally, proton magnetic resonance spectra[4] (D_2O) are consistent with the proposed structures and contain no extraneous signals.

The ultraviolet absorption spectra at pH 7 permit estimates of the extinction coefficient: 21.1 mM^{-1} cm^{-1} at 261 nm for benzoyl-CoA; 21.3 mM^{-1} cm^{-1} at 262 nm for 2-hydroxybenzoyl-CoA; 21.3 mM^{-1} cm^{-1} at 261 nm for 3-hydroxybenzoyl-CoA.[11] 4-Hydroxybenzoyl-CoA exhibits ϵ of 21.4 mM^{-1} cm^{-1} at 262 nm, pH 6.[11] In contrast, phenylacetyl-CoA has a lower ϵ: 14.6 mM^{-1} cm^{-1} at 260 nm, pH 7.0.[12] The spectrum of each hydroxybenzoyl-CoA thioester exhibits a longer wavelength absorption

[8] With Eastman cellulose plates (nonfluorescent) (#13255) and 1-butanol : acetic acid : water (10 : 3 : 5) as solvent, the following R_f values are observed: acyl CoA, 0.5–0.7; CoASH, 0.3; oxidized CoA, 0.1; free acids, >0.9.

[9] E. Beutler, O. Duron, and B. M. Kelly, *J. Lab. Clin. Med.* **61**, 882 (1963).

[10] P. Talalay, *Methods Biochem. Anal.* **8**, 119 (1960).

[11] L. T. Webster, Jr., J. J. Mieyal, and U. A. Siddiqui, *J. Biol. Chem.* **249**, 2641 (1974).

[12] J. J. Mieyal, K. S. Blisard, and V. A. Siddiqui, *Bioorg. Chem.* **5**, 263 (1976).

band that corresponds to the phenolic moiety and undergoes a progressive red shift of 40 to 45 nm as the pH is raised.[11] The phenolate absorbance of 2-hydroxybenzoyl-CoA and 4-hydroxybenzoyl-CoA both have large extinction coefficients (ϵ of about 7 mM^{-1} cm^{-1} at 365 nm and 24 mM^{-1} cm^{-1} at 330 nm, respectively); this feature may be exploited for discontinuous enzymatic assays involving these compounds. At high pH, difference spectra (i.e., before esterolysis minus after esterolysis) of benzoyl-CoA and 2- and 3-hydroxybenzoyl-CoA indicate thioester bond absorbances at 269 to 273 nm with ϵ of 6.4 to 6.9 mM^{-1} cm^{-1}.[10] Under the same conditions 4-hydroxybenzoyl-CoA does not show a spectral difference attributable to thioester absorbance.

Rates of esterolysis of benzoyl-CoA and the hydroxybenzoyl thioesters have been estimated by ultraviolet difference spectroscopy.[11] For benzoyl-CoA, 3-hydroxybenzoyl-CoA, and 4-hydroxybenzoyl-CoA, these rates are directly proportional to hydroxide ion concentration; the estimated second order rate constants at 38° are 10, 1.8, and 0.11 M^{-1} min^{-1}, respectively. This is considered consistent with a nucleophilic attack of the hydroxide ion on the carbonyl carbon of the benzoate moiety. In contrast, the rate of esterolysis of 2-hydroxylbenzoyl-CoA at 38° does not vary with hydroxide ion concentration. The rate of hydrolysis is nearly constant over a pH range of 9.6 to 12.1; the pseudo-first order rate constant in this pH range is ~0.02 min^{-1}. In this case, phenolate-assisted water attack on the carbonyl carbon of the benzoyl moiety is thought to be the predominant mechanism. Such a mechanism is compatible with glycine being a better nucleophile than hydroxide ion or water for 2-hydroxybenzoyl-CoA whereas the reverse holds true for benzoyl-CoA and 4-hydroxybenzoyl-CoA.

Proton magnetic resonance studies of coenzyme A and several of its thioester derivatives indicate that these molecules, under physiological conditions, exist in different conformations in solution.[4] Thus, benzoyl-CoA and the hydroxybenzoyl-CoA thioesters exist predominantly in intramolecular adenyl–benzoyl complexes whereas phenylacetyl-CoA and coenzyme A are present as extended molecules with little or no complex formation. These differences in conformation in solution may account, in part, for the differences in acyl-CoA substrate specificity of the several acyl-CoA : amino acid N-acyltransferases.[13]

Purity and Properties of Bile Acid-CoA Thioesters[6]

The purified coenzyme A thioesters of bile acids appear homogeneous by paper chromatography in two systems and by repeat chromatography

[13] L. T. Webster, Jr., this volume, Article [40].

on Sephadex LH-20. Independent estimates of bile acid, coenzyme A, and thioester bond exhibit the predicted stoichiometry; contamination by reduced coenzyme A is less than 2%. Biological activity of these purified compounds was confirmed by demonstrating greater than 94% conversion to taurine conjugates of the bile acids following incubation in the presence of rat liver bile acid-CoA amino acid N-acyltransferase.

The ultraviolet absorbance spectrum of each of the bile acid-CoA thioesters exhibits maximal absorbance at 259 nm; ϵ is $\sim$15.0 mM^{-1} cm^{-1}. Comparison of the absorbance spectra of the bile acid-CoA thioesters to that of bile acid plus coenzyme A indicates increased absorbance at 232 nm because of the thioester bond; ϵ is 4.12 mM^{-1} cm^{-1}.

Hydrolysis of the thioester bond occurs at alkaline pH; rates of esterolysis are proportional to hydroxide ion concentration and vary with the number of hydroxyl groups on the steroid nucleus. Thus, at 38° the approximate second order rate constant with respect to hydroxide ion concentration for hydrolysis of choloyl-CoA is 130 M^{-1} min^{-1} whereas the constants for deoxycholoyl-CoA and chenodeoxycholoyl-CoA are 24 and 26 M^{-1} min^{-1} and for lithocholoyl-CoA, 8.9 M^{-1} min^{-1}. The bile acid-CoA thioesters are stable in neutral and acid solution; extrapolated half-lives for hydrolysis of the thioester bond under physiologic conditions, 37°, pH 7.42, range from 185 days for choloyl-CoA to 1580 days for lithocholoyl-CoA.

Author Index

Numbers in parentheses are reference numbers and indicate that an author's work is referred to although the name is not cited in the text.

L

M

Subject Index

A

G